GW00671939

Mechanical Engineering Series

Frederick F. Ling
Series Editor

Springer
New York
Berlin
Heidelberg
Barcelona
Hong Kong
London
Milan
Paris
Singapore
Tokyo

Mechanical Engineering Series

Introductory Attitude Dynamics
F.P. Rimrott

Balancing of High-Speed Machinery
M.S. Darlow

Theory of Wire Rope, 2nd ed.
G.A. Costello

Theory of Vibration: An Introduction, 2nd ed.
A.A. Shabana

Theory of Vibration: Discrete and Continuous Systems, 2nd ed.
A.A. Shabana

Laser Machining: Theory and Practice
G. Chryssolouris

Underconstrained Structural Systems
E.N. Kuznetsov

Principles of Heat Transfer in Porous Media, 2nd ed.
M. Kaviany

Mechatronics: Electromechanics and Contromechanics
D.K. Miu

Structural Analysis of Printed Circuit Board Systems
P.A. Engel

Kinematic and Dynamic Simulation of Multibody Systems:
The Real-Time Challenge
J. García de Jalón and E. Bayo

High Sensitivity Moiré:
Experimental Analysis for Mechanics and Materials
D. Post, B. Han, and P. Ifju

Principles of Convective Heat Transfer
M. Kaviany

(continued after index)

Prabir Basu   Cen Kefa   Louis Jestin

Boilers and Burners

Design and Theory

With 250 Figures

 Springer

Prabir Basu
Department of Mechanical Engineering
Technical University of Nova Scotia
P.O. Box 1000
Halifax, Nova Scotia B3J 2X4, Canada

Cen Kefa
Institute of Thermal Engineering
Zhejiang University
Zheda Road
Hangzhou, Zhejiang 310027, China

Louis Jestin
Department Equipment
Electricite de France
12-14 Avenue Dutrievoz
F-69628 Villeurbanne, France

Series Editor
Frederick F. Ling
Ernest F. Gloyna Regents Chair in Engineering
Department of Mechanical Engineering
The University of Texas at Austin
Austin, TX 78712-1063, USA
   and
William Howard Hart Professor Emeritus
Department of Mechanical Engineering,
   Aeronautical Engineering and Mechanics
Rensselaer Polytechnic Institute
Troy, NY 12180-3590, USA

Library of Congress Cataloging-in-Publication Data
Basu, Prabir, 1946–
    Boilers and burners : design and theory / Prabir Basu, Cen Kefa,
Louis Jestin.
        p.   cm. — (Mechanical engineering series)
    Includes bibliographical references.
    ISBN 0-387-98703-7 (hardcover : alk. paper)
    1. Steam-boilers—Design and construction.   2. Oil burners—Design
and construction.   I. Cen, Kefa.   II. Jestin, Louis.   III. Title.
    IV. Series : Mechanical engineering series (Berlin, Germany)
TJ290.B37   1999
621.1'8—dc21   99-17360

Printed on acid-free paper.

© 2000 Springer-Verlag New York, Inc.
All rights reserved. This work may not be translated or copied in whole or in part without the written permission of the publisher (Springer-Verlag New York, Inc., 175 Fifth Avenue, New York, NY 10010, USA), except for brief excerpts in connection with reviews or scholarly analysis. Use in connection with any form of information storage and retrieval, electronic adaptation, computer software, or by similar or dissimilar methodology now known or hereafter developed is forbidden.
The use of general descriptive names, trade names, trademarks, etc., in this publication, even if the former are not especially identified, is not to be taken as a sign that such names, as understood by the Trade Marks and Merchandise Marks Act, may accordingly be used freely by anyone.

Production managed by Timothy Taylor; manufacturing supervised by Jerome Basma.
Typeset by TechBooks, Fairfax, VA.
Printed and bound by Maple-Vail Book Manufacturing Group, York, PA.
Printed in the United States of America.

9 8 7 6 5 4 3 2 1

ISBN 0-387-98703-7 Springer-Verlag New York Berlin Heidelberg   SPIN 10707311

Series Preface

Mechanical engineering, an engineering discipline borne of the needs of the industrial revolution, is once again asked to do its substantial share in the call for industrial renewal. The general call is urgent as we face profound issues of productivity and competitiveness that require engineering solutions, among others. The Mechanical Engineering Series features graduate texts and research monographs intended to address the need for information in contemporary areas of mechanical engineering.

The series is conceived as a comprehensive one that covers a broad range of concentrations important to mechanical engineering graduate education and research. We are fortunate to have a distinguished roster of consulting editors on the advisory board, each an expert in one of the areas of concentration. The names of the consulting editors are listed on the facing page of this volume. The areas of concentration are: applied mechanics; biomechanics; computational mechanics; dynamic systems and control; energetics; mechanics of materials; processing; production systems; thermal science; and tribology.

I am pleased to present this volume in the series: *Boilers and Burners: Design and Theory*, by Prabir Basu, Cen Kefa, and Louis Jestin. The selection of this volume underscores again the interest of the Mechanical Engineering Series to provide our readers with topical monographs as well as graduate texts in a wide variety of fields.

Austin, Texas   Frederick F. Ling

Mechanical Engineering Series

Frederick F. Ling
Series Editor

Advisory Board

Applied Mechanics	F.A. Leckie University of California, Santa Barbara
Biomechanics	V.C. Mow Columbia University
Computational Mechanics	H.T. Yang University of California, Santa Barbara
Dynamical Systems and Control	K.M. Marshek University of Texas, Austin
Energetics	J.R. Welty University of Oregon, Eugene
Mechanics of Materials	I. Finnie University of California, Berkeley
Processing	K.K. Wang Cornell University
Production Systems	G.-A. Klutke Texas A&M University
Thermal Science	A.E. Bergles Rensselaer Polytechnic Institute
Tribology	W.O. Winer Georgia Institute of Technology

Preface

Modern society owes a great deal to fossil fuels, which have accelerated man's progress from the cave to the present age of jets and computers. Utilization of this precious gift of nature is central and critical to our lives. This book is about rational utilization of fossil fuels for generation of heat or power. The book is intended for undergraduate and graduate students with an interest in steam power plants, burners, or furnaces. For researchers, it is a resource for applications of theory to practice. Plant operators will find solutions to and explanations of many of their daily operational problems. Designers will find this book filled with required data, design methods, and equations. Finally, consultants will find it useful for design evaluation.

This book uses a format of theory-based practice. Each chapter begins with an explanation of a process. Then it develops equations from first principles and presents experiment-based empirical equations. This is followed by design methodology, which in many cases is explained by worked-out examples. Thus, the book retains the interest of the reader and help remove doubts if any on the theory.

The present monograph is a joint effort of writers on three continents—Asia, Europe, and North America—and a marriage of two scientific traditions–Eastern and Western. In the West, boilers and burners are designed for high levels of performance, but their design methods and data are buried under commercial secrecy. Very few books or even research papers are available giving exact design standards. Thus, there is limited opportunity for information exchange on design data and methodologies. The energy industry in the Eastern European countries and China, on the other hand, did not have much commercial impetus for burying their design standards under secrecy. So their design methods and data were freely exchanged, and debated in the scientific community. Eventually, they formed into national standards. These elaborate thermal design standards were developed on the basis of experience in the design and operations of thousands of boilers and burners. However, the difference in language and scientific conventions prevented easy access of Western readers to this wealth of information available in China and Eastern European countries. The present book synthesizes design methods and data from both sides of the scientific world and presents the same in a simple Western format.

The book greatly benefitted from diverse professional backgrounds of three authors. Professor Cen Kefa, Head of the Institute of Thermal Power Engineering of Zhejiang University, is holder of the coveted distinction of Academician for his lifelong contribution to the Chinese boiler industry. He received his early education in boilers and power plants in the former Soviet Union. Thereafter, he worked closely with Chinese boiler and burner industry manufacturers perfecting their design standards. The research and development work he and his colleagues have carried out in the past four decades on boilers and burners greatly contributed to this book.

Dr. Louis Jestin, Head of the Heat Exchanger Section of the Fossil Fuel Division of the Electricite de France, is a specialist in heat transfer. He is involved in the design review of all new circulating fluidized bed boilers commissioned by this company. His team also carried out in-depth technical analyses of several supercritical boiler projects. His intimate contact with modern fossil fuel boilers and familiarity with European designs greatly enriched this book.

Dr. Prabir Basu, Professor in Mechanical Engineering in Dalhousie University and President of Greenfield Research Inc., is a specialist in fluidized bed boilers. He carried out extensive design and development work in government research laboratory, boiler manufacturing company, and universities. He participated in the development of the Indian boiler standard. His research and design experience in fluidized bed boilers contributed much to this book.

We thank Mr. S.S. Kelkar, Ex-Vice President, Deutsche Babcock Power System Ltd., India, who used his 30 years experience of boiler design and familiarity with German, British, Indian, and U.S. boiler codes to write Chapter 17 on pressure part design.

Our special thanks go to Prof. Jianren Fan, Prof. Qiang Yao, and Dr. Zuohe Chi for their significant contributions to a number of chapters in this book. Their tireless effort in data collection, draft preparation, and figure drafting provided critical support to this project. Dr. Jayson Greenblatt and Prof. David Mackay proofread many versions of the manuscript. Dr. Leming Cheng, Mrs. Sanja Boskovic, and Mr. Animesh Dutta greatly helped with preparation of the final manuscript.

Finally the authors thank their waves for their support to this project.

Prabir Basu
Halifax, Nova Scotia, Canada
Cen Kefa
Hangzhou, Zhejiang, China
Louis Jestin
Lyon, France

Contents

1
Introduction

To generate steam or hot water fossil fuel boilers use the chemical energy from fuels. A nuclear boiler uses energy from nuclear fission. A waste heat recovery boiler uses the sensible heat of hot gases from a process, and a solar boiler uses energy from the sun to generate steam. Although the name *boiler* or *steam generator* implies conversion of water into steam, boilers used for space heating do not necessarily generate steam. Some of them heat cold water to desired temperatures.

The earliest reference to boilers is seen in Hero's aelopile of 200 B.C. (Fig. 1-1). After being little used for two thousand years, boilers became an integral part of the industrial revolution in Europe. Since those early days boilers have come a long way, providing about 83% of the world's electricity supply. Table 1-1 shows the chronological development of boilers.

In early-nineteenth-century boilers, the steam pressure was slightly above the atmospheric pressure. This was largely due to the difficulty of building large pressure vessels from rivetted plates. The invention of the water tube boiler removed this barrier. Boiler pressure began to rise steadily, reaching supercritical levels (Fig. 1-2). Between the seventies and nineties utility industries operated conservatively by bringing down the steam pressure used in boilers. Of late, there has been a renewed interest in the use of high-efficiency supercritical boilers. This interest arose from the environmental need to attain higher power generation efficiencies. A dividend of higher efficiencies would be decreasing CO_2 greenhouse gas emissions.

1-1 Principles of Boiler Operation

Fuel burns in the furnace of a boiler, generating heat, which is then absorbed by the heating surfaces located around it and further downstream. Figure 1-3a shows a schematic of a boiler, where steam is generated. This process is best explained by using the temperature–heat content (enthalpy) diagram of water. This diagram (Fig. 1-3b) shows the effect of addition of heat to a unit mass of water. Water is pressurized to the required pressure by a feed water pump. It is then preheated in a heat exchanger called an *economizer*. On the temperature–enthalpy diagram

TABLE 1-1. Historical development of boilers.

Year	Developer	Application
200 B.C.	Hero	Aelopile as power source; not used any more
1600 AD	Branca	First steam turbine
1680	Denise Papin	Steam digester for food processing; use of safety valve
1720	Haycock	Shell-type boiler made of copper plates
1730	James Allen	Internal flue furnace; use of bellow for combustion air
1766	William Blakey	Patent on water in tube and fire outside
1803	John Stevens	A pseudo-water-tube design used in a steamboat
1804	Richard Trevithick	First high-pressure boiler with cast iron cylindrical shell
1822	Jacob Perkins	Once-through boiler using cast iron (fire) bars
1856	Stephen Wilcox	Inclined tube boiler with water-cooled enclosures
1880	Allan Stirling	Bent tubes connecting drums
1881	Brush Electric Light Co.	Advent of present age of electric power supply system
1920s		Pulverized coal fired boiler; size limiation is removed
1957	Ohio Power Co.	Supercritical boiler
1970s		Introduction of bubbling fluidized bed boilers
1980s		Introduction of circulating fluidized bed boilers

FIGURE 1-1. Hero's turbine and steam generator

we see that as water moves from state (A) to state (B), it gains in both heat and temperature. However, the water is still below its boiling point (Fig. 1-3b). The preheated water then enters the *evaporator* section of the boiler, which forms the vertical walls of the boiler furnace shown in Figure 1-3a. These walls absorb heat from the combustion of fossil fuels in the boiler furnace. While traveling through the evaporative tube the water picks up heat, but does not necessarily rise in temperature because the heat is used in transforming water (liquid) to steam (gas). This process is represented by the horizontal line BC.

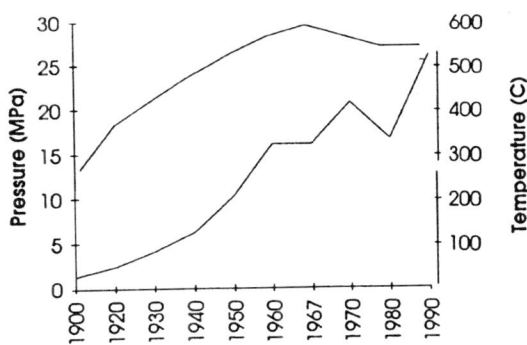

FIGURE 1-2. Historical rise of pressure and temperatures of boilers

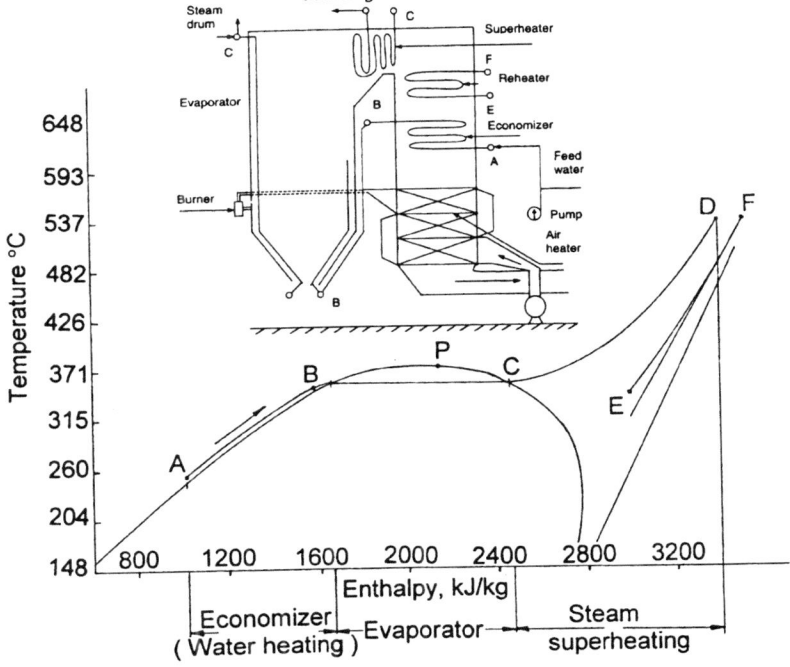

FIGURE 1-3. Temperature enthalpy diagram of water showing the heating and phase transformation of water in a boiler

The amount of heat required to convert a unit mass of water into steam is called the *latent heat of evaporation* or specific enthalpy of vaporization; this is represented by the length of BC in Figure 1-3b. We also notice in this figure that the higher the water pressure, the shorter the length of the horizontal section, and the line moves up the diagram with rising pressure. It is interesting to note that above the highest point (P) in the phase diagram there is no horizontal portion of this heating curve. This is the case for the supercritical boiler, where there is a continuous rise in temperature even when water is transformed into steam. Most of the present discussion relates to subcritical boilers, which operate below the critical pressure. The mixture of water and steam at any point in the line BC goes to a drum, where steam is separated from water. The water is recirculated through the evaporative section once again. The steam from the drum goes to another heat exchanger called a *superheater*. Here water, in its vapor state, is heated further at a constant pressure. This is shown by the line CD in Figure 1-3b. Steam at this point goes to either a steam turbine or a process heater, depending on the intended use. In a power boiler, the steam goes to a steam turbine.

After transforming some of its thermal energy into mechanical work in the turbine, a part of the steam returns to the boiler for further heating at a lower pressure. This process is called reheating, and the part of the heat exchanger that performs the task is called the *reheater*. The process is shown by line EF in Figure 1-3b.

A schematic of a large boiler plant is shown in Figure 1-4. Coal, the most common fuel, is stored in a bunker. A belt conveyor conveys the coal to a storage pile or crushing facility. The coal is crushed and screened to the required size in the crushing facility. After that it goes to another storage hopper for direct feed into a *pulverizer*, which grinds the crushed coal extremely finely, and transports it to the burner.

Combustion air and pulverized coal are injected into the furnace, where the coal is burnt in a flame. The boiler may use other firing systems like a fluidized bed or stoker without a flame to burn the fuel. The ash in the fuel is extracted partly from the bottom of the furnace. The other part is conveyed by the flue gases. This ash, known as *fly ash*, is collected either by a bag-house or electrostatic separators. The

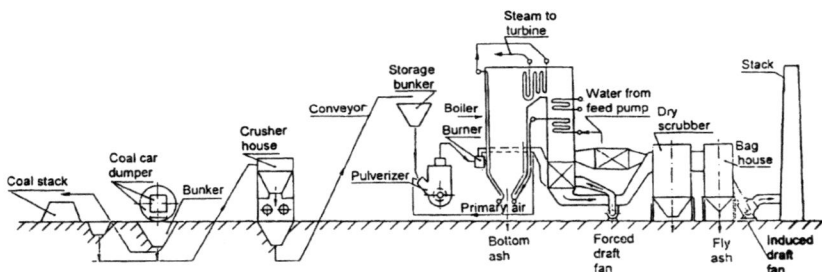

FIGURE 1-4. Schematic of a large boiler plant firing coal

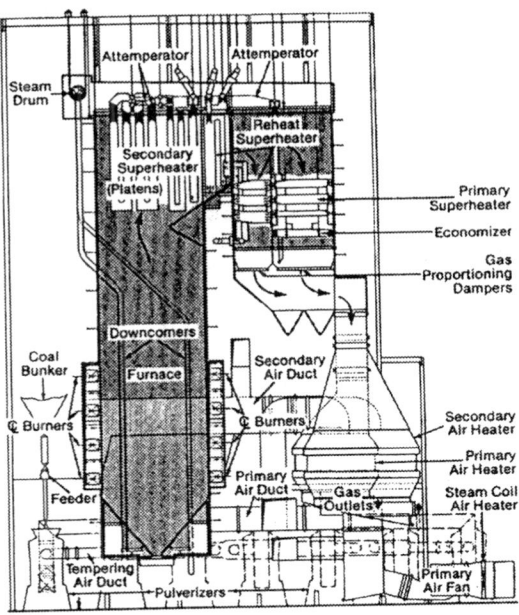

FIGURE 1-5. Cross section of a typical pulverized coal boiler [Reprinted with permission from Stulz and Kitto (1992) "Steam: Its Generation and Use," 40th edition. Babcock & Wilcox, Barberton, Ohio, p. 18-2, Fig. 3]

combustion air is supplied by a fan. This air is preheated by the flue gas in the *air preheater* section of the boiler. A part of the combustion air travels through the pulverizer, while the other part goes directly into the burner in the boiler.

The walls of a boiler furnace are usually made of evaporator tubes (Fig. 1-5). These heating surfaces absorb heat directly from the flame or the combustion products by radiation. Further up in the furnace are the superheater and reheater surfaces. These surfaces also absorb heat from the flue gas. The flue gas leaves the furnace, but is still hot. This gas enters the convective section of the boiler, called *back-pass*. Remaining sections of both the reheater and superheater cool the flue gas further. Finally, the flue gas flows over the *economizer* section of the boiler. Here, the water is preheated before entering the drum and the evaporative section of the boiler. After this section, the flue gas enters the air preheater. Further downstream, the flue gas is cleaned in a gas–solid separator (bag-house or electrostatic precipitator). For reducing the nitric oxide and sulfur dioxide content of the flue gas, the boilers may use postcombustion cleaning devices such as a selective catalytic reducer and a flue gas desulfurizer. Finally, an induced draft fan draws the flue gas out of the system and delivers it to the stack.

TABLE 1-2. Classification of boilers.

Firing method	Energy source	Use of steam	Water circulation	Steam pressure	Construction
Stoker	Coal	Utility	Natural	Atmosphere	Packaged (Shell)
Front firing burner	Liquid fuel	Industrial	Forced	Subcritical	
Tangential firing burner	Gas	Domestic	Once-through	Supercritical	Packaged (water tube)
Opposed firing burner	Solid wastes	Marine	Combined	Sliding pressure	Field erected
Downjet firing burner	Biomass	Naval			
Cyclone firing					
Bubbling fluidized	Recovery	Hot gas boiler			
Circulating fluidized	Waste heat	Cogeneration			
No firing method	Nuclear fuel				

1-2 Classification of Boilers

The boiler, the most commonly used piece of energy conversion equipment, varies widely in design depending on the firing method used, fuel fired, field of application, type of water circulation employed, and pressure of the steam. Table 1-2 gives a list of different types of boilers.

1-3 Description of Boilers

A detailed description of gas, oil, and coal flame fired boilers is presented in Chapters 5 and 6. The two types of fluidized bed boilers are described in Chapter 11. Chapter 13 includes a description of the once-through boilers. The following section presents a brief discussion on some of the other types of boilers.

1-3-1 Packaged Boilers

Packaged boilers are usually smaller in capacity. They are preassembled units with all components mounted on a shippable mount. Thus these boilers do not need assembly at site. Smaller units are shell type, larger ones are water-tube type.

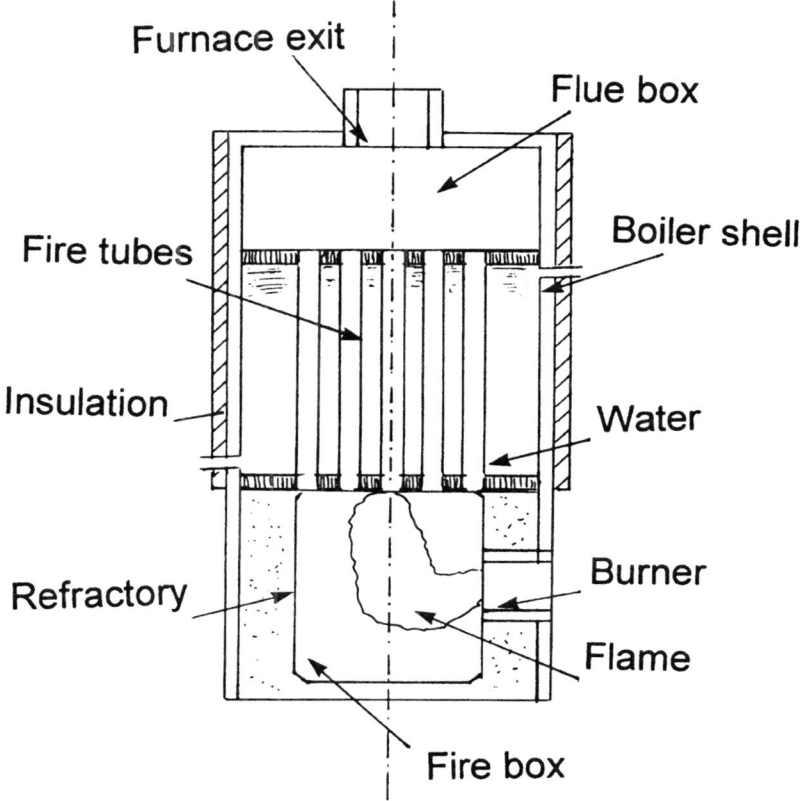

FIGURE 1-6. Domestic hot water boiler

a) Shell Type

Domestic hot water boilers are the most common example of this type of boiler, especially in colder regions. Figure 1-6 shows a view of a typical oil fired domestic boiler. The main requirement of this type of boiler is absolute reliability and safety. These boilers are often left unattended for months in the basement of a house, being turned on and off in response to signals from a simple thermostat. The burner and blower are one combined unit. The oil is ignited by a high-energy electric spark. The flame is usually short and enclosed in a small refractory chamber below a boiler shell containing water. The hot gas rises through vertical fire tubes passing through the shell. Sometimes spiral ribbons are used to enhance the gas side heat transfer coefficient. The furnace operates in negative pressure created by the chimney of the house. These boilers produce hot water at low pressures. The thermal efficiency

of these boilers is usually low — 50–65%. Some boilers use electricity instead of
an oil/gas flame for heating.

Larger shell-type packaged boilers are used for heating building complexes or
for supplying process steam to small industries. Most of these are of the fire tube
type, with combustion on a grate inside a cylindrical furnace located within the
shell. The hot flue gas exits from the far end of the furnace tunnel and returns
to pass through a number of fire tubes located around the furnace tube. Space
heating does not need a high-pressure steam. However, a higher specific volume
or low pressure makes the transport of steam over some distance uneconomical.
For this reason some industries use a standard pressure of 8.6 bar gauge (125 psig).
Pressure reducers lower the pressure at the location of steam use.

b) Water-Tube Type

Water-tube packaged boilers are usually built in capacities up to 25 kg/s. The
pressures can go up to 72 bar and temperatures to 440°C depending on the fuel and
steam rating. The capacity of these boilers is restricted by the shipping dimensions
of a rail car or truck. However, barge-mounted units can be assembled at the dock
in capacities up to 75 kg/s. These bottom-supported boilers are typically in the
shape of a "D" (Fig. 1-7). The bulk of the evaporator tubes are grouped in a bank
between drums. The oil or gas burner fires into the furnace, which is separated from
the tube bank by a baffle. So the gas travels to the rear of the furnace and returns
through the boiler bank. The furnace operates under positive pressure. Hence it is

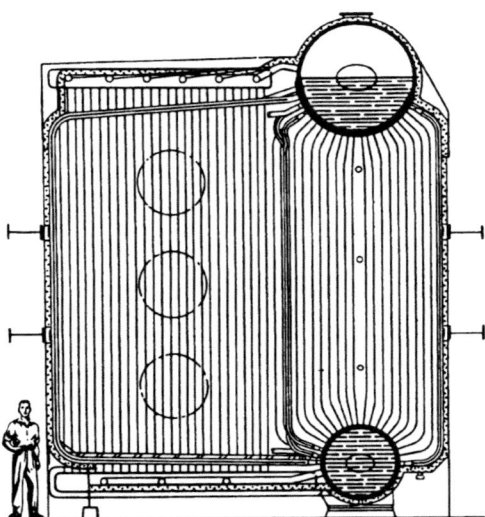

FIGURE 1-7. Packaged water tube boilers [Reprinted with permission from Stulz and Kitto
(1992) "Steam: Its Generation and Use." 40th edition. Babcock & Wilcox, Barberton, Ohio,
p. 25-9, Fig. 12]

made gas-tight by water tubes welded together. This is called the *membrane wall.* These boilers are very tightly designed for compactness and operate with a very high volumetric heat release rate.

1-3-2 Marine or Naval Boilers

Marine boilers are extremely compact. They are built to maximize the power-to-weight and power-to-volume ratio. As a result they use furnace heat release rates up to 10 MW/m^3 in naval vessels and 1 MW/m^3 in merchant vessels. Practically all marine boilers are oil fired. Now a days, many ships use diesel propulsion. Gas turbine and nuclear-powered ships are available in very limited numbers. For these ships, a waste heat boiler and or auxiliary package boiler is used to supply auxiliary steam for heating, etc.

1-3-3 Congeneration Boilers (Combined Heat and Power Plant)

Cogeneration boilers supply process steam as well as electricity. Utility boilers designed exclusively for electricity generation waste more than 60% of the combustion heat in the condenser. Even the most efficient plant designed on the Rankine cycle cannot avoid it. This heat is discharged either to the atmosphere through a cooling tower or to a river or lake. Some processes, like sugar mills, paper mills, etc., use a sizable amount of process steam as well as electricity. Cogeneration boilers are extensively used in these plants. Steam is expanded in the steam turbine only to a modest pressure. Then the low pressure steam is used in the process. Overall energy utilization efficiency of cogeneration may exceed 80%, while that for utiliy plant is less than 40%. The decision to build a cogeneration plant will depend on the relative price of electricity, investment cost, and the steam requirement.

The design of cogeneration boilers is the same as that of most industrial boilers. If the steam demand is high, it is economical to drive as many boiler auxiliaries as possible on steam. The capacity of a cogeneration plant is usually dictated by the steam requirement. Recent changes in tariff regulations in some countries allow cogeneration plants to sell excess power to the local utility. In such cases the cost of generation and the price received may influence the decision on capacity of the plant.

1-3-4 Solid Waste Fired Boilers

Many cities are finding it increasingly difficult to dispose of their garbage owing to the paucity of space for traditional landfill. Recycling and composting are two environment-friendly options. Even after that, however, a certain amount remains that must be incinerated. This figure was 15% in 1990 and is projected to rise to 50% (Stultz & Kitto, 1992, p. 27-2) in the 21st century. Furthermore some governments have made it easier to sell the electricity generated by waste-to-energy plants to the local utility. Thus the option of generating electricity by solid waste fired boilers is

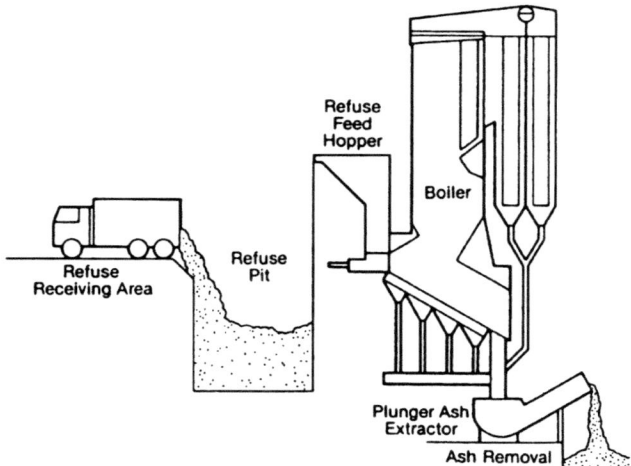

FIGURE 1-8. Mass-burn-type solid waste fired boiler [Reprinted with permission from Stulz and Kitto (1992) "Steam: Its Generation and Use," 40th edition. Babcock & Wilcox, Barberton, Ohio, p. 27-2, Fig. 3]

gaining acceptance. One type is the mass burn design, where, except for very bulky items, everything is dumped in a stepped grate (Fig. 1-8). These boilers use grate heat release rates in the range of 0.9 to 1.1 MW/m^2. The other type of boiler is the refuse derived fuel (RDF) type. In the latter type boiler, items that are economically recyclable or salable are removed. The remaining solids are shredded to <38 mm (1.5 in.) and injected into the furnace. The ash is collected on a traveling grate and dropped into a hopper. The RDF fired boilers typically use a grate heat release rate of 2.3 MW/m^2.

A common problem with all solid waste fired boilers is the corrosion of boiler tubes. Unlike coal or oil, solid waste fuel contains high levels of corrosive elements like chlorides, sodium potassium, zinc, lead, vanadium, etc. So, the lower part of the boiler is covered with corrosion resistant refractory. Presently, an overlay of inconel is also being used successfully. To protect against high-temperature corrosion in the superheaters, a parallel flow pass is used where the hottest gas hits the coldest steam. Additional protection is provided by making the hottest section of the superheater of inconel alloy. To meet the emission limits of particulates, nitric oxide, sulfur dioxide, hydrochloric acid, volatile organic compound, dioxin, furan, etc., a dry scrubber along with a bag-house is used at the back end of the boiler.

1-3-5 Biomass Fired Boilers

Biomass comes from plants and vegetation. This fuel is also a renewable source of energy. Wood chips, saw dust, bagasse, rice hull, etc., are examples of biomass fuels. A common feature of most of these fuels is that they contain smaller amounts of ash, higher amounts of oxygen, a negligible amount of sulfur, and high moisture. As a result, biomass requires less combustion air and produces less nitric oxide and sulfur oxides. A large part of the combustible is in the form of volatile matter. So, a biomass fired boiler would require a large amount of overfire air. The fuel is spread or swept by air into the furnace. Conventional wood fired boilers use stepped grate, vibratory grate, water cooled grate with holes, or traveling grate. Owing to its high moisture, biomass fuel is sometimes predried. Preheating of the combustion air is used in most of the wood fired boilers to improve their ignition and combustion efficiency. Modern biomass boilers are using fluidized bed firing with increasing success.

1-3-6 Recovery Boilers

These boilers have the unique objective of production of both steam and a chemical feed stock. Pulp and paper industries use this type of boiler extensively. Figure 1-9 shows a typical recovery boiler in a paper mill. Black liquor, an intermediate product of the pulp-making process (Kraft process), has a high heating value and is used as a fuel in this boiler. It is sprayed in the furnace such that liquid drops are large enough to drop to the floor of the furnace yet small enough to be dry before hitting it. Sub-stoichiometric primary air reduces the sodium sulfate (Na_2SO_4) of the black liquor into sodium sulphide (Na_2S). This solid product, called *smelt*, is extracted from the bottom of the boiler in molten form, and is used in subsequent steps of the pulping process. The char residue of the organic component of the black liquor burns, releasing heat in the furnace. Any entrained solids are returned to the black liquor for subsequent conversion into smelt. The secondary and tertiary air is added further up to complete the combustion. The capacity of the recovery boiler is often driven by the production rate of smelt or black liquor processing. The furnace is usually designed on the basis of grate heat release rate of 2.5–2.7 MW/m^2.

1-3-7 Waste Heat Recovery Boilers

The basic oxygen furnace, cement kiln, open hearth steel furnace, and petroleum refinery produce hot gas at temperatures between 400°C and 1900°C (Ganapaty, 1991). The heat of their flue gas can be economically recovered by generating steam in a waste heat recovery boiler. The steam generated may be used either for process heating or for generation of electricity. In combined cycle or large diesel based power plants the exhaust gas from the gas turbine or diesel engine generates heat in a heat recovery steam generator for the steam turbine. These boilers do not usually fire any fuels. Thus, they are essentially counter-current heat exchangers. Figure 1-10 shows the schematic of a typical steam generator used in a combined cycle power plant. Here, the boiler heating surfaces are made vertical to facilitate natural circulation. In recovery boilers, the gas side pressure drop across boiler is generally kept within 2.5–3.7 kPa and the minimum temperature difference

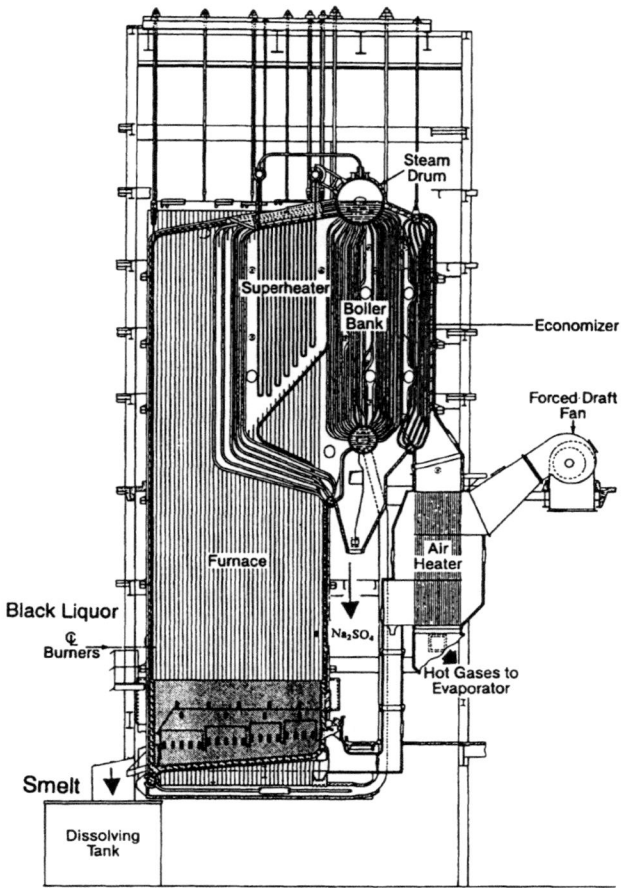

FIGURE 1-9. A black liquor fired recovery boiler [Reprinted with permission from Stulz and Kitto (1992) "Steam Its Generation and Use," 40th edition. Babcock & Wilcox, Barberton, Ohio, p. 26-9, Fig. 10]

between gas and steam water (Pinch point) is kept at 11–28°C (Stulz & Kitto, 1992, p. 31-4). A higher resistance through the boiler will adversely affect the efficiency of the gas turbine.

For highly dust-laden flue gases such as those from open hearth furnaces and cement kilns, a three-drum design is suitable. Solids separate from the gas as it flows horizontally over the vertical boiler tubes. Dust hoppers kept underneath collect the dust, which can be removed as needed. In case steam is required at all

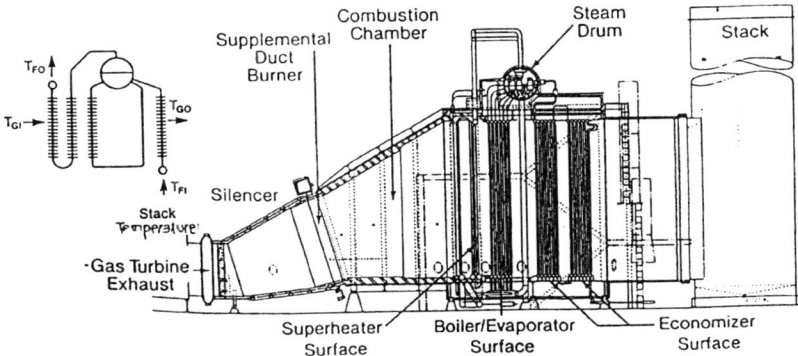

FIGURE 1-10. A heat recovery steam generator used for generation of steam in a combined cycle power plant [Reprinted with permission from Stulz and Kitto (1992) "Steam Its Generation and Use." 40th edition. Babcock Wilcox, Barberton, Ohio, p. 31-3, Fig. 3]

times, even when hot gas is not available supplemental gas or oil firing may be added.

1-3-8 Nuclear Steam Generators

These boilers are substantially different from the previously discussed boilers. Like waste heat recovery boilers, they do not contain a furnace. Heat is generated by the nuclear fission reaction in a nuclear reactor. This heat is absorbed by a coolant, which could be heavy water (CANDU reactor) or pressurized water in the pressurized water reactor (PWR). In PWR reactors, the coolant is cooled to a much higher pressure (2200 psig) such that it is in a liquid state even at 625°F. In one design, this subcooled water enters the steam generator from one side of the bottom (Fig. 1-11). The coolant (also called primary water) flows upward through a large number of parallel tubes contained in a pressure vessel. This tubes bundle is bent 180°. So the primary water flows downward and exits from the other side of the shell. The feed water for steam generation (called secondary water) enters at midlevel and travels downward through the annulus between the steam generator shell and the tube bundle. It then rises up through the core of the pressure vessel shell. The steam is generated at a modest pressure of about 1060 psi and with a superheat of about 30°F superheat. The relatively low temperature of the primary water is the limiting factor on the temperature of the secondary steam. The steam water separation takes place at the top of the generator. This steam is sent to the turbine for electricity generation.

Another type of steam generator, used in the nuclear power plant, is the boiling water reactor (BWR). Here the cooling water directly generates steam in the reactor, and this steam is used to drive the turbine.

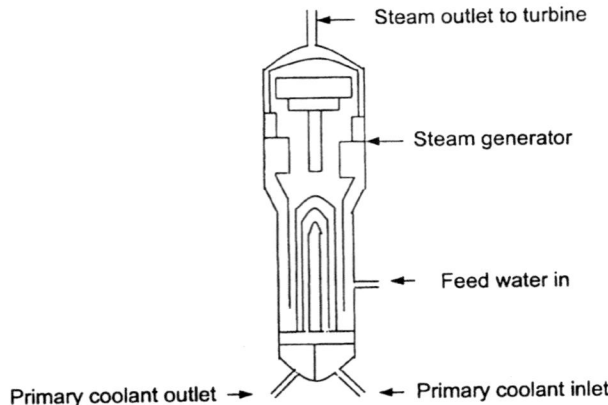

FIGURE 1-11. A simplified figure of a nuclear steam generator

References

Ganapaty, G. (1991) "Waste Heat Boilers." Prentice Hall, Englewood, New Jersey.
Stulz, S. C. & Kitto, J. B. (1992) "Steam—Its Generation and Uses," 40th edition. Babcock & Wilcox, Barberton, Ohio, pp. 25-9, 26-9, 31-3, 31-4.

2
General Design Considerations

A boiler, whether used in a process or utility plant, is a major piece of equipment. So, the purchaser applies much care to its procurement. The degree of sophistication and the extent of detail of the procurement process may vary depending on the purpose and size of boiler, but the process generally follows a common format. Typically, it starts with the preparation of a detailed specification, invitation and selection of bids, contract award, project execution, and finally the acceptance of the plant. The boiler supplier, on its part, ensures that all expectations of the purchaser are met and that the price is competitive. The vendor goes through different stages of design, construction, and commissioning. The design stages are discussed later in section 2-2. The following section discusses the scope of the boiler specifications.

2-1 Boiler Specifications

The user assesses the need for steam, keeping in view the near- and long-term requirements. The steam specification is governed by the selection of other equipment of the plant, for example, the turbine in a steam power plant or ancillary equipment in a process plant. However, one does not specify a boiler for the exact amount of steam required. Depending on the type of use and the criticality of steam availability the user may allow for excess or standby capacity. For example, if a process plant needs 100 kg/s steam, of which 20 kg/s is the minimum requirement, the user may choose to purchase six boilers of 20 kg/s capacity or two boilers of 100 kg/s capacity. Larger boilers enjoy the advantage of lower capital and operating cost per MWt heat generated. The standby capacity can be as high as 100%. However, it is less efficient to operate a large capacity boiler at a lower load than operating several smaller units at full load. Keeping such overall strategic factors in mind, a detailed specification of the steam generating plant is produced.

Table 2-1 gives a list of parameters that a boiler specification may include. The first column lists a set of general parameters common to most boilers. The other four columns indicate special conditions needed for the specific type of boiler. To avoid any confusion or doubt, a specification should be as complete and as

TABLE 2-1. A list of some boiler specification parameters.

General	Special parameters			
1 General fossil fuel fired boiler	2 Utility (burner firing)	3 Industrial (burner or stoker firing)	4 Fluidized bed (utility or industrial)	5 Heat recovery steam generator
1. Steam a) Flow rate b) Pressure c) Temperature	Max. continuous flow Max. peak flow Temperature variation	Load pattern Minimum steam flow	Same as 1	Same as 1
2. Feed water Economizer inlet temperature	Source Chemical analysis	Quality		
3. Fuel a) Type b) Ultimate analysis c) Heating value d) Physical properties	Alternate fuel & cost Delivered fuel size Ash composition Fouling & sintering properties Reactivity		Attrition (mechanical and combustion assisted if possible)	Flue gas amount & composition Gas temperature at inlet
4. Sorbent	Reactivity (if used)	Type & size Max. extent of sulfation Attrition Ca/S ratio		
5. Emissions a) Sulfur dioxide b) NO_x limit c) Particulate				
6. Solid Waste	Ash disposal system		Land fill restrictions if any	
7. Pressure drop	Gas side Steam side		Min. drop through grid or air distributor	Back-pressure of turbine Pinch point
8. Auxiliaries	Auxiliary power	Auxiliary equipment	Auxiliary power	
9. Site	Site plan & topography —ambient temperature, pressure & humidity —access of material and personnel			
10. Personnel	Operating personnel	Operating personnel		
11. Evaluation criteria				

precise as possible, but at the same time the plant should not be overspecified. This would unnecessarily increase the cost. Sometimes a purchaser specifies a feature of the boiler thinking that it might be useful if a need should arise in the future. The vendor would of course then add the cost of that feature in the design even if the purchaser does not use it at all.

The above list is not comprehensive. The actual specification may be more detailed or brief depending on the need and the buyer's preference. This list gives some indication of the type of information or guidance the boiler designer receives.

The next section deals with different stages of a custom boiler design. Sometimes, especially for smaller *packaged boilers*, a suitable boiler may be available off-the-shelf from the manufacturer's catalogue. The elaborate design process, as laid out below, is not necessary for those boilers.

2-2 Design Steps

To deliver the right boiler, satisfying all of the terms of the contract at a competitive price, the boiler manufacturer puts in a team effort. The design team, supporting development engineers, and technical sales staff work together to meet the customer's specifications and special requirements.

The entire design exercise is carried out in different phases, each serving different purposes. The details of each vary with its purpose. Examples of some are given in Table 2-2:

2-2-1 Preliminary Design

At this stage the designer often carries out a mass and material balance to match with the given steam condition and the fuel analysis. The designer also attempts to match this with any plant built in the past to interpolate or extrapolate overall size, weight, and the cost of the plant. Some commercial software packages, such as CFBCAD© (1995) are available for preliminary size and performance analysis of specific types of boilers.

2-2-2 Detailed Proposal Design

The first step in this type of design is to review three main specified parameters of the boiler with respect to steam, fuel, and environment. After that the designer may carry out the followings in appropriate sequence:

TABLE 2-2. Stages of boiler design.

Design stage	Purpose
Preliminary proposal design	Prepares a budget price
Detailed proposal design	Prepares the detailed competitive bid
Final design	Prepares manufacturing drawings

- Stoichiometric calculation for the designed fuel. Among other things, this step, as shown in Chapter 3, determines the amounts of air required, flue gas produced, and heat released per unit weight of fuel burnt.
- Heat balance of the boiler: Heat losses in different parts of the boiler are assessed here. Further details and calculation procedures are given in Chapter 3.
- Heat duty allocation: The heat duty allocation among different sections of the boiler is found here. A design optimization is carried out to optimize the boiler cost and maximize its reliability while meeting the steam parameters given in the specification.

a) Thermal Design

The thermal design determines the size of the boiler with combustion and heat transfer considerations in mind. The followings are some steps of the thermal design process:

1. Design of the furnace: This step involves determining the size and configuration of the furnace. More details on the designs of gas fired, oil fired, and pulverized coal fired furnaces are given in Chapters 5 and 6.
2. Design of heat transfer surfaces: Heating surfaces of different sections of the furnace are designed using the thermal load computed earlier. The superheater, reheater, and evaporator are often spread around the boiler. This is governed by surface area optimization and the material cost. Determining the optimum heating surface arrangement is an important part of this step. Once that is done the actual surface areas are calculated from heat transfer analysis. Chapter 7 discusses this aspect in detail.
3. Steam temperature control: A suitable strategy for safe and economical control of steam temperature is developed at this step. Design and selection of an attemperator is a part of this design. This aspect is especially important for utility and industrial boilers.
4. Design of the burner: Suitable burners must be designed or selected if the boiler is oil, gas, or pulverized coal (PC) fired. In the latter case, pulverizers must be designed or selected to suit the burners. These designs are to be carried out with due regard for NO_x control in the furnace. Burner design is discussed in detail in Chapters 8–10.
5. Fluidized bed or stoker fired boilers do not use burners. The main fuel burns on grates or distributors, which have a different design strategy integrated with the furnace design. They are discussed in Chapter 11 in greater detail.

b) Fluid Mechanical Design

1. For selection of fans and blowers a complete analysis of draft losses in the gas and airstream is carried out. Details of this are given in Chapter 16. Stoichiometric and heat balance calculations provide data on the volumetric capacity of the fans, while the draft loss calculations give the head requirement.

2. Water circulation design. Except for once-through boilers, all other boilers work under natural circulation. An analysis of the natural circulation pattern of steam–water mixture through the design evaporator circuit is essential to protect the tubes from failure owing to burn out. Design procedures for the same are given in Chapter 12. The assessment of pressure drop in steam and steam water mixture in once through circuits is also discussed in Chapter 12.

c) Equipment Selection

1. Capacity and specifications of feeders, pulverizers, bunkers, hoppers, etc., are firmed up in this step. In case the fuel handling and postcombustion flue gas cleanup are within the scope of supply, these are to be chosen at this stage. At this stage the designers attempt to maximize the use of standard available equipment. To utilize the standard equipment, the designer may have to accept some overcapacity. However, it could save engineering and manufacturing time. Also it has the benefit of proven reliability. However, the additional capital cost and power consumption should be balanced against the cost of custom design.
2. While selecting fans, blowers, and pumps according to the specifications prepared by the fluid mechanics design, an effort should be made to chose standard equipment, because custom-sized equipment, especially in smaller sizes, may prove extremely expensive. In selecting the fans and blowers, spare capacity desired by the buyer is also an important consideration.

d) Mechanical Designs

1. Pressure parts calculations: the mechanical design gives the heating surface required for different sections of the furnace. It specifies the outer size of the tube, but does not say anything about the material or thickness of the same. Also the thermal design of step a does not specify the headers and drums. This step would lay out the details of the tube circuit inside and outside the boiler. Design and selection of headers (manifold) and drums are made. The tube thickness, tube support, header, and drum details are calculated by considering the imposed stress and allowed stress of the chosen material. These calculations need to follow the governing boiler or fired pressure vessel code of the country where the boiler will be used. One of the most widely used codes is that of the American Society of Mechanical Engineers. The buyer will specify the applicable code for this design. An introduction to this is given in Chapter 17.
2. Design of steel structure and furnace enclosures: Many boilers, especially those used in utility industries, are very tall. Thus the design of the boiler support and access platform and steel enclosures is an important step in the design. This design is generally carried out as the last step because the boiler configuration and size must be decided before designing the support and enclosure. The wind load, seismic factors, and soil conditions are some of the parameters taken into consideration for this design.

2-2-3 *Final Design*

This design is usually carried out after a firm order is placed with the boiler supplier. Detailed engineering begins at this stage. Detail design is essentially the same as that described above for the detailed proposal design, except that each design has to be carried out more thoroughly. The manufacturer's drawings are to be prepared on the basis of this design. So, much care and all contributing factors are considered at this stage. A detail discussion between the purchaser or its representative may help avoid costly last-minute changes. Often the part load operation is verified at this stage to ensure that no unsafe operation is involved. A boiler performance calculation at different loads and or fuel conditions is carried out by designers to check the acceptability of boiler operation under specific conditions. This is especially important when the boiler supplier is required to meet emission levels at all operating conditions.

References

CFBCAD© (1995) An Expert System for Design of Circulating Fluidized Bed Boilers. Greenfield Research, Inc., Box 25018, Halifax B3M 4H4, Canada.

Stulz, S.C. & Kitto, J.B. (1992) "Steam." Babcock & Wilcox, Barberton, Ohio, 40th edition.

3
Fuel and Combustion Calculations

Fuel is the single most important contributor to the cost of steam generation. It also governs the design, operation, and performance of the boiler. Even the most fuel-flexible boilers, e.g., fluidized bed boilers, are fuel dependent, albeit to a lesser degree. For this reason any design or even design planning, must start from a consideration of the fuel to be used. From the fuel receiving yard to the waste disposal pond and the stack, everything depends on the characteristics of the fuel. For this reason a typical boiler design starts with combustion calculations. The combustion calculations are based on the stoichiometry of combustion reactions. So, this step is often called stoichiometry calculations. It provides specifications of most major items of a power plant like fans, blowers, fuel, waste handling plants, solid conveyors, stack size, air pollution control equipment, and finally, the size of the boiler.

The following section briefly describes the physical and combustion characteristics of different fossil fuels used in power plant boilers. It also presents design formulae for combustion calculations. Such calculations are often based on unit weight of fuel burnt.

3-1 Features of Fuel

Fossil fuels are found in gaseous, liquid, or solid form. The following section presents a discussion of the essential features of these three types of fuels.

3-1-1 Gaseous Fuel

Fuel gases are either natural or man-made. Their compositions vary widely. Typical compositions of some fuel gases are given in Table 3-1.

a) Natural Gas

Natural gases come from either gas or oil fields. Methane is the principal component of natural gas. It is accompanied by some other hydrocarbons (C_nH_{2n+2}) and

TABLE 3-1. Analysis of some typical gaseous fuels.

Fuel type	Origin of fuel	Composition by volume, %						Higher heating value kJ/Nm³	Density kg/Nm³	Combustion air requirement Nm³/Nm³
		CH_4	C_mH_n	H_2	CO	CO_2	N_2			
Natural gas	Gas field	97.3	2.4				0.3	40600	0.74	9.7
Synthetic	Water gas	15.5	7	34	32	4.3	6.5	18,950	0.77	4.27
	Coke oven	24.9	3.2	50.2	9	3.2	9.5	19,900	0.56	4.3
	Cracked	31.8	1.1	38.5	4	8.9	15.7	18,900	0.7	4.2
	Producer	3	$O_2 = 0.2$	14	27	4.5	50.9	5260	1.1	1.0
	Blast Furnace		$H_2O = 3.4$	2.4	23.4	14.4	56.4	3122	1.3	0.63

Commercial gases	Composition by volume, %						kJ/Nm³	kg/Nm³	Nm³/Nm³
	C_4H_{10}	C_3H_8	C_3H_6	C_2H_6	C_2H_4	C_4H_8			
Propane	2	65.5	30	2	0.5		102200	2.04	23.8
Butane	68.6	6.4			3.2	21.8	131400	2.65	30.4

incombustible gases. The gases from gas fields contain up to 75–98% methane. Natural gases from oil fields contain only 30–70% methane. Natural gas has a high heating value. Its lower heating value (LHV) is 36,600–54,400 kJ/Nm³.

b) Synthetic Gas

Coal gas and blast furnace gas are two principal types of synthetic gas. Those produced from coal include coke-oven gas, cracked gas, water gas, and producer gas. Blast furnace gas is a by-product of iron extraction in a blast furnace. Its main constituents are CO and H_2. Owing to its high CO_2 and N_2 content, the heating value of blast furnace gas is very low (3800–4200 kJ/Nm³). Furthermore, it contains large amounts of low melting point ash particles. It is, therefore, classed as a low rank fuel and is often burned in conjunction with heavy oil or pulverized coal.

Coke-oven gas is a by-product of coke. It contains impurities such as ammonia, benzene, and tar. So coke-oven gas should be refined before it is burned. Various producer gases (air producer gas, water-gas, and mixed producer gas) can be obtained by coal gasification in a coal gas generator. They are used as chemical raw materials and fuel. In general, their heating value range is from 3700 to 10,000 kJ/Nm³.

c) Commercial Gas

Propane or butane are the most important commercial gases. They are by-products of the petroleum refining processes and have high heating values. Therefore, these are excellent fuels for both domestic and industrial use. Unlike natural gas, the petroleum based gases are rich in propane or butane.

TABLE 3-2. Characteristics of some liquid fuels.

ASTM oil designation→	#1	#2	#3	#5	#6
Elements↓	%	%	%	%	%
S	0.01–0.5	0.05–1.0	0.2–2.0	0.5–3.0	0.7–3.5
H	13.3–14.1	11.8–13.9	$(10.6–13.0)^a$	$(10.5–12.0)^a$	$(9.5–12.0)^a$
C	85.9–86.7	86.1–88.2	$(86.5–89.2)^a$	$(86.5–89.2)^a$	$(86.5–90.2)^a$
N	0–0.1	0–0.1			
A			0–0.1	0–0.1	0.01–0.5
Specific Gravity	0.825–0.806	0.887–0.825	0.996–0.876	0.972–0.922	1.002–0.922
Viscosity in Cs (38°C)	1.4–2.2	1.9–3.0	10.5–65	65–200	260–750
Higher heating value (kJ/kg)	45,637–46,079	44,478–45,823	42,413–45,011	41,995–44,130	40,394–44,060

Specific heat is generally in the range of 1.8 kJ/kg K for 50–100°C

Volume of theoretical combustion air requirement is in the range of 10.5–10.4 Nm^3/kg fuel

^{a}estimated value, Cs = centistoke.

3-1-2 Liquid Fuels

Liquid fuels are generally distillation fractions of crude oil. For example, heavy oil, used by utility boilers, is the liquid residue left after the crude is distilled at atmospheric pressure. Light diesel oil, used for ignition, is a subsequent distillation product. Compositions of several grades of fuel oils are listed in Table 3-2. These oils are classified by American Society of Testing and Materials (ASTM) numbers.

Viscosity, flash point, pour point, sulfur content, and ash content are important properties of fuel oils.

a) Viscosity

Viscosity influences both the transport and the atomizing quality of crude oil. It depends on temperature and pressure. The influence of temperature is, however, greater. Higher oil temperatures give lower oil viscosities. However, if a heavy oil is heated above 110°C, carbon will be produced and it may block the atomizers.

Viscosity can be measured by an Engler viscometer. It is expressed as the ratio of time taken by 200 ml oil to flow through an Engler viscometer to that taken by 200 ml of distilled water. This measure of viscosity is identified by °E. For smooth transportation pipelines the viscosity of the oil should be in the range 50–80°E, but for good atomization it should be less than 3–4°E. Different types of oil require different heating temperatures to attain this viscosity.

b) Flash Point and Ignition Point

The flash point and ignition point are two important combustion properties of liquid fuels. The flash point is the temperature at which a liquid fuel will vaporize and, when it comes in contact with air and flame, flash. This flame does not burn continuously. Flash point can be measured in an open or a closed apparatus but is 20–40°C higher in the closed state.

Ignition is said to take place if the oil, ignited by an external flame, keeps burning for more than 5 seconds. The lowest temperature at which this happens is defined as the ignition temperature.

c) Solidifying (Pour) Point

Solidifying point refers to the temperature at which oil stops flowing, while pour point is the temperature at which oil will just flow. To measure this, oil is placed in a test tube tilted at 45°. The solidifying point is the highest temperature at which the oil does not flow within 5–10 seconds. The pour point is the lowest temperature at which this flow occurs. Paraffin wax content is closely associated with the solidifying point. The oil's solidifying point generally rises with paraffin content.

d) Sulfur

When sulfur is greater than 0.3%, metal corrosion on low temperature heating surfaces needs to be considered. So, depending on the sulfur content, oil can be divided into low sulfur (S < 0.5%), middle sulfur (S = 0.5%–2.0%), and high sulfur (S > 2%) grades.

e) Ash

The ash content of fuel oil is very small. However, it contains vanadium, sodium, and potassium, which lead to corrosion of metal surfaces. This aspect is discussed in more detail in the chapter on corrosion.

3-1-3 Solid Fuels

Coal is a heterogeneous substance formed through millions of years of transformation of plant and mineral matter underground. Depending on the length of this process (known as coalification) we get in ascending order of age solid fuels like peat, lignite, subbituminous coal, bituminous coal, and anthracite. Composition of typical solid fuels is given in Table 3-3.

a) Composition of Coal

Coal consists of an inorganic impurity known as ash (A), moisture (M), and a large number of complex organic compounds. The latter comprise five principal elements: carbon (C), hydrogen (H), oxygen (O), sulfur (S), and nitrogen (N) (Fig. 3-1). For this reason, the chemical analysis of coal is generally determined in terms of these elements. This analysis is called *ultimate analysis*. The mass fraction of these chemical elements in the fuel is determined according to ASTM standard D3176 for coal.

TABLE 3-3. Typical compositions of some solid fuels.*

	Sawdust	Peat	Lignite	Sub-bituminous	Bituminous	Anthracite	Petroleum coke
Proximate %	Dry	DAF					
Moisture	16.5	—^{a}	33.3	23.4	5.2	7.7	5
VM	78.6	68	43.6	40.8	40.2	6.4	1.3
FC			45.3	54	50.7	83.1	83.7
A	5.2		11.1	5.2	9.1	10.5	10
Ultimate (%)	DAF	DAF					
C	51.2	57.5	63.3	72	74	83.7	82
H	6.3	5.5	4.5	5	5.1	1.9	0.5
N		1.9	1	0.95	1.6	0.9	0.7
S		0.1	1.1	0.44	2.3	0.7	0.8
A			11.1	5.2	9.1	10.5	10
O		35	19	16.41	7.9	2.3	6
HHV (kJ/kg)	9880	20,950	16,491	21,376	29,168	27,656	28,377
IDT (C)		1120	1110	1149	1215		

*IDT = initial deformation temperature of the ash, HHV = higher heating value on "as received basis," DAF = dry ash-free basis.
^{a} As received moisture is 93%, but typical commercial peat is about 20% moisture.

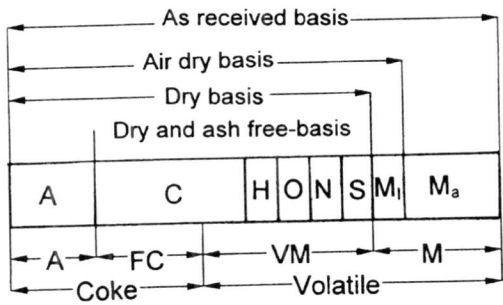

FIGURE 3-1. Composition of coal showing different bases of representation

Owing to the experimental complexity involved in ultimate analysis, another, simpler method, known as *proximate analysis*, is often used in power plants. In proximate analysis coal is considered to consist of four components: volatile matter (*VM*), fixed carbon (*FC*), ash (*A*), and moisture (*M*). It is determined following the ASTM standard D3172.

Carbon, is a major combustible element in the coal. It exists in the form of fixed carbon and volatile matter (CH_4, C_2H_3, CO). The greater the geological age of the coal, the greater the extent of carbonization and the higher the carbon content.

Hydrogen in coal, which accounts for 3–6% of its content, combines with oxygen, producing steam during combustion. This steam in the flue gas is a potential source of heat loss in a boiler.

The oxygen content of coal varies widely. Depending on the degree of carbonization it may move up from 2% for anthracite to 20% for lignite.

The nitrogen content in coal is small (0.5–2%). Coal forms nitrogen oxides during combustion and thus causes environmental pollution.

Sulfur, which is another source of air pollution, exists in three forms: organic sulfur, FeS, and sulfates ($CaSO_4$, $MgSO_4$, and $FeSO_4$). Sulfate is a constituent of ash. It cannot be oxidized. Combustible sulfur includes organic sulfur and FeS. Its heating value is about 900 kJ/kg.

Coal may have moisture in two forms: inherent and surface. The surface moisture (M_a), which is gathered by the coal during storage, etc., can be removed by air drying. However, the inherent moisture (M_i), which is trapped in coal during its geological formation, is not released except during combustion. In any case, these two forms of moisture and that formed through combustion of hydrogen in the coal, contribute to the moisture in flue gas.

b) Ash in Coal

Ash comprises the inorganic solid residues left after the fuel is completely burned. Its primary ingredients are silicon, aluminum, iron, and calcium. Small amounts of compounds of magnesium, titanium, sodium, and potassium may also be present in ash. Ash is determined by heating a sample of coal at 800°C for 2 hours under atmospheric conditions. The procedure is given in ASTM D3174.

Fusion of ash is an important characteristic of coal. It greatly influences the boiler design. Ash fusion temperatures can be measured by ASTM test D1857. When a conical sample of ash is slowly heated it goes through four stages:

1. Initial deformation temperature (IDT) is reached when a slight rounding of the apex of the cone of ash sample occurs.
2. Softening temperature (ST) is reached when the sample is fused down to a spherical lump, whose height is equal to its width.
3. Hemispherical temperature (HT) is marked by further fusion of the ash when the height of the cone is one-half the width of its base.
4. Fluid temperature (FT) is the temperature at which the ash spreads out in a nearly flat layer with a maximum height of 1.6 mm.

c) Analysis of Coal

The ultimate or proximate analyses of coal may be based on different bases depending on the situations. Generally four bases are used: *as received, air dry, dry, dry and ash-free*. A comparison of different bases of analysis of coal is shown in Figure 3-1. When an as received basis is used, the results of ultimate and proximate analyses can be written as follows:

As received basis
    Ultimate:

$$C + H + O + N + S + A + M = 100\% \qquad (3\text{-}1)$$

Proximate:

$$VM + FC + M + A = 100\% \tag{3-2}$$

The percentage amount of each constituent is represented by the constituent in italics. For example C represents the percentage of C or carbon in coal as measured by the ultimate analysis. One can convert the "as received" composition into other bases as follows:

Air dry basis:

$$C^f = \frac{100C}{100 - M_a} \% \tag{3-3}$$

where M_a is the mass of surface moisture removed from 100 kg of moist coal after drying it in air. Similarly, other constituents of coal can be expressed in this basis.

Total dry basis:

$$C^g = \frac{100C}{100 - M} \% \tag{3-4}$$

where M is the total moisture (surface + inherent) in coal, i.e,

$$M = M_a + M_i$$

Other components can be found in a similar way.

Dry and ash-free (DAF) basis:

$$C^r = \frac{100C}{100 - M - A} \% \tag{3-5}$$

where $(100 - M - A)$ is the mass of coal without the moisture and ash.

For a check one should add up percentages of all constituents on any basis to get 100%.

$$C^r + H^r + O^r + N^r + S^r = 100\%$$
$$C^g + H^g + O^g + N^g + S^g + A^g = 100\% \tag{3-6}$$
$$C^f + H^f + O^f + N^f + S^f + Mi^f = 100\%$$

3-1-4 Heating Value of Fuel

If we burn 1 kg of fuel completely and then bring the product gas and solids to the precombustion temperature of the fuel, we get an amount of heat called *higher heating value*, or *HHV*. It is also called *gross calorific value*. It can be measured in a bomb calorimeter using the standard ASTM method D2015.

The exhaust flue gas temperature of a boiler is generally in the range of 120–180°C. Hence, the product of combustion is rarely cooled to the initial temperature of the fuel, which is generally below the condensation temperature of steam. The water vapor in the flue gas does not condense, and the latent heat of vaporization is not recovered. Thus the effective heat available for use in the boiler is less than the chemical energy stored in the fuel. This lower heating value *(LHV)* is equal to the higher heating value less the condensation heat of water vapor in the flue gas. The relationship between higher heating value and lower heating value is given by

$$LHV = HHV - r\left(\frac{9H}{100} + \frac{M}{100}\right) \qquad (3\text{-}7)$$

where *LHV*, *HHV*, *H*, and *M* are lower heating value, higher heating value, hydrogen percentage, and moisture percentage, respectively, on an as received basis. Here, *r* is the latent heat of steam in the same units as *HHV*.

3-2 Stoichiometric Calculations

The amount of air required or the product gas produced in the combustion of a unit mass of fuel can be calculated from the chemical reactions involved. These amounts are the theoretical amount; they are designated by the superscript $(^0)$. Owing to imperfect mixing, combustion reactions require more than this amount of the air. So the actual amounts are higher and are expressed without the superscript. In this chapter all these amounts are expressed as volumes of gas at $0°C$ and 1 atm pressure (Nm^3) for 1 kg of fuel burned.

3-2-1 Theoretical Air Requirement and Excess Air Coefficient

Combustion is a rapid, high-temperature, and exothermic oxidation of a fuel. The reaction requires a minimum temperature (ignition temperature) and oxygen within a certain range of concentration (flammability limit). The latter requirement is less relevant for solid and liquid fuels where the oxygen and combustibles mix in the reaction zone itself.

In industrial combustion equipment, oxygen comes from air. Since the combustion is a simple oxidation process, the amount of air required for the combustion of a fuel may be easily calculated from the stoichiometry or the chemical formula of each combustible element of the fuel. The theoretical amount of air required to burn 1 kg of fuel completely, V^0 (Nm^3/kgf), is often expressed as volume at $0°C$ and 1 atm pressure since 1 kmol of any ideal gas occupies a volume of 22.4 Nm^3 at this temperature and pressure. The simple combustion reaction of carbon is

$$C + O_2 = CO_2 + 32,790 \text{ kJ/kg of carbon}$$

From elementary chemistry we know that 1 kmol of carbon combines with 1 kmol of O_2 to produce 1 kmol of CO_2. So, 12 kg carbon needs 22.4 Nm^3 of O_2 to produce the 22.4 Nm^3 of CO_2. In other words, 1 kg of carbon needs 1.866 Nm^3 O_2 to produce 1.866 Nm^3 CO_2. From Eq. (3.1) we find that 1 kg coal contains (C/100) kg carbon. Therefore, to burn the carbon in 1 kg of coal, we need (1.866 C/100) Nm^3 oxygen.

In the same way, we find that the oxygen required for combustion of the hydrogen and sulfur elements in 1 kg coal is $5.56H/100$ Nm^3 and $0.7 S/100$ Nm^3, respectively. The oxygen content of 1 kg coal can be derived in the same way as

TABLE 2-4. Recommended values of excess air coefficient α_1 at the exit of furnace.

Firing methods	Pulverized coal furnace with dry bottom	Slag-tap furnace	Bubbling fluidized bed furnace	Oil fired furnace and Gas fired furnace		Circulating fluidized bed
Types of fuel	Anthracite low-rank bituminous	Bituminous lignite	All types of coal	Negative pressure in furnace	Slightly positive pressure	All types of coal
α_1	1.20–1.25	1.15–1.20	1.3–1.5	1.08–1.10	1.05–1.07	1.2

0.7 O/100 Nm^3. This amount, however, reduces the combustion oxygen required from the air.

Thus, the theoretical volume of oxygen required to burn 1 kg of fuel completely is

$$V_{O_2}^0 = \left(1.866\frac{C}{100} + 5.56\frac{H}{100} + 0.7\frac{S}{100} - 0.7\frac{O}{100}\right) \text{Nm}^3/\text{kgf} \qquad (3\text{-}8)$$

Air contains about 21% oxygen by volume. Therefore, when 1 kg of solid, liquid, or gaseous fuel burns completely it requires an air volume of

$$V^0 = \frac{V_{O_2}^0}{0.21}\left(1.866\frac{C}{100} + 5.56\frac{H}{100} + 0.7\frac{S}{100} - 0.7\frac{O}{100}\right)$$

$$= 0.0889\,(C + 0.375\,S) + 0.265\,H - 0.033\,O \text{ Nm}^3/\text{kgf} \qquad (3\text{-}9)$$

The combustion products of carbon and sulfur are often combined as tri-atomic gases. Their reaction can be written as $R + O_2 = RO_2$, by substituting $R = C + 0.375\,S$ in Eq. (3-9).

As indicated earlier, to achieve complete combustion, an excess amount of air is added to compensate for the imperfect mixing of air and fuel. The *excess air coefficient* (α) is defined as the ratio of volumes of actual air to the theoretical air.

$$\frac{V}{V^0} = \alpha \qquad (3\text{-}10)$$

While it helps combustion, the excess air also increases the heat loss through the stack of the furnace. So, an excess air coefficient is called optimum when the heat losses are minimum. The optimum excess air value depends on the type of fuel, firing methods and furnace and burner geometry. From experience, boiler designers recommend the following values of excess air coefficient α_1 for different firing systems (Table 2-4).

In a negative pressure furnace, cold ambient air leaks into the furnace through various openings along its gas path. So, owing to the leakage, the excess air increases along the flue gas duct. The amount of air leakage, ΔV (for 1 kg fuel), is expressed as the ratio of air leakage to the theoretical air volume V^0 and is known as the air leakage coefficient, $\Delta\alpha$.

$$\Delta\alpha = \frac{\Delta V}{V^0} \qquad (3\text{-}11)$$

In any cross-section of the gas duct beyond the furnace, the excess air coefficient is the sum of leakage in upstream sections.

$$\alpha = \alpha_e + \Sigma \Delta\alpha \qquad (3\text{-}12)$$

where α and α_e are, respectively, the total excess air coefficients at the given section of the boiler and at the furnace exit.

The leakage of cold air increases the volume and reduces the temperature of the flue gas. Consequently, the heat absorption from the flue gas drops, thereby increasing the exhaust gas temperature. This adversely affects the cost of boiler operation by increasing the stack gas loss and power consumption of the induced draft fan. For example, in a pulverized coal fired boiler, when the air leakage coefficient increases from 0.1 to 0.2, the stack gas temperature rises by 3–8°C, and the boiler efficiency decreases by 0.2–0.5%. When the air leakage coefficient increases by 0.1, the FD fan and the ID fan power consumption will increase by 2 kW/MW electric power. Therefore, appropriate measures must be taken to reduce the air leakage in a boiler.

3-2-2 Calculation of Combustion Products

a) Theoretical Gas Volume

The flue gas generally consists of CO_2, SO_2, O_2, N_2, and H_2O. Theoretical flue gas volume is the volume of gaseous products of complete combustion of the fuel burning with the theoretical amount of air. From the chemical reaction equations in section 3-2-1, one can compute the flue gas volume for complete combustion of 1 kg fuel as follows.

(1) CO_2

The carbon dioxide is produced from both combustion of carbon in the fuel and the calcination of sorbents ($CaCO_3$). From the limestone decomposition reaction ($CaCO_3 = CaO + O_2$) we can find the additional CO_2 as $(0.7R'S/100)$ where R' is the molar ratio of calcium and sulfur in the fuel and sorbent combined. The amount of CO_2 produced is

$$V_{CO_2} = 1.866\frac{C}{100} + 0.7\frac{R'S}{100} \text{ Nm}^3/\text{kgf} \qquad (3\text{-}13)$$

This expression neglects the presence of magnesium in the limestone and any CaO in the coal.

(2) SO_2

Sulfur dioxide is produced by the oxidation of sulfur in the coal, but some of it is also reduced in the furnace through a reaction with the sorbent ($CaO + SO_2 + 1/2 O_2 = CaSO_4$). The limestone sorbent fed to the furnace is rarely fully utilized owing to the blockage of pores of CaO particles. What fraction can be utilized also depends on a number of factors. So, we assume that only a fraction, E_{sor}, of the sulfur of the coal is retained as $CaSO_4$. The resultant volume of SO_2 produced from 1 kg of fuel burnt is

$$V_{SO_2} = 0.7 \frac{S}{100}(1 - E_{sor}) \text{ Nm}^3/\text{kgf} \tag{3-14}$$

Sometimes, V_{RO_2} is used to represent the sum of V_{CO_2} and V_{SO_2}:

$$V_{RO_2} = V_{CO_2} + V_{SO_2} = 1.866 \frac{C + 0.375S(1 + R' - E_{sor})}{100} \text{ Nm}^3/\text{kgf}$$

(3) Nitrogen

Nitrogen comes from both air and the fuel. Therefore, theoretical nitrogen volume is

$$V_{N_2}^0 = 0.79V^0 + 0.8 \frac{N}{100} \text{ Nm}^3/\text{kgf} \tag{3-15}$$

(4) Water

Water vapor comes from

1. Combustion of hydrogen in the fuel ($11.1 \frac{H}{100}$ Nm^3/kgf),
2. Moisture in the fuel ($\frac{22.4}{18} \frac{M}{100}$ Nm^3/kgf), and
3. Moisture in combustion air ($\frac{1.293 V^0}{0.804} X_m$ Nm^3/kgf),

where 1.293 and 0.804 are the density of air and water vapor at standard conditions. The moisture content in dry air (X_m kg/kg) depends on the relative humidity and barometric pressure. A typical value of X_m is 0.013. Thus the total volume of water from 1 kg fuel is

$$V_{H_2O}^0 = 11.1 \frac{H}{100} + 1.24 \frac{W}{100} + 1.6V^0 X_m \text{ Nm}^3/\text{kgf} \tag{3-16}$$

When liquid fuel burns, steam is often used for oil fuel atomization. Atomized steam becomes a part of water in the flue gases. If 1 kg of oil needs G_{as} kg of atomizing steam, the water vapor volume is

$$V_{H_2O}^0 = 11.1 \frac{H}{100} + 1.24 \frac{W}{100} + 1.6V^0 X_m + 1.24G_{as} \text{ Nm}^3/\text{kgf} \tag{3-17}$$

Therefore, theoretical flue gas volume is the sum of volumes of tri-atomic gases, nitrogen, and the water vapor.

$$V_g^0 = V_{RO_2} + V_{N_2}^0 + V_{H_2O}^0 \text{ Nm}^3/\text{kgf} \tag{3-18}$$

b) Actual Flue Gas Volume

In practice a boiler operates with a certain amount of excess air. So, the total flue gas volume (V_g) is the sum of the theoretical flue gas volume (V_g^0) the excess air $(\alpha - 1)V^0$, and the water vapor [$1.6(\alpha - 1)V^0 - X_m$] carried with the excess air.

$$V_g = V_g^0 + (\alpha - 1)V^0 + (\alpha - 1)V^0 X_m \text{ Nm}^3/\text{kgf} \tag{3-19}$$

In Eq. (3-17) we add the additional moisture in the excess air α to get the total water vapor volume in flue gas volume as

$$V_{H_2O} = 11.1\frac{H}{100} + 1.24\frac{M}{100} + 1.24G_{as} + 1.6\alpha V^0 X_m \text{ Nm}^3/\text{kgf} \qquad (3\text{-}20)$$

The actual volume of dry flue gas is obtained by subtracting the total water from the V_g

$$V_{dg} = V_g - V_{H_2O} = V_{RO_2} + V_{N_2}^0 + (\alpha - 1)V^0 \qquad (3\text{-}21)$$

c) Heat Transfer Parameters

Tri-atomic gases and water vapor in the flue gas contribute to the nonluminous radiation in the furnace. It is influenced by their partial pressure. The partial pressure is equal to the volume fraction of the gas in the mixture. These are to be calculated separately.

Volume fraction of CO_2 and SO_2: $r_{RO_2} = \dfrac{V_{RO_2}}{V_g}$

Partial pressure of CO_2 plus SO_2: $P_{RO_2} = r_{RO_2}P$

Volume fraction of water vapor: $r_{H_2O} = \dfrac{V_{H_2O}}{V_g}$

Partial pressure of water vapor: $P_{H_2O} = r_{H_2O}P$ (3-22)

where P is the total pressure of the flue gas.

The ash particles contribute to luminous radiation in the furnace. If x_{fa} is the fraction of total ash in fuel appearing as fly ash, the concentration of the fly ash, μ, is

$$\mu = \frac{Ax_{fa}}{100G_g}.$$

The volume fraction of fly ash in the flue gas is

$$\sigma = \frac{Ax_{fa}}{100V_g}$$

The mass of flue gas, G_g, includes the mass of fuel changed into gases $(1 - A/100)$, mass of the fuel atomizing steam, and the mass of wet air consumed in burning 1 kg fuel. So, the flue gas mass G_g is given by

$$G_g = 1 - \frac{A}{100} + G_{as} + 1.306\alpha V^0 \text{ kg/kgf} \qquad (3\text{-}23)$$

Here the density of air (1.306) includes the moisture added to the dry air. Fly ash coefficient x_{fa} is the ratio of fly ash in flue gas to total ash in the fuel.

3-2-3 Incomplete Combustion

The above flue gas calculations are based on the complete combustion of fuel producing only CO_2, SO_2, and H_2O. However, the combustion is sometimes incomplete, which may at times be allowed on purpose for the reduction of NO_x. So there is some tradeoff with increases in unburnt carbon and CO. Thus, in addition to CO_2, the combustion product may also include CO, hydrogen, and C_mH_n. In modern boilers, the H_2 and C_mH_n content of the flue is so small that it can be neglected. Therefore, CO alone is regarded as the product of incomplete combustion. The carbon combustion reactions are

$$C + O_2 = CO_2$$
$$C + \frac{1}{2}O_2 = CO \tag{3-24}$$

From this we note that 1 mol of carbon produces 1 mol of combustion product irrespective of whether the product is CO_2 or CO. Thus, the total volume of carbon combustion products is unchanged.

Under incomplete combustion, if CO is the only incomplete combustion product, the dry flue gas composition would be

$$CO_2 + SO_2 + O_2 + CO + N_2^k + N_2^r = 100 \tag{3-25}$$

where N_2^k = nitrogen carried with air in dry flue, % volume; and N_2^{-r} = nitrogen released from fuel itself, % volume; and CO_2, SO_2, O_2, and CO are volume concentrations in percentage of the CO_2, SO_2, O_2, and CO in the flue gas.

An oxygen balance gives the incomplete combustion equation as (Afanasayev, 1984)

$$RO_2 + 0.605CO + O_2 + \beta(RO_2 + CO) = 21 \tag{3-26}$$

where $RO_2 = CO_2 + SO_2$. The coefficient β is given by

$$\beta = 2.37 \frac{H - \dfrac{O}{8} + 0.038N}{C + 0.375S} \tag{3-27}$$

This equation may be used to estimate the amount of CO in the flue gas from measured values of oxygen, carbon dioxide, and sulfur dioxide in the flue gas and the analysis of the fuel.

The above analysis is particularly important for oil fired systems. However, for coal fired systems solid carbon is the major source of incomplete combustion. During the design stage the unburnt carbon, as a percentage of the fuel feed, is taken from experience. For performance tests the unburnt carbon is measured by taking samples of fly ash and bottom ash. Fluidized bed boilers, using sorbents, would have carbonates in the ash. The CO_2 from the decomposition of carbonate is measured (ASTM method D1756) to separate this source of carbon from the rest in the solid residues. The total carbon mass is corrected to unburnt carbon mass by subtracting the carbon in the carbonate which is

$$CO_2 \frac{12}{44}$$

Thus,

$$\text{Unburnt carbon} = \frac{X_c(A + X_s)}{100(100 - X_c)} \text{ kg carbon/kg of fuel} \qquad (3\text{-}28)$$

where X_c is percentage measured carbon in residue and X_s is the number of kilograms of spent sorbent per 100 kg of fuel burnt, and A is the ash percentage in the fuel.

3-2-4 Determination of Excess Air Coefficient

For an operating boiler, the excess air coefficient is often determined from the analysis of the flue gas. From Eq. 3-10, excess air coefficient is defined as

$$\alpha = \frac{V}{V^0} = \frac{V}{V - \Delta V} = \frac{1}{1 - \dfrac{\Delta V}{V}} \qquad (3\text{-}29)$$

where ΔV — excess air Nm³/kgf.

In the case of complete combustion, the excess air can be expressed by oxygen volume content in dry air.

$$\Delta V = V - V^0 = \frac{V_{O_2}}{0.21} = \frac{O_2}{21} V_g \text{ Nm}^3/\text{kgf} \qquad (3\text{-}30)$$

where V_{O_2} is the oxygen (Nm³) in the flue gas from 1 kg fuel burnt.

The actual amount of air entering into the furnace can be determined from the nitrogen content of the dry flue gas.

$$V = \frac{V_{N_2} - 0.8\dfrac{N}{100}}{0.79} \qquad (3\text{-}31)$$

where N is the nitrogen percentage in the fuel and V_{N_2} is the volume (Nm³/kgf) of nitrogen produced by burning 1 kg of fuel. The factor 0.8 converts weight percentage into volume percentage of nitrogen in the fuel.

Since the nitrogen content of a fuel is very small the entire nitrogen in the flue gas can be taken to be from the combustion air. So, Eq. (3-31) is simplified as

$$V = \frac{V_{N_2}}{0.79} = \frac{\dfrac{N_2}{100} V_g}{0.79} = \frac{N_2}{79} V_g \qquad (3\text{-}32)$$

where N_2 is percentage of nitrogen in the flue gas volume, V_g per unit weight of fuel burnt.

Substituting Eqs. (3-31) and (3-32) into Eq. (3-29), we get

$$\alpha = \cfrac{1}{\left(1 - \cfrac{\cfrac{O_2}{21}V_g}{\cfrac{N_2}{79}V_g}\right)} = \cfrac{1}{1 - \cfrac{79}{21}\cfrac{O_2}{N_2}} = \cfrac{1}{1 - 3.76\cfrac{O_2}{N_2}} \tag{3-33}$$

When CO is the only incomplete combustion product, the oxygen from flue gas analysis includes oxygen in excess air and the oxygen that is not used by carbon owing to incomplete combustion. In the reaction $C + \frac{1}{2}O_2 = CO$, the oxygen unused by carbon accounts for $0.5CO$ in dry flue gas. Therefore, oxygen content in dry flue gas should be $O_2 - 0.5CO$. So excess air volume is

$$\Delta V = \frac{O_2 - 0.5CO}{21}V_g \tag{3-34}$$

From Eq. (3-26), nitrogen content of the flue gas is

$$N_2 = 100 - (RO_2 + O_2 + CO) \tag{3-35}$$

Substituting ΔV and N_2 into Eq. (3-29), and using the definition of excess air coefficient, we get an expression for the latter for incomplete combustion

$$\alpha = \cfrac{1}{\left[1 - \cfrac{\cfrac{O_2 - 0.5CO}{21}V_g}{\cfrac{100 - (RO_2 + O_2 + CO)}{79}V_g}\right]} \tag{3-36}$$

Using Eq. (3-27), we can write this expression in the form of measurable parameters:

$$\alpha = \frac{21(79 + \beta RO_2)}{(79 + 100\beta)RO_2} \tag{3-37}$$

The excess air coefficient has a direct influence on the combustion of fuel in the furnace and heat loss of stack gases. An accurate and fast measurement is a major requirement in ensuring the economic operation of a boiler.

The quality of coal supplied to a power plant often varies. When the fuel composition varies, its combustion characteristics also change. So, even if the RO_2 or $(CO_2 + SO_2)$ content of the flue gas is kept constant during operation, the excess air coefficient may still change. This indicates that the use of RO_2 to monitor the excess air coefficients is not very reliable. The same value of RO_2 may represent different excess air coefficients for anthracite, lignite, heavy oil, and natural gas. Figure 3-2 shows the variation of CO_2 and O_2 with excess air for different fuels. However, variation of fuel has little effect on the dependence of excess air on the oxygen content of the flue gas. Therefore, for a pulverized coal fired boiler, if excess oxygen is used to monitor excess air coefficient, coal variation would have a minor influence on the excess air coefficient.

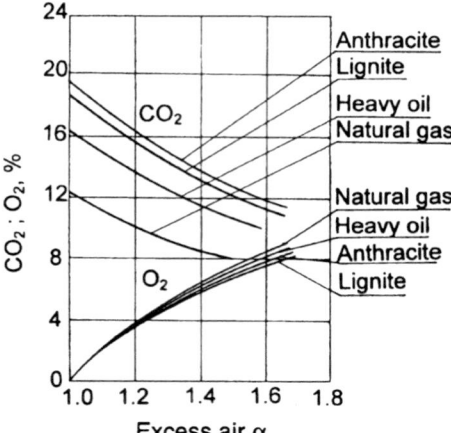

FIGURE 3-2. Variation of CO_2 and O_2 for different values of excess air and (1) anthracite, (2) lignite, (3) heavy oil, and (4) natural gas

3-3 Enthalpy Calculation of Air and Combustion Products

The calculation of the enthalpy of a gas is an important part of the design or performance analysis of a boiler. The enthalpy is the thermal energy of unit mass above a certain base. Here we will use the standard condition (0°C at 1 atm pressure) as the baseline condition. Enthalpy calculations for air or flue gas are on the basis of 1 kg fuel fired. For example, the enthalpy of flue gas, H_g, is the thermal energy of the entire flue gas produced from the combustion of 1 kg of fuel. It is not the energy per unit mass of the flue gas.

3-3-1 Enthalpy Calculation of Air

The enthalpy of total air required for combustion of 1 kg fuel is

$$H_a = \alpha H_a^0 = \alpha V^0 (CT)_a \text{ kJ/kgf} \tag{3-38}$$

The $(CT)_a$ is the enthalpy of wet air per standard cubic meter at temperature T °C. It may be found from an enthalpy table. Alternately, it may be calculated as the product of specific volume heat C (kJ/Nm^3) at T °C, and the temperature T (°C). The specific volume heat, which is the product of density and specific mass heat, is a function of temperature. Here their values are to evaluated at the mean temperature of 0 to T °C.

3-3-2 Enthalpy of Flue Gas

The enthalpy of combustion products produced from the combustion of 1 kg of fuel, denoted as H_g, is found by adding the enthalpy of each component of the flue gas. Since the specific heats of various gases are different, the enthalpy of each component is calculated separately. For instance, the enthalpy of the flue gas at a temperature $T\,°C$, is

$$H_g = H_g^0 + (\alpha - 1)V^0(CT)_k + \frac{A}{100}x_{fa}(CT)_a \text{ kJ/kgf} \qquad (3\text{-}39)$$

where enthalpy for theoretical amount of flue gas, H_g^0, is given as

$$H_g^0 = V_{RO_2}(CT)_{RO_2} + V_{N_2}^0(CT)_{N_2} + V_{H_2O}(CT)_{H_2O} \text{ kJ/kgf} \qquad (3\text{-}40)$$

This includes enthalpies of V_{RO_2} Nm^3 of CO_2 and SO_2, $V_{N_2}^0$ Nm^3 of nitrogen, V_{H_2O} Nm^3 of water vapor, $(\alpha - 1)V^0$ Nm^3 of wet excess air and $(\frac{A}{100}x_{fa})$ kg of fly ash. If the percentage of ash is very low, as in low ash fuels, the fly ash enthalpy may be ignored.

3-4 Heat Balance

Heat absorbed by the steam comes from the heat released by fuel combustion. For a variety of reasons, the fuel does not burn completely; and also the heat released cannot be fully utilized. Heat loss is unavoidable. A heat balance shows how much heat is effectively used and how much of it is wasted. The purpose of a heat balance is to identify the sources of heat loss and to find means to reduce them and thereby improve boiler efficiency.

3-4-1 Concept of Heat Balance

Figure 3-3 schematically shows the heat balance of a pulverized coal (PC) fired furnace. In a steady state the heat input, heat utilization, and heat losses in 1 kg fuel may be given as

$$Q = Q_1 + Q_2 + Q_3 + Q_4 + Q_5 + Q_6 \qquad (3\text{-}41)$$

where Q = available heat of fuel fired
Q_1 = heat absorbed by steam (utilized by the boiler)
Q_2 = heat loss through stack gas
Q_3 = heat loss by incomplete combustion
Q_4 = heat loss owing to unburned carbon
Q_5 = heat loss owing to convection and radiation from the furnace exterior
Q_6 = heat loss through the sensible heat of ash and slag

To express the heat losses in percentages we divide both sides of Eq. (3-41) with Q:

$$100 = q_1 + q_2 + q_3 + q_4 + q_5 + q_6 \qquad (3\text{-}42)$$

where $q_i = \frac{Q_i}{Q}100$ represents the percentage of available heat.

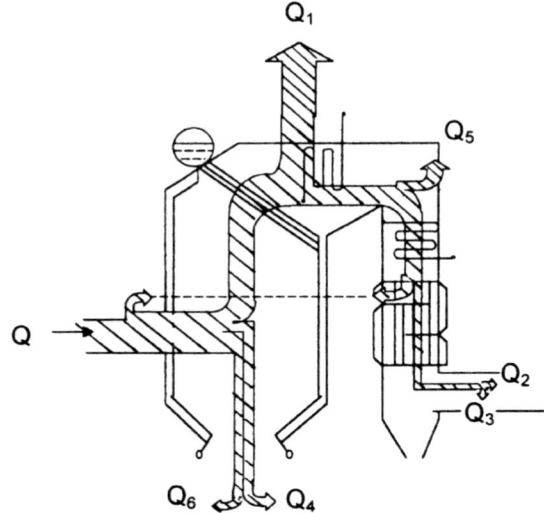

FIGURE 3-3. Schematic representation of heat balance of a boiler

Boiler efficiency, η_{bo}, is defined by the ratio of heat absorbed by the boiler and the heat provided by the fuel. It is given by

$$\eta_{bo} = q_1 = \frac{Q_1}{Q} 100 = 100 - (q_2 + q_3 + q_4 + q_5 + q_6) \qquad (3\text{-}43)$$

3-4-2 Input Heat to Boiler

The total heat input to coal fired or oil fired boilers is given by

$$Q = LHV + H_f + Q_{wr} + Q_{wh} - Q_{cd} \text{ kJ/kg} \qquad (3\text{-}44)$$

where LHV = lower heating value of fuel, as received basis, kJ/kg
  H_f = sensible heat of fuel
  Q_{wr} = sensible heat carried by the air when heated by external air heater
  Q_{wh} = total heat of the fuel atomizing steam, kJ/kg
  Q_{cd} = heat spent in decomposition of carbonates in oil shale or limestone fed as sorbent, it is proportional to the amount of CO_2 released by decomposition

The sensible heat of fuel is given by

$$H_f = C_{pf} T_f \text{ kJ/kg} \qquad (3\text{-}45)$$

where C_{pf} = specific heat of fuel, as received basis
  T_f = fuel temperature at burner or feeder exit

TABLE 3-5. Specific heat (kJ/kg °C) of coal on dry basis.

Anthracite	Bituminous	Lignite	Oil-shale	Heavy oil (C_{pf})
0.92	1.09	1.13	0.88	$0.415 + 0.0006\, T_f$

The sensible heat of coal, H_f, may be neglected if the coal is not preheated, but if its moisture percentage, M, exceeds ($LHV/1500$), it is necessary to consider the H_f.

Specific heat of fuel is calculated based on the equation

$$C_{pf} = C_{pf}^g \frac{100 - M}{100} + \frac{M}{100} 4.186 \text{ kJ/kg} \qquad (3\text{-}46)$$

where C_{pf}^g = specific heat, as dry basis, kJ/kg.

For different types of coal, C_{pf}^g is chosen from Table 3-5:

The specific heat of fuel, C_{pf}, is sometimes approximated as 2.09 kJ/kg°C.

The sensible heat carried by the air when heated by an external air heater is calculated as

$$Q_{wr} = \alpha(H_b^\circ - H_a^\circ) \text{ kJ/kgf} \qquad (3\text{-}47)$$

where H_b° and H_a° are enthalpies of theoretical air (per kg fuel burnt) at burner entrance and at ambient conditions, respectively.

When the combustion air is not preheated by an external air heater, $Q_{wr} = 0$. The enthalpy of atomizing steam is calculated from

$$Q_{as} = G_{as}(H_{as} - 2500) \text{ kJ/kgf} \qquad (3\text{-}48)$$

where G_{as} = steam consumption for oil atomization (usually 0.05-0.1 kg steam/kg oil)

H_{as} = enthalpy of atomization steam (kJ/kg)

2500 = enthalpy of steam at average exit conditions of flue gas (kJ/kg)

3-4-3 Heat Losses

Different heat loss terms in Eq. (3-42) are analyzed below.

a) Waste Gas Losses (Q_2)

The enthalpy of flue gas escaping from the boiler is higher than that of combustion air entering the boiler. So, there is a net heat loss. This loss is given by

$$q_2 = \frac{Q_2}{Q_r} 100 = \frac{\left(I_g - \alpha_{ah} I_a^0\right) \dfrac{100 - q_4}{100}}{Q} 100 \text{ \%} \qquad (3\text{-}49)$$

where I_g = flue gas enthalpy at the exit of air heater, kJ/kg fuel

I_a^0 = theoretical cold air enthalpy entering boiler, kJ/kg fuel

α_{ah} = excess air coefficient at the exit of air heater
$(100 - q_4)/100$ = correction factor owing to the difference between
calculated and actual fuel consumption

This loss (q_2) increases with increase in exit flue gas temperature. Generally q_2 increases by 1% when the exit flue gas temperature increases by 10°C. So it is desirable to reduce the exit gas temperature as much as possible. However, when the exit flue gas temperature is below the dew point, the sulfur dioxide of the gas deposits as sulfuric acid. So, the corrosion of metals in the air heater sets in. The flue gas from the combustion of a high sulfur fuel would have a higher dew point. Therefore, a boiler designed for this fuel would require a higher exit gas temperature. For large boilers, the stack gas temperature is chosen in the range of 110–180°C.

b) Heat Loss by Incomplete Combustion of Gaseous Components (Q_3)

Heat loss owing to incomplete combustion is caused by escape of combustible gases, viz., CO, H_2, CH_4, with the flue gas. Heat loss by incomplete combustion per 1 kg fuel is the sum of combustible gas volume multiplied by heating value. It is also corrected for unburnt solid carbon loss, q_4:

$$q_3 = Q_3\ 100/Q = V_g(126.4\,CO + 108.0\,H_2 + 358.2\,CH_4)\frac{100 - q_4}{100} \cdot \frac{100}{Q} \quad (3\text{-}50)$$

where CO, H_2, and CH_4 are volume percentages of carbon monoxide, hydrogen, and methane in the flue gas.

A good mixing between the fuel and air and proper furnace aerodynamics can reduce the combustible loss. For reduction of CO loss, the furnace temperature should not be too low. The incomplete combustion loss (q_3) is generally small. For example, in a pulverized coal furnace, $q_3 = 0$; gas or liquid fuel fired furnace, $q_3 = 0.5\%$; stoker firing of coal, $q_3 = 0.5–1.0\%$.

c) Heat Loss Owing to Unburned Carbon (Q_4)

Unburned carbon is present in the bottom (bed drain) ash and in the fly ash. So the total heat loss owing to unburned carbon (Q_4) is the sum of carbon in fly ash (Q_{fa}) and bottom ash (Q_{ba}) multiplied by their heating values. The heat loss q_4 as percentage of heat input is

$$q_4 = 100\frac{Q_4}{Q} = \frac{32{,}866}{Q}\left(\frac{G_{ba}C_{ba} + G_{fa}C_{fa}}{100B}\right)100\ \% \quad (3\text{-}51)$$

where G_{ba} = amount of bottom ash produced over one hour operating time, kg/h

G_{fa} = amount of fly ash generated over one hour operating time, kg/h

C_{ba}, C_{fa} = carbon contents in bottom ash and fly ash respectively, %
B = actual fuel consumption during the above one hour period, kg/h

Ash leaves the bed either through bed drain or fly ash. We define x_{ba} and x_{fa} as the mass fraction of total ash exiting through the bed drain and fly ash, respectively. An ash balance is carried out to calculate the amounts of fly ash:

$$x_{fa} + x_{ba} = 1 \tag{3-52}$$

The individual ash fraction can be found as follows:

$$x_{ba} = \frac{G_{ba}(100 - C_{ba})}{BA}$$

$$x_{fa} = \frac{G_{fa}(100 - C_{fa})}{BA} \tag{3-53}$$

where A is the ash percentage in fuel.

Substituting Eq. (3-51), we get

$$q_4 = \frac{32{,}866\,A}{Q} \left[\frac{x_{ba}C_{ba}}{100 - C_{ba}} + \frac{(1 - x_{ba})C_{fa}}{100 - C_{fa}} \right] \tag{3-54}$$

At the design stage of the boiler the following values of q_4 may be taken from experience, as shown below.

For a pulverized coal fired boiler:

> bituminous coal and slag-tap furnace, $q_4 \leq 2\%$
>
> bituminous coal and dry bottom furnace, $q_4 \leq 3\%$
>
> anthracite, $q_4 \leq 4\%$
>
> For gas fired and oil fired boilers, $q_4 = 0$

The heat loss owing to unburned carbon depends on the types of fuel, furnace and firing equipment construction, boiler load, operating conditions, furnace temperature, and the air–fuel mixture.

d) Heat Loss Owing to Surface Radiation and Convection (Q_5)

When a boiler is in operation, the external surface temperature of the furnace, flue gas ducts, steam tubes, and headers is higher than that of the ambient. The heat loss is caused by heat transfer from the surfaces to ambient through convection and radiation. The heat loss primarily depends on the surface area of furnace wall, insulating layer of tubes, and ambient temperature. The heat loss is calculated by

$$Q_5 = \frac{\Sigma F_{sb}}{B}(h_c + h_r)(T_{sb} - T_0) \text{ kJ/kgf} \tag{3-55}$$

$$\text{and} \quad q_5 = \frac{Q_5}{Q} 100 \%$$

where ΣF_{sb} = external surface area of boiler exposed to the ambient, m^2
h_c = coefficient of heat transfer by convection, kW/(m^2°C)

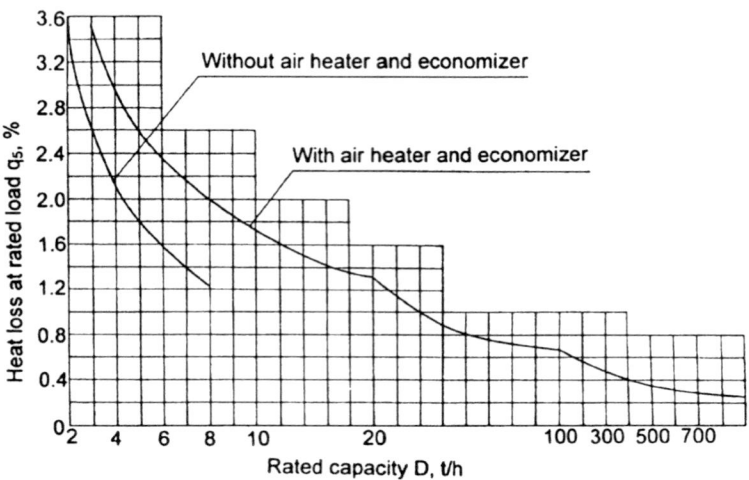

FIGURE 3-4. Radiative and convective heat losses from the external surface of typical pulverized coal fired boilers

h_r = coefficient of heat transfer by radiation, $kW/(m^2 °C)$
T_{sb} = average temperature of the boiler surface, °C
T_0 = average ambient temperature, °C
B = fuel consumption in kg/s

Fuel consumption increases in direct proportion to boiler capacity, but the exterior surface of the boiler does not increase at that proportion. So, the convection and radiation loss from the boiler as percentage of the heat input q_5 decreases with increase in boiler capacity. Since the heat loss is very difficult to measure, q_5 is obtained from experimental data. Figure 3-4 shows typical values for PC boilers. Similar values for circulating fluidized bed boilers are given in Basu and Fraser (1991). Modern water wall cooled circulating fluidized bed boilers have a lower radiation loss than that given in Basu and Fraser (1991).

When a given boiler operates at low load, the skin temperature of the outer casing of the boiler does not change much. Thus, while the heat loss Q_5 does not change much, the heat input Q decreases. So, the part load heat loss q_5' increases with decreasing boiler load.

$$q_5' = q_5 \frac{D}{D'} \qquad (3\text{-}56)$$

where q_5, q_5' = heat loss at rated and operating load, respectively
D, D' = boiler rating and operational capacity, respectively, kg/s

In boiler calculations, heat losses owing to surface radiation and convection in each section of the heating surface should be considered. To simplify the calculation, the difference in design and ambient temperature in different sections of flue gas ducts is ignored, and the heat loss is assigned in proportion to the amount of heat transfer from each section. Under these conditions, a coefficient φ, known as the HEAT PRESERVATION COEFFICIENT, is used to calculate the heat loss.

$$\varphi = \frac{\text{Heat transferred from heating surface to steam or water}}{(\text{Heat transfer from surface to steam or water} + \text{Heat transfer from surface to ambient})}$$

The heat preservation coefficient is also equal to the ratio of heat released by flue gas to heat absorbed by steam or water. If we assume it to be constant for all steam/water sections of the flue gas passage, φ is given by

$$\varphi = \frac{Q_1 + Q_{ah}}{Q_1 + Q_{ah} + Q_5} \tag{3-57}$$

where Q_{ah} = heat absorbed by air heater
Q_1 = heat absorbed by water and steam

When a boiler has no air heater, or the ratio of heat absorbed by the air heater is very small compared to that by steam and water, φ can be shown to be given by

$$1 - \varphi = \frac{q_5}{\eta_{bo} + q_5} \tag{3-58}$$

where η_{bo} = boiler efficiency.

e) Heat Loss of Ash and Slag (Q_6)

When solid fuel is burned, ash and slag leave the furnace at a rather high temperature (about 600–800°C). This results in sensible heat loss in ash and slag. The heat loss depends on fuel ash content, fuel heating value, and slag deposition method. For a high ash content and low heating value fuel, this loss, q_6, is large. The sensible heat loss of a slag-tap furnace is larger than that of a furnace with dry bottom. A fluidized bed boiler using high ash coal or sorbents also loses heat through bottom ash or bed drain. So, for these boilers, this heat loss should be considered. For a pulverized coal fired furnace this loss needs to be considered only for high ash coal.

$$A > \frac{LHV}{419}$$

For a slag-tap furnace, sensible heat loss can be calculated by

$$Q_6 = x_{ba} \frac{100}{100 - C_{ba}} C_{pas} \frac{A}{100} T_{slag} \tag{3-59}$$

where C_{pas} = specific heat of ash and slag, kJ/kg C
x_{ba} = fraction of bottom ash
C_{ba} = carbon contents in bottom ash
A = ash percentage in fuel

If the slag temperature is not known, one could take the following as a first approximation:

$$T_{slag} = 600°C \text{ for dry bottom PC boiler (Parilov \& Ushakov, 1989)}$$
$$= T_m + 100°C \text{ for slagging bottom PC boiler}$$
$$= 300°C \text{ for CFB or bubbling boiler with heat recovery ash cooler}$$
$$= 800°C \text{ for CFB or bubbling bed boiler without ash cooler}$$

3-4-4 Thermal Efficiency and Fuel Consumption of Boilers

The combustion heat of the fuel is partly absorbed by the water and steam. The heat utilized, Q_1, is calculated from

$$Q_1 = \frac{D_{sup}(H''_{sup} - H_{fw}) + D_{rh}(H''_{rh} - H'_{rh}) + D_{bw}(H' - H_{fw})}{B} \tag{3-60}$$

where
B = fuel consumption, kg/s
D_{sup} = superheated steam flow rate, kg/s
H''_{sup} = enthalpy of superheated steam, kJ/kg
H_{fw} = enthalpy of feed water, kJ/kg
D_{rh} = flow rate of reheated steam, kg/s
H'_{rh}, H''_{rh} = enthalpy of steam at inlet, outlet of reheater, kJ/kg
D_{bw} = flow rate of blow-down water, kg/s
H' = enthalpy of saturated water at the pressure of steam drum, kJ/kg

The blow-down steam may sometimes be up to 5–10% of the main steam flow in a small boiler. So, the heat taken away by the blow-down water should be considered. However, in a condensing plant the blow-down water of a boiler is no more than 1–2% of the main steam flow rate. Then the heat loss through blow-down can be neglected.

If the heat picked up by the steam (Q_1), input heat (Q), and fuel consumption (B) is known, the efficiency of the boiler can be calculated. Alternately, if the input heat (Q), heat picked up by steam (Q_1) and its efficiency (η_{bo}) are known, the fuel consumption of boiler B can be obtained.

$$\eta_{bo} = \frac{100}{BQ}[D_{sup}(H''_{sup} - H_{fw}) + D_{rh}(H''_{rh} - H'_{rh}) + D_{bw}(H' - H_{fw})] \tag{3-61}$$

$$B = \frac{100}{\eta_{bo}Q}[D_{sup}(H''_{sup} - H_{fw}) + D_{rh}(H''_{rh} - H'_{rh}) + D_{bw}(H' - H_{fw})] \tag{3-62}$$

In combustion calculations, fuel is considered to burn completely. But, owing to unburned carbon, only $(1 - q_4/100)$ kg out of 1 kg fuel fed actually generates combustion heat. Therefore, the total air volume needed for actual combustion and total flue gas volume are correspondingly reduced. Thus, in these volume calculations, fuel quantity should be corrected. That is, actual fuel consumption B is known from the fuel burnt, B_j.

$$B_j = B\left(1 - \frac{q_4}{100}\right) \tag{3-63}$$

The calculation of fuel supply should be on the basis of actual fuel consumption rather than the fuel burnt.

3-4-5 Performance Test of Utility Boilers

Heat balance is the focus of performance or boiler acceptance test. Many codes also specify the environment performance in the acceptance test. The objectives of the heat balance test are

(1) to determine the thermal efficiency of the boiler;
(2) to determine each component of the heat loss to find the reasons for, and thereby identify the means to reduce, heat loss;
(3) to determine various working limits for different parameters, such as the excess air coefficient, CO_2 volume content in dry flue gas, waste gas temperature, and superheated steam temperature at different loads.

There are two methods of measuring boiler efficiency:

(a) Input-output method and
(b) Heat-loss method.

The heat-loss method is based on calculation of various heat losses. The input-output method measures the heat absorbed by the water and steam and compares it to the total energy input based on the higher heating value of the fuel. Both methods are mathematically equivalent and would give identical results if the required heat balance factors were measured without error. When accurate instruments and testing techniques are used, there is reasonably good agreement between the two calculation procedures. However, for practical boiler tests with limited instrumentation, comparisons between the two methods are generally poor. This variation results primarily from inaccuracies associated with the measurement of the flow and energy content of the input and output streams.

The input-output method requires accurate measurements of the fuel feed rate. Also, accurate data must be available on steam flowrate, pressure, temperature, air temperature, and stack temperature to complete the energy balance calculations. Because of the many physical measurements required on the boiler and potential for significant measurement errors, the input-output method is not feasible for field measurements in plants where precision instrumentation is not available. Large errors are possible.

The heat-loss method subtracts individual energy losses from 100% to obtain percent efficiency. It is recognized as the standard approach for routine efficiency tests, especially in industrial sites where instrumentation is minimal. In addition to being more accurate for field testing, the heat loss method identifies exactly where the heat losses are occurring. Thus it aids energy-saving efforts. The method requires determination of the exit flue gas; excess O_2, CO, RO_2, combustibles, and unburned carbon in refuse; exit flue gas temperature; combustion air temperature; etc. The method is much more accurate and is the more accepted method of

TABLE 3-6. ASME Performance Test Codes. (Note: the code is revised periodically).

Code No.	Titles of the code
PTC 1	General instructions
PTC 2	Code on definition and values
PTC 3.1	Diesel and burner fuels
PTC 3.2	Solid fuels
PTC 3.3	Gaseous fuels
PTC 4.1	Steam-generating units
PTC 4.2	Coal pulverizers
PTC 4.3	Air heaters
PTC 21	Dust-separating apparatus
PTC 27	Determination of dust concentration in a gas stream

determining boiler efficiencies in the field, provided that the measurements of the flue gas conditions are accurate.

The American Society of Mechanical Engineers (ASME) has produced standard methods for testing the performance of the boiler and other components of the power plant. These test codes are periodically updated and serve as a guide for performance testing in the West. Table 3-6 lists test codes for different units:

3-5 Generation of SO_2 and NO_x

During the combustion of coal, the sulfur in it is oxidized to sulfur oxides, and the nitrogen in the fuel or air is oxidized to nitrogen oxides. Release of these gases to the atmosphere gives rise to acid rain precipitation, smog formation, and low level ozone formation. The following section briefly describes the processes involved in the generation and destruction of these gases.

3-5-1 Sulfur Dioxide Control

Sulfur dioxide is the main product of sulfur oxidation, though a small amount (1% for coal firing and 2% for oil firing) is converted into sulfur trioxide. Limestone ($CaCO_3$) is an effective absorbing materials for sulfur dioxide. In the case of a PC boiler fine limestone can be injected into the furnace, or the SO_2 laden flue gas can be washed by lime or limestone slurry. In a fluidized bed boiler limestone is used as the bed material. It calcines to CaO, which reacts with SO_2, producing $CaSO_4$. Thus, instead of leaving the boiler as a gaseous pollutant, sulfur is discharged as a solid residue. The molar volume of $CaSO_4$ is greater than that of CaO. As a result the reaction product $CaSO_4$ blocks passages to the internal pores of CaO. For this reason 30–50% of the CaO remain unutilized (Basu & Fraser, 1991, p. 149). The level of sorbent utilization depends on the reactivity of the sorbent, its size, temperature, and cyclone efficiency. For engineering calculations of CFB boilers one can use the following simplified expression (Basu & Fraser, 1991):

$$\frac{F_{sor}}{F_c} = \frac{3.12 E_{sor}\rho_{bav} H_{fur}.S - 100 P'U_g[A]E_c \ln(1 - E_{sor})}{E_c[\delta_{max} X_{caco3}\rho_{bav} H + 100 P'U_g \ln(1 - E_{sor})]} \tag{3-64}$$

The amount of SO$_2$ produced in a boiler can be estimated from the following equation:

$$\text{SO}_2 \text{ produced}, G_s = 2.F_c \left(\frac{S}{100}\right) K \text{ kg/s}$$

where $K = 0.95$ for cyclone furnace, 0.97 for dry bottom PC firing, and 0.99 for oil firing (Stulz and Kitto, 1992). The sulfur dioxide emission is generally expressed as concentration in the flue gas.

SO$_2$ concentration $= 100\, G_s/M_g \%$ by weight $= G_s \times 10^6/M_g$ ppm by weight

$$= \frac{100.G_s}{\left[G_s + \dfrac{64}{M_{fg}(M_g - G_s)}\right]} \% \text{ by volume}$$

$$= \frac{10^6.G_s}{\left[G_s + \dfrac{64}{M_{fg}(M_g - G_s)}\right]} \text{ ppm by volume}$$

where M_g is the weight of flue gas produced per second; and M_{fg} is the molecular weight of flue gas $= 30.2$ for coal firing and 29 for oil firing.

3-5-2 Nitrogen Oxide Control

The nitrogen of the combustion air as well as that in the fuel may oxidize to nitric oxide during combustion, giving two sources of nitric oxide. The first source is called thermal NO$_x$; the second, fuel NO$_x$. The rate of formation of NO$_x$ is influenced by the temperature and the oxygen available in the combustion zone. Lower flame temperature and excess air reduces the thermal NO$_x$ generation. The fuel NO$_x$ is reduced by reducing the oxygen availability in the combustion zone. A more detailed description of the NO$_x$ control by a low NO$_x$ burner is given in Chapter 9.

In a fluidized bed boiler the nitrogen in air is oxidized insignificantly at the low combustion temperature (800–900°C), but the fuel nitrogen does oxidize to an appreciable degree. The formation and destruction of NO in fluidized bed combustors is complex, and involves a large number of complex equations (Kilpinen et al., 1992). However, a simpler predictive model (Talukdar & Basu, 1995) using the overall reactions can give a closed form solution. This equation is useful for engineering calculations for CFB boilers.

Nitrous oxide, which is a 200 times (Winter, 1995) more damaging greenhouse gas than CO$_2$, is another major product of fuel burning. It is formed by the oxidation of HCN (Kilpinen and Hupa, 1991).

$$\text{HCN } (+\text{O}) \rightarrow \text{NCO} - (+\text{NO}) \rightarrow \text{N}_2\text{O}$$

At higher temperatures NCO is rapidly removed by competing reactions (Winter, 1995). So, N$_2$O formation drops sharply at elevated temperatures. For this reason

TABLE 3-7. Conversion chart for sulfur dioxide and nitrogen oxide emission from boilers.[*]

				g/GJ			lb/10⁶ btu		
To convert from↓	mg per Nm³	ppm NO$_x$	ppm SO$_2$	Coal[a]	Oil[b]	Gas[c]	Coal[a]	Oil[b]	Gas[c]
mg/Nm³	1	0.487	0.350	0.350	0.280	0.270	8.14×10^{-4}	6.51×10^{-4}	6.28×10^{-4}
ppm NO$_x$	2.05	1		0.718	0.575	0.554	1.67×10^{-3}	1.34×10^{-3}	1.29×10^{-3}
ppm SO$_2$	2.86		1	1.00	0.801	0.771	2.33×10^{-3}	1.86×10^{-3}	1.79×10^{-3}
(ng/J) Coal[a]	2.86	1.39	1.00	1			2.33×10^{-3}		
or Oil[b]	3.57	1.74	1.25		1			2.33×10^{-3}	
(g/GJ) Gas[c]	3.70	1.80	1.30			1			2.33×10^{-3}
lb/10⁶ Coal[a] btu	1230	598	430	430			1		
Oil[b]	1540	748	538					1	
Gas[c]	1590	775	557			430			1

[*]From IEA (1986). Basis of concentration indicated by superscripts.
[a]Coal-flue gas dry 6% excess O$_2$ @ 350 Nm³/GJ.
[b]Oil-flue gas dry 3% excess O$_2$ @ 280 Nm³/GJ.
[c]Gas-flue gas dry 3% excess O$_2$ @ 270 Nm³/GJ.

the high-temperature PC or oil/gas fired furnaces produce much less nitrous oxide than the low-temperature fluidized bed furnaces.

3-5-3 Unit Conversion

The sulfur dioxide and nitric oxide emissions are often expressed in different units. Table 3-7 gives conversion factors for these.

Nomenclature

[kg/kgf *stands for kilogram of the quantity per kilogram of fuel burnt*]

$(CT)_k$	enthalpy of wet air, kJ/Nm³
A	ash percentage in fuel, %
B	actual fuel consumption in 1 hour operating time, kg/h
B_j	fuel burnt, kg/s
C	carbon percentage in fuel, %
C_{ba}	carbon contents in bottom ash, %
C_{fa}	carbon contents in fly ash, %
C_{pf}	specific heat of fuel, as received basis, kJ/kg °C
C_{psa}	specific heat of ash and slag, kJ/kg °C
D	boiler rating, kg/s
D	operational capacity, kg/s
D_{bw}	flow rate of blow-down water, kg/s
D_{rh}	flow rate of reheated steam, kg/s
D_{sup}	superheated steam flow rate, kg/s
$E_{c=}$	average cyclone efficiency, %
E_{sor}	sulfur capture efficiency, %

F_c	coal feed rate, kg/s
F_{sor}	sorbent feed rate, kg/s
F_{sb}	external surface area of boiler, m^2
G_{as}	steam consumption for oil atomization, kg steam/kg oil
$G_{ba,}$	amount of bottom ash during one hour operating time, kg/h
G_{fa}	amount of fly ash during one hour operating time, kg/h
G_g	flue gas mass per unit mass of fuel burnt, kg/kg
H'	enthalpy of saturated water at the pressure of steam drum, kJ/kg
H_{fur}	height of furnace above secondary air level, m
H	hydrogen percentage in fuel, %
H''_{rh}	enthalpy of steam at outlet of reheater, kJ/kg
H''_{sup}	enthalpy of superheated steam, kJ/kg
H_{fw}	enthalpy of feed water, kJ/kg
HHV	higher heating value of fuel, kJ/kg
H_a	enthalpy of actual combustion air for 1 kg fuel, kJ/kg fuel
H_b^0	enthalpy of theoretical amount of combustion air at burner entrance, kJ/kg fuel
H_{ab}^0	enthalpy of theoretical air at ambient conditions, kJ/kg fuel
h_r	coefficient of heat transfer by radiation, kW/(m^2°C)
H_f	sensible heat of fuel, kJ/kg
h_c	coefficient of heat transfer by convection, kW/(m^2°C)
H'_{rh}	enthalpy of steam at inlet of reheater, kJ/kg
H_{wh}	enthalpy of atomization steam, kJ/kg
H_y	enthalpy of total flue gas, kJ/kg fuel
H_y^0	theoretical enthalpy of total gas, kJ/kg fuel
I_a^0	theoretical cold air enthalpy entering boiler, kJ/kg
I_g	flue gas enthalpy at the exit air heater, kJ/kg
LHV	lower heating value of fuel as received basis, kJ/kg
M	moisture percentage in fuel, %
M_a	surface moisture in fuel, %
M_i	inherent moisture in fuel, %
N	nitrogen percentage in fuel, %
N_2^k	nitrogen carried with air in dry flue, % volume
N_2^r	nitrogen released from fuel itself, % volume
O	oxygen percentage in fuel, %
P	total pressure of flue gas, Pa
P'	proportionality constant in pore plugging time (s) $= P' \times$ (SO$_2$ concentration, kmol/m^3)$^{-1}$
P_{H_2O}	partial pressure of water vapor, Pa
P_{RO_2}	partial pressure of tri-atomic gases, Pa
Q	available heat of fired fuel, kJ/kg
Q_1	heat absorbed by steam or utilized by the boiler, kJ/kgf
Q_2	heat loss owing to waste heat, kJ/kgf
Q_3	heat loss by incomplete combustion, kJ/kgf
Q_4	heat loss owing to unburned carbon, kJ/kgf

q_5	heat loss at operating part load, %
q_5	heat loss at rated load, %
Q_5	heat loss owing to heat transfer from the furnace wall to ambient, kJ/kgf
Q_6	heat loss with physical heat of ash and slag, kJ/kgf
Q_{ah}	heat absorbed by air heater, kJ/kgf
Q_{cd}	heat spent in decomposition of carbonates in oil shale or limestone. kJ/kgf
Q_{fa}	heat loss owing to unburned carbon in fly ash, kJ/kgf
Q_{ba}	heat loss owing to unburned carbon in bottom ash, kJ/kgf
q_i	ratio of available heat and heat losses
Q_{as}	total heat of the fuel atomizing steam, kJ/kgf
Q_{wr}	sensible heat carried, kJ/kgf
r	latent heat of steam, kJ/kgf
r_{H_2O}	volume fraction of water vapor
r_{RO_2}	volume portion of tri-atomic gases
S	sulfur percentage in fuel, %
T_0	average ambient temperature, °C
T_r	fuel temperature at burner, °C
T_{sb}	average temperature of the boiler surface, °C
U_g	gas velocity, m/s
V_{dg}	actual volume of dry flue gas, Nm³/kgf
V_g	total flue gas volume, Nm³/kgf
V_g^0	theoretical flue gas volume, Nm³/kgf
$V_{H_2O}^0$	vapor volume, Nm³/kgf
$V_{N_2}^0$	theoretical volume of nitrogen, Nm³/kgf
V^0	air volume, Nm³/kgf
$V_{O_2}^0$	theoretical volume of oxygen, Nm³/kgf
V_{RO_2}	sum of V_{co2} and V_{so2}, Nm³/kgf
V_{SO_2}	gas volume of sulfur dioxide, Nm³/kgf
x_{ba}	fraction of bottom ash, %
X_{CaCO_3}	mass fraction of calcium carbonate in sorbent
V	actual air, Nm³/kgf
V_{CO_2}	gas volume of carbon dioxide, Nm³/kgf
X_c	percentage measured carbon in residue, %
x_{fa}	fraction of total ash in the fly ash, %
X_m	moisture content in dry air, kg/kg
X_s	kg of spent sorbent per 100 kg of fuel

Greek Symbols

σ	mass concentration of fly ash
$\Delta\alpha$	air leakage coefficient
α	excess air coefficient
α_{ah}	excess air coefficient at the exit of air heater
α_e	excess air coefficient at furnace exit

β	fuel characteristic index (Eq. 3-27)
η_{bo}	boiler efficiency %
μ	nondimensional mass concentration of fly ash
δ_{max}	maximum extent of sulfation
ε	voidage of the bed
ρ_{bav}	average bed density, kg/m^3

Superscripts

f	air dry basis
g	total dry bases
r	dry and ash basis
0	theoretical amount

References

Afnasayev, V. (1989) "Testing of Boilers" (translated from the Russian). Mir Publishers, Moscow, ISBN5-030000024-0, p. 59.

Basu, P., and Fraser, S. (1991) "Circulating Fluidized Bed Boilers, Design and Operation." Butterworth-Heineman, Stoneham, Massachusetts, p. 169.

D1756 (1981) "Standard Test Method for Carbon Dioxide in Coal." Annual Book of ASTM Standards, Part 26.

D1857 (1981) "Standard Test Method for Fusibility of Coal and Coke Ash." Annual Book of ASTM Standards, Part 26.

D2015 (1981) "Standard Test Method for Gross Calorific Value of Solid Fuels by the Adiabatic Bomb Calorimeter." Annual Book of ASTM Standards, Part 26.

D3172 (1981) "Standard Method for Proximate Analysis of Coal and Coke." Annual Book of ASTM Standards, Part 26.

D3174 (1981) "Standard Test Method for Ash in the Analysis of Coal and Coke." Annual Book of ASTM Standards, Part 26.

D3176 (1981) "Standard Method for Ultimate Analysis of Coal and Coke." Annual Book of ASTM Standards, Part 26.

Kilpinen, P., and Hupa, M. (1991) Homogeneous N$_2$O chemistry at fluidized bed combustion conditions—a kinetic modeling study. Combustion and Flame 85:94–104.

Kilpinen, P., Glarborg, P., and Hupa, M. (1992) Reburning chemistry at fluidized bed combustion conditions—a kinetic modeling study. Industrial Engineering and Chemistry Research, 31:1477–1490.

Parilov, V. and Ushakov, S. (1989) "Testing and Adjustment of Steam Boilers." Mir Publishers, Moscow, p. 69.

PTC 4.1 American Society of Mechanical Engineers, Performance Test Code 4.1. 1991.

Talukdar, J., and Basu, P. (1995) A simplified model of nitric oxide emission from a circulating fluidized bed combustor. Canadian Journal of Chemical Engineering, 73: 635–643.

Winter, F. (1995) "Single fuel particle and NO$_x$/N$_2$O—emission characteristics under circulating fluidized bed combustor conditions." Ph.D. Thesis, University of Technology, Vienna, Austria.

4
Coal Preparation Systems for Boilers

Coal can be burned in a number of ways. Depending on the characteristics of the coal and the particular boiler application, the designer may chose pulverized coal (PC) firing, cyclone firing, stoker, or fluidized bed firing methods. Whatever method is applied, raw coal should be prepared before being fed into the furnace. The preparation process has a major impact on the combustion in the furnace. Different types of boilers have different requirements as to the characteristics (size and moisture content, etc.) of the fuel fed into the furnace. Thus there is a variety of coal preparation systems.

Pulverized coal (PC) firing is the dominant method used in modern power stations. So, most of this chapter will focus on coal preparation systems for PC boilers and their design. Fluidized bed (FB) firing is developing very quickly and finding its way to wider applications in large-scale commercial boilers. Hence a brief discussion on coal preparation for fluidized bed boilers is also presented here.

4-1 Coal Preparation Systems

Coal is usually delivered to the site by barge, train, truck, or belt conveyor. It is first stacked for reserve. Many plants keep an emergency stockpile as a strategic reserve, which can be used in the event of extended interruption of fuel delivery. From the stack, coal is transported to the coal preparation plant. Figure 1-4 in Chapter 1 shows a typical coal handling and preparation plant. This part of the plant is common to most types of coal firing systems. Differences in preparation come further on; these are described below for two types of firing systems: PC and FB firing.

4-1-1 Pulverized Coal Boilers

In a PC boiler, the coal is first crushed in a crusher. Its size is further reduced to <200 μm in a pulverizer; it is then dried, heated, and blown into the burners of the boiler. Three systems for this process have been developed: storage or

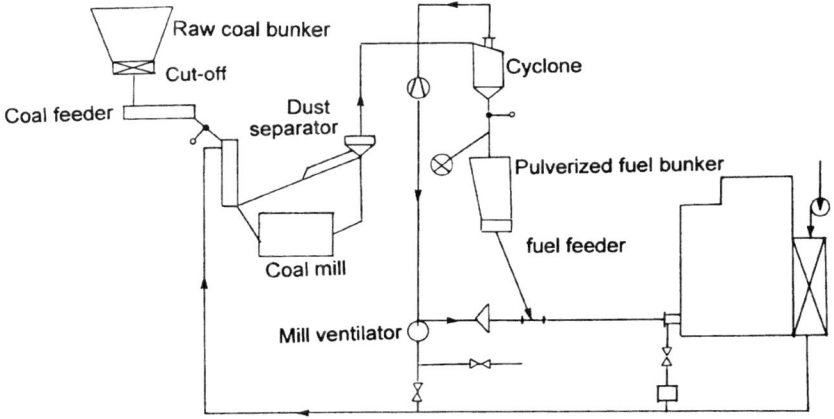

FIGURE 4-1. Coal storage system

bin-and-feeder system, direct fired system, and semidirect system. These methods differ on the mode of drying, feeding, and transportation characteristics. Although these three are all widely used, the most popular is the direct fired system.

a) Storage System

In a storage system (Fig. 4-1), coal is pulverized and conveyed by air or gas to a suitable collector where the carrying medium is separated from the coal, which is then transferred to a storage bin. Hot air or flue gas, used in a pulverizing mill for drying, is often used as the carrying medium. The carrying gas is vented after separation from the fuel. From the storage bin, the pulverized coal is fed into the furnace as required.

b) Direct Fired System

In a direct fired system (Fig. 4-2), coal is pulverized and transported, with air, directly to the furnace. Hot air or diluted flue gas is supplied to the pulverizer for both drying and transporting the pulverized fuel to the furnace. This air also serves as the primary air. It is a part of the total combustion air.

c) Semidirect System

In a semidirect system (Fig. 4-3), a cyclone collector located between the pulverizer and furnace separates the conveying medium from the coal. The coal is fed directly from the cyclone to the furnace in a primary-air stream that is independent of the milling system. The drying hot medium is returned to the mill. In this system, the pulverizing system itself serves as a storage.

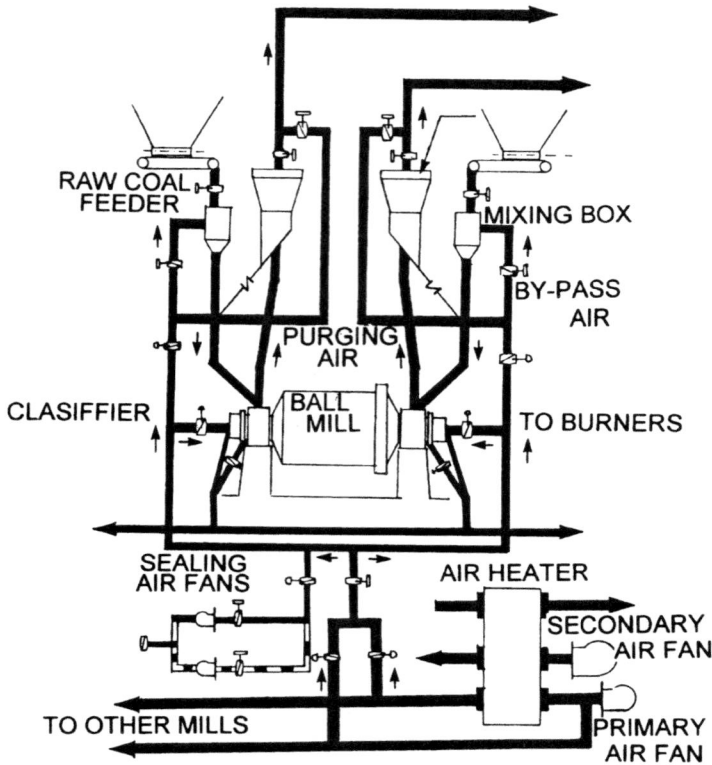

FIGURE 4-2. Direct-fired system

d) Comparison Between the Storage System and the Direct-Fired System

A semidirect system is much like a direct fired system except for the primary air source. So, we will present only a comparison between the direct system and the storage system.

a) A direct fired system is much simpler and has fewer components. Thus the capital cost and power consumption for a direct fired system are lower than those for a storage system.

b) In a storage system, the bunkers store a large amount of coal. So, in case of a breakdown of the pulverizers the interruption in coal supply is less compared to that for the direct fired system. In this regard, a storage system is more reliable. This is why it is used more widely in places where sustained and reliable operation is more important than efficiency.

c) A storage system can easily respond to the load change of the boiler by adjusting the feeders. A direct fired system, on the other hand, must adjust the whole system starting from the raw coal bunker.

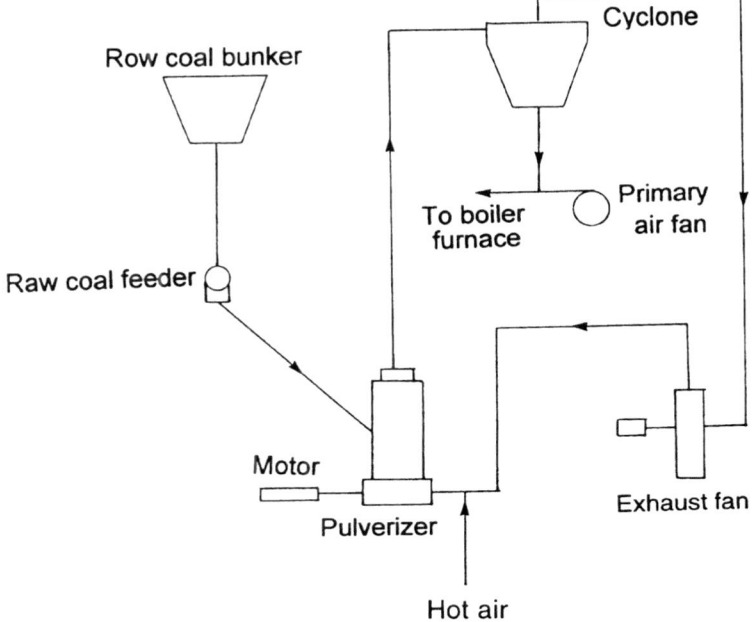

FIGURE 4-3. Semi-direct system

Nowadays, control techniques for boilers are well developed. Hence the advantages of the storage system are not as appreciable as before. Also, the cost of a storage system is much more than that of a direct fired system. For these reasons, the direct corrected fired system is in more use.

4-1-2 Fluidized Bed Boilers

Since the feed-stock (coal and limestone) used for a fluidized bed (FB) boiler is coarser ($<6,000$ μm) than pulverized coal (<200 μm), a crusher-based coal preparation system is sufficient for the former. It is much simpler and less costly than that of a PC boiler. The fuel preparation plant of fluidized bed boiler comprises a crusher and a drier. Unlike PC combustion fluidized bed firing does not require drying of fuels. Some fluidized bed boilers burn coal slurry directly. The drying is required only for transportation of the fuel to the furnace. For this reason, the drier is particularly important if pneumatic transport is considered for conveying the feed-stock to the furnace. The dryer may use the hot flue gas from the boiler for driving away the surface moisture of the feed. Most fluidized bed boilers are of the indirect firing type, which means the crushed coal and sorbents are stored in an intermediate bunker, whence thus are conveyed to the feeder for feeding. The design of a crusher is described is later.

a) Feed Size

Singer (1991) recommends crushing coal and sorbents to the following sizes for use in fluidized bed boilers:

Circulating fluidized bed (CFB) firing:
Coal   Top size = 3–9 mm and d_{50} = 0.5–1.0 mm
Sorbent   Top size = 1.0 mm and d_{50} = 0.2 mm

Bubbling fluidized bed firing:
Coal   Top size = 25–50 mm (over-bed feeding); <25% below 1.0 mm
          Top size = 10 mm (under-bed feeding); <20% below 0.5 mm

The size distribution of the crushed coal often plays an important role in the furnace hydrodynamics, especially for high ash coal that does not require limestone addition. Thus in many cases the crusher size distribution becomes a requirement for the boiler performance guarantee. In the case of circulating fluidized bed (CFB) boilers the size distribution and inventory of solids in the furnace is greatly influenced by the size spectrum of the crushed coal and or sorbent. In one large CFB boiler, where an incorrectly specified crusher gave a wider spread of the limestone size distribution, the furnace had to be run at a very high temperature in order to maintain the steam output, raising the sorbent use and nitric oxide emission.

The amount and nature of ash have some bearing on the size to which coal is to be crushed. High ash coal generally contains more extraneous ash; i.e., the ash remains in discrete lump form. Unless coal is crushed to finer sizes the carbon is not burnt and the resulting bed materials would not be sufficiently small to give the required hydrodynamic conditions in the bed. In low ash coal, the ash is generally of an intrinsic nature; i.e., it is uniformly dispersed as fines. So, a low ash coal should be crushed more coarsely, such that coal fines do not escape unburnt immediately upon feeding into the bed.

In the case of CFB boilers, if the feed is too fine an excess amount of bed-solids will escape the CFB loop through the cyclone. This would leave insufficient materials in the system. On the other hand, a very coarse feed will produce fewer solids in the upper furnace. This would result in a reduced level of heat absorption, leading to a higher furnace temperature, higher emission, and a lower steam generation rate. At the same time it would result in more solids in the lower bed, adding to the furnace pressure drop. Fuels like high ash anthracite are crushed to 3-mm top size, while low ash bituminous can be crushed to 6-mm top size (Singer, 1991).

4-2 Pulverizing Properties of Coal

Coal properties have a great impact on the performance of the coal pulverizing system and the whole boiler. To discuss the coal preparation system and their components, it is necessary to introduce some important coal properties first.

TABLE 4-1. Typical pulverized coal fineness requirements.

Fuel	Fixed carbon,%			Heating value, kJ/kg		
	97.9–86	85.9–78	77.9–69	>30,240	30,240–25,590	<30,240
	High-rank coal (% <74 μm)			Low-rank coal (% <74 μm)		
Water-cooled furnace	80	75	70	70	65	60
Cement kiln	90	85	80	80	80	—

4-2-1 Coal Fineness

For suspension burning of solid fuels, rapid ignition requires a minimum amount of fines in the fuel–air mixture. Conversely, to obtain maximum combustion efficiency, a minimum amount of coarse particles in this fuel–air mixture is desirable. So, a PC mixture should contain a minimum amount of coarse particles and a maximum amount of fine particles. The function of the coal pulverizing system is to get the ideal combination of these.

The fine particle content of the fuel is usually expressed as the percentage through a 200-mesh screen (74 μm). The coarseness is designated as the percentage retained on a 50-mesh screen (297 μm). The following recommendations (Table 4-1) on the fineness of pulverized coal can be made (after Stulz & Kitto, 1991, p.12-8).

The number of openings per linear inch designates the mesh of a screen. Thus, a 200-mesh screen has 200 openings to the inch, or 40,000 holes per square inch. The diameter of the wire used in making the screen governs the size of the openings. The U.S. Standard and W.S. Tyler are the most common screen sieves. The mesh and openings of these screens are shown in Table 4-2.

4-2-2 Grindability

To predict the performance of a pulverizer on a specific coal, the ease with which the coal can be pulverized must be known. The *Hardgrove grindability index* (HGI) provides a measure of the ease of pulverizing a coal. The HGI is not an inherent property of the coal. Rather, it represents the relative ease of grinding coal when tested in a particular type of apparatus. The HGI can be applied to find a particular size and, to a lesser degree, type of pulverizer. The HGI is measured in a Hardgrove grindability machine. This value shows how much coal can be milled to a certain fineness in a test mill consuming a certain amount of power. Hence the HGI is approximately (not directly) proportional to the mill capacity. This is because of the difference between a commercial pulverizer and a grindability test machine that, with no provision for continuous removal of fines, is of the batch type rather than the continuous type. The crushing pressure of the test equipment is also considerably less.

To overcome the above-mentioned discrepancies, some correction factors for commercial equipment are developed by pulverizer manufactures.

TABLE 4-2. Comparison of the sieve openings.

Mesh, U.S. Standard sieve	Inches	Millimeters
20	0.0331	0.84
30	0.0234	0.595
40	0.0167	0.420
50[a]	0.0117	0.297
60	0.0098	0.250
100[a]	0.0059	0.149
140	0.0041	0.105
200	0.0029	0.074
325	0.0017	0.044
400	0.0015	0.037

Mesh, U.S. Tyler sieve	Inches	Millimeters
20	0.0328	0.833
28	0.0232	0.589
35	0.0164	0.417
48[a]	0.0117	0.295
60	0.0097	0.246
100[a]	0.0058	0.147
150	0.0041	0.104
200[a]	0.0029	0.074
325	0.0017	0.043
400	0.0015	0.037

[a]Commonly used screens in pulverized coal practice for combustion purposes.

4-2-3 Moisture

Usually a reference to moisture in coal pertains to the total moisture content, which includes the equilibrium moisture and the surface or free moisture. Equilibrium moisture varies with the coal type or rank and mine location. It is also called *"bed"* or *"seam"* moisture. The surface moisture is the difference between total moisture and bed moisture.

The surface moisture adversely affects both pulverizer performance and the combustion process. The surface moisture produces agglomeration of the fines in the pulverizing zone. This reduces pulverizer drying capacity because of its inability to remove the moisture efficiently. Agglomeration of fines has the same effect as coarse coal has on the combustion process. Here the surface available for the chemical reaction is reduced. Since drying in the mill is an accepted method of preparing coal, sufficient hot air at certain temperature is necessary for the mill. Figure 4-4 shows that the air temperatures required to dry coal of different moistures vary with different coal–air mixtures. The capacity utilization of a pulverizer depends on the availability of sufficient hot air to dry the coal. If there is a deficiency of hot air, the mill output will be limited to the "drying capacity" and not the "grinding capacity."

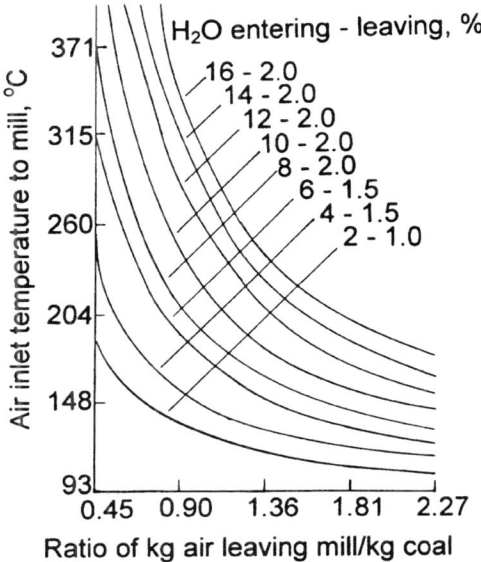

FIGURE 4-4. Air temperatures of drying coal at different moistures and coal–air mixtures

4-2-4 Abrasion

Coal is pulverized by contact with the balls, rolls, rings, races, and liners of the pulverizer. Though these are made of much harder materials, eventually all of them experience wear owing to erosion, abrasion, and metal displacement in the grinding process. Thus, the power for grinding and the maintenance of the grinding elements make up the major costs of the pulverizing operation.

The abrasiveness of a particular coal can be represented by its *abrasion index*. The index, which is used in the estimation of mill wear during grinding, is expressed in milligrams of metal lost from the blades of the test mill per kilogram of coal ground.

4-3 Pulverizing Air System

Drying makes the fuel easier to grind and increases the mill capacity. Furthermore, a dried and heated fuel is easier to ignite. So, the fuel must be dried and heated before it enters the furnace. However, to avoid spontaneous combustion and explosion, the temperature to which the fuel is heated must not exceed certain values. Thus, the pulverizing air system should be designed to provide the required amount of air to dry and transport the fuel, but prevent overheating of the fuel. The air temperature at the mill inlet depends on the drying requirement of the coal. It is therefore decided by the type of coal being pulverized and its surface moisture.

4-3-1 Source of Air

The best source of hot air for mill drying is either a regenerative or recuperative air heater, which extracts heat from the flue gas. The air heater used in large boilers can usually provide adequate heating for almost any fuel moisture condition. While drying a high-moisture coal, hot and cold flue gas mixed with the preheated air is used. The use of a mixture of gas and preheated air has some advantages:

1. The risk of explosion is reduced since the oxygen concentration in the flue gas is low.
2. The use of flue gas can reduce the temperature in the burner region so as to avoid slag formation,
3. The mixing proportion of hot and cold gases can be adjusted to respond to a large variation in coal moisture with the changing temperature of the primary air.

4-3-2 Storage Pulverizing Systems

In the coal-storage pulverizing system, a cyclone collector separates the air used in the pulverizer for drying, classifying, and conveying. The pulverized coal is conveyed either mechanically or pneumatically to intermediate storage or bunkers. Controllable feeders at the bunker outlets deliver the required quantity of coal to the fuel lines, whence conveying air carries the fuel to the burners.

a) Primary Air

In some installations, the primary air is taken from the room or a preheated air duct, or both. In other cases, the vented air from the pulverizer system is used for all or parts of the primary-air source. In installations of the first kind, the only air used is that needed to carry the coal to the furnace and to provide the required velocity of coal–air mixture at the burners. The amount of air, therefore, depends on the type of firing system and the type of piping system used. In medium-speed mills the coal-to-air weight ratio is in the range of 1.7 to 2.2, while in the tube mill it is 1.2 to 1.5 (Clapp, 1991). The air requirement decreases with coal throughput into the mill. For example, in a vertical spindle type mill the air-to-coal ratio decreased from 2.5 to 1.68 when the coal rate per pulverizer increased from 26 to 54 t/h (Stulz & Kitto, 1992, p. 12-7).

b) Vented Air

For coal to be pulverized and classified economically and properly, it must either be dried when delivered to the pulverizer or during the pulverizing operation.

This drying is accomplished by the drying air or gas. In a storage system, either all or a portion of the drying medium is discharged from the system by venting. This removes the moisture that has been evaporated from the fuel. The drying medium supplies a major portion of the heat to evaporate the fuel moisture, and, in so doing, it approaches saturation. Some of this saturated mixture must be removed or vented from the system. The amount vented depends on the pulverizer output, the moisture removed from the fuel, the initial temperature of the drying medium, the temperature of the vented material, and the efficiency of the drying system.

The vented mixture must be disposed of by venting to the atmosphere directly or through the boiler stack. Alternately, it can be used as a part of the air supply to the furnace. In the first case, since the vent contains some extremely fine coal (up to 2% of the amount being pulverized), economically it cannot be vented directly without collecting the coal dust in cyclone concentrators, bag filters, air washers, or a combination of these. In the second case, the air need not be cleaned.

c) Air Temperature

Pulverized-coal feeder and fuel-piping arrangements determine the primary-air temperature, which may be as high as 315°C. If the preheated air is too hot to be used directly, tempering with cold air at the suction side of the fan can be carried out. This is done by inserting a resistance in the preheated air and placing an adjustable cold-air opening between this damper and the fan inlet. Any tempering air admitted to the system reduces the quantity of air passing through the air heater, which reduces the overall unit efficiency. So, minimum amount of tempering air should be used in the primary-air system.

The dried coal while being transported by air should avoid self-ignition as well as condensation of moisture. The Table 4-3a gives some typical values of air temperature required for different types of coal. Table 4-3b shows the allowable mill outlet temperature for different coals and mills as recommended by Singer (1991, pp. 11–18).

If vented air is to be used, the temperature will depend on that of the vent, the position of the vent in the return air line, and on whether preheated or cold air is added to the primary air at the fan inlet. The vented air will usually have a relative humidity of 70–90%, and a temperature of 40–70°C, and it will contain coal fines from the collector.

TABLE 4-3a. Temperature of hot air (°C).

| Fuel | Anthracite | Lean coal, inferior bituminous | Brown coal | | Bituminous coal |
			Hot air drying	Fuel drying	
Temperature of hot air (°C)	380–430	330–380	350–380	300–350	280–350

TABLE 4-3b. Allowable mill outlet temperature, °C.

System	Storage	Direct	Semidirect
High-rank, high-volatile bituminous	54	77	77
Low-rank, high-volatile bituminous	54	71	71
High-rank, low-volatile bituminous	57	82	82
Lignite	43	43–60	49–60
Anthracite	93	–	–
Petroleum coke (delayed)	57	82–93	82–92

4-3-3 Direct-Fired Arrangements

In direct-fired arrangements, two methods are utilized for supplying the air requirements and overcoming resistance. One method utilizes a fan behind the pulverizer, while the other has a fan ahead of it. The last is a suction system and the fan handles coal-dust-laden air, while the latter is a pressure system and the fan handles relatively clean air.

The volume requirement of the exhauster or blower depends on the pulverizer size. It is usually fixed by the base capacity of that pulverizer. The pressure requirement is a function of the pulverizer and classifier resistance, the fuel distributing system, and burner resistance. This resistance is in turn affected by the system design, the required fuel-line velocities, and density of the mixture being conveyed.

In a suction system, the coal feeder discharges against a negative pressure, whereas in the pressure system, the feeder discharges against a positive pressure of 4500 to 5300 Pa. No coal feeder can act as a seal. Thus, the head of coal above the feeder inlet must be utilized to prevent back-flow of the primary air.

a) Suction System

The suction system has a number of advantages. It is easy to keep the area around the pulverizer clean. It is easy to control the air flow through the pulverizer by adjusting the flow rate of the constant-temperature coal–air mixture with a positioning device. The temperature of the coal–air mixture is controlled by a single hot-air damper and a barometric damper through which a flow of room air is induced by the suction in the pulverizer. The fan is designed for a constant, low-temperature mixture and has a low power consumption.

The main disadvantage of the suction system is the maintenance required on the exhauster. Another disadvantage is that since the system works under negative pressure, some ambient air leaks into the system, which reduces the amount of air through the preheater. This reduces the overall efficiency of the boiler.

b) Pressurized System

This system retains the advantages of the suction system in the design of the fan for constant, low-temperature mixture and the relative ease of airflow control. Two dampers, one in the hot-air duct to the mill and one in the cold-air duct to the mill, control the amount of pulverizer airflow. Biasing the hot and cold air dampers controls the temperature of the mixture leaving the pulverizer.

Since the pressurized system works under positive pressure, the ambient air will not leak into the system. So the overall efficiency of the boiler is not reduced. The wear on the fan is also lower. Another advantage is that the low pressures in the pulverizers reduce the problem of sealing the head of coal over the raw-fuel feeder as with pulverizers under direct blower pressure. The disadvantage of the system is that the pulverizers and the ducts must be made leakproof so as to prevent explosion and dust related pollution of the surrounding area.

4-4 Size-Reducing Machines

Raw coal should be crushed to a size acceptable to pulverizers. A CFB boiler does not need to grind the coal to a very fine size, but it is still necessary to crush the coal to a reasonable size so as to achieve uniform feed of the fuel. A PC boiler requires coal to be finely ground. Thus, crushers are key components in both PC boilers and CFB boilers, while pulverizers are important components in the PC coal preparation system. The types of pulverizers and their performance greatly affect the coal preparation system.

Pulverizers use either one, two, or all three of the basic principles of the particle communition process: namely, impact, attrition, and crushing. With respect to speed, these machines may be classified as low, medium, and high. The three most commonly used pulverizers are ball-tube, ring-roll or ball-race, and impact or hammer mill, which belong to the low, medium, and high speed categories, respectively.

Since the pulverizers are more important than the crushers, most of this chapter is focused on the three types of the pulverizers in the order of low speed to high speed. After examining pulverizers, we discuss crushers briefly.

4-4-1 Ball-Tube Mill Pulverizers

The oldest pulverizer design still in frequent use is the ball-tube mill. It (Fig. 4-5) is essentially a horizontal cylindrical chamber, rotated on its axis. Its length is comparable to its diameter. The cylinder is filled with wear-resistant material contoured to enhance the action of the tumbling balls, and the balls fill 25–30% of the cylinder volume. Grinding is caused by the tumbling action of the cylindrical chamber. Coal particles are trapped between balls as they impact.

Ball-tube mills may be either single- or double-ended. In the former, air and coal enter through one end and exit opposite. Double-ended mills are fed with coal and air at each end and ground-dried coal is extracted from each end. In both types, classifiers are external to the mill and oversize material is injected back to the mill with the raw feed.

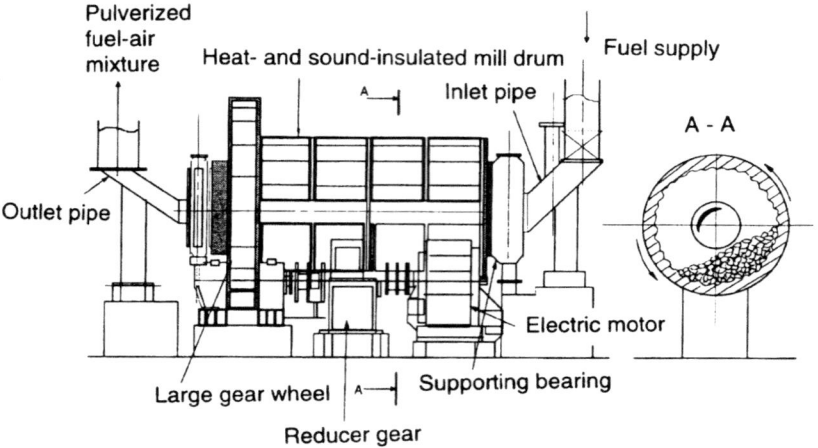

FIGURE 4-5. Arrangement of ball-tube mill

a) Speed of a Ball-Tube Mill

The rotational speed of a ball-tube mill has a great influence on its performance. Solid particles are swept to the wall by centrifugal force. After moving up some distance the particles tumble owing to gravity. The speed at which the grinding media stop tumbling and adhere to the internal walls of the mill is called *critical speed*. The critical speed is reached when the centrifugal force balances the gravitational force on the balls. From force balance (Fig. 4-6) we get

$$m\omega_c^2 R = mg \tag{4-1}$$

where m is the mass of the ball and ω_c is the critical angular speed of the mill in radians per second.

If we define n_c as the critical speed of rotation in revolutions per minute, we have

$$\omega_c = \pi n_c/30 \tag{4-2}$$

Introducing the above equation into Eq. (4-1), we obtain

$$n_c = 30/R^{0.5} = 42.3/D^{0.5} \tag{4-3}$$

The mill must operate at a value less than n_c. The optimum theoretical speed is determined by a study of the internal movements of the balls.

When the mill rotates below the critical speed, the particles detach themselves when the radial component of the gravitational force equals the centrifugal force.

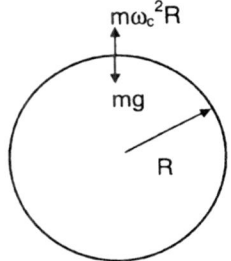

FIGURE 4-6. Force acting on the ball

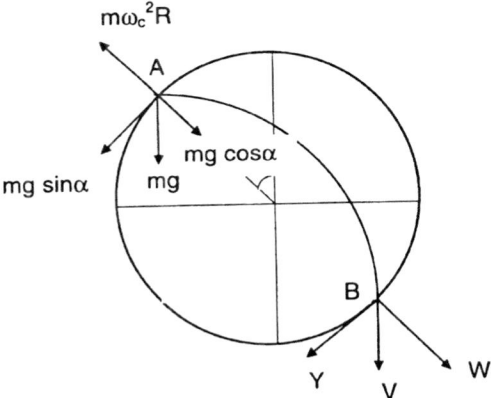

FIGURE 4-7. Trajectory of a ball

At separation point A (Fig. 4-7), we have the following equation if we consider that the ball is free in its trajectory and is unaffected by other balls:

$$m\omega^2 R - mg \cos\alpha = 0 \qquad (4\text{-}4)$$

If a is defined as the ratio of operating and critical speed,

$$a = \omega/\omega_c = n/n_c \qquad (4\text{-}5)$$

Eq. (4-4) can be written as

$$\cos\alpha = a^2 \qquad (4\text{-}6)$$

The above equation shows that we can determine the position of point A from the speed of the mill.

Since the greater the relative velocity w, the higher the grinding efficiency, the optimum speed will be that which leads to a maximum w. The expression of w as a function of α is (Fontanille, 1996)

$$w = 4\cos\alpha(1 - \cos^2\alpha)(g/\omega_c) \tag{4-7}$$

For the maximum efficiency we equate $dw/d\alpha = 0$. This gives

$$\cos\alpha = 0.577 \tag{4-8}$$

Hence,

$$a = (\cos\alpha)^{0.5} = 0.76 \tag{4-9}$$

Thus, maximum efficiency for a given mill is achieved at a rotation equal to 76% of the critical speed.

In practice, ball mills rotate between 66% and 80% of their critical speed. This discrepancy from the theoretical speed is affected by such parameters as type of lifters, the ball filling rate, and the mill type.

b) Determination of Power Requirement

The power requirement of a ball-tube mill depends on the following factors:

• rotational speed of the mill
• rate of filling of the mill

For a given mill, the required power can be written as (Fontanille, 1996)

$$P = KLD^{2.5}X \tag{4-10}$$

where K is a constant factor related to the characteristics of the coal, L is the internal length of the mill, D is the diameter, and X is the power factor. The value of K varies slightly with the change of coal characteristics. Thus, from Eq. (4-10), we can see that the power consumption of a ball-tube mill will not change much when the load changes. Therefore, it is not economical for the mill to work under low load.

For a mill of given diameter and length, the power factor is a function of the rotational speed of the mill and of the rate of filling:

$$X = f(a, T) \tag{4-11}$$

c) Pulverizing Capacity

Pulverizing performance basically depends on two factors:

• size reduction capacity of the coal
• geometrical characteristics of the mill

We can write:

$$Q = f(H_C, H_R, F_C, P_B, D_B, n_B) \tag{4-12}$$

where

> Q = flow rate of pulverized coal
> H_C = moisture content of coal
> H_R = hardgrove grindability
> F_C = coal fineness at mill outlet
> P_B = weight of balls
> D_B = inside diameter of mill
> n_B = rotational speed of mill

The most important advantage of the ball-tube mill is its high availability. These mills are well suited for hard and abrasive coals. They also have such advantages as a large fuel reserve, rapid response to rating change, very good fineness, and great operating flexibility. However, they typically require larger building space and higher specific power consumption than the vertical medium-speed pulverizer. They also have higher metal wear rates. Such characteristics make their capital and operation costs higher than those other mills.

4-4-2 Ring-Roll and Ball-Race Mill Pulverizer

Ring-roll and ball-race mills are the most commonly used pulverizers for coal grinding. They are of medium speed. For size reduction these machines primarily utilize crushing, attrition, and some amount of impact. The widely used B&W E mills, EL mills, MPS mills, and C-E roller mills and bowl mills all belong to this family. Since the shafts of the ring-roll and ball-race mills are all vertical and the air sweeps in the vertical direction, they are also called vertical air-swept mills (Fig. 4-8).

In the vertical air-swept mills, the grinding action takes place between two surfaces. The roller may be either a ball or a roll, while the surface over which it rolls may be either a race or a ring. The movement of the roller causes motion between particles, while the roller pressure creates compressive loads between particles. The movement of particle layers under pressure causes attrition, which is the dominant size reduction mechanism. As grinding proceeds, fine particles are removed from the process to prevent excessive grinding, power consumption, and wear. Figure 4-8 shows the essential components of a vertical mill and coal circulation in this type of mills. As shown in the figure, an upward flow of air fluidizes and entrains the size-reduced coal. The rising air flow, mixed with the coal particles, creates a fluidized bed just above the throat. The air velocity is low enough to entrain only the smaller. The air–solids mixture leaving the bed forms the initial stage of size separation or classification. As the air–solids mixture flows upward, the flow area increases and velocity decreases, returning larger particles directly to the grinding zone. The final stage of size separation is provided by the classifier located at the top of the pulverizer. This device is a centrifugal separator. In the classifier, the coarse particles come out of the suspension and fall back into the grinding zone. The finer particles remain suspended in the air mixture and exit to the fuel pipe.

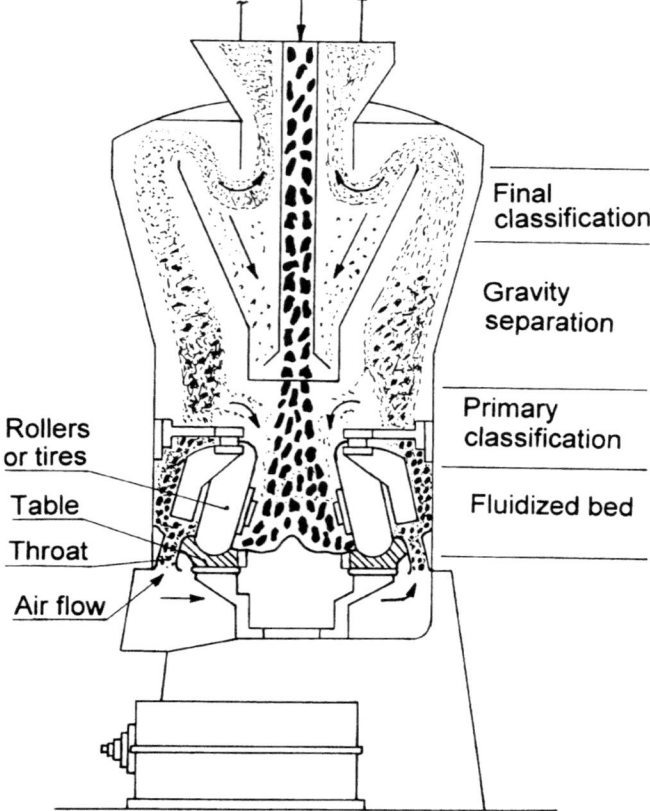

Final
classification

Gravity
separation

Primary
classification

Fluidized bed

Rollers
or tires

Table

Throat

Air flow

FIGURE 4-8. Ring-roll and ball-race mill

Since there is sufficient residence time of the coal in the pulverizer, it is easy
for the heated air to dry the coal to the desired level of humidity. Provided with
sufficient air at a temperature to produce a satisfactory mill outlet temperature, it
can handle very wet coals with only a small reduction in capacity.

These mills are compact and occupy a relatively small amount of floor space per
unit of capacity. Their power consumption is also lower than that of the low-speed
pulverizer. Thus, the total cost of a ring-roll or ball-race mill is much lower than
that of a ball-tube mill. These mills also have the advantage of operating quietly.
However, they cannot adapt to hard and abrasive coals very well.

4-4-3 Impact Mill (Hammer Mill)

An impact mill consists of a series of hinged or fixed hammers revolving in an
enclosed chamber lined with cast wear-resistant plates. Grinding results from a

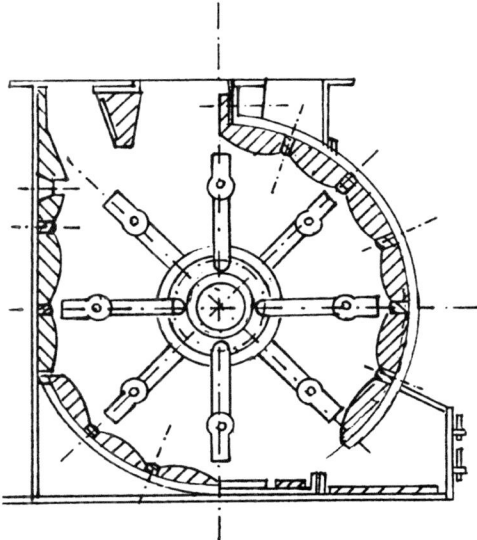

FIGURE 4-9. Impact (or hammer) mill

combination of hammer impact on the larger particles and attrition of smaller particles by particle-to-particle and particle-to-grinding-surface abrasion. An air system with the fan mounted either internally or externally on the main shaft induces a flow through the mill. An internal or external type of classifier can be used. Figure 4-9 shows a hammer or impact mill.

This type of mill is simple, compact, and inexpensive. It may be built in very small sizes, and its ability to handle high inlet-air temperatures, plus the return of dried classified rejects to the incoming raw feed, makes it an excellent dryer. However, the high-speed design results in high maintenance and high power consumption when grinding fuels. Progressive wear on the grinding elements produces a rapid drop-off in product fineness, and it is difficult, if not impossible, to maintain fineness over the life of the wearing parts. An external classifier permits maintenance of fineness, but only at the expense of a considerable reduction in capacity as parts wear. The maximum capacity for which such mills can be built is lower than that of most other types.

4-4-4 Coal Crushers

Numerous types of crushers are commercially available. The type generally used for smaller capacities is the swing-hammer type. This crusher has proved satisfactory for overall use and has demonstrated reliability and economy. The swing-hammer crusher consists of a casing enclosing a rotor to which pivoted hammers or rings are attached. Coal is fed through a suitable opening in the top of the casing

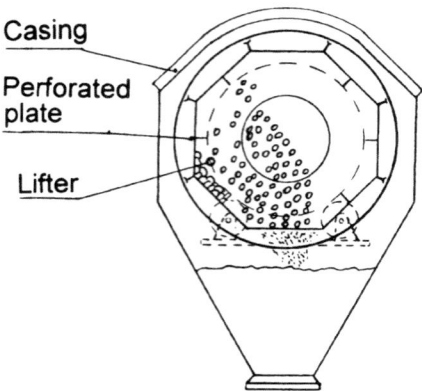

FIGURE 4-10. Bradford breaker

and crushing is effected by impact of the revolving hammers or rings directly on, or by throwing the coal against, the liners or spaced grate bars in the bottom of the casing. The degree of size reduction depends on the hammer type, speed, wear, and bar spacing. The bar spacing is usually a little larger than the desired coal size. These crushers produce a uniform coal sizing and break up pieces of wood and foreign material with the exception of metallic objects. Foreign material that is too hard to crush is caught in pockets.

The most satisfactory crusher for large capacities is the Bradford breaker. This design (Fig. 4-10) consists of a large-diameter, slowly revolving (about 20 rpm) cylinder of perforated steel plates. The size of the perforations determines the final coal size. In diameter, these opening are usually 30 mm to 38 mm. The breaking action on the coal is accomplished as follows: the coal is fed in at one end of the cylinder and carried upward on projecting vanes or shelves. As the cylinder rotates, the coal cascades off these shelves and breaks as it strikes the perforated plate. As the coal drops a relatively short distance, coal crushing occurs with the production of very few fines. Coal broken to the screen size passes through the perforations to a hopper below. Rocks, wood, slate, tramp iron, and other foreign material are rejected. This breaker produces a relatively uniform product and uses little power.

4-5 Other Components for Coal Preparation Systems

In addition to the crusher and pulverizer several other components are needed to convey the coal finally to the furnace. They are discussed below.

4-5-1 Cyclone Collector

Cyclone collectors are used in storage systems to separate the pulverized coal from the coal–air suspension. They are also applied to separate the fines entrained by the vented air so that the fines can be returned to the pulverized-coal bunker and increase the overall efficiency of the system.

To adapt to the coal preparation system, the cyclone collector used in the system is a little different from the widely used tangential-inlet cyclone. After entering the cyclone, the coal–air suspension swirls down the cyclone chamber. Owing to the centrifugal force, the coal particles move to the wall of the cyclone, then go slowly down to the hopper. The coal is then transported to the pulverized coal bunker.

4-5-2 Coal Feeders

A coal feeder is a device that supplies the pulverizer with an uninterrupted flow of measured amount of coal to meet system requirements. It is especially important in a direct-fired system. There are several types of feeders, including the belt feeder and the overshot feeder.

a) Belt Feeder

The belt feeder uses an endless belt running on two separated rollers receiving coal from above at one end and discharging it at the other. Varying the speed of the driving roll controls the feed rate. A leveling plate fixes the depth of the coal bed on the belt. The belt feeder (Fig. 4-11) can be used in either a volumetric or a gravimetric type of application. The gravimetric type has gained wide

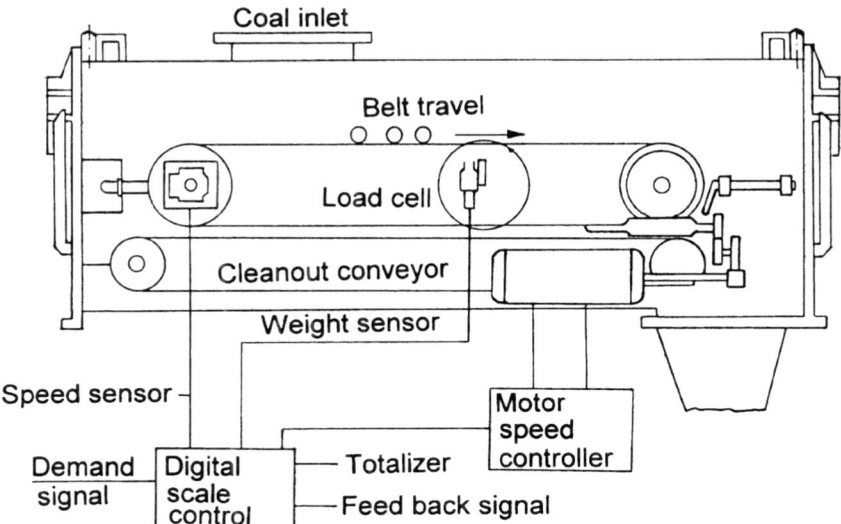

FIGURE 4-11. Belt feeder

popularity in the industry for accurately measuring the quantity of coal delivered to each individual pulverizer. Generally, it is applied to PC and CFB boilers
having combustion-control systems requiring individual coal metering to the fuel
burners.

There are two accepted methods of continuously weighing the coal on the feeder
belt. One method uses a series of levers and balance weights; the other, a solid
state load cell across a weigh span on the belt. Both are very accurate mechanisms
and well accepted by utilities. This same belt feeder design can also be used for
volumetric measurement.

b) Overshot Feeder

The overshot roll feeder (Fig. 4-12) has a multibladed rotor that turns about a fixed,
hollow, cylindrical core. This core has an opening to the feeder discharge and is
provided with heated air to minimize wetcoal accumulation on surfaces and to aid
in coal drying. A hinged, spring-loaded leveling gate mounted over the rotor limits
the discharge from the rotor pockets. This gate permits the passage of oversize
foreign material. Feeders of this type may be separately mounted, or they may be
integrally attached to the side of a pulverizer.

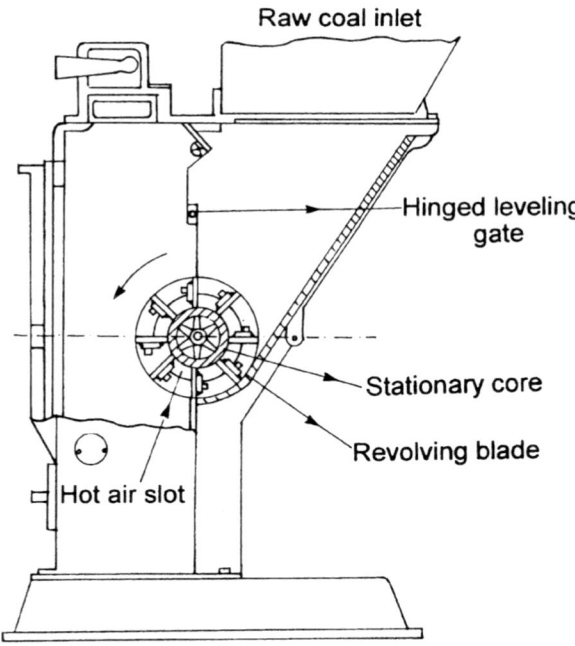

FIGURE 4-12. Overshot roll feeder

4-6 Design of Coal Preparation System for Pulverized Coal Boiler

Coal preparation systems for a PC boiler should be designed according to the capacity of the boiler and the type of fuel used by the boiler. These two factors determine what kind of system and what kind of pulverizer are to be applied and, furthermore, the whole coal preparation system.

4-6-1 Design Procedure

The most important component in the coal preparation system is the pulverizer. Thus, after the capacity of the boiler and the type of the coal are given, one should first select the type of the pulverizer system, then the pulverizer, and finally select other components of the system. The detail of the design procedure will be described afterwards. A broad outline of the design procedure is described first.

1. Select the type of coal preparation system. Out of the three kinds of coal prepa- ration systems: the storage system, direct-fired system, and semi-direct system. The most widely used is the direct-fired system. When choosing a direct-fired system, one should also determine whether to use a suction air system or a pressure system.
2. Select the type of pulverizers from the different types of pulverizers are given for different types of coal; therefore, the pulverizers should be chosen according to the coal specification.
3. Choose the capacity of the pulverizers and the number of pulverizers. In a large boiler, two to five pulverizers are used in the coal preparation system.
4. Calculate the air flow rate and temperatures. Normally a pulverizer has some limits on the maximum exit temperature and flow rate of the drying air. From these limits, the feed rate of the fuel, and from the energy balance, one can calculate the flow rate and temperatures of the air.
5. On the basis of the calculated air flow rate and the air temperature, one can select the fans and design the air-pipe system.
6. Select the coal feeders according to the pulverizers selected.
7. Select the coal crushers.
8. Design the coal bunkers.

4-6-2 Example

> Design a coal preparation system for a power station boiler burning bituminous coal whose characteristics are shown in Table 4-4. The capacity of the boiler is 200 MW and the total coal feed rate is 97 t/h under standard load.

TABLE 4-12. Characteristics of the coal*.

Moisture (%)	Fixed carbon (%)	Volatile matter (%)	Ash (%)	Heating value (kJ/kg)	HGI
10	43.3	35.9	10.8	22,080	50

*Characteristics are based on a proximate analysis.

SOLUTION

1. *Select the type of coal preparation system.* We choose the direct-fired system with pressurized pressure primary air, which is most widely used in power station boilers.
2. *Select the type of pulverizers.* We choose a ball-tube BBD mill, which is good for a direct-fired system. For this kind of coal, we can choose either a vertical air-swept mill or a ball-tube mill. However, for certain kinds of coal, only one of the two types of mills will be suitable. For example, a ball-tube is well suited for very hard, abrasive coal, while a vertical air-swept mill is not good for this situation. On the other hand, a ball-tube mill cannot work well for coal with high moisture content, say, 30%, whereas a vertical mill is well suited for this situation.
3. Table 4-5 shows the capacities of different types of BBD ball-tube mills. From the table, one can see that two BBD 40–60 mills, whose total standard capacity is 116 t/h, are suitable. The mills can be adjusted to fit the requirement of different boiler fire powers by simply changing mill air flow rate.
4. *Determine the drying air flow rate and the air temperatures.* The drying air flow rate is determined by the pulverizers. Every type of pulverizer has its own recommended air/coal weight ratio. For a BBD ball-tube mill, a ratio of 1.75 is good. Hence, under standard boiler capacity,

Total drying air flow rate $(Q_a) = 1.75 \times 97 = 170$ t/h
Drying air flow rate per mill = $170/2 = 85$ t/h

The mill outlet mixture temperature is determined by the types of fuel and the coal preparation system. Referring to Table 4-3b, we can choose 71°C as the outlet temperature.

The required air inlet temperature can be calculated on the basis of energy conservation. According to energy conservation,

Heat given by the hot air (H_a) = heat required to dry the coal (H_c)

Heat given by the hot air can be expressed as

$$H_a = Q_a(T_{inlet} - T_{outlet})C_a \qquad (4\text{-}13)$$

TABLE 4-12. Capacities of BBD mills.

Type of mill	34–48	40–60	47–60	47–72
Capacity of mill[a] (t/h)	30.5	58	79	99

[a]Capacities are for coal whose moisture is 10% and HGI is 50.

where

$$Q_a = \text{total air flow rate,}$$
$$T_{inlet}, T_{outlet} = \text{inlet and outlet temperature of the air,}$$
$$C_a = \text{heat capacity of the air}$$

Heat used to dry the coal can be divided into the following parts:

1. For heating the dry coal to the outlet mixture temperature
2. For heating the moisture to the outlet temperature
3. For evaporating the moisture. The moisture in the coal cannot be totally evaporated. For coal with 10% moisture, only 50–60% of the total moisture will be evaporated. We can choose a value of 55% for our calculation. Thus, the heat required to dry the coal.

$$H_c = Q_{dc}(T_{outlet} - T_{rc})C_c + Q_m(T_{outlet} - T_{rc})C_m + Q_m R_e E_e$$
$$= Q_c[(1 - Mo)(T_{outlet} - T_{rc})C_c + Mo(T_{outlet} - T_{rc})C_m + Mo R_e E_e]$$
$$(4\text{-}14)$$

where

$$Q_{dc} = \text{feed rate of the dry coal}$$
$$T_{rc} = \text{temperature of the raw coal}$$
$$C_c = \text{heat capacity of the dry coal}$$
$$Q_m = \text{feed rate of the moisture of the coal}$$
$$C_m = \text{heat capacity of the moisture}$$
$$R_e = \text{percentage of moisture evaporated}$$
$$E_e = \text{evaporating enthalpy of water}$$
$$Q_c = \text{feed rate of the coal}$$
$$Mo = \text{moisture content of the coal}$$

Substituting the following values, $C_a = 1.009$ kJ/(kg°C), $T_{rc} = 25°C$, $C_c = 0.92$ kJ/(kg°C), $C_m = 4.174$ kJ/(kg°C), $R_e = 55\%$, $E_e = 2334$ kJ/(kg°C), $Mo = 10\%$, and combining Eqs. (4-13) and (4-14), we get $T_{inlet} = 176°C$.

Considering that about 5% of the energy released by the hot air may be lost to the surroundings through conduction, we should choose an inlet temperature about 10°C higher than the calculated value. Thus we can let T_{inlet} be 185°C. In a ball-tube mill system, the drying air temperature can be easily adjusted to fulfill the requirement of the drying purpose.

The drying air is usually a mixture of the preheated air and the room air or a mixture of the preheated air, the flue gas, and the room air. Here we use the first combination, which is simpler. Normally, the temperature of the preheated air is about 300°C. From the energy balance, we have

$$(T_{inlet} - T_r) = R_h(300 - T_r)/100 \qquad (4\text{-}15)$$

where

$$T_r = \text{room air temperature} = 25°C$$
$$R_h = \text{percentage of preheated air in the drying air}$$

TABLE 4-15. Parameters for the coal preparation system.

	Symbol	Unit	Value
Total coal feed rate	Q_c	t/h	97
Type of pulverizer	—	—	Ball-tube BBD 40–60
Standard capacity per pulverizer	—	t/h	58
Number of pulverizers	—	—	2
Moisture in raw coal	Mo	%	10
Moisture in pulverized coal	Mo_{pc}	%	4.5
Heat capacity of dry coal	C_c	kJ/(kg°C)	0.92
Drying air flow rate	Q_a	t/h	170
Air inlet temperature	T_{inlet}	°C	185
Mixture outlet temperature	T_{outlet}	°C	71
Hot air percentage of drying air	R_h	%	58

Substituting the values of T_{en} and T_{inlet}, we get that $R_h = 58\%$.

All parameters for the coal preparation system are given in Table 4-6.

4-7 Fuel Feeding in Fluidized Bed Boilers

Three types of fuel feed systems are used in fluidized bed boilers.

• under-bed system
• over-bed system
• in-bed system

Bubbling fluidized bed boilers usually use the under-bed and over-bed systems, while a circulating fluidized bed boiler uses an in-bed system of feeding.

4-7-1 Bubbling Fluidized Bed Boilers

An under-bed feed system (Fig. 4-13) is essentially a pneumatic transport system. It transports the coal from a storage silo to the bottom of the bubbling bed. The feed points usually pass through the distributor or the grate of the boiler furnace. The under-bed system cannot spread the coal as evenly as the over-bed system does. However, it allows a longer residence time for the fuel and its volatile in the hot combustion zone. This allows for good mixing of the fuel-volatile with the air. Hence it was first developed for bubbling fluidized bed boilers burning less reactive fuels to increase their combustion efficiency. These less reactive fuels require a longer time to burn. However, the under-bed system is complicated and very expensive, especially for large boilers where the coal has to be evenly split among large numbers of fuel feed points. For average coal one under-bed feed point can serve 1.5–2.8 m² of bed area. For reactive coal the area served by one point will go down and for nonreactive coal it will go up.

An over-bed feeder is simpler in construction. Coal is sprayed over the bed by means of a spreader feeder (Fig. 4-14). Thus fewer feeders are required. For

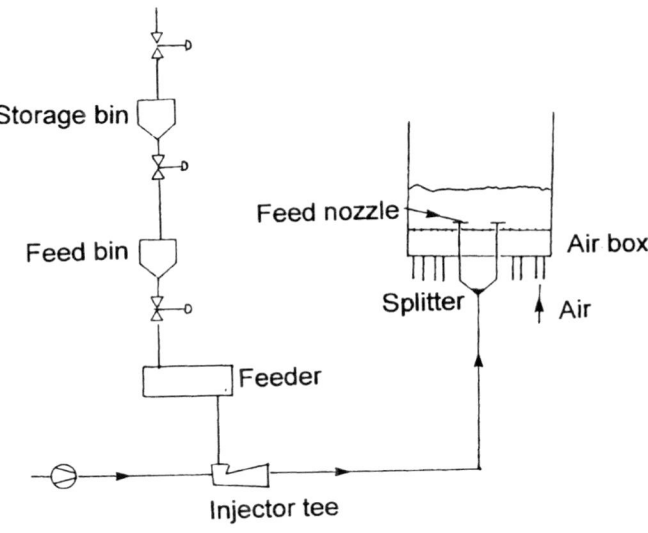

FIGURE 4-13. Under-bed feed system

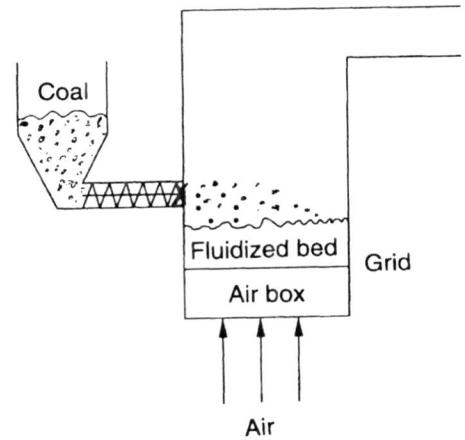

FIGURE 4-14. Over-bed feed system

example, one over-bed feeder can serve about 15 m^2 of bed area (Basu & Fraser, 1991). A major disadvantage of this type is the loss of unburnt carbon fines and unused sorbents. This puts a limitation on the amount of fines tolerated in the crushed coal. On the other hand, the moisture, especially the surface moisture, is an advantage to over-bed feeding. It reduces the carbon loss owing to the entrainment.

4-7-2 Circulating Fluidized Bed

In circulating fluidized bed (CFB) boilers, there is no definite bed depth. The entire furnace is in a state of refluxing. So, the fuel–air mixing is very intense. The under-bed system is not required here as it cannot enhance the combustion efficiency any further. Coal is generally fed from the sides.

In a typical system a belt feeder carries the coal from the bunker. It may also measure the feed rate of the coal. It drops the coal into hoppers on the side of the furnace. The coal is then injected into the bed with the assistance a feed screws or pressurized air. Because the furnace pressure at the feed point can be as high as 1270 mm of water gauge, the feed system must seal against this pressure to allow the coal to enter the furnace. The pressure seal is accomplished by the use of an isolation valve and a head of coal between the coal silo and a belt feeder. Because an in-bed feed system is much simpler and inexpensive than an under-bed system, it is widely used in CFB boilers.

In some CFB boilers the fuel is fed into the loop seal. This coal enters the furnace along with the hot recycled solids. Thus it gives better mixing of fuel with the hot bed solids.

A CFB boiler often uses sorbent for sulfur capture. It is pneumatically conveyed from the storage bunkers and injected at various points around the periphery of the lower part of the furnace. The sorbents are slow-reacting solids, so their mixing is not critical as it is for coal.

Nomenclature

a	ratio of rotational speed (Eq. 4-5)
C_a	heat capacity of the air, kJ/kg°C
C_c	heat capacity of the dry coal, kJ/kg°C
C_m	heat capacity of the moisture, kJ/kg°C
D	diameter, m
D_B	inside diameter of mill
E_e	evaporating enthalpy of water, kJ/kg°C
F_C	coal fineness at mill outlet
H_a	heat released by the hot air, kJ/kg
H_c	heat to dry coal, kJ/kg
H_C	moisture content of coal
H_R	hardgrove grindability
K	constant factor (Eq. 4-10)
L	length, m
m	mass of the ball, kg
Mo	moisture content of the coal, %
n_B	rotational speed of mill
n_c	critical speed of rotation, 1/min
P	power, MW
P_B	weight of balls, kg

Q flow rate of pulverized coal, t/h
Q_a total air flow rate, t/h
Q_c feed rate of the coal, t/h
Q_{dc} feed rate of the dry coal, t/h
Q_m feed rate of the moisture of the coal, t/h
R_e percentage of moisture evaporated
R_h percentage of preheated air in the drying air
T ratio of filling
T_{inlet} inlet temperature of the air, °C
T_{outlet} outlet temperature of the air, °C
T_r room air temperature, °C
T_{rc} temperature of the raw coal, °C
w function of α (Eq. 4-7)
X power factor

Greek Symbols

α angle (Fig. 4-7)
ω_c critical angular speed, rad/s

References

Basu, P., and Fraser, S.A. (1991) "Circulating Fluidized Bed Boilers, Design and Operation." Butterworth-Heinemann, Stoneham, Massachusetts.

Clapp, R.M. British Electricity International (1991) "Modern Power Station Practice, Vol. B: Boilers and Ancillary Plant." London, Pergamon Press.

Fontanille, D. (1996) Recent developments in ball tube mills in thermal power plants. GEC Alsthom, Stein Industrie, Report No. DTB/FAI/LOE.

Kakac, S. (1991) "Boilers, Evaporators and Condensers." John Wiley and Sons, Inc., New York.

Singer, J, ed. (1991) "Combustion: Fluidized-Bed Steam Generator." ABB Power Corporation, Windsor, Connecticut, Chapter 9.

Singer, J.G. (1981) "Combustion Fossil Power System." Combustion Engineering Inc., Windsor, Connecticut.

Stultz, S.C., and Kitto, J.B. (1992) "Steam, Its Generation and Use." Babcock & Wilcox, Barberton, Ohio, p. 12-8.

5
Design of Oil Burners

Fuel oil is the most commonly used liquid fuel. Easy transportation and simple and less expensive design make oil fired boilers very attractive. Fuel oil is used wherever the fuel cost does not outweigh other advantages of oil fired boilers. Liquid fuels burn in a vapor state. To facilitate fuel vaporization a major goal of oil burner design is to increase the contact surface area of the oil with air. For this reason, oil is always atomized in small droplets on entry to the boiler. The present chapter presents an overview of oil supply systems and some procedures for selection and design of oil burners.

5-1 Design of Oil Supply System

A typical oil supply system includes an oil tank, oil strainer, oil supply pump, heater, and connecting pipelines to the boiler. Accordingly the system may be divided into several parts: oil tank, inlet header, heating loop, and exit header as shown in Figure 5-1. General considerations in the design of the oil supply system are safety, cost, convenience, and emergency provision.

5-1-1 Heating Loop

For economic reasons most power and industrial boilers use less expensive heavy fuel oil. It is difficult to atomize this fuel owing its high viscosity at ambient temperatures, so the oil is heated prior to burning. Heating is an important part of the oil burners. The heating loop consists of strainer, pump, and heater. Usually there are two types of heating systems: the *single unit heating loop* and the *centralized heating loop*. In the former the strainer and the heater form a set with the pump, which connects the strainer and the heater (Fig. 5-1). At the end of the loop, it is connected with a common header (also called mother pipe) to send the oil to the boiler. Sometimes, for equipment arrangement and economic considerations, the capacity and the number of heaters cannot be matched with those of oil supply pumps. Therefore, the centralized heating system is needed where a common

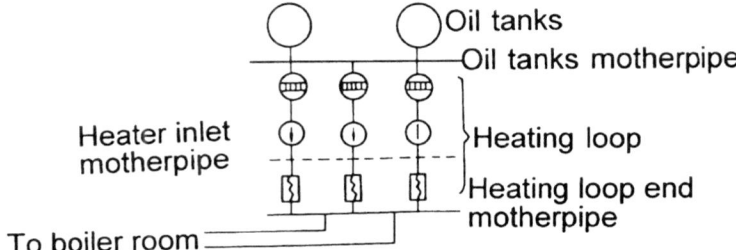

FIGURE 5-1. Schematic diagram of an oil supply system, including the heating loop

header (mother pipe) is placed between the heater and oil pump. The header collects the oil from individual the pumps and then distributes to boilers.

For practical reasons the oil supply builds in some extra capacity. Instead of chosing one pump of 100% capacity, one may choose several pumps of smaller capacities. For example, 3 nos × 50% capacity; 4 nos × 33% capacity; 4 nos × 50% capacity; 6 nos × 25% capacity, and one or two pumps are left as spares.

The oil heaters are arranged in no fewer than three groups. When the largest group of heaters is off, the remaining heaters should meet the heating demands of oil flow for the maximum continuous steam rates of all the boilers in the plant. So, the heater systems may be chosen as 3 nos × 50% capacity, 4 nos × 33% capacity, and 5 nos × 25% capacity; and one spare group.

5-1-2 Oil Supply System with Two Stages of Pumps

In the previous arrangement only one pump is used to transfer the oil from the oil tank to the boiler room. In this system, the oil pump is located between the oil tank and the heaters. Therefore, the heater must bear the oil pressure at the pump exit, which is very high. Burner oil pressure increases with boiler capacities. For example, the burner oil pressure is 2.0 MPa for a 120 t/h steam output boiler, 3.5 MPa for a 220 t/h boiler, 5.0 MPa for 410 t/h and 1000 t/h boilers, and 7.0 MPa for 670 t/h and 2000 t/h boilers. As the oil pressure at the pump exit should be higher than these values, the design of the heaters increasingly complex for higher capacity boilers. Furthermore, since the pumps are located upstream of the heaters, the viscosity of the pumped oil is limited by the maximum allowed temperature of the oil tank. This restricted temperature gives a higher oil viscosity. This lowers the pump efficiency and therefore more energy is consumed by the pump. To overcome the above shortcomings, a two stage pump system is employed.

In the two stage system, the first stage pump transfers the oil to the heaters from oil tanks. Then the oil is heated, and the second stage pump transfers the oil to the boiler room. Thus the second stage heater pump has to bear only the burner pressure. However, it benefits from the lower viscosity, and so the heater design is much simpler.

5-2 Oil Atomizers

An oil burner, used in an oil fired boiler, would generally consist of an oil gun, an air register, and some supplementary equipment, such as ignition apparatus, etc.

Oil is injected through the oil gun. It atomizes into droplets and then burns in the furnace. The tip of the oil gun is its main component and is known as an *atomizer*. The atomizer plays a decisive role in the quality of atomization.

The combustion air is injected through the *air register* (discussed in section 5-3) in such a way that it provides the best combustion condition for atomized oil spray.

The main determining factors for good combustion in an oil fired boiler are good atomizing quality and complimentary air supply. In this section, the principles of operation, construction, and calculation of the oil atomizers are discussed. Air is supplied through air registers. This will be discussed later in section 5-3.

The oil enters the furnace through the atomizer and it breaks up into droplets. Good atomizing quality is a basic requirement of good combustion. The finer the atomized oil droplets, the faster they burn. Therefore, the main requirement of the oil atomizer is that the droplets be as small as possible and that the distribution of the spray meet the air register's demand. This requires an appropriate atomizing angle and spray flow density distribution. In addition, the design of the atomizer should be simple for reliable, convenient, and adjustable operation, and the design must permit easy cleaning and overhauling.

5-2-1 Types of Oil Atomizers

Following five main types of oil atomizers are used in oil fired boilers.

1. Mechanical atomizer (simple mechanical and circle mechanical atomizer)
2. Steam or compressed air atomizer
3. Low pressure air atomizer
4. Rotary cup atomizer
5. Supersonic atomizer

These are described below:

5-2-2 Mechanical Atomizer (Pressure Jet)

The mechanical atomizer is one in which the pressure of the fuel itself atomizes the fuel. It is also called pressure jet atomizer. There are several types of mechanical atomizers. The one most widely used is shown in Figure 5-2. There are three main parts of the burner:

- Atomizer (1) (Fig. 5-2)
- Swirler (2)
- Flow distributor (3)

These are arranged along the burner axis. The oil first flows through a set of small holes of the flow distributor, then it flows through the tangential grooves of the swirler, which make the oil swirl. Then the oil enters the swirl room of the atomizing slice. From here the oil swirls rapidly in the swirl room and spurts out from the atomizing nozzle, forming droplets. Since the mechanical design of the flow passage makes the oil atomize, these types of atomizers are called *mechanical atomizers*.

There are two commonly used mechanical atomizers:

1. Uniflow, or simple mechanical atomizer
2. *Return-flow*, also called *circle mechanical atomizer*

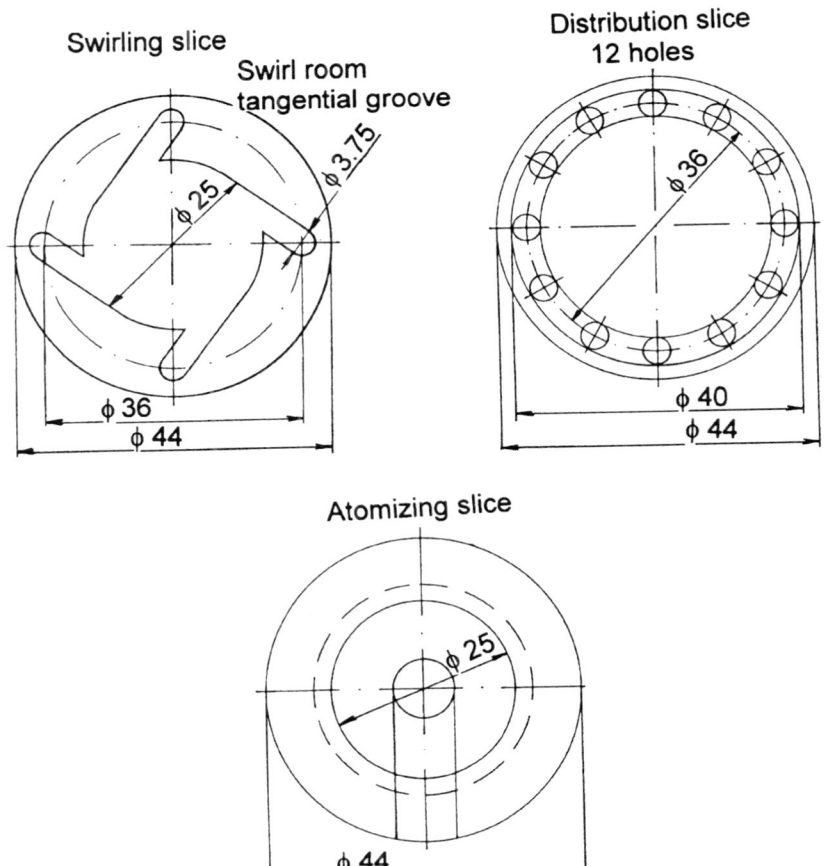

FIGURE 5-2. Cross section views of a simple mechanical atomizer of capacity 1700–1800 kg/h. All dimensions are in mm

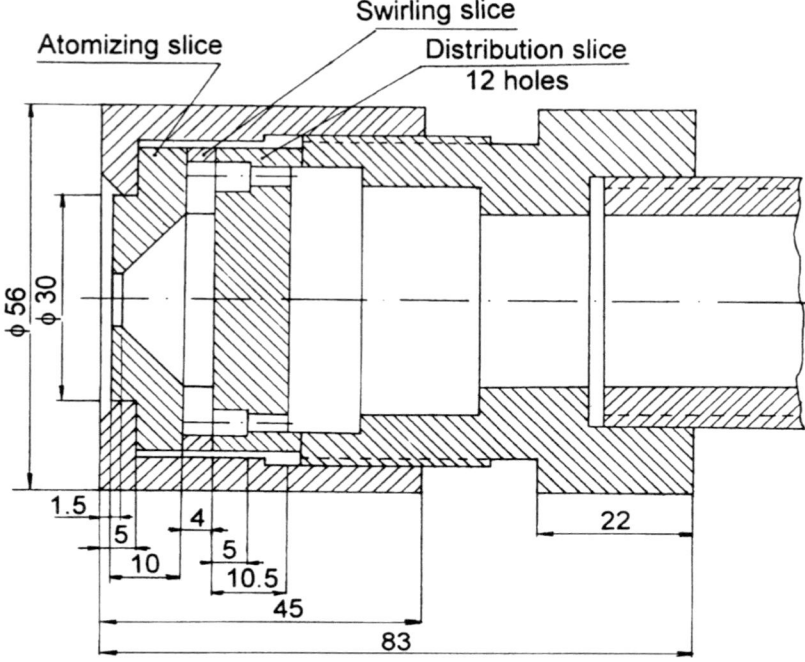

FIGURE 5-2. (*Continued*)

In the uniflow atomizer, oil enters the atomizers at a pressure above 2 MPa gauge for a heat input of 20–24 MW. These atomizers are generally used in small, medium, and marine boilers.

In the return-flow atomizer, a part of the oil returns to the oil delivery system. It is used when there is a need for capacity variation, e.g., in large industrial and utility boilers. Oil pressure is in the range of 4–7 MPa gauge and the maximum capacity can go up to 60 MW.

Other mechanical atomizers used are the tangential-hole-type cylinder atomizer and the tangential-hole-type spherical atomizer. Details of these burners are available in the literature (Junkai, 1992; Xiezhu, 1976).

Now we will discuss the design method for different types of mechanical atomizers. For the simple mechanical atomizer we review first the theory and then the design method.

a) Mechanical Atomizer

The main indicators of the atomizing quality are atomizing angle, droplet size, and distribution of oil flux.

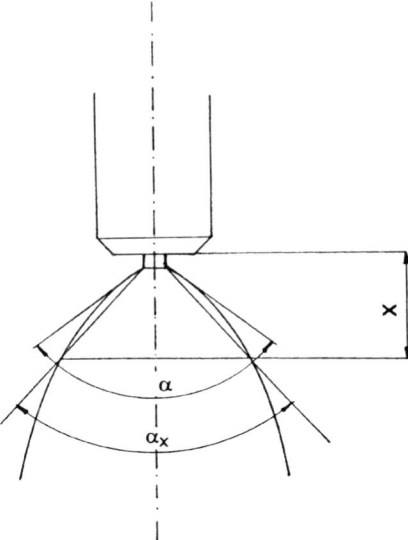

FIGURE 5-3. Definition of the angle of atomization

(1) Atomizing Angle

The atomized oil spurts out of the nozzle, forming a cone. The angle of this cone, known as *atomizing angle*, plays an important part in the design of the air register. In the absence of any external force, the oil leaving the nozzle would appear as a conical sheet. In the center of the oil spray, a certain pressure head is formed. The pressure around the spray cone is higher than the center pressure which makes the oil sheet rupture into spray. Therefore, the edge of the oil spray is not a straight line. There are several ways to describe the atomizing angle as described below and in Figure 5-3.

(1) Exit atomizing angle (α): Draw a line at the exit of the atomizer nozzle tangent to the edge of the spray. The angle between the two tangents is called the exit atomizing angle and is represented by α. The value of the exit atomizing angle is nearly the same as its theoretically calculated value (Fig. 5-3).

(2) Conditioned atomizing angle: At a certain distance X from the atomizer nozzle, a line is drawn perpendicular to the axis of the oil spray. If we draw two lines from the atomizer inlet to the points of intersection of perpendicular line with the spray profile, we get a smaller cone. The angle of this cone is called the *conditioned angle* (Fig. 5-3).

It is apparent that the conditioned angle is less than the exit atomizing angle. The difference can be 20° or greater. The conditioned angle changes with the distance X selected.

(2) Spray Atomization Size (Size of Droplets)

Droplet size is the main index of atomizing quality. Usually, the mean diameter of droplets is used as a parameter. The usual practice is to use the area mean diameter. In this book mean diameter would refer to the area mean diameter. The size of the droplet may vary with the position in the spray. So the drop sizes are sampled at specific locations. There are two sampling methods:

(1) Sample along the radius at a certain distance X as shown in Figure 5-3, then measure mean diameters.
(2) Sample at the position of greatest concentration of oil droplets, then determine the mean diameter. Since the droplets are larger at all other positions, the mean diameter calculated from this sample will also be larger.

When the oil flow and air collide at a very high velocity, the oil can also break up into small droplets. Experimental results (Xiezhu, 1976) have shown that the largest droplet diameter, d_m, depends on the relative velocity, W_{rel}, between the oil droplet and the air as follows:

$$d_m = \frac{K}{\rho W_{rel}^2} \tag{5-1}$$

where ρ is the density of the air in kg/m^3 and K is a coefficient depending on the characteristics of oil, mainly the viscosity and surface tension. For moderately viscous common fuel oil, K is about 600.

It has been also shown (Weinberg, 1953) that the Sauter mean diameter of droplet size is proportional to $(FN^3/\Delta P)^{0.143}$, where FN is *flow number* defined by the ratio of $(G/\Delta P)^{0.5}$. There are several emipirical expressions (Dombrowski & Munday, 1968) for droplet size, including the one shown above. The droplets will break up further if the diameter of the droplets is larger than the value calculated with the above-mentioned relationship.

(3) Oil Flux Density

Oil flux density is defined as the volume of fuel flowing through unit area perpendicular to the direction of oil spray in unit time. The distribution of oil flux density has an important effect on the air registration. The oil flux density should be uniform around the periphery of the spray. Otherwise there would be poor combustion. Also a very high flux density is not desirable at the spray center where there is a recirculation zone.

b) Design Methods

(1) Calculation of Oil Flow Rate

The working principle of the mechanical atomizer is shown in Figure 5-4. For the theoretical analysis, the oil is assumed an ideal liquid without any frictional resistance.

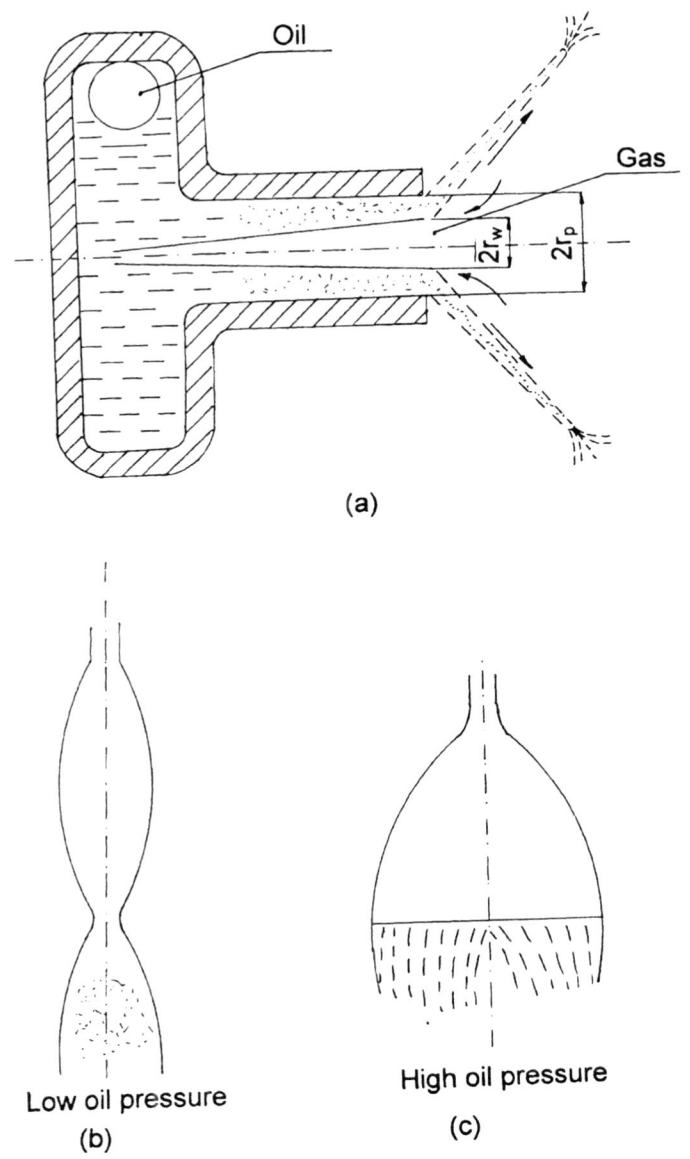

FIGURE 5-4. A sketch showing the operation of the atomizing process of the mechanical atomizer. (a) Oil enters the atomizer tangentially and leaves it in the shape of a conical jet. (b) High oil pressure produces a wide cone. (c) Low oil pressure produces a twisted narrow jet

The oil enters at a radius R and then it swirls through the nozzle (radius r_p), forming a film of radius r_w. So, a centrifugal force acts on the liquid.

By applying Bernoulli's equation and conservation of mass, the flow rate of oil can be derived, as (Xiezhu, 1976, pp. 62–65).

$$G = 3600 \,\mu\pi \cdot r_p^2 \sqrt{2\rho P_0} \ \ (\text{kg/h}) \tag{5-2}$$

where μ is the *flow coefficient* given by

$$\mu = \cfrac{1}{\sqrt{\cfrac{A^2}{1-\varphi} + \cfrac{1}{\varphi^2}}} \tag{5-3}$$

where the ratio of oil flow area and the nozzle area is $\varphi = 1 - \frac{r_w^2}{r_p^2}$

r_p = nozzle radius, m
r_w = radius of oil film vortex inside the nozzle (Fig. 5-4), m
P_0 = oil pressure at the inlet of nozzle above that at the exit, Pa.
ρ = the oil density, kg/m³

Here, A is a *characteristic coefficient* of the nozzle, which is calculated as below:

$$A = \frac{\pi r_p R}{\sum f} \tag{5-4}$$

where $\sum f$ is the total area of tangential grooves, and R is the swirl radius.

For maximum flow rate the flow coefficient μ should be maximum. So, using $d\mu/dA = 0$, we get

$$A = \frac{1-\varphi}{\sqrt{\dfrac{\varphi}{2}}} \tag{5-5}$$

Substituting this into Eq. (5-3),

$$\mu = \sqrt{\frac{\varphi^3}{2-\varphi}} \tag{5-6}$$

(2) Calculation of the Atomizing Angle

At the exit of the nozzle, the oil spins out with a certain axial (w) as well as with a tangential (u) velocity. The atomizing angle, α (Fig. 5-3) formed by the outer edge of the oil spray cone would depend on the ratio of these two velocities. It is derived as (Xiezhu, 1976, p. 66)

$$\tan\frac{\alpha}{2} = \frac{u}{w} = \frac{(1-\varphi)\sqrt{8}}{\left(1 + \sqrt{1-\varphi}\right)\sqrt{\varphi}} \tag{5-7}$$

The actual cone angle is slightly different from that calculated above.

(3) Empirical Method of Calculation of Flow Coefficient, μ

Empirical coefficient method

Extensive tests carried out by Chinese scientists (Xiezhu, 1976) established the following relationships:

$$\mu_p = 0.88\,\mu \qquad (5\text{-}8)$$

$$\alpha_p = 0.87\,\alpha \qquad (5\text{-}9)$$

where μ_p = practical flow coefficient,
$\quad\quad\ \ \alpha_p$ = practical exit atomizing angle
$\quad\quad\ \ \mu$ = theoretical values calculated from Eq. (5-6)
$\quad\quad\ \ \alpha$ = theoretical values calculated from Eq. (5-7)

Empirical formula method

The above coefficient still requires the knowledge of theoretical flow coefficient and the atomization angle. To further simplify the design calculation, experimenters (Xiezhu, 1976) found the following relationship, but for a limited range of design conditions:

$$\mu_p = 0.37 A^{-1.1} \qquad (5\text{-}10)$$

$$\alpha_{R=200} = 57 A^{0.37} \text{ degree} \qquad (5\text{-}11)$$

The above relationship (Eq. 5-10, 5-11) is valid for conditions $A = 0.8 \sim 1.8$; inlet oil pressure $= 2.0$ MPa, oil viscosity $= 3°$E (see Chapter 3, Section 3-1-2 for definition of °E); and nozzle diameter <5 mm.

Besides the characteristics coefficient of the nozzle (A), the ratio of the radius of the nozzle and the vortex radius (r_p/R) has an important effect on the operation of the atomizer. The smaller the ratio r_p/R, the larger the vortex radius.

The nozzle characteristics may also be obtained to allow calculation of more practical values of the flow coefficient μ using Eq. (5-3).

$$A_p = 2.094\left(\frac{r_p}{R}\right)^{0862} A \qquad (5\text{-}12)$$

where A_p is a modified practical value of the characteristics coefficient, A.

Oil flow rate

By using the practical flow coefficient and considering the friction loss in the atomizer, we can calculate the oil flow rate of the atomizer as shown below:

$$G = 3600\,\mu_p r_p^2 \sqrt{2\rho(P_0 - \Delta P)} \ \text{(kg/h)} \qquad (5\text{-}13)$$

The friction loss of the atomizer can be calculated as

$$\Delta P = \sum \xi(\rho v^2/2)\,\text{Pa} \qquad (5\text{-}14)$$

where the velocity in the swirl chamber is

$$v = \frac{G}{3600\,\rho \cdot g \sum f} \qquad (5\text{-}15)$$

where G is the oil flow rate in kg/h, ρ is oil density in kg/m^3, and $\sum f$ is the total flow area of the groove. The combined coefficient of hydrodynamic resistance, $\sum \xi$, is the sum of individual resistance coefficients, $\xi_1, \xi_2, \xi_3, \xi_4$, which are the resistance coefficients of swirl room inlet, cylinder wall, tip wall, and the nozzle, respectively $\sum \xi = \xi_1 + \xi_2 + \xi_3 + \xi_4$. Here, the $\sum \xi$ valve may be taken from Figure 5-21, which gives the combined resistance coefficient for both tangential groove atomizer and tangential hole cylinder atomizer against the log $(A \cdot Re)$. The Reynolds number, Re, is based on the groove dimension.

$$Re = \frac{Wd}{\gamma} \qquad (5\text{-}16)$$

where W, the oil velocity in the groove, is given by $(2 P_0/\rho)^{0.5}$ and hydraulic diameter of groove, $d = [2 hb/(h + b)]$; γ is the kinematic viscosity of oil.

EXAMPLE:

A simple mechanical atomizer has the following characteristics. The required oil flow rate is 1000 kg/h. The main design parameters are as follows: radius of the nozzle, $r_p = 2.05$ mm; diameter of the swirl room, $D_x = 14.8$ mm; width of the tangential groove, $b = 2.5$ mm; depth of the tangential groove, $h = 2.4$ mm with 4 grooves; inlet oil pressure, $P_0 = 3.0$ MPa; density of oil, $\rho = 900$ kg/m^3; kinematic viscosity of oil, $\gamma = 20 \times 10^{-6}$ m/s.
Find if this design can deliver the required oil flow rate.

SOLUTION:

(a) Calculation of friction loss in the nozzle.
    First the vortex radius is calculated:

$$R = D_x / 2 - h / 2 = (14.8 - 2.4)/2 = 6.15 \, \text{mm}$$

The nozzle characteristics coefficient is calculated from Eq. (5-4):

$$A = \frac{\pi \cdot R \cdot r_p}{f} = \frac{3.14 \times 6.15 \times 2.05}{4 \times 2.5 \times 2.4} = 1.65$$

The oil velocity in the groove, W, is calculated from the inlet oil pressure using Eq. (5-16):

$$W = \sqrt{\frac{2 P_0}{\rho}} = \sqrt{\frac{2 \times 3,000,000}{900}} = 81.6 \, \text{m/s}$$

The hydraulic diameter of the tangential groove is calculated from Eq. (5-16).

$$d = \frac{2 \times 2.4 \times 2.5}{(2.4 + 2.5)} = 2.45 \, \text{mm};$$

$$Re = \frac{W \cdot d}{\gamma} = \frac{81.6 \times 0.00245}{20 \times 10^{-6}} = 9996$$

To find the friction coefficient from Graph 5-21 we can use

$$\log(A\ Re) = \log(1.65 \times 9900) = 4.214$$

So the friction coefficient is read from Figure 5-21 against log $(A.Re)$ as $\sum \xi = 3.31$.

The velocity of oil in the tangential groove is calculated from Eq. (5-15):

$$v = \frac{G}{3600\,\rho\sum f} = \frac{1000}{3600 \times 900 \times (4 \times 2.5 \times 2.4 \times 10^{-6})} = 12.8 \text{ m/s}$$

The friction loss of the atomizer is obtained from Eq. (5-14):

$$\Delta P = \sum \xi \rho \frac{v^2}{2} = 4.41 \times 900 \times \frac{12.8^2}{2} = 240000 \text{ Pa} = 0.24 \text{ MPa}$$

(b) Calculation of flow coefficient.

The modified practical flow coefficient is calculated from Eq. (5-10) for $A = 1.65$ as follows:

$$\mu_p = 0.37 A^{-1.1} = 0.37 \times 1.65^{-1.1} = 0.21$$

(d) The oil flow rate, G, of the atomizer is calculated from Eq. (5-13):

$$G = 3600\,\mu_p \pi r_p^2 \sqrt{2\rho(P_0 - \Delta P)}$$
$$= 3600 \times 0.21 \times 3.14 \times (2.05/1000)^2$$
$$\times (2 \times 900 \times (3,000,000 - 240,000)^{0.5} = 703 \text{ kg/h}$$

The computed flow rate 703 kg/h is about 30% less than the required flow rate of 1000 kg/h. So, the nozzle diameter and or other parameters have to be changed to meet this requirement.

c) Circle Mechanical (Return-Flow) Atomizer

In a simple mechanical atomizer, the oil flow rate is directly proportional to the square root of the oil pressure. So, the oil flow rate does not change much with the change in the oil pressure. Therefore, to increase the oil flow rate, the oil pressure must be increased greatly. A circle mechanical atomizer or return-flow atomizer provides an alternative to this problem. Here, adjustable atomizers are used to control the flow rate instead of increasing the oil pressure. A schematic of this system is shown in Figure 5-5. Two common types are discussed below.

(1) Inlet Area Changing Atomizer

The flow rate of the atomizer is adjusted by changing the inlet area of the tangential groove.

(2) Circle Atomizer

The oil enters the swirl chamber through tangential holes in the swirl section (Fig. 5-6). One part of the oil swirls out from the atomizer nozzle; the other part

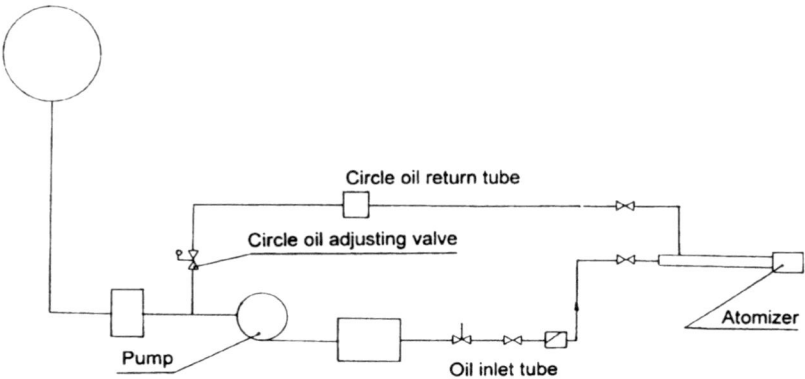

FIGURE 5-5. The oil pipe arrangement for a return oil or circle oil atomizer

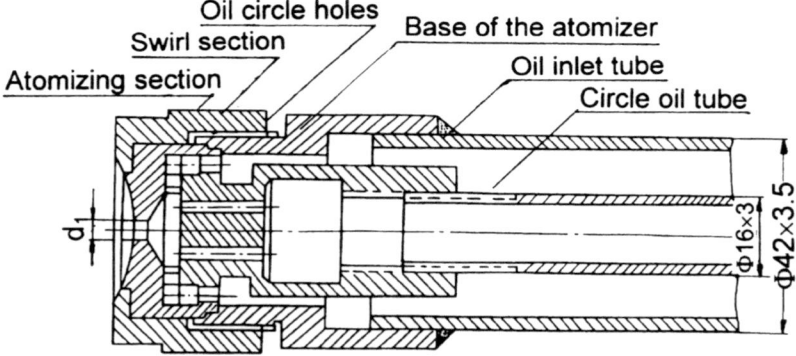

FIGURE 5-6. Cross section view of a return oil inner circle type atomizer

returns from the swirl chamber through a number of circle oil holes on the back of
the swirl chamber. This oil returns to the pump. It is called *inner circle atomizer*.
The inner oil circle atomizers are widely used in power plants.

There is another design, known as the *outer circle atomizer*, where oil enters
through the central hole and returns through peripheral holes.

The oil supply system of the oil circle atomizer is shown in Figure 5-5. One part
of the oil entering the burner returns to the pump through the oil circle holes on
the back of the swirl. The flow rate of the circle or returning oil is adjusted with
the circle oil adjusting valve. When the circle oil adjusting valve opens up, the oil
pressure in the circle oil pipe decreases. So the return or the circle oil flow rate
increases and therefore the oil flow rate to the atomizer decreases.

The following three methods are employed to control the oil flow rate through the atomizer.

1. Keep the pressure difference between the inlet oil and the return or circle oil constant. So, when the circle oil pressure is decreased by opening the circle oil adjusting valve, the inlet oil pressure must decrease correspondingly. There is some decrease in the oil entering the atomizer but the range of adjustment is relatively low here.
2. Keep the oil flow rate at the inlet constant. When the circle oil pressure decreases, the inlet pressure decreases. So, at low load the inlet pressure is low. Thus the adjustment range in this method is higher than that in the above described method.
3. A combination of two methods above. When the oil flow rate is high, the first method of keeping the inlet oil pressure constant is adopted. When the oil flow rate is low, the method to keep the inlet oil flow rate constant is adopted. The adjustment range is wider than that of either method separately.

Multiple holes for the return oil can be drilled in a circle on the back plate of the swirl chamber. The diameter of individual return oil holes is selected according to the required adjustment range. When the diameter of the oil hole is 1.3 times the diameter of the atomizer nozzle, the maximum adjustment ratio

$$\left(n_{max} = \frac{G_{max}}{G_{min}}\right)$$

can be about 3. The circle on which the return oil holes are located is called the *oil circle*. The diameter of this cannot be larger than $(D_x - 2h)$, where D_x is the diameter of swirl room and h is the depth of tangential groove through which the oil enters the swirl room. The diameter of the oil circle can be as large as 0.5–0.6 times larger than the diameter of the swirl room, but it should be less than $D_x - 2h$. The diameter of the individual return oil hole can be 2–3 mm.

Instead of having multiple holes for oil return there can be a single large central hole. When the cross section area of this return oil hole is close to that of the nozzle, the maximum adjustment ratio, n_{max}, can be about 3. Under the above-described condition, the maximum ratio of the pressure of return oil and that of the inlet oil is about 0.5–0.6. In case of multiple small holes this ratio is 0.65–0.7.

5-2-3 Steam (or Compressed Air) Atomizer

The steam atomizer appeared on the market before the mechanical atomizer did. However, because of the high consumption of steam (0.4–0.6 kg steam/kg oil), the use of simple steam atomizers declined in the industry. Recently, the demand for higher capacity and low excess air operation prompted the reappearance of steam atomizers. Newer steam atomizers, however, have much lower steam consumption (0.1–0.02 kg/kg oil). Steam (air) atomizers have fewer difficulties with repairs and

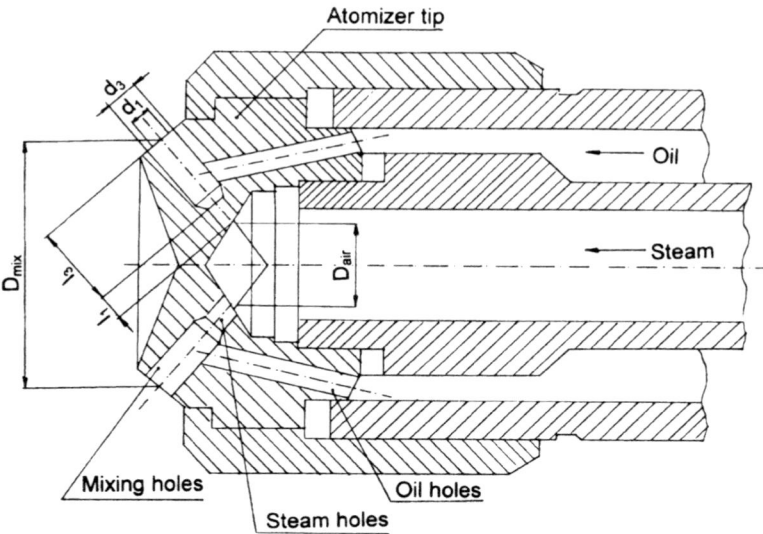

FIGURE 5-7. A Y-type steam oil atomizer

maintenance owing to their oil pressure lower requirements. Also, they produces much finer fuel droplets.

In a steam atomizer (Fig. 5-7), the atomization is assisted by either high-pressure air or high-pressure steam. Accordingly, they are of either the *steam-mechanical atomizer or pure steam atomizer* type.

a) Pure Steam Atomizer

The oil pressure of the steam-only atomizer is very low. The oil sheet of the spray cone is broken up by the high-speed steam jet emerging from a steam nozzle. The steam consumption rate of this burner is therefore large. It is generally used in small industrial boilers.

The design of the pure steam atomizer is simple (Fig. 5-8). Oil flows in the central tube while the steam flows in the annulus. The exit section area of steam can be adjusted by the hand wheel. The oil pressure is so low (0.2–0.25 MPa) that even a raised oil tank can provide the necessary head for the burner. The flame length is, however, thin and long (about 2.5–7.0 m). The oil flows out from a central tube and the steam flows in the annulus. They meet at the nozzle tip.

b) Steam-Mechanical Atomizer

In a simple mechanical atomizer (Fig. 5-2), no steam is consumed, but its capacity regulation is limited and at low load the atomizing quality deteriorates rapidly. In a steam atomizer, on the other hand, the adjustment ratio is high and at low load the

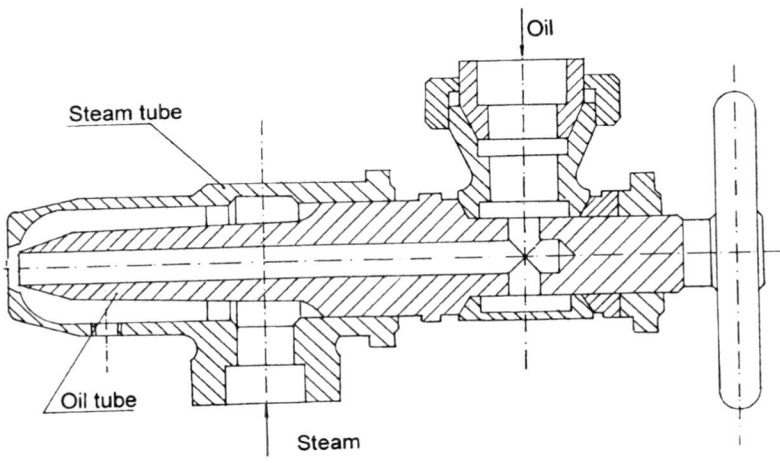

FIGURE 5-8. Pure steam atomizer. Low-pressure oil in the central tube is atomized by steam from the annulus

atomizing quality is good, but the steam consumption is high. The advantages of these two types of atomizers are combined in steam-mechanical atomizers. These atomizers use oil pressure lower than that used for pure mechanical atomizers. Also, the steam consumption is much lower than that for pure steam atomizers. Yet these can provide a very large capacity modulation of the burner.

The steam-mechanical atomizer is extensively used in utility boilers. The oil pressure of this type of atomizer is higher (0.5–2.0 MPa) than that of pure steam atomizers. The steam pressure is higher than the oil pressure, but the difference is no more than 0.1 MPa (Stulz & Kitto, 1992). The oil spurts out from the oil nozzle and the high velocity steam flow assists its atomization. Figure 5-7 shows one type of steam-mechanical atomizer, called Y-type. The steam, admitted through the central tube, enters small but long steam holes drilled at an angle on end of the central tube. The oil admitted through the annular tube enters oil holes. The oil and steam meet at a larger hole called the *mixing hole*. Then they spurt into the furnace, forming fine oil droplets. The steam pressure is about 0.6–1.0 MPa. The steam consumption is relatively low (~0.02 kg steam/kg oil). The capacity of these burners can go up to 10 t/h of oil. Design procedure is given later.

Steam-mechanical atomizers can be of two types:

- Inner mixing type
- Outer mixing type

(1) Inner Mixing Type

In inner mixing type atomizers steam and oil meet in a mixing chamber at the tip of the burner and then jet out of the chamber through a series of holes. Table 5-1

TABLE 5-1. Test results of an inner mixing steam-mechanical atomizer.

No.	Steam pressure (MPa)	Oil pressure (MPa)	Oil flow rate (kg/h)	Steam consumption rate (kg steam/kg oil)
1	0.8	0.75	1650	0.0887
2	0.8	0.65	1260	0.107
3	0.7	0.65	1470	0.09
4	0.7	0.55	1220	0.097
5	0.6	0.55	1270	0.078

shows test results for a typical inner mixing type of atomizer. The atomizer used here has 12 steam holes with diameter $\phi 3$ mm; 6 steam grooves with depth and width 3×3 mm; 6 oil holes with diameter $\phi 3$ mm; and 8 nozzle holes with diameter $\phi 3.2$ mm in the mixing chamber. The angle, α, between the central lines of the holes in the mixing chamber $= 60°$.

(2) Outer Mixing Type

Figure 5-9 shows an outer mixing (RG-W-1) type steam-mechanical atomizer. The steam first enters the annulus of the burner. Then it flows through the swirl vane and steam orifices to impact on the oil atomizing it. Since the oil meets the steam outside the atomizer the chances of the high-pressure oil entering the steam to contaminate the steam water system is minimum.

c) Design Method for Y-type Atomizer

There is no standard design procedure for this type of atomizer. A tentative design method is provided below for reference. It refers to the Y-type atomizer shown in Figure 5-7. The design involves calculations for the sizing of the atomizer,

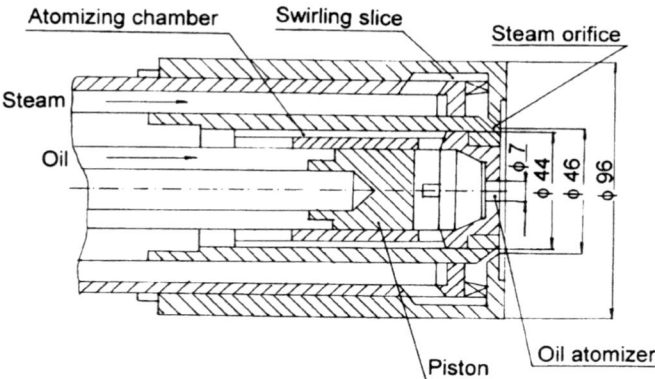

FIGURE 5-9. Cross section view of an outer-mixing-type steam mechanical atomizer (RG-W-1)

the steam consumption rate, and the mean droplet diameter under conditions of known oil flow rate, oil pressure, steam pressure, oil temperature, and steam temperature.

(1) Number of Holes, n

The number of holes should be between 5 and 30 per atomizer. A larger number of small-diameter holes can be used only if the oil flow rate is constant.

(2) Oil Hole Diameter, d_1

The oil at pressure P_0 enters the mixing hole through the oil holes (Fig. 5-7). For the chosen number of holes, n, the hole diameter, d_2 can be calculated from the total cross section area of the oil holes, F_1. The required oil flow rate, G, would be proportional to the square root of the pressure drop across the oil hole and its density.

$$G = 3600 \, \mu F_1 \sqrt{2\rho \cdot \Delta P} \qquad (5\text{-}17)$$

where G = oil flow rate of the atomizer, kg/h,
μ = flow coefficient. Usually, $\mu = 0.7$
F_1 = total cross section area of the oil holes, m^2
ρ = density of oil, kg/m^3
ΔP = difference between the oil pressure P_0 before the atomizer and that P_3 at the mixing point, Pa
$\Delta P = P_0 - P_3$

(3) Steam or Air Hole Diameter, d_2

The steam (air) hole diameter, d_2, is based on the steam (air) consumption rate, m_2, which should be in the range of 0.015–0.02 kg steam (air)/kg oil to ensure proper atomization quality.

From fluid mechanics we note that the steam (air) consumption rate is proportional to the square root of its pressure at the entrance to the atomizer. So, its consumption per hole can be calculated as

$$m_2 = 3600 \, \mu F_2 \psi \sqrt{P_2 \rho_s} \; (kg/h) \qquad (5\text{-}18)$$

where μ = flow coefficient,
F_2 = total cross section area of air holes, m^2
ψ = 2.09 for superheated steam, 1.99 for saturated steam, 2.14 for compressed air
P_2 = absolute pressure of the atomizing media at the inlet of the atomizer, Pa
ρ_s = density of the steam/air at inlet of the atomizer, kg/m^3

Owing to the resistance of the steam (air) holes the pressure in the mixing chamber P_3 will be lower than that before it. We take this pressure P_3 as a fraction of the upstream pressure, P_2. So, we can write

$$P_3 = \beta P_2$$

where β is a coefficient. It is about 0.94 for oil pressure greater than steam pressure.

From Figure 5-7 we find that after the atomizing media enters the mixing hole from the steam hole, it is disturbed by the oil flowing out from the oil holes. So, it is difficult to know the flow coefficient μ precisely. It also depends on the ratio of areas of air holes, oil holes, and the mixing holes and on the exit velocity in the oil holes. For first approximation one can take a value from linear approximation of the experimental data showing its dependence on the oil exit velocity (Fig. 5-10).

The ratio of steam hole diameter and the oil hole diameter is a function of the steam flow rate, which is generally constant for the Y-type atomizer. We can take the ratio in the range of $(d_2/d_1 = 1.0–1.4)$ for a first guess.

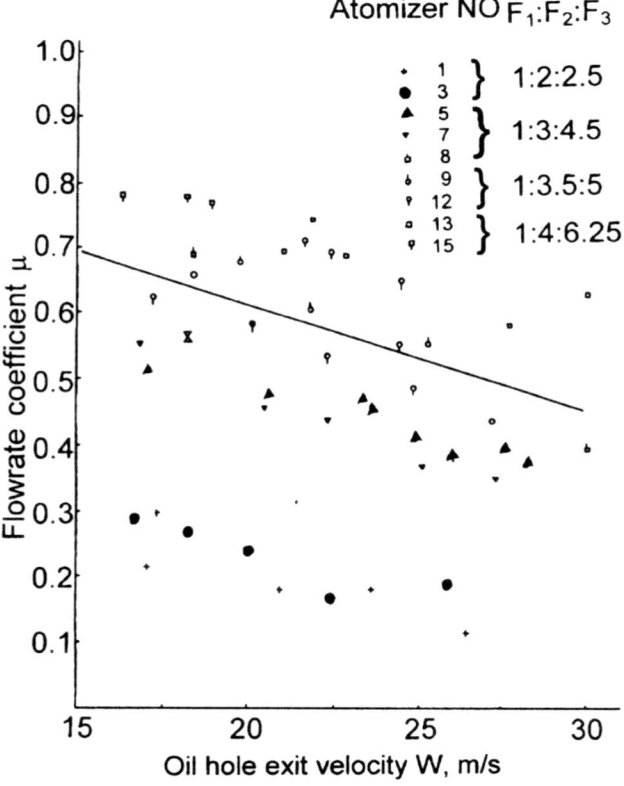

FIGURE 5-10. Flow coefficient, μ, for the flow of oil through oil holes plotted as a function of oil hole exit velocities and different ratios of areas of air/steam hole (F_1), oil hole area (F_2) and mixing hole area (F_3) for Y-type atomizer

(4) Mixing Hole Diameter, d_3

The area of the mixing hole is close to the sum of cross section areas of the steam holes and oil holes. From this it follows that the ratio of the mixing hole diameter and the steam hole diameter (d_3/d_2) is in the range of 1.4–1.8.

(5) Length of the Mixing Hole

The length of the mixing hole (l_3) is the distance between the exit of the oil hole and the exit of the mixing hole. The distance between the exit of the oil holes and the exit of the steam holes (length, l_1) is also an important design parameter. To allow good mixing of steam and oil and a complete transfer of the energy of the steam to the oil, the following ratios may be used:

$$l_1/d_3 = 0.7$$
$$l_3/d_3 = 4.0 - 5.0$$

(6) Angle between the Oil Hole and the Steam Hole

The oil holes and steam holes are set at an angle with each other. The recommended range of this angle is in the range of 50–65°. A lower value may give higher oil velocity and thinner oil film.

(7) Atomizing Angle, θ_a

The preferred range of this angle is 70–90°. The actual atomizing angle may be about 15–25° larger than this designed value of θ_a.

(8) Diameter, D_{air}, of the Steam Hole Center Line

The steam holes are arranged on a circle of diameter D_{air}. If the minimum gap between adjacent holes is 1 mm, the diameter of the center line circle containing n holes is around the circumference, then

$$D_{air} = \frac{n(d_2 + 1)}{\pi} \qquad (5\text{-}19)$$

(9) Annular Diameter, D_{mix}, of the Mixing Hole

From geometry (Fig. 5-7) it is given by

$$D_{mix} = D_{air} + 2(l_1 + l_2 + l_3) \sin \frac{\theta_a}{2} \qquad (5\text{-}20)$$

The ratio of the length of steam hole, l_2, and the hole diameter, d_2 can be chosen in the range of 2–10 depending on the mechanical design and the convenience of manufacturing process.

(10) Droplet Weighted Mean Diameter

The following empirical relation may be used for estimation of droplet sizes (Wigg, 1964).

$$d_{drop} = \frac{200 v^{0.5} m^{0.1} \left(1 + \dfrac{1}{q}\right)^{0.5} d_3^{0.1} \sigma_{oil}^{0.2}}{\rho_{steam}^{0.3} W_r} \text{ mm} \qquad (5\text{-}21)$$

where
v_{oil} = viscosity of oil, centistoke (Cst)
m_{oil} = oil flow rate per hole, g/s
q = steam consumption rate, g steam/g oil
d_3 = diameter of the mixing hole, cm
σ_{oil} = surface tension of the oil, dyne/cm
ρ_{steam} = steam density at the steam pressure in the mixing hole, g/cm^3
w_{cr} = relative velocity between oil and the atomizing media. (For
steam we can approximate it as $w_{cr} = k(P_{steam} v_{steam})^{0.5}$ m/s.)
k = 3.33 for superheated steam
= 3.22 for saturated steam
= 3.38 for compressed air
P_{steam} = pressure of atomizing media, N/m^2
v_{steam} = specific volume of steam, m^3/kg

EXAMPLE:

The oil flow rate of a Y-type atomizer is 2500 kg/h. The maximum oil pressure is 1.6 MPa. The steam pressure is 0.8 MPa. The steam temperature is 250°C. The oil is preheated to 140°C. Calculate the sizes of the parts of the atomizer and estimate the effects of the operating parameters on the atomizing quality.

SOLUTION:

1. The properties of heavy oil at 140°C are taken from oil data book: ρ_{oil} = 980 kg/m^3, viscosity v_{oil} = 13.8 Cst; surface tension σ_{oil} = 28 dyne/cm.
2. Properties of steam at the given pressure and temperature are taken from steam table:

$$\rho_{steam} = 3.35 \text{ kg/m}^3 = 3.35 \times 10^{-3} \text{ g/cm}^3; \quad v_{steam} = 0.299 \text{ m}^3/\text{kg}$$

3. Selection of number of oil holes. We chose $n = 11$ for the high oil flow rate.
4. The oil flow rate per hole:

$$m_{oil} = \frac{G}{3600\,n} = \frac{2500 \times 1000}{(3600 \times 11)} = 63.4 \text{ g/s}$$

5. Leaving some design margin, we take the operating oil pressure as 80% of the maximum oil pressure. So, the pressure difference between the oil and steam

holes is

$$\Delta P_{oil} = 0.8 P_{oil} - P_{steam} = 0.8 \times 1.6 - 0.8 = 0.48 \text{ MPa}$$

6. The oil velocity is calculated using a simple orifice equation with an orifice coefficient of 0.7:

$$V_{oil} = 0.7 \sqrt{\frac{2 \Delta P_{oil}}{\rho_{oil}}} = 21.8 \text{ m/s}$$

7. The cross section area of an oil hole is therefore

$$f_{oil} = \frac{m_{oil}}{\rho_{oil} \cdot V_{oil}} = \frac{0.0634 \times 10^6}{980 \times 21.8} = 2.9 \text{ mm}^2$$

The oil hole diameter d_2 is

$$\sqrt{\frac{4 f_{oil}}{\pi}} = 1.92 \text{ mm}$$

8. Following the suggestion give in step 3 of the Design method of Y-type atomizer, we take the oil and steam hole diameter ratio $d_2/d_1 = 1.2$; From step 4 of this method we take the steam and mixing hole diameter ratio $d_3/d_1 = 1.4$, then we obtains $d_1 = 1.6$ mm, $d_3 = 2.24$ mm.

9. From steps 6 and 9 the remaining geometric parameters are calculated by making a suitable choice:

$$l_2/d_2 = 2; \quad l_2 = 3.2 \text{ mm}; \quad l_1/d_1 = 2; \quad l_1 = 3.2 \text{ mm}; \quad \theta = 50°$$

The premixing length $l/d_1 = 0.75; l = 1.2$ mm
The mixing length $l_3/d_1 = 4; l_3 = 6.4$ mm
The atomizing angle $\varphi = 90°$
The annular diameter, D_{air}, of the air hole:

$$D_{air} = \frac{n(d_1 + 1)}{\pi} = \frac{11.(1.6 + 1)}{\pi} = 9.1 \text{ mm, (take 10 mm)};$$

The annular diameter $D_{mix} = D_{air} + (l_1 + l + l_3) = 25.3$ mm.

10. Calculation of steam consumption rate, m_j, per hole.
We find the value of flow coefficient μ from Figure 5-13 against the oil velocity 21.8 m/s as $\mu = 0.65$.
From Eq. (5-18) we get:

$$m_j = \mu x F_1 x \psi \sqrt{P_{steam} v_{steam}}$$
$$= 0.65 \times (\pi/4) \times (1.6/1000)^2 \times [0.8 \times 1,000,000 \times (1/0.299)]^{0.5}$$
$$= 0.00446 \text{ kg/s}$$

11. The number of steam and oil holes are equal. So, the steam consumption per kg of oil is

$$q = \frac{m_j}{m_{oil}} = \frac{0.00142}{0.0634} = 0.07 = 7\%$$

12. The calculation of the spray droplets weighted mean diameter requires the critical velocity of atomizing medium, w_{cr}

$$w_{cr} = 3.33\,(800,000 \times 0.299)^{0.5} = 1628\ \text{m/s}$$

Now, using Eq. (5-21) we get

$$
\begin{aligned}
d_{drop} &= \frac{200 v_{oil}^{0.5} g_{oil}^{0.1} \left(1 + \dfrac{1}{q}\right)^{0.5} d_3^{0.1} \sigma_{oil}^{0.2}}{\rho_{steam}^{0.3} W_{cr}} \\[2mm]
&= \frac{200 \times 13.8^{0.5} \times 63.4^{0.1} \left(1 + \dfrac{1}{0.07}\right)^{0.5} \times 0.224^{0.1} \times 28^{0.2}}{(3.35 \times 10^{-3})^{0.3} \times 1628} \\[2mm]
&= 0.124\ \text{mm} = 124\ \mu\text{m}
\end{aligned}
$$

If the steam consumption rate and the mean droplet diameter are acceptable, the first step of calculations is completed. Otherwise, the oil pressure, the steam pressure, and the ratios of the design are changed and recalculated until the requirement is satisfied.

The method above can also be repeated to calculate the operation under different loads. Results from a set of similar calculations for different conditions are shown below:

Oil pressure, MPa	1.2	1.4	1.6
Steam pressure, P_{steam}	0.8	0.8	0.8
Oil flow rate, G_{oil}, kg/h	1380	2000	2500
Load ratio %	50.5	80	100
Mean diameter d_{drop}, μm	76.5	109	135

5-2-4 Low-Pressure Air Atomizer

A typical low-pressure air atomizer is shown in Figure 5-11. The oil exits at a low velocity from the center of the atomizer. Higher-velocity air comes from the annulus and impinges on the oil jet at a very high velocity (70–100 m/s). This impact atomizes the oil. The low-pressure air atomizer is usually used in small industrial boilers with an air pressure of 2–7 kPa. Some low-capacity (150–200 kg/h) burners use oil at about 0.03 MPa–0.15 MPa. Usually the entire combustion air is used as the atomizing media. So, the mixing of air and oil is good. The excess air coefficient, $\alpha = 1.1$–1.15.

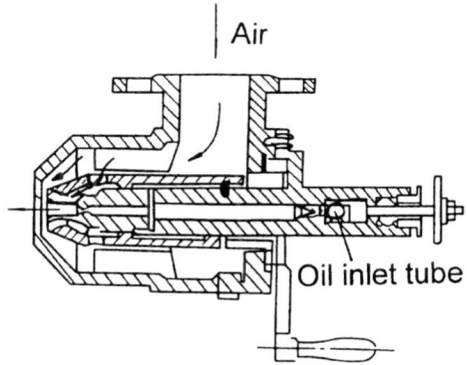

FIGURE 5-11. A low-pressure air-assisted oil atomizer

a) Design Procedure

A simple design procedure for a low-pressure air atomizer is given below

1. The diameter, D_{oil}, of the tube through which the oil passes can be found from mass balance.

$$D_{oil} = [(4 \times G)/(3600\,\pi\rho\,W)]^{0.5}\ \text{m}$$

$$= 18.8(G/\rho \cdot W)^{0.5},\ \text{mm} \qquad (5\text{-}22)$$

where G = oil flow rate, kg/h
        ρ = oil density, kg/m³
        W = oil velocity, about 0.1–1.0 m/s

2. The cross section area A of the oil nozzle can be found from Eq. (5-2), which is written in the form, $G = 3600\,\mu\,A_{oil}(2\rho\,P_0)^{0.5}$

$$A_{oil} = 196\frac{G}{\mu\sqrt{P_0\rho}}\ \text{mm}^2 \qquad (5\text{-}23)$$

where μ = oil nozzle flow coefficient = 0.2–0.3,
        P_0 = oil pressure before the nozzle in Pa, N/m²
        ρ = oil density, kg/m³

3. The diameter, D_a, of the air inlet tube is found as

$$D_a = 18.8\sqrt{\frac{G_a}{\rho_a W_a}}\ \text{mm} \qquad (5\text{-}24)$$

where G_a = air consumption rate, kg/h
        ρ_a = air density, kg/m³
        W_a = air velocity, 10–15 m/s

4. The exit section area, A_a, of the air nozzle is

$$A_a = 196\frac{V_a}{\mu}\sqrt{\frac{\rho_a}{P_a}}\ \text{mm}^2 \qquad (5\text{-}25)$$

where V_a = air flow rate, m³/h
          P_a = air pressure before the nozzle, Pa
          μ = air nozzle exit flow rate coefficient = 0.6–0.8

5. The theoretical size (radius) of the spread of atomized oil droplets, r (Wang, 1996) is

$$r = \frac{(200\text{--}300)}{P_a}\ \text{mm} \qquad (5\text{-}26)$$

6. The approximate flame length is given by Wang (1996) as

$$L = 2\left(42 + \frac{60}{V_{air}}\right)d_{noz}\ \text{m} \qquad (5\text{-}27)$$

where L = the flame length, m
          V_{air} = the air flow rate per unit of fuel, m³/kg
          d_{noz} = the diameter of the atomizer nozzle, m

b) Other Types of Atomizers

In addition to the examples above, there are several other types of air atomizers. The basic designs and working principles are similar to those described above. They differ only in mechanical details of the design. They are

1. K-type atomizer
2. R-type low-pressure oil atomizer
3. RK-type low-pressure atomizer
4. TB-type low-pressure atomizer

5-2-5 Supersonic Atomizer

The supersonic atomizer is another type of steam-mechanical atomizer. Here, steam or compressed air enters the steam (air) chamber, and then spurts out through an annular gap between the nozzle and the shell. Then the steam (or air) hits the atomizer tip (resonant cavity) and produces a supersonic wave. The oil flows radially from a central tube, meeting the steam before the resonant cavity. The supersonic sound wave breaks up the oil drops further to small and even sizes. The capacity adjustment ratio is very large, about 20–50, and the oil pressure requirement is usually low (0.5–0.7 MPa).

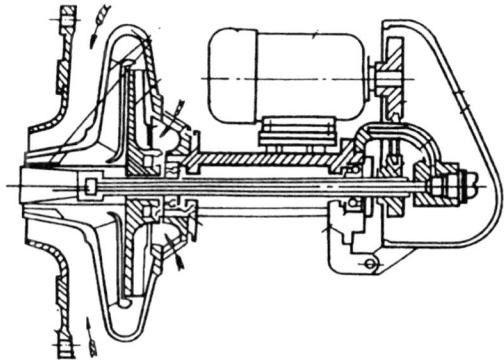

FIGURE 5-12. Cross section view of a rotary cup oil atomizer

5-2-6 *Rotary Cup Atomizer*

The rotary cup is a type of mechanical atomizer (Fig. 5-12). The oil is atomized by the rotary motion of the swirling chamber. A motor rotates the swirling cup at a very high (3000–6000 rpm) speed. The oil enters the inner wall of the swirling rotary cup from a hollow axis. Under the effect of centrifugal force, the oil is thrown out to the truncated conical wall of the rotary cup. Because the velocity of this oil is very high, it is atomized while leaving the rim of the cup. There is also a high-velocity air flow from the primary air fan around the rotary cup, which can further improve the atomization. The rotary cup is more difficult to manufacture and operate because the cup is a moving component rotating at high speed. Also, it would need a higher level of maintenance. These atomizers are usually used in industrial boilers.

a) Merits and Demerits of the Rotary Cup Atomizer

The rotary cup atomizer has the following positive and negative features:

1. It can be operated at a low oil pressure of 0.05–0.2 MPa. This allows it to work without a pump. A gravitational head of less than 2 m of the oil tank is sufficient to provide the required oil pressure.
2. There is no fine nozzle holes for spray. This relaxes the standard of filtration and viscosity requirements. Since a viscosity of 14°E is adequate, the heavy oil needs to be heated only to 0°C–70°C against 95°C–100°C for other mechanical atomizers.
3. The air and oil droplets are mixed very well; so the spray is easy to ignite and burns completely at a low excess air coefficient.
4. The oil pump, the fan, and the atomizer are combined in a single, compact lightweight unit, which is particularly suitable for packaged or mobile units.
5. The turndown is easy and the adjustment ratio can be as high as 1:20.

6. The rotary cup is heated by combustion radiation, which helps the oil film preheat, evaporate, atomize, and mix with air as it spurts out. So, the combustion is complete and the flame is short and wide. The atomizing angle is in the range of 50°–120°.
7. The electricity consumption for the supply oil, air, and for atomization is small.
8. However, the noise is greater than that for mechanical atomizers, but is less than that for steam atomizer.
9. The cup rotates at a high speed in a high-temperature environment; therefore a high degree of precision in its manufacture is needed. Also, the cup materials must withstand high-temperature erosion.

b) Operating Principles and Design Calculation

When the oil swirls inside the rotary cup, a rotating oil film is formed by the action of friction and centrifugal forces (Fig. 5-13). From theory (Fraser et al., 1963), the radial velocity u_m of the oil film along the cup wall is calculated as

$$u_m = \left[\frac{0.667 Q^2 N^2 \sin\theta}{\nu D_{cup}} \right]^{1/3} \tag{5-28}$$

where Q = oil flow rate, m^3/s
D_{cup} = inner diameter of the rotary cup, m
N = rotary speed of the cup, rpm
ν = kinematic viscosity of oil, m^2/s
θ = slope of inner surface of the cup

Assuming the thickness, h, of the oil film to be small compared to the cup diameter, we can write

$$Q = \pi D_{cup} \, h \cdot u_m$$

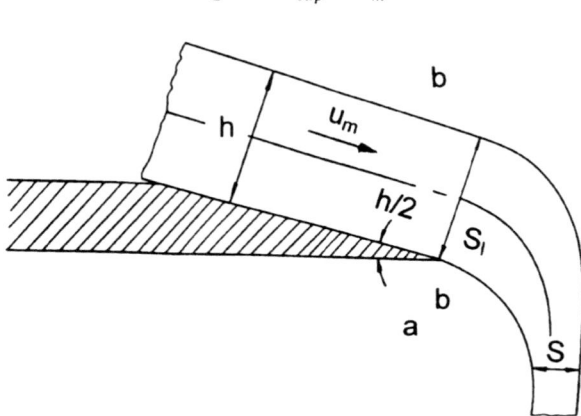

FIGURE 5-13. Formation of oil film at the exit of a rotary oil cup atomizer

Substituting the value of u_m from Eq. (5-28), we get

$$h = \left[\frac{0.0484\,Qv}{D^2 n^2 \sin\theta}\right]^{1/3} \tag{5-29}$$

(1) Oil Film Thickness

After the oil film leaves the rotary cup, it disperses in a tangential direction under the action of centrifugal force. An annular oil film is formed around the rotary cup. The further from the cup, the thinner the oil film. This thickness, S (Fig. 5-13) is calculated using Eq. (5-30) (Fraser et al.,1963):

$$S = \frac{Q}{2\pi V_T}\left[\frac{D_L^2 u_m^2}{4 V_T^2} + a D_L + a^2\right]^{-1/2} \tag{5-30}$$

where D_L = diameter of the cup mouth
             V_T = tangential velocity of the oil
             a = radial distance from the cup mouth from the axis of the cup

The change of oil film thickness from the cup mouth affects the atomizing quality directly. The oil film thickness decreases rapidly after leaving the cup. For example, for a 225 kg/h atomizer rotating at 4500 rpm the film thickness of a 45 Cst viscosity oil film at the exit of the cup is about 283 μm thick. At a distance of 3.2 mm, the oil film thickness decreases to 74 μm. Thereafter, the change in thickness is small. At a distance of 25 mm, the thickness changes only to 30 μm.

When the oil flow rate is very low, a continuous oil film cannot be formed at the exit of the cup. Spiral flow lines are formed instead at the. The distribution of droplets without an oil film is very uneven. The flame is difficult to stabilize. This sets the lowest limit of heat load on the burner. The critical condition to form a continuous oil film is (Fraser et al., 1963)

$$\frac{\rho^{0.81}N^{0.67}Q^{1.14}v^{0.19}}{\sigma \cdot D_L^{0.81}} > 0.363 \tag{5-31}$$

where ρ = oil density,
             Q = critical oil flow rate,
             σ = surface tension (=2.5 kg/m when ρ = 900 kg/m^3)
             N = revolution
             v = viscosity

(2) The Tapering Angle, θ, of the Rotary Cup

This is the angle of taper at the edge of the rotary cup. It has a great effect on the oil film thickness, S, at the exit of the oil cup. To avoid large droplet size, θ is selected to be larger than 35°.

*(3) The Effect of the Air Nozzle Position and the Impact Point
on the Atomizing Quality*

From experiments, the droplet size (mean diameter) at 75 mm from the atomizer
can be calculated with the following empirical formula (Fraser et al.,1963):

$$d_s = 6 \times 10^{-4} + \frac{0.59\,\sigma^{0.5}V_r^{0.21}}{\rho_a^{0.5}(aD_L + a^2)^{0.25}}\left[1 + \frac{0.065}{M_r^{1.5}}\right]\left[\frac{Q}{V_T^3(0.5V_r^2 - V_r + 1)}\right]^{0.5} \text{cm}$$

$$(5\text{-}32)$$

(valid for $a \geq 5$ mm)

where σ = surface tension, g/cm
             ρ_a = air density, g/cm^3
             M_r = weight ratio of air and oil
             V_r = velocity ratio of air and oil
             V_T = tangential velocity of oil, cm/s (all units are c.g.s.)

It can be observed that further the impact point is from the cup mouth, the
smaller the droplet size. It is very difficult to control the location of the impact
point of primary air with the oil. It is related to the following:

a) The distance between the oil cup and the air nozzle exit. Generally the cup is
   moved forward from the air nozzle. Usually this distance is about 10 mm or
   more.
b) The swirl strength and flow rate of the primary air. The larger the swirl strength
   and the larger the flow rate, the larger the radial distance a.

5-3 Air Registers

The air register supplies combustion air to the burner just as the atomizer supplies
fuel oil to it. Both are key components of the burner system. These two components
greatly affect the combustion properties of an oil fired boiler. The combustion air
is generally fed into the boiler furnace through the air register around the atomizer.
The air register includes an air tube, ignitor, and flame holder. The ignitor ignites the
flame and the flame holder helps anchor the flame in place, preventing a blow-out.

5-3-1 Types of Air Registers

Air registers are divided into two types: swirl type and advection-flow type. The
following section discusses these two types in details.

a) Swirl Air Registers

Figure 5-14 shows a swirl air register. The combustion air is split into two streams.
One stream enters the boiler through a central tube and then exits through a flame-
holder with vanes at its end. The other stream enters from the annulus around the

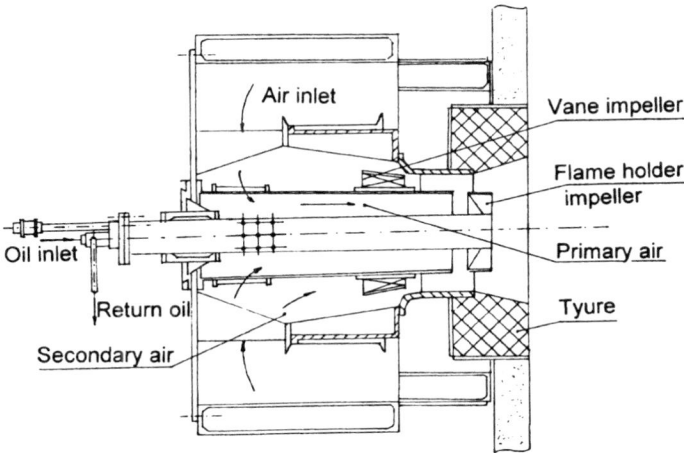

FIGURE 5-14. A swirl-type air register

central tube. This stream also passes through a vaned impeller that gives a swirl to the annulus air. After passing through the flame-holder, the core air forms a recirculation zone in front of the burner. The recirculating zone provides a steady source of ignition. The volume of air carried by the two air streams is adjusted with the tubular air doors. The oil gun is inserted through the central tube. This may also carry some air.

The swirl air register is further classified into following groups according to the design of the air swirler in the air registers:

Turbo air registers
Tangential moveable vane air registers
Axial movable impeller air registers, etc.
Fixed tangential impeller air registers
Axial fixed-vane air registers

b) Advection-Flow Air Registers

A typical advection flow air register is shown in Figure 5-15. Here the air enters the air register from a large wind box through an air door. Most of the air flows into the boiler directly and a small amount flows through the flame-holder (Fig. 5-16). The flame-holder with vanes also adds some swirl to the air.

Because only a small amount of air is swirled in the center, the characteristics of the air flow at the burner exit approach those of the direct flow to a great extent. Hence advection-flow air registers are also referred to as direct-flow air registers.

Advection-flow air registers are used increasingly for their simple design, small resistance, and low NO_x formation during combustion. Design considerations are explained in this section.

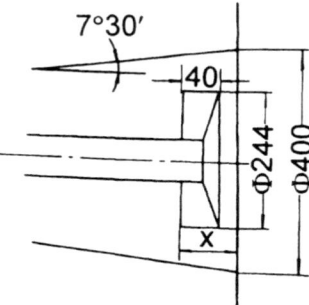

FIGURE 5-15. A venturi-type advection-flow air register

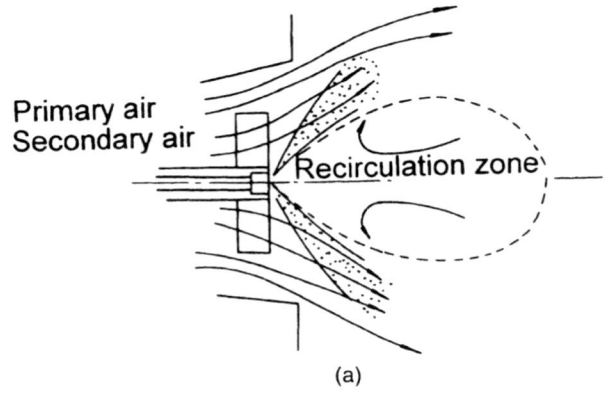

(a)

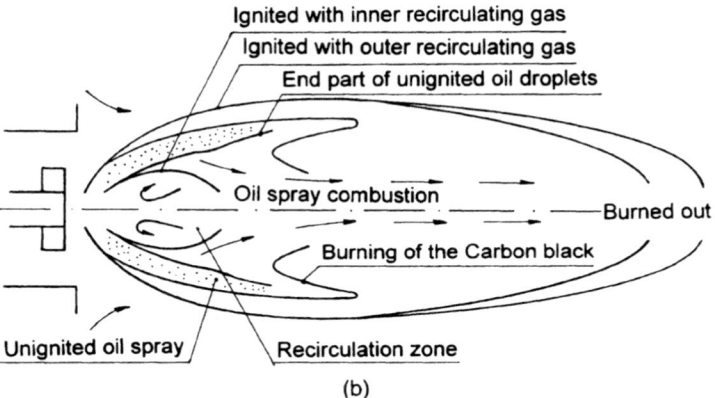

(b)

FIGURE 5-16. Recirculation zone in advection-type air register (a) without root air where the primary air passes through the flame holder around the atomizer and (b) with root air where the primary air is premixed with the oil in the oil gun

5-3-2 Basic Principles of the Air Register

Good atomization and good air arrangement are primary objectives of the air register. Some of the guiding principles for the design of air registers are discussed below.

a) Root Air is Necessary

Root air is a part of the primary air required to be mixed with the oil before it is ignited at the root of the atomized oil. Oil burners often produce unburnt carbon particles, which come from two sources: residual coke left after the oil is burned, and carbon black particles produced from the thermal decomposition of oil–gas. Coke formation is inevitable when heavy oil or crude oil is fired. To reduce the coke formation the design should ensure complete combustion of the fuel in the boiler. The time required for complete burning of coke particles is proportional to the square of their diameter, which in turn depends on the size of the oil droplets. Therefore, good atomization is the key to the complete burning of coke.

Thermal decomposition of oil can be reduced by premixing a part of the combustion air with the oil in advance. The primary air mixes with oil before the fuel reaches the atomizer. Since the oil is now oxygenated, the thermal decomposition does not occur, even though complete combustion may not still be obtained.

Figure 5-16 shows how the ignition and combustion of the oil takes place in the recirculation zone with or without root air. In an absence of root air, the atomized oil does not burn well in the recirculation zone (Fig. 5-16a) as it is very hot but deficient in oxygen. So, the thermal decomposition occurs easily. If the entire primary air is passed as root air through the oil gun (Fig. 5-16b) the oil burns better as it benefits from both the high temperature of the recirculating zone and oxygen of the root air.

Theoretically, steady combustion is possible if the entire air is mixed with the oil before the outlet of the burners. However, it is difficult to form a recirculating zone if the entire air is fed before ignition. The air velocity will be too high to allow stable ignition. Therefore, the amount of root air must be restricted to 15–30% of the combustion air.

b) Early Mixing Should Be Intense

The remaining part of the combustion air (primary or secondary) must be mixed with the atomized oil evenly as soon as the oil is sprayed from the burner before ignition. The following principles may help avoid the problem of early mixing.

1. For multiple burners the oil and air capacity of each burner must match. The excess oxygen from one burner zone may not be of much use to a neighboring oxygen-deficient burner. So, their deviations should not exceed ±1% and ±3% for oil and air capacities, respectively.
2. As air leaves the register it forms a cone of a certain angle called the *expanding angle* (Fig. 5-16a). Usually, the expanding angle of air is less than the atomizing

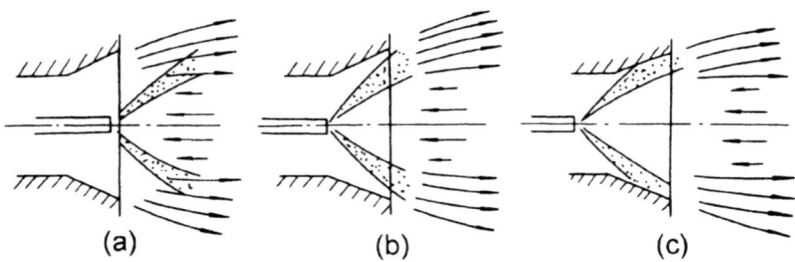

(a)   (b)   (c)

FIGURE 5-17. Effect of position of the oil nozzle with respect to the air register exit on the mixing of air with the fuel

angle of oil. In a swirl burner, the intensity of the swirl determines the expanding angle of air. A high swirl intensity gives a larger expanding angle and vice versa. When the swirl intensity is very low, the expanding angle of air and the recirculating zone is small. This causes delayed ignition, flame elongation, and excessive and coarse soot particles in the flue gas. The soot also increases when the swirl intensity, and consequently the expanding angle of the air is very large. Since the early mixing of oil and air is not vigorous in this case, soot is formed by thermal decomposition of oil. Thus an optimum value of swirl intensity and the expanding angle is necessary.

3. In advection-type air registers, there is an option for swirling both primary and secondary air. To enhance the early mixing it is more efficient to use swirling of the primary rather than the secondary air.

4. The position of the oil nozzle or the atomizer in relation to the swirl-type air register is very important. The location of the spray of the atomized oil and air jet must be arranged with respect to each other to ensure good early mixing of oil and air. Figure 5-17 shows the effect of different locations of the nozzle in a swirl-type air register. The nozzle head, stretched too far out into the recirculating zone (Fig. 5-17a), would make the mixing weak. If it is drawn too far back (Fig. 5-17c) the atomized oil will be sprayed on either the tuyere or the water wall. Figure 5-17b shows the ideal location.

An advection-type burner also needs an appropriate distance between the nozzle and the flame-holder.

c) Size and Location of the Recirculation Zone Should Be Appropriate

The recirculation zone in front of the burner serves as a flame stabilizer. It provides a constant ignition source to the fuel–air mixture and helps anchor the flame in place. This zone of hot gas should not be so large as to touch the burner nozzle as it might burn the center tube. At the same time it should not be so small especially in advection-type registers, as it may adversely affect the early mixing of primary and secondary air. The exact shape of the recirculation zone should be obtained by experiments in hot conditions. As an alternative, computational fluid dynamic analysis may also help predict its shape and location for a given burner geometry at a given throughput.

d) Later Stage Mixing Should Be Intense Too

The fuel air mixing in the later stage, i.e, further downstream is as important as the early mixing. With poor later-stage mixing, the unburnt gas and soot from the base of the flame would not burn completely in the tail of the flame, as it is usually oxygen deficient. The air swirl can increase the disturbance at the base of the flame, which is the exit of the burner. But the disturbance dissipates rapidly. Increasing the air velocity is one way to increase the mixing in the entire length of the flame. Since the mixing is intensified by increasing the air velocity, a relatively low excess air combustion is now possible at a high air velocity. Therefore, the air velocity at the exit of the burner in oil-fired boilers has been increased from the past practice of 20–35 m/s to 30–60 m/s or even more in newer systems. However, higher air velocity means larger resistance in the burner and therefore higher electrical power consumption. So this would necessitate a reduction in the resistance coefficient in the burner.

Special burner arrangements like opposed jet or tangentially fired systems can also enhance the downstream mixing in the burner. These will increase the mixing in the furnace and, therefore, enhance the combustion of unburnt species. In conclusion, a good practical design of air register should have the following:

a) Root air
b) A flame-holder in the center air to help stabilize the flame and increase the early mixing
c) Improved adjustment of the secondary air and reduction in its swirl intensity

5-3-3 Design of Other Types of Swirl Air Registers

Besides the main type described above, there are several other types of swirl air registers, which are discussed below.

a) Turbo Air Registers

Both primary and secondary air is swirled through the turbo air register. The air flows into the register from one side tangentially. The swirl intensity can be increased by increasing the tangential velocity through a reduction in its flow cross section area. Both primary and secondary air can be swirled through such a tangential entry.

b) Tangential Vane Air-Registers

Figure 5-18 shows a tangential-vane air register, where the air is swirled through a flow swirler made of tangential vanes, which can be either fixed moveable. A tangential-vane air register splits the air into two streams: primary and secondary. The secondary air is swirled through the tangential moveable vanes. The primary

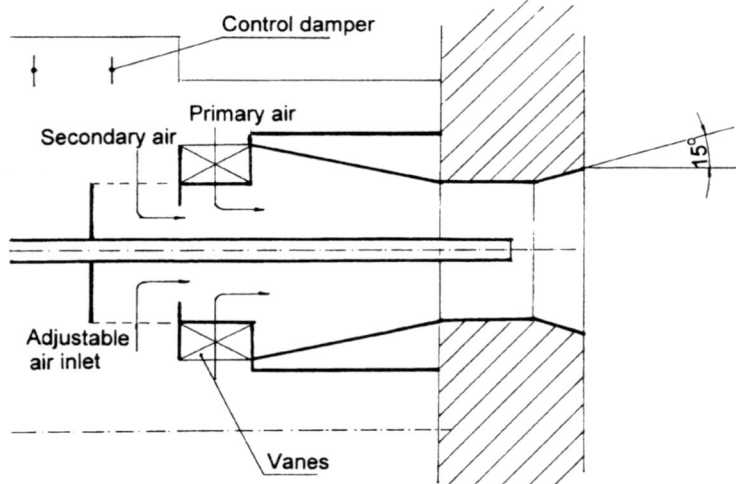

FIGURE 5-18. Tangential-vane-type swirl air register

air is fed from the center through an annular tube with many apertures. In a fixed tangential impeller air register control dampers are installed at the inlet of the air register for adjusting the air capacity. Also, an extender moves back and forth to change the exit area of the impeller vane.

In a fixed tangential impeller air register, the peripheral air at the exit is distributed better than that in the turbo air register but the controllability of the former is worse and the air resistance is higher.

c) Axial Vane Air Registers

An axial moveable impeller air register is shown in Figure 5-19. The air axially enters a set of moveable impellers, which add swirl to the flow. When the impellers are drawn out, a part of the air can be fed directly into the furnace through the annular space around the impeller. The impellers can be moved back and forth to adjust the capacity ratio of the swirl-flow air and the straight-flow air, thereby adjusting the swirl intensity. In an axial fixed-vane air register, air flow is separated into three streams. One flows in from the outer annulus, which is a straight-flow air; one is swirl-flow air through the axial fixed vanes; the other is central air through the center tube.

5-3-4 Advection-Flow Air Registers

Advection-flow air registers are widely used in low-excess-air burners. In this type, the capacity ratio of the air through the flame-holder (primary air) and the total air cannot be adjusted. This ratio is dictated by the geometric dimensions of the register.

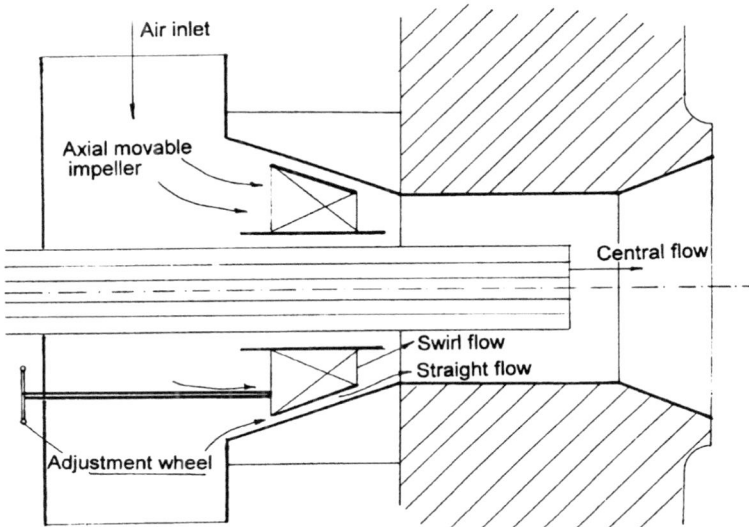

FIGURE 5-19. An axial moveable swirl-type air register

There are two types of advection-flow air registers.

1. Straight-tube
2. Venturi type

A venturi-type advection-flow air register is shown in Figure 5-15. Here both primary and secondary air enters the air register. The secondary air passes straight through the annular passage around the flame-holder, while the primary air passes through the flame-holder, receiving some swirl. As far as combustion is concerned there is no great difference between the straight-tube and venturi-type air register. The venturi type allows easy measurement of the air capacity in the air register, which is important for low oxygen combustion.

Design for Advection Air Register

In a venturi-type air register the inlet diameter is close to that at the outlet or the tuyere of the register. The diameter of the throat may be 0.7–0.75 the diameter of the tuyere. By applying the Bernoulli's equation between the throat and the annular inlet and considering the friction losses, we can derive an expression for flow through the venturi air register, q.

$$\frac{P_1}{\rho} + \frac{V_1^2}{2} = \frac{P_h}{\rho} + \frac{V_h^2}{2} + \xi\frac{V_h^2}{2}$$

$$Q = V_h \times F_h \times 3600 = 3600 \times F_h \sqrt{\frac{2(P_1 - P_h)}{\rho\left[1 + \xi - \left(\frac{F_h}{F_{in}}\right)^2\right]}} \qquad (5\text{-}33)$$

where Q = air flow rate through the register, m³/h
          ξ = resistance coefficient
          F_{in} = annular section area at the inlet, m²
          F_h = section area of the throat, m²
          $P_1 - P_h$ = pressure difference between the inlet and throat, Pa

Near the outlet (tuyere) a part (Q_w) of the total flow (Q) passes through the flame-holder (Fig. 5-14), and the remainder passes through the annular open area around it. Taking the resistance coefficient of the open area as 1.0 and that through the flame-holder ξ_w, one can calculate the fraction of the total flow passing through the flame holder.

$$\frac{Q_w}{Q} = \frac{1}{1 + \sqrt{\xi_w\left[\left(\frac{D_d}{d_{fh}}\right)^2 - 1\right]}} \qquad (5\text{-}34)$$

where d_{fh} = diameter of the flame-holder, m
          D_d = diameter of the duct where the flame-holder is located, m

Considering the resistance of parallel flow through the annulus and the flame-holder the overall resistance coefficient of the advection register can be derived as

$$\xi = \xi_r + \frac{1}{\left[1 - \left(1 - \frac{1}{\sqrt{\xi_w}}\right)\left(\frac{d}{D}\right)^2\right]^2} \qquad (5\text{-}35)$$

where ξ_r = resistance coefficient at the inlet of the advection-flow air-register
          ξ_w = resistance coefficient of the flame-holder

For venturi-type advection-flow air registers the resistance coefficients should refer to the area of the section chosen. The outlet area of the tuyere is chosen for convenience because the mean axial velocity is usually given for the tuyere section. The resistance coefficient corresponding to the tuyere outlet section is

$$\xi_0 = \xi\left(\frac{F}{F_0}\right)^2 = \left[\xi_r + \frac{1}{\left[1 - \left(1 - \frac{1}{\sqrt{\xi_w}}\right)\left(\frac{d}{D}\right)^2\right]^2}\right]\left(\frac{F_0}{F}\right)^2 \qquad (5\text{-}36)$$

where F = calculating section area, m²
          F_0 = outlet area of the tuyere, m²

The calculating section is usually the section of air duct coinciding with the inlet of the flame-holder. Some typical values of the resistance coefficient of the advection air register are given in Table 5-2.

TABLE 5-2. Resistance coefficients of advection flow air register ($d/D = 0.6$).

ξ_r	0.7	0.7	0.7	0.2	0.2	0.2
ξ_w	3	5	10	3	5	10
ξ	2.09	2.26	2.44	1.59	1.76	1.94

EXAMPLE:

The dimensions of a venturi-type advection-flow air register are shown in Figure 5-19. The resistance coefficient at the inlet $\xi_r = 0.18$, and the resistance coefficient of the flame-holder $\xi_w = 5.3$. Evaluate the percentages of primary air and the resistance coefficients under the conditions that the flame-holder is stretched in or not.

SOLUTION:

When the flame-holder is not stretched in the distance x between the inlet of the flame-holder and the tuyere or the outlet of the register is equal to the width of the flame-holder, 40 mm.

1) Diameter of the air register at the calculating section $D = D_0 - 2 \times \tan\varphi = 400 - 2 \times 40 \times \tan 7.5° = 389$ mm

2) Diameter ratio flame-holder and calculating section, $d/D = 244/389 = 0.626$

3) Cross section area of the tuyere, $F_0 = (\pi/4)D_0^2 = 0.785 \times 0.4^2 = 0.125$ m^2

4) Calculating section area, $F = (\pi/4)D^2 = 0.785 \times 0.389^2 = 0.119$ m^2

5) Overall resistance coefficient of the register is calculated using Eq. (5-36)

$$\xi_0 = \left[\xi_r + \frac{1}{\left[1 - \left(1 - \frac{1}{\sqrt{\xi_w}}\right)\left(\frac{d}{D}\right)^2\right]^2} \right] \left(\frac{F_0}{F}\right)^2$$

$$= \left[0.18 + \frac{1}{\left[1 - \left(1 - \frac{1}{\sqrt{5.3}}\right)(0.626)^2\right]^2} \right] \left(\frac{0.125}{0.119}\right)^2 = 2.0$$

6) The fraction of primary air flow through the flame-holder is calculated from Eq. (5-34):

$$\frac{Q_w}{Q} = \frac{1}{1 + \sqrt{\xi_w}\left[\left(\frac{D}{d}\right)^2 - 1\right]} = \frac{1}{1 + \sqrt{5.3}\left[\left(\frac{1}{0.626}\right)^2 - 1\right]} = 0.219$$

$$= 21.9\%$$

7) When the flame-holder is pulled 100 mm in the register, the inlet of the flame-holder will be $(100 + 40)$, or 140, mm away from the tuyere.

So, $D = 400 - 2 \times 140 \tan 7.5° = 363.4$ mm. Now repeating the calculation above, we can show that the primary air flow through the flame-holder (Q_w) is now increased to 26.3% of the total flow.

5-4 Design Principles of Oil Fired Boilers

The design of oil fired boilers involves the design of atomizers, air registers, the furnace, heat transfer surfaces, and auxiliary equipment. The design of the atomizer and the air register has been discussed earlier. The design of heat transfer surfaces and auxiliary equipment will be discussed in other chapters. The present section will discuss the design of the furnace.

5-4-1 Volume Heat Release Rate in Furnace

The volumetric heat release rate is an important design parameter of the design of furnaces. It defines the space required to complete the combustion of fuels.

a) Combustion Volume

The primary function of the furnace is to ensure that the fuel has enough residence time in the furnace to complete its combustion. The fuel or air does not necessarily flow uni-directionally in a straight line as in a plug flow reactor. Thus, to quantify the residence time of fuel air in the furnace, an overall approach is taken by the use of an index called *Furnace volume heat release rate* (kW/m³). It is represented by q_v:

$$q_v = \frac{B \cdot LHV}{V_f} \ (\text{kW/m}^3) \tag{5-37}$$

where B = fuel flow rate, kg/s
           LHV = heating value, kJ/kg wet
           V_f = furnace volume, m³

Generally B and LHV are input parameters in the design. The volumetric heat release rate, q_v is selected from Table 5-4, given later and the volume of furnace is then calculated from the above equation.

Typical fuel oil requires 7.46 lb of theoretical air per 10^4 btu (0.000231 nm³/kJ) of the higher heating value of oil (Stultz & Kitto, 1992, p. 9-8). Taking account of excess air and the other products released during combustion, one can derive the following rule of thumb to estimate the volume of combustion product per unit mass of fuel.

$$V_{fg} = (1.1\alpha_x + 0.08)\left(\frac{LHV}{4186}\right) \text{Nm}^3/\text{kg} \tag{5-38}$$

where *LHV* is the lower heating value of oil in kJ/kg and α_x is the excess air coefficient.

The volume of flue gas will depend on the temperature and pressure in the furnace. So for a given temperature, T_f, and pressure, P_f, in the furnace, one obtains the gas volume in the furnace per unit mass of fuel as

$$V_g = V_{fg} \times \frac{T_f \times P_{atm}}{273 \times P_f} \; (m^3/kg) \qquad (5\text{-}39)$$

where T_f = average temperature of the gas in the furnace, °K
    P_f, P_{atm} = pressures of gas in furnace and in standard atmosphere, respectively, Pa

The main gas flow does not go through some dead spaces in any furnace. Therefore, the entire furnace volume is not effective for combustion. The effective utilization volume V_{eff} of the furnace is a fraction (φ) of the geometric furnace volume:

$$V_{eff} = \varphi_{eff} V_f \qquad (5\text{-}40)$$

where φ_{eff} = volume effectiveness.

The furnace volume effectiveness factor (φ_{eff}) is in the range of 0.6–0.8. It is low for front firing and high for opposed jet and tangential firing. A higher effectiveness factor gives a conservative design.

One obtains the retention time of the gas in the furnace by dividing the effective furnace volume by the gas volume. It is also the retention period of the fuel in the furnace. Using Eqs. (5-37) to (5-40), we get

$$\tau = \frac{\varphi_{eff} V_f}{V_g B} = \frac{\varphi_{eff} P_f}{T_f (1.1\alpha_x + 0.08) q_v} \left[\frac{273 \times 4186}{P_{atm}} \right] (s) \qquad (5\text{-}41)$$

where q_v is the volumetric heat release rate and T_f is the average temperature of the gas in the furnace. This would lie somewhere between the adiabatic flame temperature, T_{th}, and the furnace exit gas temperature, T_{ou}. Blokh (1988) suggests the following experience-based empirical relation to estimate this.

$$T_f = 0.925\sqrt{T_{th} \times T_{ou}} \; °K \qquad (5\text{-}42)$$

where T_{th} = theoretical adiabatic flame temperature °K
    T_{ou} = gas temperature at the furnace exit, °K

The adiabatic flame temperature is a theoretical value depending on the excess air and the air preheat temperature. It can also be approximated using an experience-based empirical relation of Chinese designers (Xiezhu, 1976, p. 40):

$$T_{th} = \frac{2730}{(\alpha_x + 0.3)} + 0.7 T_{pre} + 273 \; °K \qquad (5\text{-}43)$$

where T_{pre} = preheat temperature of combustion air, °C.

In conventional boilers, the furnace pressure is very low. So the effect of pressure on the gas volume or the volume heat release is negligible.

EXAMPLE:

> Calculate the average retention time of the fuel in the furnace of a boiler. The gas pressure in the furnace is 1.03 atm. The furnace volume heat release rate is 0.298 MW/m^3. The excess air coefficient is 1.1. The theoretical adiabatic combustion temperature is $2353°K$. The gas temperature at the exit of the furnace is $1405°K$ and volume effectiveness, φ_{eff}, is 0.6.

SOLUTION:

First, the average gas temperature is calculated from Eq. (5-42):

$$T_f = 0.925\sqrt{2353 \times 1405} = 1682°K$$

The furnace pressure, $P_f = 1.03 \times 0.101 = 1.04$ MPa; and the volumetric heat release rate $q = 298\,kW/m^3$.

Then, the residence time is calculated from Eq. (5-41):

$$\tau = \frac{\varphi_{eff}P_f}{T_f(1.1\alpha_x + 0.08)q_v}\left[\frac{273 \times 4186}{P_{atm}}\right]$$

$$= \frac{0.6 \times 0.104 \times 4186 \times 273}{298 \times 1682 \times 0.101(1.1 \times 1.1 + 0.08)} = 1.09\,s$$

b) Heat Transfer Requirement of an Oil Fired Furnace

The secondary function of the furnace is to cool the gas temperature by transferring heat to the water-cooled walls around it. The flue gas needs to be cooled to a desired temperature. In an oil fired furnace, the gas should not be cooled below $1000°C$ at the furnace exit. The burning rates of fuel drops are very low at such low temperatures. So any unburnt fuel will not complete its combustion and therefore the gas temperature at the exit of the furnace for large-scale oil fired boilers is kept in the range of $1350-1400°C$. At higher temperatures the fouling problem in the superheater tubes could be serious. Depending on the oil quality, the inlet gas temperature to the superheater section may have to exceed $1050-1100°C$ to avoid fouling from fuel oil. The mechanism of fouling will be discussed in detail in Chapter 14.

The metal temperature of the superheater tube is also not allowed to exceed values set by the permissible stress (see Chapter 17). This imposes a limit on the maximum gas temperature at the exit of the furnace. The high gas temperature at the exit of the furnace raises the average temperature in the furnace, which increases the heat absorption by radiative heat transfer to surfaces. This is damaging, especially

TABLE 5-3. The shape coefficient of a cubical volume.

Length of one edge of the cube	1	2	3
Volume, m^3	1	8	27
Surface area, m^2	6	24	54
Shape coefficient, 1/m	6	3	2

TABLE 5-4. Furnace volume heat release rates of different types of oil-fired boilers.

Boiler types	Furnace volume heat release rate (q_v)	
	(MW/m^3)	(MJ/m$^3 \cdot$ h)
Utility boiler	0.23–0.35	840–1250
Industrial boiler	0.35–0.58	1250–2100
Marine boiler	0.77–1.00	2700–3600
Naval vessel boiler	~8.13	~29,300 × 10^3

to high-pressure boilers. So there is a need to maintain the furnace temperature within limits by allowing the furnace wall to absorb appropriate fraction of the combustion heat.

While the amount of combustion heat released in the furnace is proportional to the furnace volume, the amount of heat absorption is proportional to the surface area of the enclosing walls. The furnace volume and its wall surface area do not increase in the same ratio as the size of the furnace increase. The surface area increases at a lower rate than does the volume. *Shape coefficient*, which is the ratio of enclosing surface and enclosed volume, decreases with increasing volumes (Table 5-3).

Therefore, the surface area per unit steam output is larger in small boilers. The furnace volume is decided on the combustion condition. Sometimes it may be necessary to have some area without water-wall coverage. However, for the large boiler, the shape coefficient is small. Even with the water walls arranged on all the furnace walls, the gas temperature at the exit of furnace is still higher. To achieve the required cooling, heat transfer surfaces are arranged inside the furnace as platens.

c) Selection of Furnace Volume Heat Release Rate

Volumetric heat release rates are not same for all boilers. They depend on the type of application. For example, the furnace volume heat release rate of the boilers in naval vessels is 20–30 times higher than that of the utility boilers, because space is more important than thermal efficiency in warships. The furnace volume heat release rate of utility oil-fired boilers is in the range of 0.23–0.35 MW/m^3. It is limited by both furnace heat absorption considerations as well as by the need to restrict the NO$_x$ emission. Table 5-4 lists the range of furnace volume heat release rates (q_v) for four types of oil fired boilers.

A higher furnace volume heat release rate will increase the aerodynamic loss in the burner. The consequent higher power consumption by the fan reduces the overall plant efficiency. Based on the above information one could statistically correlate the data of burner pressure drop with volumetric heat release rate for all types of the oil fired boilers in the following form:

$$\Delta P = K'q_v^m \qquad (5\text{-}44)$$

where ΔP = pressure loss of the burner
K' = coefficient; its value depends on the design and arrangement of the burners
m = empirical constant; its value lies in the range 0.9–1.1

5-4-2 Furnace Cross Section Area

The volume of the furnace is decided from the furnace volume heat release rate, but its plan or cross section area depends on another characteristic parameter known as *furnace grate heat release rate*, q_F, which is defined as

$$q_F = \frac{B \cdot LHV}{F_{sec}} \ \text{kW/m}^2 \qquad (5\text{-}45)$$

where B = fuel flow rate, kg/s
LHV = lower heating value of the fuel, kJ/kg
F_s = cross section area of the furnace, m^2

The grate heat release rate, q_F, often increases with the capacity of the boiler. Some typical values of grate heat release rates for oil fired boilers are given in Table 5-5 below:

If the furnace grate (or plan) area reduces, the furnace will be narrow and tall. In a narrow and tall furnace, it is easy to increase the flame length, and the furnace volume effectiveness factor will be better. However, if the local heat release rate is too high in the area near the burners of a narrow furnace, oil droplets will hit the furnace walls. The temperature level in the burner area will increase, making the local heat transfer too high. On account of all these, the furnace section area heat release rate cannot be too high.

As shown below, the ratio of the grate and the volumetric heat release rate gives the furnace height, h.

$$\frac{q_f}{q_v} = \frac{\dfrac{B \cdot LHV}{F_{sec}}}{\dfrac{B \cdot LHV}{V}} = \frac{V_f}{F_{sec}} = H_{fur} \qquad (5\text{-}46)$$

TABLE 5-5. Average grate heat release rates for oil or gas fired boilers.

Steam capacity (t/h)	130	220	400	670–950	1000–1600
q_F (MW/m^2)	<4.0	4.07–4.77	4.19–5.23	5.23–6.16	<9.0

Burner zone heat flux is another parameter that is also used at times for design. This index q_s is defined as

$$q_b = \frac{B \cdot LHV}{S_r} \, \text{kW/m}^2 \qquad (5\text{-}47)$$

where S_r = radiative heat transfer area in the burner zone, m^2.

The heat absorption in the furnace varies in direct proportion to q_b. We have seen earlier that in large capacity boilers, the volume of the furnace is decided also from heat transfer considerations. This is mainly to avoid excessive heat absorption by any part of the boiler heating surface. So the designer tries to limit the heat flux on the heat transfer areas to 0.58 MW/m^2. Generally, in an oil fired boiler, the maximum heat flux on the heat transfer area is 2.5 times its average value. Therefore, the average heat absorption by heat absorption surfaces should not exceed (0.58/2.5 = 0.23 MW/m^2). Also, the average heat flux in the furnace is about half that of the burner zone heat flux, (q_H). So, the limit of the furnace radiative heat absorption rate is about 0.46 MW/m^2.

5-4-3 Arrangement of the Burners

Depending on which wall accommodates the oil burners, there can be different burner arrangements e.g., front wall, opposite wall, tangential, and bottom arrangements.

Front wall arrangements are used when swirl-type air registers are used and the capacity of the boiler is not large. The piping for oil and ducting for air is simple for this arrangement. The opposite wall arrangements are used for large capacity boilers. For burners in opposed jet firing arrangements flames may impinge on each other, producing a secondary disturbance, which is favorable to later-stage mixing in boilers. To avoid flame interference in swirl-type burners, a minimum distance is maintained between the burners. A similar clearance is also allowed to avoid impingement on the side wall and bottom of the furnace. These minimum distances are

- The distance between center lines of adjacent burners (2.5–3.0) d_0.
- The distance between the burner and the side wall (3.0–3.5) d_0.
- The distance between the burner and the furnace bottom 3.0 d_0.

Here, d_0 is the diameter of the mouth of the burner.

Some typical minimum burner spacings for advection-type air registers are shown in Figure 5-20. When the burners are arranged on the front wall, the furnace depth should consider the flame length. The flame length depends on many factors related to the combustion process.

In an oil fired boiler, there are many advantages to arranging the burners in corners, but the air and oil pipes systems are more complex and the furnace cross section needs to be square. In very large capacity boilers this may pose some problems.

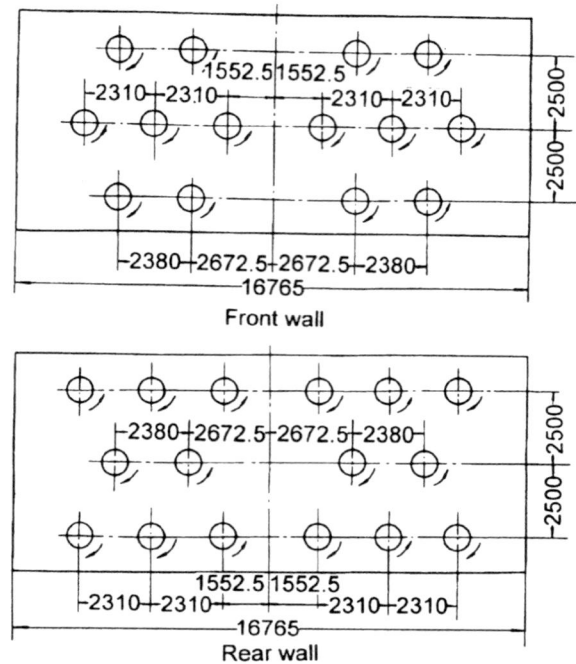

FIGURE 5-20. Arrangement of burners on the furnace wall of a 500 MW oil fired boiler, the direction of swirl in individual burner is shown by the arrow

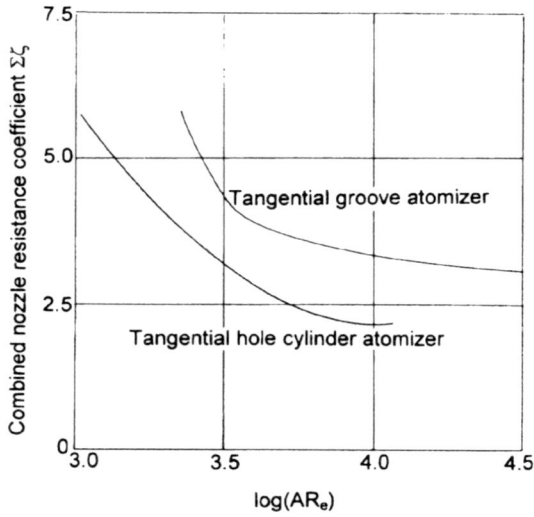

FIGURE 5-21. A plot of total resistance coefficient of a mechanical atomizer against log (ARe), where A is the characteristic nozzle coefficient and Re is the Reynolds number on the basis of the groove: (1) Tangential groove atomizer; (2) Tangential hole cylinder atomizer

Nomenclature

$\mu_{r=200}..k$

A	characteristic coefficient (Eq. 5-4)
A_a	Cross section area of air nozzle, m^2
A_{oil}	cross section area of oil nozzle, m^2
a	radial distance from the cup exit of the oil film, m
A_p	practical value of the characteristics coefficient A
b	width of tangential groove, m
B	fuel flow rate, kg/s
D_d	diameter of the duct where the flame-holder is located, m
D_{oil}	inlet tube diameter, m
d	droplet diameter, m
d_{fh}	diameter of the flame-holder (Eq. 5-34), m
d_m	largest droplet diameter, m
d_{noz}	diameter of the atomizer nozzle, m
d_0	diameter of the burner mouth, m
d_1	oil hole diameter, m
d_2	air hole diameter, m
d_3	diameter of the mixing hole, m
D_a	diameter of the air inlet tube, mm
D_{air}	diameter of the circle joining the centers of steam hole, m
D_{cup}	inner diameter of rotary cup
D_{mix}	annular diameter, m
D_T	diameter of the cup mouth, m
D_x	diameter of swirl room, m
F	calculation section area, m^2
F_{in}	annular section area at the inlet (Eq. 5-33), m^2
F_1	total section area of oil holes, m^2
F_2	total section area of air holes (Eq. 5-18), m^2
F_h	section area of the throat, m^2
F_{sec}	cross section area of the furnace, m^2
F_o	outlet area of the tuyere, m^2
F_{sec}	cross section area of the furnace, m^2
FN	flow number
G	flow rate of oil, kg/h
G_a	air consumption rate, kg/h
g	acceleration due to gravity, m/s^2
H_{fur}	height of furnace, m
h	thickness of the oil film in the cup (depth of groove), m
K	coefficient (Eq. 5-1)
k	coefficient (Eq. 5-21)
K'	empirical constant in Eq. (5-44)
L	approximate flame length, m
LHV	lower heating volume of the fuel, kJ/kg

l_2	length of steam hole, m
l_2	distance of the exit of steam hole from the exit of the oil hole, m
l_3	length of mixing hole, m
m	exponent in Eq. (5-44)
m_2	consumption of atomizing medium (steam or air), kg/h
m_j	steam consumption rate, kg/h
m_{oil}	oil flow rate per hole, g/s
N	rotary speed of cup, RPM
n	number of holes
n_{max}	maximum adjustment ratio
P_a	air pressure before entering the nozzle, Pa
P_f	inlet oil pressure, Pa
P_0	oil pressure at the inlet of nozzle above that at the exit, Pa
P_1	pressure difference the inlet, Pa
P_2	pressure of the atomizing media at the inlet of the atomizer, Pa
P_3	pressure at the mixing point, Pa
ΔP	pressure drop in nozzle due to hydrodynamic resistance, Pa
Q	oil flow rate, m^3/s
q	steam consumption rate for atomization, kg/kg oil
q_b	burner zone heat release rate, KW/m^2
q_F	grate heat release rate, kW/m^2
q_v	volumetric heat release rate, kW/m^3
q_H	heat flux on burner zone, kW/m^3
R	swirl radius (Fig. 5-4), m
r	theoretical radius of oil droplet, mm
r_p	inner radius of nozzle (Fig. 5-4), m
r_w	inner radius of the swirling oil film (Fig. 5-4), m
S_r	radiation heat transfer area, m^2
S	thickness of oil film, m
u	tangential velocity of oil at nozzle exit, m/s
V_a	air flow rate, m^3/h
V_{air}	air flow rate per unit of fuel, m^3/kg
V_f	furnace volume, m^3
V_T	tangential velocity of the oil, m/s
W	oil velocity in the groove or inner tube, m/s
W_{rel}	relative velocity between the oil droplet and the air, m/s
w_{cr}	critical velocity of the atomizing media
w	axial velocity of oil at nozzle exit, m/s

Greek Symbols

ρ	density of oil, kg/m^3
ρ_a	density of air, kg/m^3
ρ_s	density of atomizing medium, steam or air, kg/m^3
γ	kinematic viscosity of oil, m^2/s

μ	orifice flow coefficients
μ_p	practical or modified flow coefficient
$\mu_{r=200}$	practical or modified flow coefficient 200 mm away from the nozzle
ν_{oil}	viscosity of oil, centistoke (Cts)
σ_{oil}	surface tension of the oil, dyne/cm,
ρ_{steam}	steam density at the steam pressure in the mixing hole, g/cm^3,
α	enclosed angle of spray cone (Fig. 5-3)
α_p	modified cone angle of spray
β	pressure drop factor
$\sum f$	total area of tangential grooves, m^2
$\sum \xi$	sum of individual flow resistance coefficients
ψ	flow rate factor in Eq. (5-18)
θ	slope of rotary cup's wall
θ_a	atomizing angle of Y-type atomizer
φ	ratio of oil flow area and the nozzle area (Eq. 5-3)

References

Blokh, A.G. (1988) "Heat Transfer in Steam Boiler Furnaces." Hemisphere Publishing Corporation, Washington, p.178.

Dombrowski, N., and Munday, G. (1968) "Biochemical and Biological Engineering Science." Academic Press, New York.

Fraser, R.P., Domrowski, N., and Routley, J.H. (1963) Performance characteristics of rotary cup blast atomizers. Journal of the Institute of Fuel, pp. 316–327.

Junkai, Fen (1992) "Principle and Calculation of Boilers" (in Chinese)." 2nd edition. Science Press, Beijing.

Stultz, S.T and J.B. Kitto (1992) "Steam—Its Generation and Use." Babcock & Wilcox, Barberton, Ohio.

Weinberg, S. (1953) Proceeding of the Institution of Mechanical Engineers 18:240.

Wigg, L.D. (1964) Drop-size prediction for twin-fluid atomizers, J. Institute of Fuel 37 (Nov.):500–505.

Xiezhu, B. (1976) "Design and Operation of Oil-Fired Boilers" (in Chinese). Hydro-Electric Press, Beijing.

6
Boiler Furnace Design Methods

The furnace is the most important part of a boiler. Its primary function is to provide adequate space for fuel particles to burn completely and to cool the flue gas to a temperature at which the convective heating surfaces can be operated safely. How this can be done depends on the type of firing method employed. For example, in the case of stoker fired boilers, the combustion takes place mostly on the grate. Here the heat transfer takes place beyond the combustion zone. In pulverized coal (PC) or circulating fluidized bed (CFB) fired boilers, on the other hand, the combustion and heat transfer to the water walls take place simultaneously in the furnace. As a result, the furnace of a stoker fired boilers will be designed differently from that of the latter two boilers. Even PC and CFB boilers vary in design considerations as the thermal and hydrodynamic conditions in a CFB boiler are different from those of a PC boiler.

An introduction to the design of a CFB boiler is given in Chapter 11. A more comprehensive treatment of the design of CFB boilers is available elsewhere (Basu & Fraser, 1991). Oil and gas fired furnaces are discussed in Chapter 5. Stoker fired boilers are not widely use in modern plants. So this chapter concentrates on pulverized coal (PC) fired boilers.

6-1 General Design Principles

A typical oil or PC fired boiler furnace (Fig. 6-1) is enclosed by a water-cooled wall, known as a *water wall*. An array of burners fires the fuel in flames that radiate heat to the walls of the furnace. Some boiler furnaces also accommodate additional heating surfaces like superheater or reheater surfaces. These surfaces cool the flue gas leaving the furnace to an appropriate temperature for the downstream heat exchange surfaces in the back-pass or convective section.

General requirements of boiler furnaces irrespective of their firing method are as follows:

1) The furnace should provide the required physical environment and the time to complete the combustion of coal or any other fuel particles.

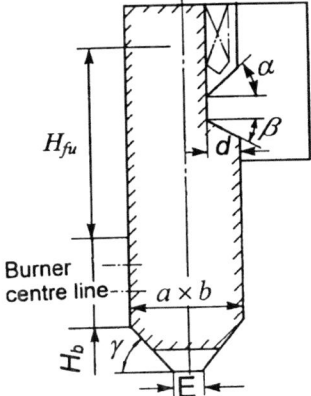

FIGURE 6-1. A schematic of an oil, gas, or pulverized coal fired furnace showing the definition of its volume by hatched lines

2) The furnace should have adequate radiative heating surfaces to cool the flue gas sufficiently to ensure safe operation of the downstream convective heating surface.
3) Aerodynamics in the furnace should prevent impingement of flames on the water wall and ensure uniform distribution of heat flux on the water wall.
4) The furnace should provide conditions favoring reliable natural circulation of water through water wall tubes.
5) The configuration of the furnace should be compact enough to minimize the amount of steel and other materials used for construction.

6-1-1 Determining the Furnace Size of a Boiler

The shape of the furnace depends on several factors, including flame shape, firing method, and arrangement of burners. A typical shape of a pulverized coal (PC) fired furnace with a dry bottom is shown in Figure 6-1. While working with the furnace, the designer needs to define the boundaries of the furnace. These boundaries are defined by the central line plane of water wall and roof tubes, the central lines of the first-row superheater tubes or slag screen, and the horizontal plane passing through half-height of ash hopper. Typical proportions of a PC boiler are given below with reference to Figure 6-1 (Lin, 1991).

$$\alpha = 30 \text{ to } 50°; \quad \beta > 30°; \quad \gamma = 50 \text{ to } 55°;$$
$$E = 0.8 \text{ to } 1.6 \text{ m}; d = (0.25 \text{ to } 0.33)b$$

For uncooled or refractory lined furnaces the furnace boundary is defined by the surface of the furnace wall. If the furnace bottom is horizontal, its lower boundary would be the horizontal furnace floor itself. In stoker fired boilers, the effective furnace volume excludes the volume of the fuel layer.

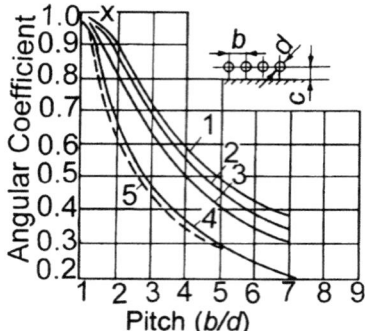

FIGURE 6-2. Angular coefficients of water wall tubes at different distances from the refractory wall: (1) $c > 1.4 d$; (2) $c = 0.8 d$; (3) $c = 0.5 d$; (4) $c = 0$; (5) fully cooled membrane wall

The *projected surface area* of the water wall surface, F, is the area of the plane defining the boundary of the effective furnace volume. Thus, the effective projected area of a section between two tubes of the water wall is calculated as shown below:

$$F = p \cdot l \ \text{m}^2 \tag{6-1}$$

where l is the length (or height) of the water wall and p is pitch or the distance between two adjacent tubes of the water wall. In a boiler furnace the radiation received by a tube is not uniform over its circumference. Furthermore, sometimes the entire furnace wall is not covered by tubes. Thus to simplify the heat transfer calculation, the radiation heat absorbing surface is thought of as a continuous flat surface that has the same heat absorption as clean spaced-out tubes. Instead of the effective projective surface area of wall, F, an *effective radiative heat absorbing surface area*, S_{eff}, is used by multiplying the former by an angular coefficient, x.

$$S_{eff} = \sum (Fx) \ \text{m}^2 \tag{6-2}$$

The *angular coefficient*, x, of tubes located before furnace wall depends on their pitch, p, and their distance from the refractory wall (Fig. 6-2). The angular coefficient also takes account of the radiation from the refractory wall behind the tubes. For a membrane-type water wall where tubes touch each other, $x = 1.0$. For pendant superheater slag screen and boiler tube bundles, $x = 1.0$. For sections of the furnace wall without water walls such as the burner zone and man-hole, $x = 0$.

6-1-2 Design of PC Boilers

There are two aspects of the design of a furnace. The first part is concerned with the generation of heat; the second part involves absorption of the heat in the furnace. The first part ensures that the designed amount of fuel can be burnt in the given furnace volume, liberating the required amount of heat. Practical designs are based on a permissible firing rate, which depends on a number of factors; these are

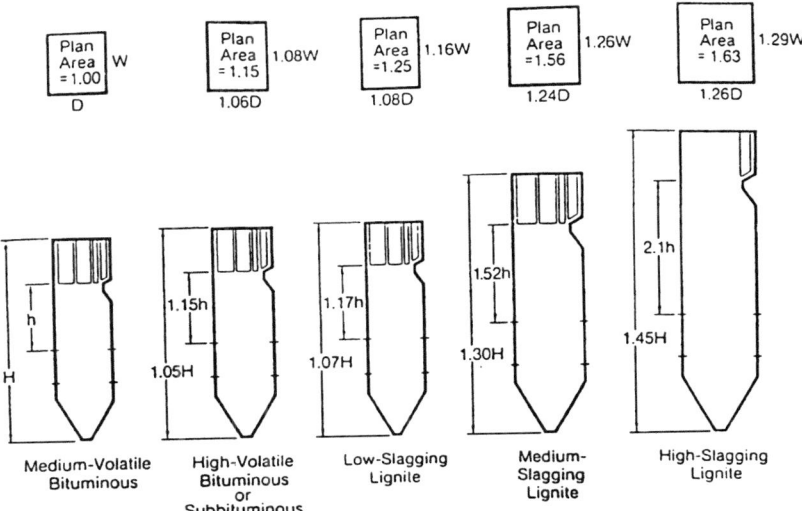

FIGURE 6-3. Effect of coal rank on the size of PC furnace for a constant heat output (Reprinted with permission from Singer, J.G., (1981), Fossil Power Systems, Combustion Engineering, Inc., Windsor, USA, P. 7-3, Figure 1)

explained below. The second part ensures that the enclosing furnace walls absorb the required fraction of the generated heat. In case of PC boilers this involves complex heat transfer analysis, which is elaborated in Section 6-3.

a) Firing Rates of PC Furnaces

Heat release rates and furnace exit gas temperatures are some of the important parameters used for the design of the size of the furnace. The following section presents methods for selection and calculation of these parameters. Unlike volume in a circulating fluidized bed furnace, the volume of a PC or oil fired furnace is greatly influenced by the type of fuel fired (Lafanechere et al., 1995). For each type of fuel there is a specified heat release rate. The furnace cross section area and its height are influenced by the rank of the coal. Figure 6-3 shows how the furnace size and shape changes for a different types of coal though the heat input is constant.

The heat release rate is expressed on three different bases, furnace volume (q_v), furnace cross section area (q_F), and water wall area in the burner region (q_b).

(1) Heat Release Rate Per Unit Volume, q_v

The volumetric heat release rate (q_v) is the amount of heat generated by the combustion of fuel in a unit effective volume of the furnace. It is given as

$$q_v = \frac{B\ LHV}{V}\ kW/m^3 \qquad (6\text{-}3)$$

where B = Designed fuel consumption rate, kg/s
           V = Furnace volume, m^3
           LHV = Lower heating value of fuel, kJ/kg

A proper choice of volumetric heat release rate will ensure that

• Fuel particles are burnt substantially.
• The flue gas is cooled to the required safe temperature before leaving the furnace. This temperature, which is known as furnace exit gas temperature (FEGT), is critical for safe operation of downstream heat exchanger surfaces.

For small-capacity boilers (>220 t/h steam) a volumetric heat release rate, chosen on the basis of complete combustion of fuel, is close to that chosen on the basis of the cooling of the flue gas. However, that is not the case for larger-capacity boilers because with increasing boiler capacity, the area of the enclosing furnace walls does not increase at the same rate as does the furnace volume. For example, if the boiler capacity is doubled, the furnace height cannot be doubled because of cost considerations. So the available furnace wall area would not be twice as much. The volumetric heat release rate also depends on the ash characteristic, firing method, and arrangement of burners. Some typical values of q_v are shown in Table 6-1.

(2) Heat Release Rate Per Unit Cross Sectional Area (q_F)

This is the amount of heat released per unit cross section of the furnace. It is also at times called grate heat release rate. It is given by

$$q_F = \frac{B.LHV}{F_{grate}} \text{ kW/m}^2 \qquad (6-4)$$

where F_{grate} is the cross sectional or grate area of the furnace m^2.

The grate heat release rate, q_F, reflects the temperature level in the furnace to some degree. When q_F increases, the temperature in the burner region rises, which helps stability of the pulverized coal flame, but it increases the possibility of slagging in the furnace. For small boilers (steam capacity $D \leq 220$ t/h) it is not essential to check q_F because compared to its volume, the cross section area of

TABLE 6-1. Typical values of volumetric heat release rate (q_v) in MW/m^3.

Coal type	Dry-bottom furnace q_v(MW/m^3)	Wet (slagging) bottom furnace q_v(MW/m^3)		
		Open furnace	Half-open furnace	Slagging pool
Anthracite	0.110–0.140	≤0.145	≤0.169	0.523–0.598
Semi-anthracite	0.116–0.163	0.151–0.186	0.163–0.198	0.523–0.698
Bituminous	0.14–0.20			
Oil	0.23–0.35			
Lignite	0.09–0.15	≤0.186	≤0.198	0.523–0.640
Gas	0.35			

* The lower limit is for lignite when softening temperature (ST) is lower than 1350°C.

TABLE 6-2. Upper limits of q_F for tangential fired furnace.

Boiler capacity (t/h)	Upper limit of q_F in MW/m^2		
	$ST^a \leq 1300°C$	$ST = 1300°C$	$ST \geq 1300°C$
130	2.13	2.56	2.59
220	2.79	3.37	3.91
420	3.65	4.49	5.12
500	3.91	4.65	5.44
1000	4.42	5.12	6.16
1500	4.77	5.45	6.63

^{a}ST = Softening temperature of ash, °C.

the furnace is very large. Since this is not the case for large-capacity boilers, the q_F must be carefully chosen. Thus it depends on both the capacity of boilers and the softening temperature (ST) of the fuel. Typical values of the upper limits of q_F are given in Table 6-2 for a wide range of coal in boilers up to 1500 t/h capacity. Beyond this capacity the limit of q_F changes only marginally.

The volume of the furnace can be determined from the chosen value of q_v, while the height of the furnace is calculated from the q_F.

(3) Heat Release Rate Per Unit Wall Area of the Burrner Region (q_b)

The burner region of the furnace is the zone of most intense heat. So this area is designed separately using a third type of heat release rate, q_b called *burner zone heat release*. It is based on the water wall area in the burner region. It is defined by

$$q_b = \frac{B\,LHV}{2(a+b)H_b} \text{ kW/m}^2 \tag{6-5}$$

where a and b are the width and depth of the furnace respectively, and H_b is the distance between the top edge of the uppermost burner and the lower edge of the lowest burner, m.

The parameter q_b represents the temperature level and heat flux in the burner region. This additional parameter is often used for judging the general condition of the burner region in a large boiler. It depends on the fuel ignition characteristics, ash characteristics, firing method, and arrangement of the burners. Some recommended values of q_b are shown Table 6-3.

TABLE 6-3. Recommended values of burner region heat release rate, q_b.

Fuel	q_b in MW/m^2
Brown coal	0.93–1.16
Anthracite and semi-anthracite	1.4–2.1
Lignite	1.4–2.32

TABLE 6-4. Minimum depth (b_{min}) of a furnace with swirl
burners on the front wall.

Boiler capacity (t/h)→	130	220	420	670	>670
Coal	6.0	7.0	7.5	8.0	$\geq(5\text{–}6)d_r$ [a]
Oil	5.0	5.0	6.0	7.5	

[a] d_r = maximum nozzle diameter of swirl burner.

b) Furnace Depth

Heat release rates help determine the volume as well as the cross section area
of the furnace. They do not, however, define the depth-to-breadth proportion of
the furnace, which is an important parameter from both combustion and heat
absorption standpoints. For example, in a very narrow (low depth-to-breadth ratio)
PC furnace, the flame emerging from the front wall can hit the opposite wall. This
would severely damage the tubes. So for a given furnace cross section, the depth-
to-breadth ratio of the furnace should be calculated to avoid this possibility. There
is, however, a minimum value of the furnace depth, b_{min}, which depends on the
capacity of the boiler and the type of fuel fired (Table 6-4). The arrangement of
burners, heat release rate per unit furnace area, power output of each burner, and
flame length are some of the factors that influence the breadth and depth of the
furnace, particularly when swirl burners are used.

In a corner fired furnace, where burners are located on four corners, a width-to-
depth ratio (a/b) less than 1.1–1.2 may ensure a good flow pattern in the furnace.
While choosing the depth, b, the flue gas velocity in the convective pass and the
steam velocity in superheater should also be considered. The steam–water mixture
rises in parallel through the water walls enclosure of the furnace. So, the periphery
of the furnace should also consider the steam–water velocity through these tubes.
It is especially important for boiler pressures above 9.8 MPa.

c) Furnace Height

The furnace should be sufficiently high so that the flame does not hit the superheater
tubes. The shortest distance (H_{fu}) between the burner and any heating surface is the
vertical distance from the center line of the uppermost burners to the midpoint of
the furnace exit (Fig. 6-1). So this height must exceed a minimum value, depending
on the type of coal used and the capacity of the boiler. The shorter the furnace
height, the worse the natural circulation. Thus a minimum furnace height is also
needed for the required degree of natural circulation. Some typical values are
shown below for a dry bottom PC boiler (Table 6-5).

d) Furnace Exit Gas Temperature, T_{ou}

The furnace exit gas temperature (FEGT) is an important design parameter. It defines the ratio of heat absorption by the radiant heating surfaces in the furnace and
that by the convective heating surfaces downstream of the furnace. This parameter is governed by both technical and economical factors. A higher FEGT would
make the furnace compact but the convective section larger. This also increases the

TABLE 6-5. Lower limit of H_{fu}(m) (Lin, 1991).

Boiler capacity (t/h)	65–75	130	220	420	670
Anthracite	8	11	13	17	18
Bituminous	7	9	12	14	17
Lignite					
Oil	5	8			

fouling potential of the tubes. These considerations give the optimum value of the furnace exit gas temperature in the range from 1200°C to 1400°C. However, the actual exit temperature of the furnace is often kept below the optimum value to prevent fouling of the convective heating surfaces. For boilers without platen superheaters the exit temperature is the temperature of the flue gas entering the screen tubes.

The FEGT is chosen slightly below the ash deformation temperature (DT). (See Section 3-1-3b in Chapter 3 for definitions of DT and ST.) Otherwise there would be severe fouling of the back-pass tubes by molten ash. If the difference between ash deformation (DT) and the softening temperatures (ST) is smaller than 100°C, the FEGT should be lower than (ST − 100)°C. For boilers with platen superheaters, the flue gas temperature behind the platen superheater should be smaller than (DT − 50)°C or (ST − 150)°C. The temperature in front of the platen superheater should be lower than 1250°C for coal with weak slagging propensity. For coals with strong and moderate slagging properties, this temperature should be below 1110°C, and 1200°C, respectively.

Oil flame produces much smaller amounts of ash. So the furnace exit gas temperature of an oil fired furnace can be higher than that for pulverized coal (PC) fired furnaces. The FEGT of oil fired furnaces is, however, chosen below 1250°C to prevent high temperature corrosion of the superheater tubes.

6-2 Flame Emissivity

Radiation being the dominant heat transfer process in PC, oil, and gas fired boilers, the emissivity of the flame is a very important parameter. The emissivity calculations must distinguish between gas/oil flame and pulverized coal flame. In gas or oil flame the main contributors of radiation are tri-atomic gases (e.g., CO_2, H_2O, SO_2) and soot particles. In case of solid fuel flame ash and coke particles are major contributors. The following section discusses the emission characteristics of different flame constituents and presents equations for the same.

6-2-1 Types of Radiation from Flames

There are four sources of radiation from flames:

a) Tri-atomic Gases

The thermal radiation of a flame is mainly due to three tri-atomic gases, CO_2, H_2O, and SO_2. The radiation from mono-atomic and di-atomic gases like N_2, O_2, H_2, etc., is relatively weak at temperatures around 2000°K. So it can be neglected in the calculation of gaseous radiation in the furnace. The radiation of tri-atomic gases is selective and falls mainly in the region of infrared wavelengths. Its intensity is determined by the temperature and partial pressure of each constituent of the radiative medium and the thickness of the radiant gas volume. However, under the same temperature the radiant intensity of gaseous substances is lower than that of solids like soot or ash.

b) Soot Particles

Hydrocarbons in the fuel are decomposed at high temperatures during combustion reactions, forming soot particles. These particles have strong radiation ability and can make the flame very bright. For coal fired boilers the radiation from soot particles can be neglected because the amount of hydrocarbon in coal is relatively low. But in boilers fired with heavy oil, the radiation from soot particles is significant and, therefore, should be taken into account.

c) Coke Particles

Coke are present in coal-flames. These solid particles emit strong radiation. The intensity of solid particle radiation in the flame depends on the particle size, the particle concentration in the furnace volume, and the particle properties. Coke particles are usually 10–250 μm. Their radiant intensity is close to that of a black body. But their concentration is not high (<0.1 kg/m^3), and they are mainly concentrated near the burners. Therefore, their radiation accounts for only 25–30% of the total radiation in the furnace.

d) Ash Particles

Ash particles are formed after coke particles burn out. They are mostly 30–50 μm. Ash particles also radiate heat. In a pulverized coal fired boiler the radiation intensity of ash particles constitutes 40–60% of the total radiation. The radiation intensity of ash particles is lower at a higher temperature and increases as the gas is cooled.

6-2-2 Mean Beam Length of Radiation

The flame is essentially a high temperature jet of flue gas. Some solid particles are suspended in this flue gas. It is semi-transparent, but every part of the gas emits radiation, which is transmitted through rest of the body of the gas. In that sense the flame radiation differs from the radiation from a surface of solid and liquid. Thus, the thicker the gas layer, the more intense the radiation.

In a boiler furnace the heat transfer involves heat exchange between the flame and the water wall. If the heating surface is the enclosing surface of the furnace, the total heat exchange is equal to the radiation of the flame to the enclosing surface. However, the shape of the furnace is complex. The radiation arriving at the enclosing surface from different directions travels different distances. This results in different magnitudes of the radiation. As a result the calculation of total radiation exchange is even more complex.

When radiation passes through an absorbing gas medium of thickness, S, a part of it is absorbed by the medium. The intensity of the transmitted radiation after absorption is given by the Beer's Law (Holman, 1991),

$$I_s = I_o e^{-kS} \qquad (6\text{-}6)$$

where I_o, I_s are radiation intensities before gaseous absorption and after passing through absorbing gases, respectively: and e^{-kS} is transmissivity of the absorbing gas of thickness S.

For nonreflecting gas, transmitivity + absorptivity = 1.0. So from Eq. (6-6), the absorptivity is $(1 - e^{-kS})$. Thus, if we consider that when black body radiation E_0 passes through a gas layer of thickness S, the amount of radiation absorbed by the medium will be

$$E_s = (1 - e^{-kS})E_o \qquad (6\text{-}7)$$

where E_o is black body radiation power at same the temperature as the flame, and $(1 - e^{-kS})$ is absorptivity of gas volume.

The radiant energy can travel from the source to the receiver along many paths. The shortest path will be the one perpendicular to one of them. The radiation, traveling at an angle, would go over a much longer distance; therefore more would be absorbed. To simplify the absorption calculation one uses an average thickness of the radiating gas layer (effective radiation layer) thickness, S, called *mean beam length*. It is equivalent to the radius of the hemisphere receiving radiation from its center.

Generally, the calculation of S is very complex. One may use Table 6-6 for standard geometry. In the absence of information on a specific geometry one could use the following approximation to calculate the mean beam length or the effective radiation layer thickness, S (Holman, 1991);

$$S = 3.6\left(\frac{V}{F}\right) \qquad (6\text{-}8)$$

where V is the volume of the gas and F is the enclosing surface area.

TABLE 6-6. Effective mean beam length for standard geometry.

Gas volume enclosure	Mean beam length (S)
Infinite parallel plate	1.8 × distance between plates
Circular cylinder (Infinite length), radiation to wall	0.95 × diameter
" (diameter = height), Radiation to base center	0.71 × diameter
" (diameter = height) radiation to all surfaces	0.6 × diameter
Cube, radiation to any surface	0.66 × edge length
Volume surrounding infinite tube bundles, radiation to one tube	
Equilateral triangle arrangement, pitch = 2 × diameter	3 × (pitch-diameter)
Equilateral triangle arrangement, pitch = 3 × diameter	3.8 × (pitch-diameter)
Square arrangement	3.5 × (pitch-diameter)

6-2-3 Emissivity of Pulverized Coal Flame

The presence of solid particles in the flame of a pulverized coal furnace introduces additional complexity owing to the presence of scattering by the solids. For all solid fuels the flame emissivity is calculated by the equation

$$a_{fl} = 1 - e^{-kPS} \qquad (6-9)$$

where k is the coefficient of radiant absorption in the furnace (1/mMPa); P is the pressure of gases in the furnace (MPa), and S is the effective thickness of the radiant layer (m) or mean beam length.

Flame radiation is absorbed by tri-atomic gases ($k_y r$), ash particles ($k_h \mu_h$) and burning char particles ($10c_1 c_2$). So, the combined coefficient of radiant absorption, k, is calculated by adding these terms:

$$k = k_y r + k_h \mu_h + 10 c_1 c_2 \ 1/\text{m} \cdot \text{MPa} \qquad (6-10)$$

where μ_h is concentration of ash particles in the furnace, kg/kg.

Here c_1 and c_2 are determined by the type of fuel and the firing method, respectively. For anthracite, $c_1 = 1$. For bituminous coal, lignite, peat, oil shale, and wood $c_1 = 0.5$. For a flame (PC/oil) fired boiler, $c_2 = 0.1$. For a stoker fired boiler $c_2 = 0.03$. The coefficient of radiant absorption due to tri-atomic gases (k_y) is calculated by the equation (Lin, 1991, p. 403)

$$k_y = \left(\frac{7.8 + 16 r_{H_2O}}{3.16 \sqrt{rPS}} - 1 \right) \cdot \left(1 - 0.37 \frac{T_{ou}}{1000} \right) \ 1/(\text{m} \cdot \text{MPa}) \qquad (6-11)$$

where r = total volume concentration of tri-atomic gases $r = r_{RO_2} + r_{H_2O}$
T_{ou} = gas temperature at the furnace outlet, K

The coefficient of radiant absorption owing to ash particles (k_h) is calculated by the expression

$$k_h = \frac{5990}{\left(T_{ou}^2 d_h^2 \right)^{\frac{1}{3}}} \ 1/(\text{m} \cdot \text{MPa}) \qquad (6-12)$$

where d_h = diameter of ash particles, μm
For ball-mill, $d_h = 13 \ \mu$m
For ball-race mill and beater-mill, $d_h = 13 \ \mu$m
For stoker fired boilers, $d_h = 20 \ \mu$m

6-2-4 Emissivity of Liquid and Gas Flame

Flames may be luminous or nonluminous. The flame of tri-atomic gases is nonluminous. Its absorption and radiation are in the infrared wavelength region. Therefore, the flame of tri-atomic gases is transparent. Soot particles make the flame luminous. So near the burner area of a heavy oil fired boiler, the flame is luminous owing to the presence of soot particles. The luminosity of the flame decreases along the flame length because the soot particles are burnt away (Curve 2 in Fig. 6-4). At the

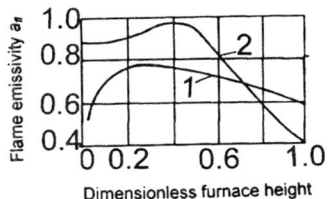

FIGURE 6-4. Variation of flame emissivity along the height of a furnace for different fuels

furnace exit the flame is often nonluminous. The luminosity of a gas fuel flame is much lower than that of a liquid fuel flame, owing to the absence of soot particles. Thus we see that the entire body of the flame is not luminous. The part or fraction (m) of the flame that is luminous gives luminous radiation, and the other part gives nonluminous radiation from the tri-atomic gases. So the total emissivity, a_{fl}, of liquid and gas flame is the sum of contribution of both luminous and nonluminous parts of the flame.

$$a_{fl} = ma_{lu} + (1-m)a_{tr} \qquad (6\text{-}13)$$

where a_{lu} = luminous flame emissivity
          a_{tr} = flame emissivity of tri-atomic gases

The fraction (m) expresses the degree of flame luminance, and it depends on the heat duty of furnace volume and the type of fuel. For liquid fuel, $m = 0.55$. For gaseous fuel, $m = 0.1$. The emissivity of tri-atomic gases is calculated by

$$a_{tr} = 1 - e^{-k_y r PS} \qquad (6\text{-}14)$$

where k_y is the coefficient of radiant absorption due to tri-atomic gases, 1/mMPa, r is the volume concentration of tri-atomic gases, P is the furnace pressure (MPa), and S is the mean beam length (m).
  The emissivity of luminous flame of gaseous or liquid fuels is determined by

$$a_{lu} = 1 - e^{-(k_y r + k_c)PS} \qquad (6\text{-}15)$$

where k_c is the coefficient of radiant absorption owing to soot particles, which is determined as follows (Lin, 1991, p. 404):

$$k_c = (2 - \alpha_{fu})[0.00016.T_{ou} - 0.5]\left(\frac{C}{H}\right) \qquad (6\text{-}16)$$

where α_{fu} = excess air ratio at furnace exit:
          T_{ou} = the furnace exit gas temperature
          C = carbon fraction in fuel—as received basis
          H = hydrogen fraction in fuel—as received basis

6-3 Heat Transfer Calculations for the PC Boiler Furnace

This section presents several procedures for the calculation of heat transfer in a PC furnace.

6-3-1 A Practical Approach to Heat Transfer Calculations for Boiler Furnaces

When fuel burns in a boiler furnace it releases a large amount of energy, which heats up the products of combustion (flue gas) to a very high temperature. This temperature may range from 1500°C to 1600°C in the flame core. Though the products of combustion are cooled by superheaters and evaporators in the furnace, their temperature at the exit of the furnace are still in the range of 1000–1250°C. The flame transfers its heat energy to the heating surface in the furnace by radiation. Because the flue gas flows through the furnace at a low velocity, the convective heat is only a small fraction (about 5%) of the total heat transferred to the walls in a conventional boiler furnace. Thus, to simplify the calculations, only radiative heat transfer is considered in the furnace.

Nonuniform temperature distribution and fouling of tubes in the furnace make the heat transfer calculations complex. Furthermore, the furnace temperature varies along its width, depth, and height. Figure 6-5 shows the temperature distribution in 220 t/h boiler furnace. Figure 6-5a shows that the temperature difference between two points at any elevation of the furnace is very large. Figure 6-5b shows the temperature distribution along the height of the furnace. This is responsible for the nonuniform heat flux along the height of the furnace walls. However, to simplify the design calculations, this variation is sometimes ignored and an average temperature and heat flux is used instead. As may be seen from Figure 6-5(b), the furnace temperature is highest near the burner region. Above this height the flue gas temperature decreases continuously.

The emissivity of the flame varies both laterally and axially. To calculate the radiative heat transfer along the height of a furnace, the variation of the emissivity along the height of the furnace should be considered. Figure 6-4 shows a typical variation of the emissivity of the flame along the height of a PC furnace.

To simplify the radiative heat transfer calculations, we use the following assumptions:

(1) The radiation heat exchange alone is considered in the heat transfer calculations for a furnace.
(2) The flame, at an average temperature T_{fl}, exchanges radiative heat with the heating surface of the water wall. The adiabatic (theoretical) flame temperature T_{th} is used to calculate T_{fl}.
(3) The flue gas temperature T_{ou} at the furnace exit (FEGT) is used as a characteristic temperature for the design.
(4) The heat absorbing surface of the water wall and the heat emitting surface of the flame are assumed to be parallel to each other with a surface area F.

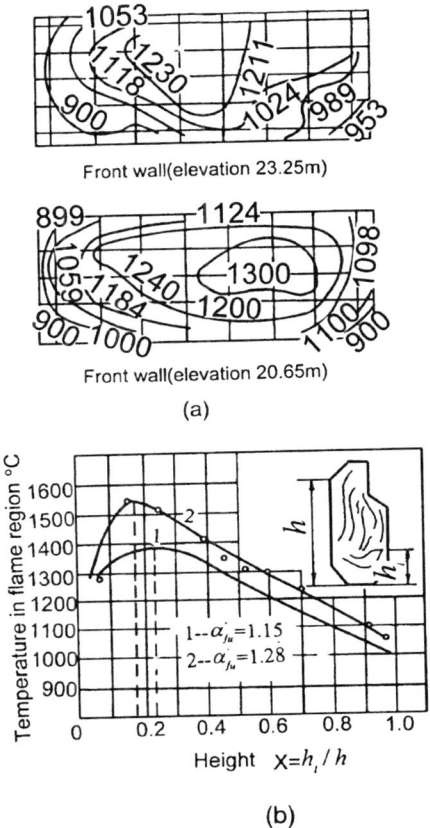

FIGURE 6-5. Temperature distribution in a 220 t/h boiler furnace: (a) variation in the horizontal plane at elevations 23.25 m and 20.65 m; (b) variation of temperature along the height of the furnace for excess air ratios (1) $\alpha = 1.15$ and (2) $\alpha = 1.18$ [From Pan Congzhen, 1986]

Using the above assumptions, we can develop the following equations to describe the heat exchange in the furnace.

a) Thermal Balance Equation

The total heat absorbed by the water (or steam) from the flue gas in a boiler furnace is a fraction, ϕ, of the difference between the total heat liberation, Q_{fu}, and the enthalpy of flue gases leaving the furnace, I_{ou}. The other part $(1 - \phi)$ is lost owing to radiation and convection from the furnace exterior. So the heat absorbed, Q_{abs}, in a furnace is

$$Q_{abs} = \phi B(Q_{fu} - I_{ou}) = \phi B \overline{VC_p}(T_{th} - T_{ou}) \text{ kW} \qquad (6-17)$$

where B = fuel consumption, kg/s;
T_{th}, T_{ou} = theoretical flame temperatures and furnace exit gas
               temperature respectively, K
       V = flue gas volume per unit mass of fuel burnt, Nm^3/kgf
       C_p = average specific heat of flue gas, $kJ/(Nm^3 K)$
       ϕ = coefficient of heat retention

Q_{fu} and I_{ou} are the heat available and the heat leaving the furnace, respectively, per unit mass of fuel burnt.

The fraction of heat retained by the water and steam, ϕ, is given by (Blokh, p. 170):

$$\phi = 1 - \frac{q_5}{q_5 + \eta} \tag{6-18}$$

where η is the boiler or thermal efficiency.

If $\overline{VC_p}$ is the average specific heat of combustion products formed by 1 kg of fuel within the temperature interval $T_{th} - T_{ou}$, then

$$\overline{VC_p} = \frac{Q_{fu} - I_{ou}}{T_{th} - T_{ou}} \text{ kJ/kg K} \tag{6-19}$$

b) Heat Available to the Furnace, Q_{fu}

Available heat is the sum of combustion heat (Q_{to}) and that brought by the preheated air per unit mass of fuel burnt. The combustion heat, Q_{to}, as used here, is the available heat from the combustion of unit mass of fuel. It is, therefore, the sum of lower heating value of fuel, heat brought into the furnace by externally preheated fuel, and steam used for steam-injected atomizers. However, some of this heat is lost. These losses include

q_3 heat loss owing to chemical incompleteness of combustion, %
q_4 heat loss owing to physical incompleteness of combustion (carbon loss), %
q_5 heat loss owing to radiation and natural convection from boiler exterior, %
q_6 heat loss owing to slag removed from the surface, %

Thus the furnace heat available, Q_{fu}, includes the useful heat released from fuel and the heat introduced into the furnace by preheated and leakage air,

$$Q_{fu} = Q_{to} \frac{100 - q_3 - q_4 - q_6}{100 - q_4} + Q_{ai} \text{ kJ/kg fuel} \tag{6-20}$$

The heat brought into the furnace by preheated and cold leakage air, Q_{ai}, is found from

$$Q_{ai} = (\alpha_{fu} - \Delta\alpha_{fu} - \Delta\alpha_{pul})I_{ha} + (\Delta\alpha_{fu} + \Delta\alpha_{pul})I_{ca} \text{ kJ/kg fuel} \tag{6-21}$$

where α_{fu} is the excess air ratio at the furnace outlet; $\Delta\alpha_{fu}, \Delta\alpha_{pul}$ are leakage air coefficients of the furnace and pulverization system, respectively; and I_{ha}, I_{ca} are enthalpies of preheated and cold air, respectively, kJ/kg fuel.

c) Radiative Heat Transfer Equation

For radiative heat transfer the furnace and the flame are considered two parallel planes with infinite areas. From the basic equation of radiative heat transfer the total net radiative heat flux, Q_r, is

$$Q_r = a_s \sigma F \left(T_{fl}^4 - T_{wa}^4\right) \text{ kW} \tag{6-22}$$

where a_s is the emissivity of the furnace wall-flame parallel planes system. It is given by

$$a_s = \cfrac{1}{\cfrac{1}{a_{fl}} + \cfrac{1}{a_{wa}} - 1} = \cfrac{a_{fl} a_{wa}}{1 - (1 - a_{fl})(1 - a_{wa})}$$

where T_{fl}, T_{wa} = flame and furnace wall temperature, respectively, K
σ = Stefan-Boltzman constant, 5.670×10^{-8} W/m^{2}K^4
a_{fl}, a_{wa} = emissivity of flame and furnace wall, respectively

To calculate the radiative heat transfer from the flame to the heating surfaces using Eq. (6-22) we need to know the flame emissivity, a_{fl}; flame temperature, T_{fl}; the emissivity, a_{wa}, and the water wall temperature, T_{wa}. All of these are not readily available. So a simplified approach as given below is used.

To express the emitted radiation flux, J_1, of the flame we use an overall furnace emissivity, a_{fu}, as follows:

$$J_1 = a_{fu} \sigma T_{fl}^4 \text{ kW/m}^2 \tag{6-23}$$

where T_{fl} is average temperature of the flame inside the furnace, K. Overall furnace emissivity, α_{fu}, is different from the flame emissivity, α_{fl}.

The water wall is not a pure black body. It is often fouled. So the heat absorbed by the water wall is only a part of the incident heat. A coefficient, Ψ, is used to define the fraction of the outgoing radiation of the flame absorbed by the heating surfaces of the water wall. It is known as the *thermal efficiency factor*. The heat absorbed q is

$$q = \psi J_1 = \psi a_{fu} \sigma T_{fl}^4 \text{ kW/m}^2 \tag{6-24}$$

Total heat absorbed by the entire water wall surface area, F, is therefore given by

$$Q_r = \psi F a_{fu} \sigma T_{fl}^4 \text{ kW} \tag{6-25}$$

6-3-2 Alternative Approach of Nondimensional Temperature

The water wall is often fouled. So using the thermal efficiency factor, ψ, we can write the heat balance where radiative heat exchange with the wall is equal to the loss in enthalpy of the flue gas. From Eq. (6-22) and (6-25) we get

$$Q_r = \phi B \overline{VC_p}(T_{th} - T_{ou}) = a_s \psi \sigma F \left(T_{fl}^4 - T_{fw}^4\right) \tag{6-26}$$

Dividing both sides by T_{th} and using $\theta_{ou} = T_{ou}/T_{th}$; $\theta_{fl} = T_{fl}/T_{th}$; and $\theta_{fw} = T_{fw}/T_{th}$, we get

$$1 - \theta_{ou} = \frac{a_s C}{B_0}\left(\theta_{fl}^4 - \theta_{fw}^4\right) \tag{6-27}$$

where C and n are empirical constants. The Boltzman number, B_0, is defined as

$$B_0 = \frac{\phi \cdot B \cdot \overline{VC_p}}{\sigma \psi F T_{th}^3} \tag{6-28}$$

The average temperature of the flame, T_{fl}, will lie somewhere between the theoretical (adiabatic) flame temperature, T_{th}, and the furnace exit temperature, T_{ou}. But its precise value is difficult to calculate. However, it is a common knowledge (experimental observations and theoretical reasoning) that the furnace exit temperature (θ_{ou} or T_{ou}) increases when the furnace temperature (θ_{fl} or T_{fl}) increases. So, an empirical approximation based on experimental data is used to relate T_{fl}, T_{ou}, and T_{th}:

$$\theta_{fl} \propto \theta_{ou}^n \tag{6-29}$$

Furthermore the wall temperature is very small compared to the furnace temperature θ_w. So $\theta_{fw}^4 \ll \theta_{fl}^4$. Thus, Eq. (6-27) can be simplified as

$$\frac{a_s C}{B_0}\theta_{ou}^{4n} + \theta_{ou} - 1 = 0 \tag{6-30}$$

From Eq. (6-30) and (6-22), therefore, the dimensionless temperature of the flue gas at the furnace exit, θ_{ou}, can be found as a function of, C, n, furnace emissivity, and the Boltzman number.

$$\theta_{ou} = f(B_0, a_{fu}, C', n) \tag{6-31}$$

The empirical coefficients C' and n depend on the properties of fuel fired, the combustion conditions, the furnace design, as well as the heat absorbing surfaces. These can be obtained from experimental data (Gurvich and Blokh, 1956) as follows.

$$\frac{\theta_{ou}}{1 - \theta_{ou}} = \frac{1}{M}\left(\frac{B_0}{a_{fu}}\right)^{0.6}$$

or

$$\theta_{ou} = \frac{T_{ou}}{T_{th}} = \frac{B_0^{0.6}}{M \cdot a_{fu}^{0.6} + B_0^{0.6}} \tag{6-32}$$

where M is the *temperature field coefficient*, discussed later in Section 6-6.

To calculate the flue gas temperature at the furnace exit one can write Eq. (6-32) as

$$T_{ou} = \frac{T_{th}}{M\left(\dfrac{\sigma \cdot a_{fu} \cdot \psi F T_{th}^3}{\phi \cdot BVC_p}\right)^{0.6} + 1} \text{ K} \tag{6-33}$$

One could rearrange Eq. (6-33) using Eq. (6-17) in the following form to find the surface area required, F, to attain a given furnace exit temperature, T_{ou}

$$F = \frac{\phi \cdot B\overline{VC_p}}{\sigma \cdot a_{fu}\psi T_{th}^3}\left[\frac{1}{M}\left(\frac{T_{th}}{T_{ou}} - 1\right)\right]^{\frac{1}{0.6}} \qquad (6\text{-}34)$$

Substituting values of VC_p from Eq. (6-19) in above, we get

$$F = \frac{B \cdot q'}{\sigma \cdot a_{fu}\psi M \cdot T_{th}^3 T_{ou}}\left[\frac{1}{M}\left(\frac{T_{th}}{T_{ou}} - 1\right)\right]^{0.66} \text{m}^2 \qquad (6\text{-}35)$$

where q' is the heat absorbed by heating surfaces in the furnace per unit mass of fuel burnt,

$$q' = \phi(Q_{fu} - I_{ou}) \text{ kJ/kg fuel} \qquad (6\text{-}36)$$

6-3-3 Calculation Procedure for Heat Transfer in a Furnace

There are two types of calculations:

- Design calculations
- Performance calculations

In design calculations the heat transfer surface area is computed for a given furnace exit temperature, θ_{ou}, while in performance calculations the furnace exit temperature, θ_{ou}, is computed for a given furnace geometry.

Many calculating parameters depend on the size and design of the furnace. This makes the design calculation more difficult. The latter type, where the furnace exit temperature is computed for a given furnace design and size, is relatively easy as the calculation parameters are known.

For design calculations first, all physical dimensions of furnace, including the arrangement of the water wall tubes and the volume of the furnace and the area of the water wall tubes, are assumed. Then the temperature, θ_{ou}, at the furnace exit is computed using Eq. (6-32) or (6-33). If the difference between the computed value of θ_{ou} and its desired value is unacceptably large, the designer would adjust the size of the furnace surfaces until the value of θ_{ou} meets the design requirement.

A typical boiler design calculation starts with the stoichiometric calculations for the fuel fired followed by the heat balance calculations (see Chapter 3). From the stoichiometric and heat balance calculations, the rated fuel consumption B, the heat retention coefficient ϕ, and the average specific heat $\overline{VC_p}$ of the flue gas can be determined. To use Eq. (6-32) or (6-32) to calculate the flue gas temperature θ_{ou} at the furnace outlet, the surface area F of the water wall tubes, the coefficient ψ of thermal efficiency, and the coefficient M and the emissivity a_{fu} of the furnace have to be determined or solved in advance. Methods for their calculation will be discussed in the following paragraphs. When the flue gas temperature θ_{ou} at the furnace exit is obtained, we can work out the radiative heat of the furnace by Eq. (6-36).

6-4 Water Wall Arrangement

The reliability of circulation of steam–water mixture in the water wall evaporative tube circuit should be considered in designing the water wall circuit. For a good design the water wall tubes are divided into several groups. Tubes in each group would have similar geometry and similar heating conditions (heat flux and temperatures, etc.).

The ratio of flow areas of the downcomer and riser tubes is an important parameter. It determines the flow resistance to water. This ratio is kept within a recommended range to reduce the flow resistance (Table 6-7).

When the boiler capacity increases, more steam has to be produced in each water wall tube so that the flow area of water wall tubes per ton of steam generated reduces. This parameter cannot be increased infinitely, as that would affect the reliability of natural circulation. So the steam generation per unit cross section area of riser tube is kept within the range listed in Table 6-8.

Boilers, with design pressures above 9.8 MPa, generally use a distributed downcomer system. In this arrangement two or more downcomer tubes in one water circulation loop are used to ensure good distribution of water in the riser tubes. For boilers with pressures less than 9.8 MPa one common downcomer tube system is generally used. The water velocity in the downcomer is chosen with care, as very high water velocity in the downcomer would increase the resistance. This may impede the natural circulation as well as make it difficult to separate steam and water in the drum. Similarly, too low a velocity will increase the diameter of the downcomer unacceptably, increasing the cost. Some recommended values of inlet water velocities in downcomer and riser tubes are given in Table 6-9.

For controlled circulation or assisted circulation boilers (see Chapter 12) it is necessary to install throttling orifices at the entrance of riser tubes. The amount of throttling would depend on the flow distribution among tubes before throttling. The riser tubes are usually divided into several groups to reduce different heat absorption among them. Controlled circulation design is also suitable for subcritical

TABLE 6-7. Flow area ratio between downcomer and riser tubes.

Drum pressure (MPa)		4–6	10–12	14–16	17–19
Area ratio	Coal firing	0.2–0.3	0.35–0.45	0.5–0.6	0.6–0.7
Downcomer	Oil firing		0.3–0.4	0.4–0.5	0.5–0.6
Riser					

TABLE 6-8. Steam production rate per unit cross section area of riser tube (t/h · m^2).

Drum pressure (MPa)		4–6		10–12	14–16	17–19
Steaming capacity (t/h)		≤75	≥120	160–420	400–670	≥850
Height of water wall tubes (m)		10–12	12–24	20–40	25–45	30–55
Steam generation	Coal fired boiler	60–120	120–200	250–400	420–550	650–800
area of riser tube	Oil fired boiler	75–150	150–250	320–480	520–680	750–900
[t/(h · m^2)]						

TABLE 6-9. Recommended values of water velocities at the inlet to the downcomer and riser tubes in m/s.

Drum pressure (MPa)		4–6	10–12	14–16	17–19
Velocity at inlet	Riser tubes entering drum	0.5–1.0	1.0–1.5	1.0–1.5	1.5–2.5
Of riser tubes (m/s)	Riser tubes entering upper header	0.4–0.8	0.7–1.2	1.0–1.5	1.5–2.5
	Waterwall tubes heated in both sides		1.0–1.5	1.5–2.0	2.5–3.5
Inlet velocity in downcomer		≤3	≤3.5	≤3.5	≤4

TABLE 6-10. Main design parameters for controlled circulation boiler.

Steam per unit cross section area of riser tube [ton/(h m^2)]	Circulation water velocity (m/s)	Inlet velocity in downcomer (m/s)	Water velocity in orifice (m/s)
1400	0.5	2.5	6–9

TABLE 6-11. Flow rates in lower radiative region of a once-through boiler (kg/m^{2}s).

Boiler pressure range	Horizontal or slightly inclined tube panel	Vertical riser tube panel		Spiral riser (Meander system) (Fig. 13-7)
		Single-pass	Multipass	
Subcritical	1200–3000	1600–2700	1200–2000	1500–2500
Supercritical	1500–3000	2100–2700	1600–2000	2000–3000

pressure boilers. Some important design parameters for controlled circulation boilers are given in Table 6-10.

For once-through boilers (see Chapter 7) the choice of flow rate of fluid in tubes depends on the hydrodynamic stability of flow, allowable tube metal temperature, and the flow resistance. When the flow rate increases, the hydrodynamic stability of flow improves and the tube metal temperature decreases, but the flow resistance increases. The heat flux on the tubes varies along the height of the furnace. So the furnace is generally divided into three regions—upper radiative region, middle radiative region, and lower radiative region. The intensity of heat flux is greatest in the lower radiative region. Therefore, the flow rate in the lower radiative region should be highest. Table 6-11 lists recommended values of flow rates in the lower radiative region. The flow rates in the middle and the upper regions should be smaller than that in Table 6-11.

6-5 Fouling and Thermal Efficiency Factors for Water Wall Tubes

Radiative heat exchange takes place between the high-temperature flame and the cooler heating surfaces of the water wall. The heat transfer coefficient inside the tube carrying water is an order of magnitude higher than that outside the tube.

Therefore, when the tube's external surfaces are clean, its surface temperature is close to that of the working fluid, and the entire radiation falling on it is absorbed by the fluid. In practice, the external surface of the water wall is often covered (fouled) by ash deposits. Furthermore, the water tube surface is not parallel to the flame. So a number of coefficients are used to find the actual radiative heat absorption.

6-5-1 Angular Coefficient

The angular coefficient of water wall, x, is a geometric factor. It is the ratio of heat absorbed by water wall tubes to that which can be absorbed by a continuous flat plane with emissivity and temperature equal to those of the tubes. For a membrane water wall the angular coefficient, x, is 1.0; for others, it is determined from Figure 6-2.

6-5-2 Fouling Factor

The thermal resistance of the ash is high. So the temperature of the external surface of ash layer deposits is much higher than that of the tube. The flame, now, exchanges heat with the hotter external surface of the ash layer on the tubes instead of the cooler metal walls of the tube. This reduction in heat absorption is taken into account by multiplying the heat absorption calculated based on the water temperature by a correction factor ξ.

The fouling factor, ξ, depends on the properties of fuel, the combustion conditions, and the design of water wall tubes. Table 6-12 gives some recommended values.

The fouling factor, ξ, of a studded water wall of a liquid-bath furnace covering the fireproof paint is calculated by the following equation:

$$\xi = b'[0.53 - 0.00025\ T_m] \qquad (6-37)$$

where T_m is the melting temperature of ash, and b' is an empirical coefficient equal to 1.0 for a normal boiler furnace and 1.2 for constricted boilers. When platen heating surfaces are placed at the exit of furnace, the factor ξ should be multiplied by another coefficient, β, which accounts for the heat exchange between the platen heating surfaces and the furnace of the boiler. So the new fouling factor is

$$\xi_p = \beta \cdot \xi \qquad (6-38)$$

TABLE 6-12. Fouling factor for water wall tubes.

Type of water wall tubes	Type of fuel	ξ
	Gas, mixture of gas & heavy oil	0.65
	Heavy oil	0.55
Water wall and	Coal	0.35–0.45
Studded water wall	All fuels burning on grate	0.60
	All fuels	0.20
Water wall with refractory backing	All fuels	0.20

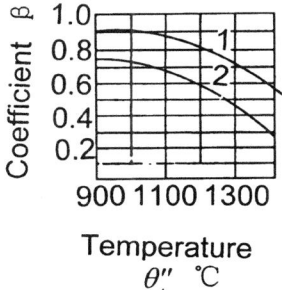

FIGURE 6-6. Values of the correction factor β to take account of radiation exchange between platen and furnace [From Congzhen, 1986]

The coefficient β depends on the properties of fuel and the temperature of the platen heating surfaces (Fig. 6-6).

6-5-3 Thermal Efficiency

If a clean water wall is a perfect black body all radiation falling on it will be absorbed. This amount is $x F a_{fu} \sigma_0 T_{fi}^4$, considering the angular coefficient of water wall tubes, x. In case of fouling, the tube is not a black body. So it absorbs only a fraction, ξ, of the falling radiation. Thus the actual heat absorption by the water wall tubes is equal to $\xi x F a_{fu} \sigma_0 T_{fi}^4$. So we define the coefficient of thermal efficiency ψ as the fraction of incident radiation absorbed by the tubes:

$$\psi = x \cdot \xi \qquad (6-39)$$

Table 6-12 gives values of the fouling factor, ξ. When the fouling factor ξ of all sections of the furnaces are not same and the angular coefficient, x, also varies one needs to use an average value of the thermal efficiency coefficient $\bar{\psi}$.

$$\bar{\psi} = \frac{\sum \psi_i F_i}{F} \qquad (6-40)$$

where $\bar{\psi}$ is the average coefficient of thermal efficiency, and ψ_i is the coefficient of individual parts of the water walls. F and F_i are the total area and the area of individual sections of the water wall tubes, respectively. The section of the furnace without water wall tubes will have a coefficient ψ_i equal to zero.

6-6 Temperature Field Coefficient, *M*

The temperature field coefficient M of Eqs. (6-32–6-35) accounts for the temperature distribution in the furnace. It is a function of the relative level of the burners and on the type of fuels burnt. This relationship is expressed as (Lin, 1991)

$$M = A' - B'(X_r + \Delta X) \qquad (6-41)$$

where X_r is the relative position of the highest temperature zone in the furnace. It is given by h_r/h_{fu}; and ΔX is a correction factor. The correction factor ΔX accounts for the actual position of the flame core. For horizontally arranged burners $\Delta X = 0$; for tilting burners directed upward or downward at $20°$, $\Delta X = \pm 0.1$; and for burners arranged on front walls or on opposing walls, $\Delta X = 0.05-0.1$.

Here, h_r is height of the burner axis above the bottom of the furnace or the midplane of the cold-ash hopper; and h_{fu} is the height of the midpoint of the flue gas exit above the bottom of the furnace or the midplane of the furnace chamber hopper. In case of several rows of burners, the average burner height, h_r, is found as follows:

$$h_r = \frac{\sum n_i B_i h_{ri}}{\sum n_i B_i} \text{ m} \tag{6-42}$$

where n_i = number of burners in row i
         B_i = fuel consumption of burners in row i, kg/s
         h_{ri} = height of burners in row i, m

The coefficients A' and B' in Eq. (6-41) depend on the properties of fuel and the type of furnace. For traveling grate or stoker type furnaces $A' = 0.59$ and $B' = 0.5$, and $X_r + \Delta X = 0.14$. For pulverized coal, oil, or gas fired furnaces A' and B' can be obtained from Table 6-13. Alternately, a average value of 0.445 can be used (Gurvich & Blokh, 1956).

6-7 Furnace Emissivity

The furnace emissivity, a_{fu}, is different from the emissivity of parallel plane system, α_s, used in Eq. (6-22). Net radiation leaving the furnace wall, J_2, and that from the flame, J_1, are given by

$$J_2 = a_{wa}\sigma T_{wa}^4 + (1 - a_{wa}) \tag{6-43}$$

$$J_1 = a_{fl}\sigma T_{fl}^4 + (1 - a_{fl}) J_2 \tag{6-44}$$

The actual furnace emissivity, a_{fu}, is determined by the flame emissivity and the coefficient of thermal efficiency of a water wall. $J_1 = a_{fu} \cdot \sigma T_{fl}^4$. From above one

TABLE 6-13. Coefficients A' and B' in Eq. (6-41).

Fuel	Open-type boiler (Fig. 1-5)		Constricted-type boiler	
	A'	B'	A'	B'
Gas, fuel oil	0.54	0.2	0.48	0
High volatile or reactive coal	0.59	0.5	0.48	0
Low reactive fuels (anthracite, high ash coal)	0.56	0.5	0.46	0

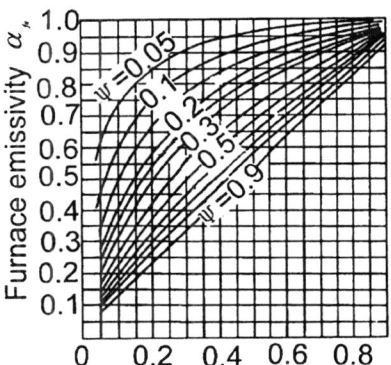

Flame emissivity $\alpha_{\prime}$

FIGURE 6-7. Overall emissivity of the furnace against flame emissivity for different values of ψ

could express J_1 as

$$J_1 = \frac{a_{fl}\sigma T_{fl}^4 + (1 - a_{fl})a_{wa}\sigma T_{wa}^4}{1 - (1 - a_{fl})(1 - a_{wa})} \qquad (6\text{-}45)$$

According to the definition, ψ is given by

$$\psi = \frac{J_1 - J_2}{J_1} = \frac{a_{fu}\sigma T_{fl}^4 - \left[a_{wa}\sigma T_{wa}^4 + (1 - a_{wa})a_{fu}\sigma T_{fl}^4\right]}{a_{fu}\sigma T_{fl}^4} \qquad (6\text{-}46)$$

Rearranging Eq. (6-46), we obtain

$$\left(\frac{T_{wa}}{T_{fl}}\right)^4 = a_{fu}\left(1 - \frac{\psi}{a_{wa}}\right) \qquad (6\text{-}47)$$

Substituting Eq. (6-47) into Eq. (6-46), the following equation is obtained:

$$a_{fu} = \frac{a_{fl}}{a_{fl} + (1 - a_{fl})\psi} \qquad (6\text{-}48)$$

We can also obtain the value of a_{fu} from Figure 6-7. For a stoker fired boiler ψ is 0.6 for all fuels (Blokh, 1987). However, the radiation of fuel layer should be taken into account in this case.

6-8 Distribution of Heat Load in Furnace

We can determine the enthalpy of flue gas at the furnace exit, I_{ou}, by using information given in Sections 6-2 to 6-7. From Eq. (6-17) we can calculate the total heat absorption by water walls $Q = \phi B_j(Q_{fu} - I_{fu})$, kJ/kg fuel burnt. Here, ϕ is the coefficient of heat retention. The average heat absorption per unit area of the water wall surface (heat load or heat flux) is

$$q = \frac{Q}{F} \text{ kW/m}^2 \tag{6-49}$$

However, the temperature and the emissivity are not uniformly distributed in the furnace. Therefore, the heat flux is also different along the furnace width, depth, and height. So it is difficult to calculate the heat load exactly. To determine the heat absorption in a certain zone in the furnace we take the following simplified approach of using a coefficient of nonuniformity, η_k

1. Chose a value of η_k for the specific zone, which is known from previous experimental data and experience.
2. Calculate the local heat flux q_i at that zone using the following equation:

$$q_i = \eta_i q \text{ kW/m}^2 \tag{6-50}$$

3. Calculate the heat absorption Q_i in that zone

$$Q_i = q_i F_i \text{ kW} \tag{6-51}$$

Using Eq. (6-53),

$$Q_i = \eta_i F_i \frac{Q}{F} \text{ kJ/kg} \tag{6-52}$$

The heat loads of individual furnace walls are different from each other. When burners are arranged on the front wall the value of η_i of the rear wall is equal to 1.1, while that of the other two are equal to each other. This value is determined by the heat balance. The heat flux varies across the furnace width with higher values at the center and lower on the sides. Hence the heat load is usually lower at the corners. The heat flux also varies along the furnace height. At the burner zone the heat flux is highest. And it is lower at the top and the bottom of the furnace. Figure 6-8 shows the heat flux distribution along the furnace height for a gas/oil fired boiler and a dry bottom PC boiler.

The radiant heat transfer from the furnace to its exit (Q_{fc}) is usually calculated in the boiler thermal calculations. This part of the heat is absorbed by the heating surfaces at the furnace exit and the convection heating surface. Q_{fc} is determined by

$$Q_{fc} = \eta_h F_c \frac{Q_{fu}}{F} \text{ kJ/kg} \tag{6-53}$$

where $Q_{fu} =$ total heat absorption in the furnace
$F =$ total area of the water wall surface
$F_c =$ area of the furnace exit

The coefficient of heat flux uniformity, η_h, at the furnace exit is determined by Figure 6-8. When there is platen heating surface at the furnace outlet, the coefficient of heat transfer, β, between the platen and the furance should be taken into account.

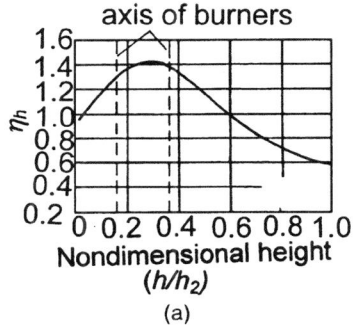

(a)

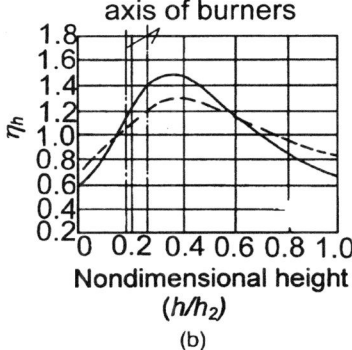

(b)

FIGURE 6-8. Distribution of heat flux along the height of the furnace in a (a) gas or oil fired furnace; (b) a dry-bottom bituminous or anthracite (------) or lignite (———) fired boiler

EXAMPLE: Heat transfer calculation of a 400 t/h boiler

Design a 400t/h reheat-type pulverized coal boiler using the following specifications:

1. Boiler evaporation capacity D_{sup}: 420 t/h
2. Reheat steam flow D_{re}: 350 t/h
3. Water-feed temperature T_{wa}: 235°C
4. Water-feed pressure p_{wa}: 15.6 MPa (gauge pressure)
5. Superheat steam temperature T_{sup}: 540°C
6. Superheat steam pressure p_{sup}: 13.7 MPa (gauge pressure)
7. Reheat inlet steam temperature T'_{re}: 330°C
8. Reheat outlet steam temperature T''_{re}: 540°C
9. Reheat inlet steam pressure P'_{re}: 2.5 MPa (gauge pressure)
10. Reheat outlet steam pressure P''_{re}: 2.3 MPa (gauge pressure)
11. Ambient temperature T_{ca}: 20°C

12. Fuel characteristics:
    (1) Fuel type: bituminous coal
    (2) Ultimate analysis, as received basis (%): $C = 47.9$; $O = 5.15$; $S = 0.45$;
        $H = 3.04$; $N = 0.86$; $M = 7.86$; $A = 34.74$
    (3) Dry and ash-free basis volatile matter: $VM = 24.8\%$
    (4) Lower heat value: $LHV = 18289$ kJ/kg
    (5) Ash melting point: $T_1, T_2, T_3 > 1500°C$
13. Pulverization system: indirect pulverization, cylinder ball mill
14. Steam drum operating pressure: 15.2 MPa (gauge pressure)

The general layout of the boiler, including its arrangement, heating surface
arrangement, superheater, and reheater flow circuit, are known. Furnace configuration and dimensions are also known.

SOLUTION

The combustion calculations, mass balance of air, and enthalpy-temperature
calculations of air and flue gas were done elsewhere. So we will use those results
directly. Now, only the furnace heat transfer calculation needs to be done. The
following table gives furnace geometry and operating parameters. The calculation
then follows.

A.

Step	Parameter	Symbol	Unit	Equation	Value
1	Area of the side wall	F_1			22.0
		F_2	m^2	given	322.8
		F_3			31.6
2	Total area of side wall	F_s	m^2	$(F_1 + F_2 + F_3)$	376.4
3	Area of the front wall	F_f	m^2	given	335.8
4	Area of the back wall	F_b	m^2	given	255.6
5	Area of the gas passage at the exit of the furnace	F_{out}	m^2	given	60.9
6	Area of the burners	F_h	m^2	given	28.8
7	Angular coefficient of the water wall	x		Membrane wall [Section 6-5-1]	1.0
8	Angular coefficient of the furnace outlet	x_{out}		Fully covered by tubes	1.0
9	Average angular coefficient of the total furnace	$\bar{x}$		$\bar{x} = \frac{F_s x + F_f x + F_b x + F_{out} x_{out}}{F_s + F_x + F_b + F_{out}}$	1.0
10	Total area of water wall in the open region of the furnace	F_{tot}	m^2	$F_{tot} = F_s + F_f + F_b + F_{out} - F_b$	999.9
11	Volume of the furnace	V_{tot}	m^3	Given	1806.7
12	Furnace cross section	F_{sec}	m^2	Given	80.4
13	Effective radiant layer thickness of the furnace	S		$3.6 V_{tot} / F_{tot}$	6.50

(Contd.)

A. (*Continued*)

Step	Parameter	Symbol	Unit	Equation	Value
14	Height of the center line of the burner group	h_r	m	Given	6.80
15	Height of the furnace	h_{fu}	m	Given	22.0
16	Relative height of the burners	h_r / H_{fu}			0.31
17	Relative height of the flame core zone	x_r		$h_r / h_{fu} + \Delta x$ as $\Delta x = 0$	0.31

Furnace design data have been obtained. Now a furnace heat transfer calculation can be done according to following steps. The stoichiometry, heat balance, and enthalpy calculations are not shown here for brevity.

B.

Step	Parameter	Symbol	Unit	Equation	Value
1	Temperature of the hot air	T_{ha}	°C	Given	320
2	Theoretical enthalpy of hot air	I_{ha}	kJ/kg	From enthalpy calculation	2114
3	Leakage coefficient of the furnace	$\Delta\alpha_{fu}$		Given	0.05
4	Leakage coefficient of the pulverization system	$\Delta\alpha_{pul}$		Given	0.06
5	Temperature of the cold air	T_{ca}	°C	Given	20
6	Theoretical enthalpy of the cold air	I_{ca}^o	kJ/kg	From enthalpy calculation	129.6
7	Excess air coefficient at the exit of the air preheater	β_{ap}''		$\alpha_{fu}'' - (\Delta\alpha_{fu} + \Delta\alpha_{pul})$ $\alpha_{fu}'' = 1.2$ given	1.09
8	Heat carried into the furnace by the air	Q_{ai}	kJ/kg	$\beta_{ap}' I_{ha} + (\Delta\alpha_{fu} + \Delta\alpha_{pul}) I_{ca}$	2318.5
9	Heat carried into the furnace by per unit kg fuel	Q_{fu}	kJ/kg	Eq. (6-20), where Q_{to} is calculated and q_3, q_4, q_6 are given	20607
10	Theoretical combustion temperature	T_{th}	K	From enthalpy calculation	2184
11	Gas temperature at the exit of the furnace	T_{ou}	K	Assumed	1493
12	Gas enthalpy at the exit of the furnace	I_{ou}	kJ/kg	From enthalpy calculation	12485
13	Average specific heat of the combustions products	$\bar{V}\bar{C}_p$	kJ/(kg°C)	Eq. (6-19)	11.75

(*Contd.*)

B. (*Continued*)

Step	Parameter	Symbol	Unit	Equation	Value
14	Volume fraction of the water vapor	r_{H_2O}		From stoichiometric calculation	0.0842
15	Volume fraction of the tri-atomic gas	r_{RO_2}		From stoichiometric calculation	0.1426
16	Volume fraction of the tri-atomic gas	r	$r = r_{H_2O} + r_{RO_2}$		0.2268
17	Furnace pressure	P	MPa	Given	0.1
18	Radiant absorption coefficient of the tri-atomic gas	k_y	1/(m · Mpa)	Eq. (6-11)	2.926
19	Radiant absorption coefficient of the ash particle	k_h	1/(m · Mpa)	Eq. (6-12)	77.4
20	Nondimensional coefficient	c_1		For Bituminous coal (6-2-3)	0.5
21	Nondimensional coefficient	c_2		For PC boiler (6-2-3)	0.1
22	Concentration of ash particles	μ_h	kg/kg	Stoichiometric calculations	0.0369
23	Radiant absorption coefficient of flame	k	1/(mMpa)	Eq. (6-10)	4.02
24	Exponent of Eq. (6-9)	$k \cdot P \cdot S$			2.61
25	Flame emissivity	a_{fl}		Eq. (6-9)	0.926
26	Total area of the furnace wall	F_{tot}	m²	From step a.10	999.9
27	Fouling coefficient of water wall	ξ		Coal fired water wall (Table 6-12)	0.45
28	Thermal efficiency coefficient of the waterwall	ψ		Eq. (6-38)	0.45
29	Furnace emissivity	a_{fu}		Eq. (6-47)	0.965
30	Coefficient M	M		Section 6-6	0.405
31	Coefficient of heat retention	ϕ		Heat balance calculation	0.996
32	Fuel consumption	B	kg/s	Heat balance calculation	19.80
33	Flue gas temperature at the exit of the furnace	T'_{ou}	K	Eq. (6-34)	1525
34	Verification of the flue gas temperature at the exit of the furnace			$[T_{ou}(assume) - T'_{ou} (calculation)]$ calculation completed	32 < 100
35	Gas enthalpy at the exit of the furnace	I_{ou}	kJ/kg	Enthalpy temperature calculation	12,805
36	Heat absorption in the furnace	q	kJ/kg	$\varphi(Q_{fu} - I_{ou})$	7771

(*Contd.*)

B. (*Continued*)

Step	Parameter	Symbol	Unit	Equation	Value
37	Heat release per unit volume in the furnace	q_V	MW/m^3	$\frac{B\,LHV}{V_{tot}}$	0.20
38	Heat release per unit cross area of the furnace	q_F	MW/m^2	$\frac{B\,LHV}{F_{sec}}$	0.45
39	Average radiant heat per unit furnace area	q_{av}	MW/m^2	$\frac{B\,Q_{fu}}{F_{tot}}$	0.15

Nomenclature

$1 - e^{-kS}$	absorptivity of gas volume
A'	coefficient (Eq. 6-41)
a	width of the furnace, m
a_{fu}	overall emissivity of furnace
a_s	emissivity of the parallel planes system
α_{wa}	emissivity of furnace wall
a_{fl}	flame emissivity
a_{lu}	luminous flame emissivity
a_{tr}	flame emissivity of tri-atomic gases
B	designed fuel consumption rate, kg/s
B_i	fuel consumption of burners in row i, kg/s
B_0	Boltzman number (Eq. 6-28)
B'	coefficient in Eq. (6-41)
b	depth of the furnace, m
b'	empirical coefficient (Eq. 6-37)
C'	empirical coefficient (Eq. 6-31)
C	carbon content of as received bases
c_1	coefficient determined by the type of fuel (Eq. 6-10)
c_2	coefficient determined by the firing method (Eq. 6-10)
C_p	specific heat of flue gas, kJ/Nm^{3}K
D	steam capacity, t/h
d_h	diameter of ash particles, μm
d_r	maximal nozzle diameter of swirl burner, m
e^{-kS}	transmissivity of absorbing gases
E_o	black body radiation power at same temperature as flame, kW/m^2
E_s	amount of radiation absorbed by the medium, kW/m^2
F	effective surface area of water wall (Eq. 6-22), m^2
F_{grate}	cross section or grate area of the furnace (Eq. 6-4), m^2
F	enclosing surface area (Eq. 6-8), m^2
F_c	area of the furnace outlet, m^2
F_i	area of the individual sections of the water wall, m^2

H	hydrogen content of as received bases
H_b	height of the burner region (Eq. 6-5), m
H_{fu}	shortest furnace height defined as the shortest distance between furnace exit and the burners (Eq. 6-5), m
h_r	average burner height above the ash hopper, m
h_{fu}	height of the center of gas exit above the ash hopper, m
h_{ri}	height of burners in row i, m
I_{ca}	enthalpy of theoretical cold air, kJ/kg
I_{ha}	enthalpy of preheated air, kJ/kg
I_o	radiation intensity before gaseous absorption, kW/m^2
I_{ou}	enthalpy of flue gases leaving the furnace, kJ/kg
I_s	radiation intensity after absorption in the medium, kW/m^2
J_1	emitted radiation of the flame, kW/m^2
J_2	radiation leaving the furnace wall (Eq. 6-43), kW/m^2
k	coefficient of radiation absorption, $1/m \cdot MPa$
k_c	coefficient of radiant absorption owing to soot particles, $1/m \cdot MPa$
k_h	coefficient of radiant absorption owing to ash particles, $1/m \cdot MPa$
k_y	coefficient of radiant absorption owing to tri-atomic gases, $1/m \cdot MPa$
LHV	lower heating value of fuel, kJ/kg
l	length of water wall, m
m	luminance fraction of a flame
M	temperature field coefficient (Eq. 6-32)
n	empirical coefficient (Eq. 6-31)
n_i	number of burners in row i
P	pressure of gases in the furnace, MPa
p	pitch distance between two tubes on water wall, m
q	average heat absorption (Eq. 6-49), kW/m^2
q	absorbed heat, kW/m^2
q'	heat absorbed by heating surfaces per unit weight of fuel burnt, kJ/kg fuel
q_3	heat loss owing to chemical incompleteness of combustion, %
q_4	heat loss owing to physical incompleteness of combustion, %
q_5	heat loss owing to radiation and natural convection from boiler exterior, %
q_6	heat loss owing to slag removed from the surface, %
Q_{ai}	heat brought into the furnace by preheated and cold leakage air, kJ/kg fuel
q_b	heat release rate per unit water wall of the burner region, kW/m^3
q_F	grate heat release rate, kW/m^2
Q_{fc}	radiant heat transfer from the furnace to its exit, kJ/kg fuel
Q_{fu}	heat entering furnace through combustion and hot air, kJ/kg fuel
Q_r	radiative heat flux, kW
Q_{abs}	total heat absorbed in furnace, kW
Q_{to}	available heat, kJ/kg
q_v	volumetric heat release rate, kW/m^3

r	total volume concentration of tri-atomic gases
r_{H_2O}	volume concentration of water in flue gas
r_{R_2O}	volume concentration of other tri-atomic gases in flue gas
S	mean beam length or effective thickness of absorbing gas layer, m
S_{eff}	total effective radiative heat absorbing surface area, m^2
T_m	melting temperature of ash, K
T_{fl}	absolute temperature of flame, K
T_{fl}	average temperature of flame (Eq. 6-23), K
T_{ou}	gas temperature at the furnace exit (FEGT), K
T_{th}	adiabatic flame temperature (Eq. 6-28), K
T_{wa}	absolute temperature of furnace wall, K
V	furnace volume (Eq. 6-3), m^3
V	volume of the gas, m^3
VC_p	average specific heat, kJ/kg · K
x	angular coefficient (Eq. 6-39)
X_r	relative position of the highest temperature zone
Δx	correction factor

Greek Symbols

β	coefficient (Eq. 6-38)
α_{fu}	excess air coefficient at furnace exit
$\Delta\alpha_{fu}$	leakage air coefficient of the furnace
$\Delta\alpha''_{fu}$	excess air ratio at the furnace outlet
$\Delta\alpha_{pul}$	leakage air coefficient of the pulverization system
Φ	coefficient of heat retention (Eq. 6-17)
η	boiler or thermal efficiency
η_η	coefficient of heat flux uniformity (Eq. 6-53)
μ_h	concentration of ash particles in the furnace, kg/kg
θ_{fl}	nondimensional temperature (T_{fl}/T_{th})
θ_{fw}	nondimensional temperature (T_{fw}/T_{th})
θ_{ou}	nondimensional temperature (T_{ou}/T_{th})
σ	Stefan-Boltzman constant, $5.678 \times 10^{-8} W/m^2K^4$
ξ	fouling factor (Eq. 6-37)
ξ_p	modified fouling factor for platens at the furnace exit
ψ	thermal efficiency factor
ψ_i	coefficient of each part of the water walls
η_k	coefficient of nonuniformity

References

Basu, P., and Fraser, S., (1991) "Circulating Fluidized Bed Boilers." Butterworth Heinemann, Stoneham, Massachusetts.

Blokh, A.G. (1987) "Heat Transfer in Steam Boiler Furnaces." R.Viskanta, ed. Hemisphere Publishing Co., Washington, p.172.

Feng, J. and Shen, Y. (1992) "Boiler Principles and Calculation" (in Chinese). Science Press, Beijing.

Gurvich, A.M. and Blokh, A.G. (1956) Concerning temperature in furnace space. Energomashiostroeniya 6:11–15.

Holman, J.P. (1991) "Heat Transfer." McGraw Hill, New York, p. 426.

Lafanechere, L., Basu, P., and Jestin, L. (1995) Effects of fuel parameters on the size and configuration of circulating fluidized bed boilers Journal of the Institute of Energy 68:184–12.

Li, Y. (1990) "Studies on Thermal Calculations in Boiler Furnaces" (in Chinese). M. E. thesis, Shanghai Institute of Machinery.

Lin, Z.H. (1991) "Boilers, Evaporators, and Condensers." S. Kakac, ed. John Wiley & Sons, New York, pp. 363–469.

Pan Congzhen (1986) "Boiler Principles" (in Chinese). Hydraulic Power and Electric Power Press, Beijing, p. 173.

Reznikov, M.I. and Lipov, Y.M. (1985). "Steam-Boilers of Thermal Power Stations." Mir Publishers, Moscow, p. 248.

7
Convective Heating Surfaces

After leaving the furnace the flue gas enters the convective section of the boiler, where it cools further by transferring heat to water, steam, and in some cases to the combustion air. The principal mode of heat transfer in this section is forced convection. Hence this section is called the *convective section*. Also, since it is located at the back of the boiler, it is called *back-pass*. Here the gas enters at about the furnace exit temperature and leaves at slightly above the stack temperature. In most utility boilers, the heat exchangers located in the convective section include the superheater, reheater, economizer, and air preheater. These are generally arranged in series. In some designs, the convective section may be split in two parallel paths with parts of the superheater and reheater located in each section. Here, the gas flow rate through individual sections can be controlled by means of dampers to adjust the superheat or reheat steam temperatures. This chapter will discuss design methods and operating issues related to the convective section.

7-1 Design of Superheater and Reheater

The superheater and reheater are two important heat exchangers in the convective section. Their designs are discussed below.

7-1-1 Arrangement of Surfaces

The function of a superheater is to heat the high-pressure steam from its saturation temperature to a higher specified temperature. The purpose of the reheater is to heat the partially expanded low-pressure steam to the designed high temperature. The reheated steam, thereafter, returns to the medium-pressure sections of the turbine for further expansion work. The pressure of the reheated steam is only about 20% that of the superheated steam, but its exit temperature is close to the final temperature of the superheated steam. The steam flow through the reheater is about 90% of that of the main steam flow rate. The flow resistance in the superheater may be 10% of the superheater exit pressure, while that of the reheater does not exceed 10% of the entry pressure of the reheater.

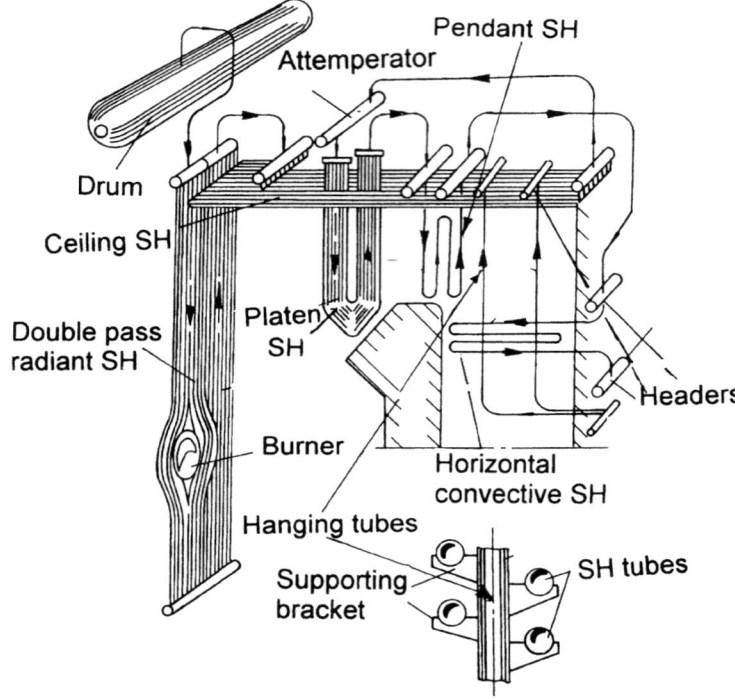

FIGURE 7-1. Flow arrangements of different types of superheaters

Operating and cost considerations require superheating and reheating to be carried out in several stages. Depending on how these stages receive the heat they are classified into three types: convective, radiant, and semi-radiant. Figure 7-1 shows a typical arrangement of superheater elements in a subcritical pressure boiler featuring all types of superheaters.

a) Convective Superheater

Convective superheaters are arranged in the back-pass or the convective section of the boiler. These are either vertical or horizontal types according to the tube orientation. The vertical one (Fig. 7-2) is always arranged in the horizontal crossover duct joining the furnace and the back-pass. It is also called a pendant superheater. Mechanical design of this type of superheater is simple, and it is easily supported from the roof. Also, these superheaters suffer less from fouling. However, it is difficult to drain water from these superheaters after a boiler shutdown.

In the horizontal arrangement (Fig. 7-3) the superheater tubes run horizontally from a set of headers. It is relatively easy to drain condensed water from horizontal superheaters. In conventional-type (two pass Π type) boilers horizontal

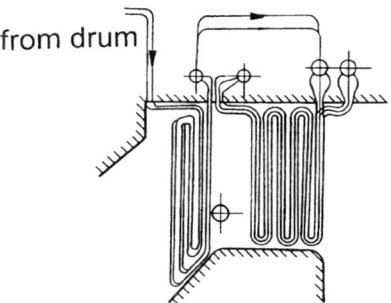

FIGURE 7-2. Pendant or vertical convective superheaters are located in the cross-over duct between the furnace and the back-pass

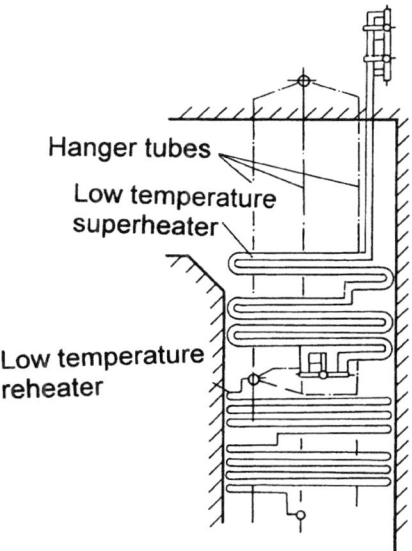

FIGURE 7-3. Horizontal superheater and reheaters are located in the back-pass of π-type boilers

superheaters are located in the back pass while in stack (or Tower)-type boilers, horizontal superheaters may be used in the furnace.

Convective superheaters are made of parallel rows of tubes (Fig. 7-3). These terminate in headers at two ends. These tubes are arranged in the "in-line" configuration. The outside diameter of the tubes is in the range of 32–51 mm, and the tube thickness is in the range of 3–7 mm. The transverse pitch (S_1/d) is between 2 and 3, but the longitudinal pitch (S_2/d) depends on the bend radius of the tubes.

TABLE 7-1. Typical flow rates of steam in superheater and reheater tubes.

Type of superheater	Pressure duty	Mass flow velocity (kg/m²s)
Convective superheater	Medium pressure	250–400
	High pressure	500–1000
Platen superheater		800–1000
Wall superheater		1000–1500
Reheater		250–400

For most common superheater tubes $S_2/d = 1.6$–2.5. In order to attain the desired steam velocity inside the tubes, the superheater tubes may be arranged in multiple parallel tubes. The desired steam velocity depends on the type of superheater and the operating range of pressure. Some recommended values of steam velocities in the superheater and reheater tubes are shown in the Table 7-1.

When the inlet gas temperature is close to 1000°C, the ash carried in the flue gas is still soft. So the front or leading rows of superheater tubes are widely spaced to avoid slagging (transverse spacing ($S_1/d \geq 4.5$) and longitudinal spacing ($S_1/d \geq 3.5$)).

b) Platen Superheater

The platen superheaters are flat panels of tubes located in the upper part of the furnace, (Fig. 7-1) where the gas temperature is high. They are, therefore exposed to very severe conditions. The tubes in the front rows of the platen tube especially receive very high radiation as well as a heavy dust burden. Inadequately cooled tubes may easily burn out. Some measures can be taken to avoid this potential hazard. For example, the outermost tubes of the platen can be made shorter than others so as to have steam mass flow rate higher than that in other tubes and therefore better cooling. In some cases; superior tube materials are used in the outer most tube.

Figure 7-4 shows some typical arrangements of platen superheaters. The outer diameter of platen superheater tubes is in the range of 32–42 mm. The platens are usually widely spaced ($S_1 = 500$–900 mm), but the tubes within the platen are spaced narrowly ($S_2/d = 1.1$). The number of parallel tubes in a platen is generally in the range of 15–35, depending on the design steam velocity.

c) Pendant Superheater

The pendant-type superheater or reheaters (Fig. 7-2) are arranged in the horizontal crossover duct connecting the furnace and the back-pass. These are also panels like the platens but more closely spread than platens. A pendant heater receives heat by convection and some radiation.

d) Reheater

The pressure drop inside the reheater tubes has an important adverse effect on the heat rate of the turbine. For example, the heat rate of a turbine can increase by 0.2–0.3% for each 0.098 MPa increase in the pressure drop in the reheater tube

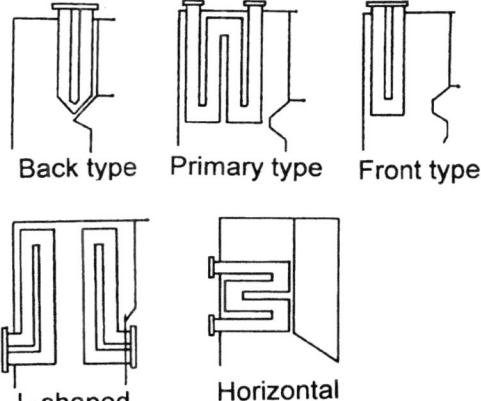

Back type   Primary type   Front type

L-shaped   Horizontal

FIGURE 7-4. Different arrangements of platen superheaters

system. So the pressure drop though the reheater tubes should be kept as low as possible. This requires the use of large-diameter (42–60 mm) tubes. As the heat transfer coefficient of low-pressure reheater steam is low, an overheating of the walls of the tubes can easily occur. So to maintain the temperature of the tube wall within safe operating limits one needs to either increase the steam velocity or reduce the local gas temperature. Higher steam velocity increases the pressure drop. So, the reheaters are located in relatively cold regions.

Vertical reheaters are located in the horizontal duct, while the horizontal reheater is arranged in the back-pass. These designs are similar to those of pendant superheaters (Figs. 7-2, 7-3). Typical mass flow velocities in reheaters are given in Table 7-1.

7-1-2 Configuration of Superheater and Reheater

The superheater and reheater banks are arranged with a view to the followings:

- Required steam parameters are met
- Steam temperature can be controlled easily
- There is no overheating of the tubes
- The design is economic

a) Superheater

The temperature of the superheated steam of a low-pressure boiler is not as high as it is in a high-pressure boiler. So here the superheaters are arranged in the convective section, where the gas temperature does not exceed 700–800°C. In a medium-pressure boiler, the steam temperature ranges from 450° to 480°C. The radiative heat released in the furnace is just sufficient for the evaporative heat requirement the water. So one could still use convective superheater. Here, the superheater is divided into low- and high-temperature sections along the direction of the steam flow. The low-temperature section is arranged behind the high-temperature section along the gas flow path. Counterflow arrangement of heat exchanger is adopted to avoid the need for expensive superheater materials. The high-temperature section arranged in a higher gas temperature region is divided into two stages. The first stage is arranged on both sides of the gas duct and a counter-current flow pattern is adopted. The second stage is arranged in the center of the gas duct and parallel flow pattern is adopted.

Superheaters are often divided into more than one stage. However, the enthalpy rise of steam in each section should not exceed 250–420 kJ/kg. For medium pressure boilers, the enthalpy rise should not exceed 280 kJ/kg. For high-pressure boilers, the value should not exceed 170 kJ/kg. The rise in enthalpy in the last section should facilitate a good control characteristic of the steam temperature.

A header located between two sections facilitates better mixing of the steam exiting from different parallel tubes. In order to reduce the nonuniform heat absorption special header arrangements are used (Fig. 7-5).

b) Reheater

Reheaters may be single stage or double stage. In a single-stage arrangement, the flow resistance is less and the system is simple, but the heating could be nonuniform. No steam flows through the reheater during the period of boiler startup. This creates

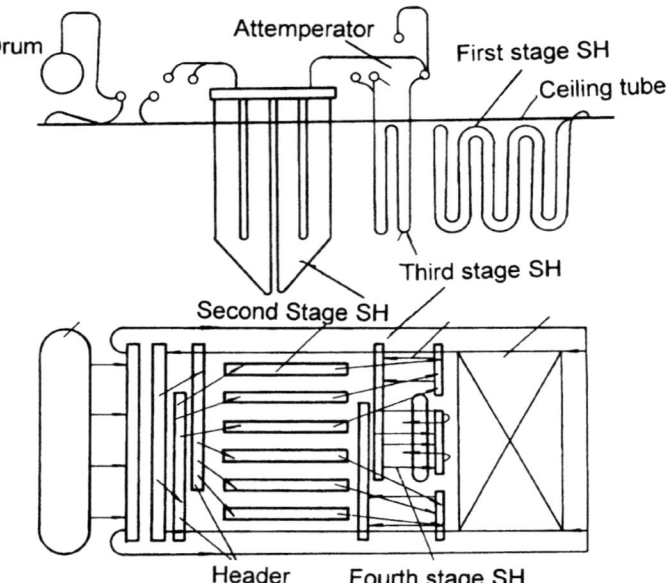

FIGURE 7-5. Superheater layout for a large high-pressure boiler

a special problem. If the reheater is located in a gas duct where the gas temperature is below 800°C and appropriate steel is used, reheater can be allowed to be heated without steam flowing through it for a short period during the boiler start-up or trip. However, if the reheater is located in a higher gas temperature region, a more reliable bypass system is necessary.

7-2 Temperature Control in Superheater and Reheater

Control and adjustment are two similar concepts, but these imply two different functions here. Adjustment refers to physical changes in the boiler hardware to achieve a specified steam temperature or heat absorption. A modification of the tube length or insertion of an orifice inside reheater tubes are examples of adjustments. These changes cannot be made while the boiler is in operation. Control, on the other hand, can be effected by making changes in the operation of the boiler. Tilting the burner and injection of water into the steam are some of the means used to control the steam temperature.

Maintenance of the steam temperature at the desired level is vital for the efficient operation of a power plant. For example, above 124 bar pressure a 20°C change in steam temperature may bring about a 1% change in the heat rate (Stultz & Kitto, 1992, pp. 18–13). Thus, in a modern power boiler utmost effort is made to maintain the steam temperature within a tolerance level of ± 6°C.

The fuel firing rate can control the steam output and the steam pressure, but it cannot directly influence the steam temperature. A number of operating parameters and the basic design of the heating surface affect the steam temperature. The following section elaborates means for control and adjustment of superheater and reheater performance.

Control of Superheat and Reheat Temperatures

a) Steam Temperature Control

The temperature of the steam leaving the superheater or reheater depends on a number of operating factors like boiler load, excess air, feed water temperature, fuel characteristics, etc. Table 7-2 shows how by changing the operating parameters one can change the steam temperature.

Superheaters are of two types: radiant and convective. The radiant superheater is located inside the radiant furnace. So it absorbs majority of the heat through radiation from the furnace. The convective superheater, on the other hand, is located in the convective section downstream of the furnace. The heat transfer to this superheater is primarily by convection from the flue gas.

When the boiler load is increased by increasing the fuel firing, a larger mass of flue gas is produced, while the furnace temperature is nearly unchanged. As a result, the radiative heat transfer, which is a function of the temperature, does not change. However, the increased mass of flue gas increases the convective heat absorption in the rest of the boiler increasing the steam flow through the superheater. With

TABLE 7-2. Effect of some operating parameters on the temperature of steam leaving a convective-type superheater.

Changes in operating parameters	Resulting changes in steam temperature (°C)
Boiler load ±10%	±10
Excess air ratio of furnace ±10%	±(10–20)
Feed water temperature ±10°C	±(4–5)
Moisture fraction in coal ±1%	±1.5
Ash fraction in coal ±10%	±5
Soot blowing	Increases owing to cleaning of superheaters surfaces, but decreases owing to cleaning of upstream surfaces.
Use of saturated steam for auxiliaries	Increases if main steam flow is maintained by increasing the firing rate

heat absorption in the radiant superheater unchanged, the increased steam flow leaves the superheater at a low temperature.

In case of convective superheaters, the heat absorption increased directly with the increase in the mass of the flue gas. Thus when the boiler load is increased, the steam temperature increases in spite of the increase in the mass of steam flow. The changes in steam temperature with the increase in boiler load or firing rate are shown in Figure 7-6. It shows that a combination of radiative and convective superheaters may help maintain the steam temperature with changes in load. A 40–60% radiative heat absorption may give good steam temperature characteristics under varying loads.

The heat absorbed by the convective reheater decreases when the boiler load decreases. This drop is even larger when the reheater is arranged in a lower gas temperature zone. When the plant works at a fixed pressure, the temperature of

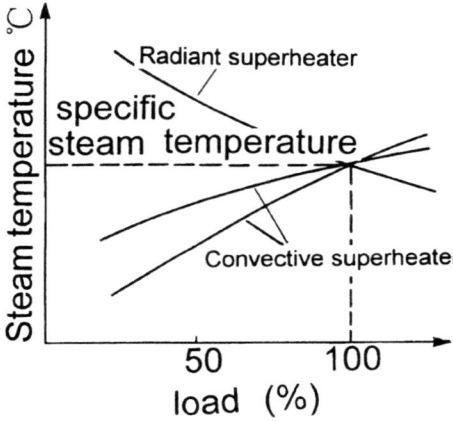

FIGURE 7-6. Variation of exit temperature of the steam with load for radiant and two convective superheaters

TABLE 7-3. Techniques for steam temperature control (Applicable techniques are marked with "X").

Medium	Means for control	Without reheater Superheater control	Boilers with reheater Superheater control	Reheater temperature control Anthracite	Bituminous	Lignite	Oil, gas
Steam side	Surface-type attemperator	X					
	Spray-type attemperator	X	X	X		X	
	Steam-steam heat exchanger			X		X	
Gas side	Gas bypass			X	X		
	Gas recirculation				X		X
	Tilting burner	X	X				

the steam at the inlet to the reheater decreases with decreasing load. So more heat should be absorbed by the reheater to maintain its exit steam temperature at the design level. To help alleviate this problem, variable pressure operation may be used.

b) Methods of Controlling Steam Temperature

The steam temperature can be controlled either by controlling the condition inside the tube or outside of it. Both means have advantages and disadvantages. Some commonly used techniques are shown in Table 7-3. For reheat boilers, a combination of these two methods is usually used.

(1) Spray-Type Attemperator

In a spray-type attemperator, high-quality pure water is injected into the superheated steam to cool it down. A typical design is shown in Figure 7-7. To reduce

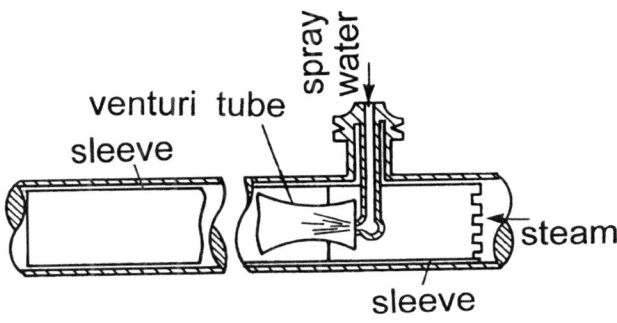

FIGURE 7-7. A spray-type attemperator

the steam enthalpy by 60–85 kJ/kg in an intermediate pressure boiler, the amount of water injection required will be 3–5% of the main steam flow rate.

A spray-type attemperator is generally arranged between two superheater stages. A very high quality of water is required for the spray. Any deposits would damage the superheater. The condensate of the feed water heater is a good source of this water. For plants whose feed water quality is not so good, a self-generating condensate system and an attemperator may be used. Here partially saturated steam is condensed by feed water, then the condensate is sprayed into the attemperator to control the steam temperature.

To design an attemperator, special care should taken to ensure that a) the quality of atomization is uniform, and b) the length of atomizing section is sufficiently long to protect the steam line from thermal shock.

If the attemperator is nozzle type, swirl type, or multi-hole nozzle type, special attention should be paid to the vibration caused by the steam stream vortex.

(2) Steam–Steam Heat Exchanger

In a steam–steam heat exchanger (Fig. 7-8), the superheated steam flows through a number of small parallel tube; and the reheated steam flows through the larger shell tubes outside these. The temperature of the superheated steam can be adjusted by varying the flow rate of the reheat steam. A steam–steam heat exchanger is more suitable for high-capacity boilers in which the radiant heat surface occupies a major fraction of total surface. In such a boiler, the superheated steam temperature increases with a decrease in the boiler load. The reheated steam is heated by excess heat of the superheated steam. The main disadvantage of this is that the design is too complicated and it is not a convenient arrangement.

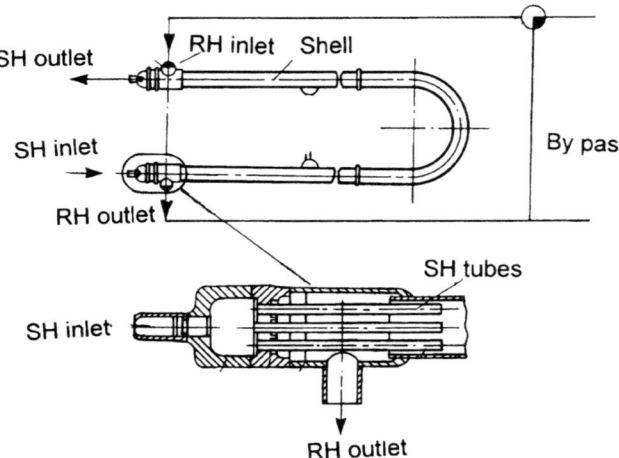

FIGURE 7-8. A shell- and tube-steam to steam-type heat exchanger for steam temperature control

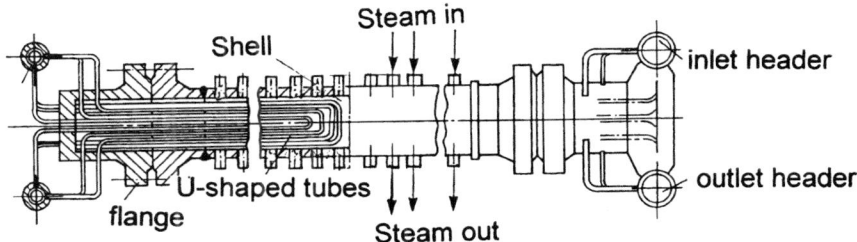

FIGURE 7-9. A surface-type attemperator

(3) Surface-Type Attemperator

A surface-type attemperator is a tubular heat exchanger. As the cooling water does not come in contact with the steam, there is no rigid limitation on the quality of the water. Its design is simple, but there is a long delay in the control response. It is generally used in low- or medium-pressure boilers. A surface-type attemperator may be installed at the inlet of the superheater or between the two superheater sections. The latter position is more commonly used today. The cooling water is taken from the economizer. By adjusting the amount of the cooling water, the heat absorption is varied and therefore the steam temperature can be controlled. The heat exchanger in the surface-type attemperator can be either a U-shaped tube (Fig. 7-9) or twin-walled tube. An U-shaped tube attemperator, arranged between two sections of the superheater, usually does not work well because of the U-shaped tube rupture caused by the thermal fatigue. Some of the measures taken to prevent it include the following:

1) To limit the minimum flow rate of the cooling water, and to vary the flow rate with stability (too large a variation is not allowed).
2) To increase the bent radius of the U-shaped tube.
3) To increase the feed water temperature.

The twin-walled tube attemperator comprises outer and inner tubes. The distance between the outer and inner tubes is 3–5 mm. Water flows into the inner tubes and flows out through the annulus. The characteristics of this type attemperator follow:

1) Inner and outer tube can expand freely. So there is no thermal stress. This prevents tube rupture, which is common in U-shaped attemperators.
2) There is no limit on the amount of the cooling water used. So the controlling range of the steam temperature is wide.
3) Parts of the twin-walled tubes are arranged in a cross-over pattern, with left and right tubes, which gives uniform temperature control characteristics, and permits a simplified design of the outlet tubes.
4) Inner tubes do not transfer heat directly. So heat transfer area and flow resistance are greater than those of U-shaped tubes.

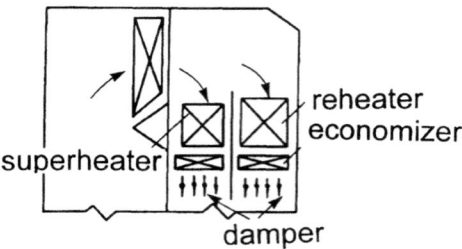

FIGURE 7-10. Gas bypassing arrangement for steam temperature control

Twin walled attemperators are generally used in the low- and medium-pressure boilers where superheater systems are relatively simple. However, one major problem is that steam may erode the outer tubes vertically.

(4) Gas Bypass

This method controls the steam temperature by varying the flow rate of flue gas over the superheater (Fig. 7-10). The heat transfer coefficient changes with the change in gas velocity, which in turn alters the heat absorption by the superheater. The design of a gas-bypass system is simple and it is easy to operate, but the control response is sluggish. Gas bypassing is effective in the range of 0–40% damper opening. To avoid damper distortion, it should be arranged where the gas temperature is below 400–500°C. However, special attention should be paid to possible erosion. The gas duct should be airtight, for which a membrane-type wall is recommended.

(5) Tilting Burner

Tilting burner control is usually used in tangential fired furnaces equipped with vertical tilting burners. By directing the nozzles of the burners upward or downward, the location of the main combustion zone can be shifted and therefore the furnace exit gas temperature can be changed. When the burner tilts upward or downward by 20–30°, the furnace exit gas temperature varies between 100–140°C. This control is precise, and no additional heating surface and additional power consumption is needed. But for low fusion point coal, the variation in temperature should be limited to avoid ash deposits on the wall.

7-3 Adjustment of Heat Absorption in Superheater and Reheater

Adjustment implies permanent hardware changes made to the superheater or reheater. It is required when the operating conditions depart for the ideal design condition. Nonuniform absorption of heat in tubes is an example. This potentially hazardous problem must be dealt with by permanent alteration of the boiler.

7-3-1 Adjustment of Steam Temperature

The steam temperature can be regulated by one or a combination of the following methods:

1. Addition or removal of superheater/reheater surfaces.
2. Removal of evaporator surfaces reduces the steam flow but increases the gas temperature, resulting in an increase a steam temperature. Conversely, an addition of evaporative surfaces would decrease the steam temperature.
3. Addition of refractory on evaporative surfaces may increase steam temperature.
4. Alteration of mass velocity of flue gas over the convective superheater by using baffles may favorably alter the steam temperature.

7-3-2 Nonuniform Heat Absorption

Parallel tubes are used in many sections of a boiler. The geometry, heat load, and mass flow rate of steam through each of these parallel tubes are not exactly the same. Thus, the increase in the enthalpy of the steam in individual parallel tubes are different. This gives rise to a nonuniformity in the heat absorption. The extent of this is defined by the ratio of enthalpy increase in the specific tube, Δh_p, and the average enthalpy increase, Δh_0, in all tubes. So the coefficient of nonuniformity of heat absorption Φ is written as:

$$\phi = \frac{\Delta h_p}{\Delta h_0} \tag{7-1}$$

Δh_p and Δh_0 can be expressed as

$$\Delta h_p = \frac{q_p A_p}{G_p} \tag{7-2}$$

$$\Delta h_0 = \frac{q_0 A_0}{G_0} \tag{7-3}$$

where $\quad q =$ heat flux per unit area, $kJ/m^2 s$
$A =$ area of heating surface, m^2
$G =$ flow rate, kg/s

Subscripts p and 0 represent conditions of the specific tube and that of average tubes, respectively.

Substituting Eq. (7-2) and Eq. (7-3) to Eq. (7-1),

$$\phi = \frac{q_p}{q_0} \cdot \frac{A_p}{A_0} \frac{1}{\dfrac{G_p}{G_0}} = \frac{\eta_q \cdot \eta_a}{\eta_g} \tag{7-4}$$

where

$$\eta_q = \frac{q_p}{q_0} , \quad \eta_a = \frac{A_p}{A_0} , \quad \eta_g = \frac{G_p}{G_0}$$

are defined as nonuniformity coefficients for heat flux, flow area, and flow rate, respectively. Flow area nonuniformity is generally very small. So, the nonuniformity in heat absorption is caused mostly by the variation in the heat flux and the flow.

7-3-3 Factors Influencing Nonuniformity

a) Nonuniformity in Heat Absorption

Uneven distribution of temperature and gas flow may cause nonuniform heat absorption in radiative and convective superheaters. The heat flux on the furnace wall may also be uneven. On a particular wall, the heat flux may vary along both width and height. Moreover, the spiral gas stream generated by the tangential firing brings about temperature variation along the furnace width. Because the gas spiral continues up to the exit of the furnace, there may be nonuniform heat absorption in the convective superheater as well. Slagging and fouling of heating surfaces often cause serious variation in heat absorption amongst tubes.

Nonuniform transverse spacing (S_1) may also cause uneven distribution of heat absorption. A wider gas passage has a lower flow resistance and therefore larger gas flow, which increases the heat transfer. Furthermore, the wider gas passage would also have a thicker gas radiation layer, giving higher gas radiation. The nonuniformity coefficient of heat absorption in the superheater can be in the range of $\eta_g = 1.2\text{--}1.3$.

b) Nonuniformity in Steam Flow

Many factors influence the flow variation between parallel tubes. These include methods of connection of headers differences in static pressure tube diameter and tube length. Furthermore, the nonuniform heat absorption by individual tubes also contributes to the flow variation.

We can derive an expression for the coefficient of flow nonuniformity. Any variations in static pressure along the header length and pressure difference between two ends of the tube loop are neglected. The pressure difference between two ends of the header is the sum of resistance in pipe fittings frictional resistance and the difference in static head. So it is given by

$$\Delta p_0 = \left(\sum \xi + \lambda \frac{l}{d} \right)_0 \frac{w_0^2}{2v_0} + \frac{h}{v_0} g \qquad (7\text{-}5)$$

where ξ, λ = coefficient of flow resistance in fittings and friction coefficient in the tube, respectively

d, l = inside diameter, and the length of the tube
h = height difference between the inlet and the outlet of the header
v, w = specific volume and flow velocity respectively

The subscript 0 represents the average value of the total tube bundle and p that of a specific tube.

For a specific tube, the pressure drop across it can be written as

$$\Delta p_p = k_p G_p^2 v_p + \frac{h}{v_p} g \qquad (7\text{-}6)$$

where

$$k_p = \left(\sum \zeta + \xi \frac{l}{d} \right)_p \frac{1}{2A_p^2}$$

G is the steam flow rate $(G = w/v)$.

If the static pressure variation along the header length is neglected, the pressure difference across each tube should be equal to their average value, ΔP_0,

$$\Delta P_0 = \Delta P_p \qquad (7\text{-}7)$$

From above

$$k_0 G_0^2 v_0 + \frac{h}{v_0} g = k_p G_p v_p + \frac{h}{v_p} g \qquad (7\text{-}8)$$

For superheated steam, the second term

$$\frac{h}{v} g$$

is very small compared to the first terms. So it is neglected. Thus, Eq. (7-9) is simplified as

$$k_0 G_0^2 v_0 = k_p G_p^2 v_p \qquad (7\text{-}9)$$

or

$$\eta_G = \frac{G_p}{G_0} = \sqrt{\frac{k_0}{k_p} \cdot \frac{v_0}{v_p}} \qquad (7\text{-}10)$$

It is apparent that if the resistance coefficient k_p of the specific tube is large, the flow nonuniformity coefficient η_G of the parallel tube bundle would be small. Also if the heat absorbed by an individual tube is high, the specific volume of steam, v_p, in it would be large, and η_G would become small.

c) Variation of Pressure along the Length of a Header

In a typical superheater or reheater steam enters the header and as it flows along the length of the header it is divided between several parallel tubes. So, the velocity of steam through inlet header decreases along its length (Fig. 7-11). From the Bernoulli equation the decrease in velocity results in an increase in the static pressure. The maximum static pressure will, therefore, occur at the end of the header where the velocity is zero. However, there is a frictional resistance in the header Δp_h, which counters the static pressure rise in the header owing to the drop

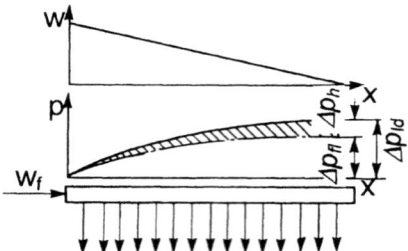

FIGURE 7-11. Variation of static pressure in the distributing header of a superheater

in velocity. Thus the maximum rise in static pressure, Δp_{fl}^{max}, occurs at the furthest end of the header:

$$\Delta p_{fl}^{max} = \Delta p_{ld} - \Delta p_h \qquad (7\text{-}11)$$

where

$$\Delta p_{ld} = \frac{\rho_f w_f^2}{2} \qquad (7\text{-}12)$$

$$\Delta p_h = \int_0^l \frac{\lambda}{d} \frac{\rho_f w_x^2}{2} \, dx \qquad (7\text{-}13)$$

Δp_{ld} = flow pressure drop, Pa
Δp_{lz} = flow resistance, Pa
w_f = steam flow velocity at the header inlet, m/s
ρ_f = steam density at the header inlet, kg/m^3
λ = friction coefficient in the header
w_x = steam flow velocity at a distance x from the entrance of the header, m/s
L = total length of the header, m

Assuming the local velocity in the header, w_x, to vary linearly with the length of the header, we can write

$$w_x = w_f\left(1 - \frac{x}{L}\right) \qquad (7\text{-}14)$$

Substituting the above equation into Eq. (7-13), we obtain

$$\Delta p_h^{max} = \frac{\lambda L}{3d} \frac{\rho_f w_f^2}{2} \qquad (7\text{-}15)$$

Substituting Eq. (7-15) and Eq. (7-12) into Eq. (7-11), we obtain

$$\Delta p_{fl}^{max} = \left(1 - \frac{\lambda L}{3d}\right)\frac{\rho_f w_f^2}{2} \qquad (7\text{-}16)$$

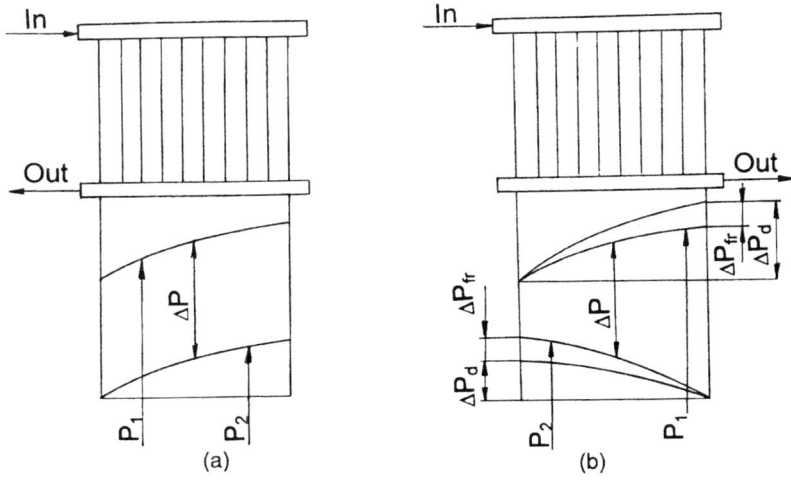

P$_1$ - Total pressure inside inlet header

P$_2$ - Total pressure inside outlet header

ΔP - Difference between pressure inside in
            and outlet headers

ΔP$_d$ - Velocity pressure increase (decrease)

ΔP$_{fr}$ - Pressure loss due to friction

FIGURE 7-12. Two types of superheater connections with the resultant static pressure variations: (a) U connections, (b) Z connection

Similarly, the maximum rise in static pressure in the collecting or the downstream header (Fig. 7-12) is

$$\Delta p_{hl}^{max} = \left(1 + \frac{\lambda L}{3d}\right)\frac{\rho_h w_h^2}{2} \qquad (7\text{-}17)$$

Figure 7-12 shows two arrangements of superheater connections. In U-type arrangements (Fig. 7-12a) the inlet and outlet connections are on the same side of the header. In Z-type arrangements (Fig. 7-12b) the connections are on the opposite side. From the above analysis we can see that in a Z-type connection the static pressure is highest on the right end in the inlet (distributing) header and lowest on the left end of the outlet (collecting header). So, tubes on the right will experience much higher pressure difference and therefore higher flow rate than the tubes on the left end of headers. Thus Z type connection gives poor flow distribution. In a U-type arrangement, however, the highest pressure in the collecting header occurs on the right end. So tubes at this end are likely to have a pressure difference that is the same as that on the right end. Thus, the U-type connection is likely to give better flow distribution.

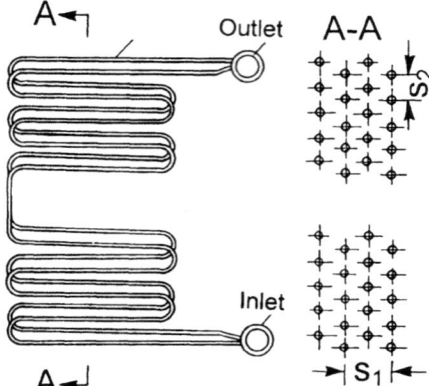

FIGURE 7-13. A typical tube arrangement in an economizer; here tubes are normal to back wall

7-4 Economizer

The economizer preheats the feed water by utilizing the residual heat of the flue gas. Thus it reduces the exhaust gas temperature and saves the fuel. Modern power plants use steel-tube-type economizers. Some old small boilers still use cast iron economizers.

Design Configuration

A typical steel tube economizer comprises two headers and rows of parallel tubes connecting them (Fig. 7-13). If the heating surface is large, instead of having one tall continuous bank of tubes, the economizer is divided into several sections (1–1.5 m high each) with intermediate headers. A clearance of 0.6–0.8 m is provided between two sections to allow access for maintenance tubes and repairs.

a) Tube Arrangement

The tube bank of an economizer may be longitudinal or transverse. In the longitudinal arrangement (Fig. 7-14a) the tubes are normal to the rear wall while those in transverse arrangement are parallel to the rear wall. The width of the back-pass is longer than its depth. As a result the length of parallel tubes in a longitudinal arrangement is shorter and there are more tubes across the header compared in a transverse arrangement. Thus the velocity of water through tubes is low, resulting in a lower pressure drop. The flue gas flowing over the tubes tends to carry more ash toward the far end of the tubes due to the centrifugal force. Thus one end of all tubes is subject to a higher level of erosion than the other end, which necessitates complete replacement of the entire tube bank for erosion damage. In a transverse

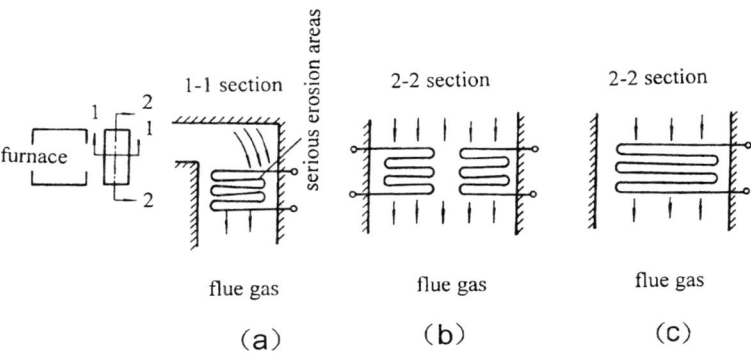

FIGURE 7-14. Gas flow pattern over three types of tube arrangements in the economizer

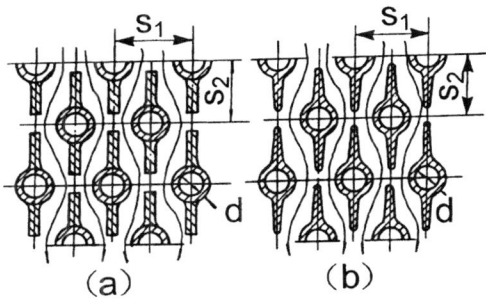

FIGURE 7-15. Extended surfaces to the economizer enhances the heat transfer: (a) rectangular fins, (b) conical fins

arrangement (Fig. 7-14b & 7-14c), where the tubes are parallel to the rear wall, only tubes away from the rear wall are subject to the higher dust concentration and consequently higher the erosion. Thus the repairs involve replacement of only affected tubes. The flow resistance of water in a transverse arrangement is, however, high because this arrangement uses smaller number of longer parallel tubes.

To make the economizer compact, rectangular (Fig. 7-15a) or trapezoidal (Fig. 7-15b) fins are welded on economizer tubes. Such fins can reduce the size of the economizer by 20–25%.

b) Tube Support

Economizer tubes can be supported either from the bottom or from the top. The top-supported economizer tube can be supported from the header located across the back-pass. The feed water circulating through the header can protect the header from the heat of the flue gas.

c) Tube Size and Spacing

The outside diameter of typical economizer tubes is 25–38 mm. Depending on the pressure of the boiler, the tubes are 3–5 mm thick. In a staggered arrangement (Fig. 7-13) the lateral spacing to diameter (S_1/d) ratio is chosen in the range of 2.5–3.0, depending on the flue gas velocity. The vertical spacing to diameter ratio (S_2/d), however, is dictated by the safe bend radius of the tube allowed by the relevant boiler code. It is generally in the range of 1.5–2.0. The water flow velocity is chosen in the range of 600–800 kg/m²s. The flow resistance becomes too large if the velocity is high. The velocity of water in the tubes used as hanging tube should be greater than that of others to avoid water downward flow. The water side resistance of the economizer should not exceed 5% of the drum pressure for high-pressure boilers 8% for mid-pressure boilers.

d) Flue Gas Velocity

The gas velocity through an economizer ranges from 7 to 13 m/s. A higher gas velocity enhances the heat transfer and saves the heating surface but it leads to higher tube erosion. Too low a gas velocity may lead to poor heat transfer and serious ash deposit on the tubes, while minimizing the erosion potential. The allowable gas velocity reduces with ash percentage and with increasing fraction ($SiO_2 + Al_2O_3$) in the ash.

e) Start-up of Economizer

During the start-up process of a boiler the water flow through an economizer is discontinuous. So when the feed water stops, the water in the economizer does not flow, but some water will evaporate owing to the heat transferred from the hot flue gas. The steam may attach itself to the inside of the tube wall or gather in the upper part of the economizer. This may lead to economizer tube burn out. So the economizer should be protected during the start up.

A recirculation tube, connecting the inlet of economizer and the bottom of the drum, allows water to flow continuously owing to the density difference between the recirculation tube and the economizer. This provides an effective cooling of the tube walls. In normal operation, the recirculation valve is closed to prevent direct flow of feedwater into the drum.

7-5 Air Heater

An air heater preheats the combustion air where it is economically feasible. The preheating helps the following:

- igniting the fuel
- improving combustion: for example (Clapp, 1991), a 22°C rise in the combustion air could improve the boiler efficiency by 1%
- drying the pulverized coal in pulverizer
- reducing the stack gas temperature and increasing the boiler efficiency

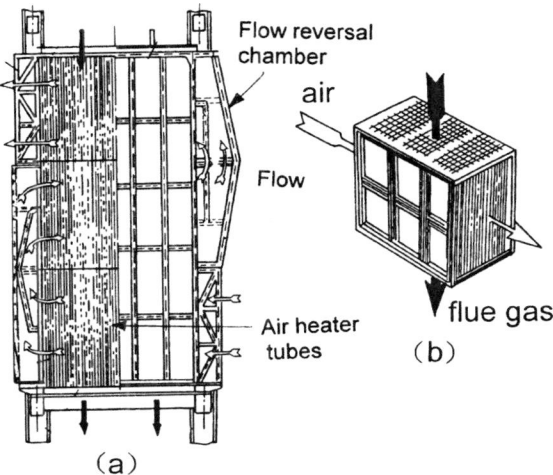

FIGURE 7-16. A Tubular air heater showing the flow directions of air and flue gas

There are three types of air heaters

1. Recuperative (tubular or plate type)
2. Rotary regenerative
3. Heat pipe.

In recuperative air heaters the heat from the hot gas flows to the cold air across plate or walls of tubes. In a regenerative air heater the heat flows directly from a packed matrix. The cold air picks up heat from the same matrix when it moves to its side later. A recuperative air heater does not involve any moving parts, but is much larger than regenerative types. For example, the volume requirement of a recuperative air heater may be as much as nine times than that of regenerative types and its weight double that of the regenerative type (Clapp, 1991). Low leakage of air into the gas stream is one advantage of a recuperative air heater.

7-5-1 Tubular Air Heater

The construction of a recuperative tubular air heater is shown in Figure 7-16. The flue gas flows longitudinally in the tube and the air flows transversely over the tube. For convenience of manufacture, transportation, and installation, tubular air heaters are generally manufactured in sections. These sections are welded together at site.

The tubular air heater is supported on the boiler frame. During operation, the tubes, shell, and the boiler frame expand to different extents due to their different temperature and construction material. So expansion joints made of thin steel flats are installed between the tube sheet and shell, and between the shell and the boiler frame. The expansion joint allows the parts to move relative to each other while ensuring airtight seal at the joints.

In a typical arrangement (Fig. 7-16) the flue gas passes straight through, but the air makes several passes over the tubes. The larger the number of passes, the closer the flow to the counterflow situation. As a result the average gas–air temperature difference air velocity are higher. This reduces the surface area requirement for a given duty. However, a larger number of flow passes increases the flow resistance. The single pass arrangement has the advantage of simplified construction, larger air flow area, and therefore lower flow resistance. However, a lower average temperature difference and lower air velocity would necessitate a larger tube area. So in large-capacity boilers, multiple pass arrangements are used.

a) Choice of Major Design Parameters

The following section presents a brief discussion on design considerations for tubular air heaters.

(1) Tube Diameter and Spacing between Tubes

Tubes are generally arranged in staggered fashion. Steel tubes of 37–63 mm diameter and 1.5 mm thickness are used. To extend the life of the low-temperature section one could use a 38 mm tube of 2 mm thickness or a 42 mm diameter tube of 2 mm thickness. In order to reduce corrosion potential, glass coating of tubes in the low temperature zone may be considered. The staggered line arrangement is usually adopted. The tube pitches used are in the range of $S_1/d = 1.5$–1.9, $S_2/d = 1.0$–1.2. The exact pitch depends on factors like heat transfer, flow resistance, and vibration.

(2) Height of Air Heater Chamber

The height of the high-temperature section may not exceed 12 m (BEI, 1991). The chamber or the air heater casing should be sufficiently strong and have good access for cleaning the ash from the casing. The height of the low-temperature section is usually 1.5–4.5 m, which is convenient for repair and maintenance.

(3) Gas and Air Flow Velocity

The gas velocity in a vertical tubular air heater is in the range of 10–16 m/s for solid fuel. The air flow velocity is usually half the gas velocity. Relatively clean fuels like oil or gas permit the use of higher velocity.

7-5-2 Plate Recuperators

Instead of tubes, parallel plates are used to form the gas and air passage in plate-type recuperators. The gas passage is 12–16 mm wide, while the air passage is about 12 mm wide. An open channel recuperative air preheater (BEI, 1991) provides a good alternative to a regenerative air preheater.

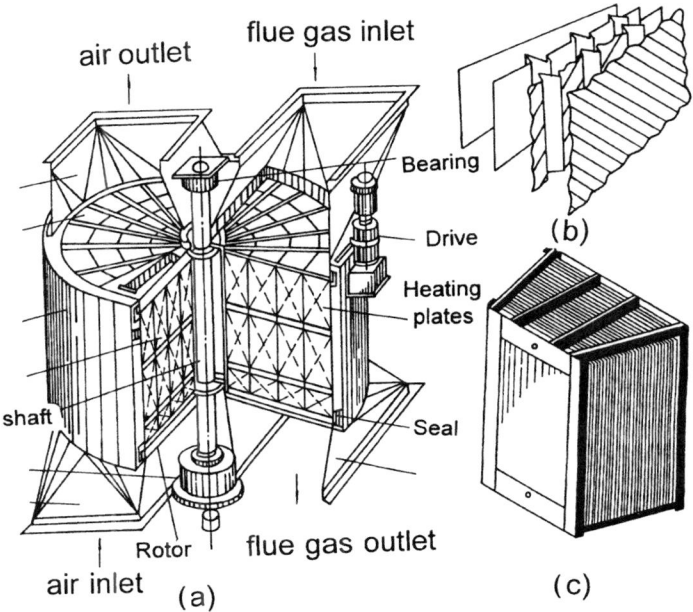

air outlet   flue gas inlet

Bearing

Drive   (b)

Heating
plates

shaft

Seal

Rotor   flue gas outlet
air inlet   (a)   (c)

FIGURE 7-17. Cut-view of a rotary-plate-type regenerative air heater with vertical shaft, it shows the (a) arrangement of the entire air heater, (b) shape of the heating elements, and (c) shape of an individual heating plate

7-5-3 Regenerative or Rotary Air Heater

In a regenerative-type air heater the hot gas and cold air flow from a set of heat storage surfaces (matrix) which move between the hot gas and cold air alternately. The heat transfer surfaces absorb heat from the hot flue gas and give it up to the cold air. The heat flows from the hot surface of the matrix to the cold air. The direction of heat flow reverses when these surfaces pass through hot gas.

Regenerative air heaters are of the rotating-plate (Lungstrom) and stationary-plate types (Rothemuhle).

a) Rotating-Plate-Type Air Heater

In the rotating-plate-type air heater (Fig. 7-17) heat storage plate elements rotate with the rotor at a low speed of about 0.75 rpm. A part of the heat storage plate elements passes through the hot flue gas. The elements are heated and flue gas is cooled. Another part of heat storage plate elements is swept by cold air, where the air is heated. The rotating component of this type of air heater can be as heavy as 500 tons, but its construction is compact.

The rotating-plate-type air heater consists of a rotor, sealing apparatus, shell, etc. The rotor is divided into 12 or 24 sectors with 12 or 24 radial division plates.

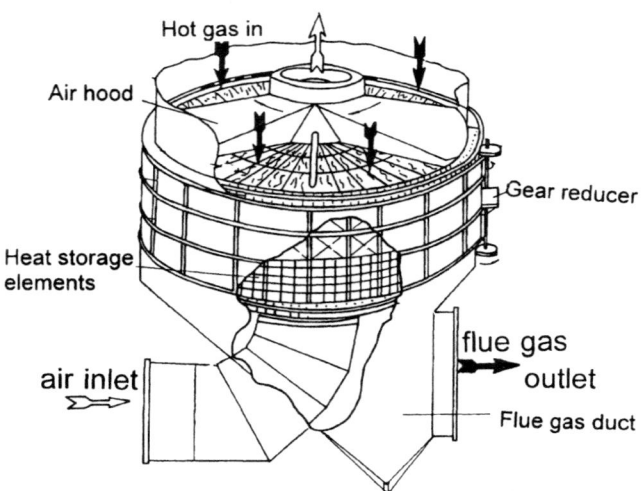

FIGURE 7-18. A stationary-plate-type regenerative air heater. The direction of the flue gas is shown by solid arrows, while that of air is shown by blank arrows

Each sector is divided into several trapezoidal sections with transverse division plates. Heat storage plates are placed in it. The low-temperature section of the rotor is made such that it allows easy replacement of the heat storage elements. The weight of the rotor is supported by a thrust bearing. Sealing is critical for the control of air leakage into the low pressure gas side. The air heater would normally use radial, axial, and circumferential seals. Among these the radial seal has the greatest influence on the air leakage. The radial seal system consists of a radial seal piece and a section seal board. Seal adjusters are installed at the sector seal board. The peripheral seals system is usually installed at the central shaft and the outer periphery of the rotor.

b) Stationary-Plate-Type Air Heater

This regenerative-type air heater is slightly different from the Lungstrom type (Fig. 7-18). Here the heat storage elements are static but the air/gas flow section rotates. The storage plates are placed in the stator. When the gas flows through the stator outside the air hood, the heat storage elements are heated. The air is in counterflow in the air hood. It absorbs heat from the heat storage elements. When the rotary air hood rotates a full circle, two heat-cooling periods are completed. The rotating parts of this type of air heater are light, but its design is complicated.

The stationary-plate-type air heater consists of a stator, seal system, and air hood. The stator's design is similar to that of the rotor mentioned above. The stator is cut into many sectors. Sealing is still a major problem. An adjustable seal system located between the air hood and the rotor is used.

TABLE 7-4. Comparison of rotating plate with stationary-plate-type air heater.

Type of air heater	Rotating-plate type	Stationary-plate type
Revolution in rpm	1.5–4	0.75–1.4
Gas flow area as % of total	40–50	50–60
Air flow area as % of total	30–45	35–45
Seal section area as % of total	8–17	5–10

The U-type seal comprises sealing plates, sealing frame, U-type sealing pieces, and the adjustable components. The sealing plates are fixed on the sealing frame and the sealing frame is connected with bottom of the hood by the U-type sealing pieces. When the hood rotates, the sealing frame also rotates. The weight of the seal system is balanced by the weight of the U-type seal board and the elastic force of the spring. The seal board contacts with the ends of stator with a small contact force. The U-type seal piece can follow the relative motion between the seal system and the air hood. As a result, it gives a good sealing effect.

An adjustable annular seal system is another type. It is located between the air hood and the fixed air duct. It consists of a cast iron seal block and seal-adjusting equipment. The annular seal system is fixed on the air flow duct. The sealing effect can be adjusted by using the spring in the adjusting equipment.

c) Major Design Considerations

Typical rotation and the ratio of flow areas are given below in Table 7.4.

(1) Velocity and Size of Heaters

The gas and air velocities are generally in the range of 8–12 m/s. The gas velocity is close to that of air. The diameter of the air heater is typically 4.8, 8.0, 8.5, and 9.5 m. The angle of the sector cabin is generally 7.5°, 15°, and 30°.

(2) Type of Heat Storage Elements

Figure 7-17b shows one type of heat storage element. Several other types of plates are shown in Figure 16-4. For elements in the high-temperature section, more attention should be paid to the effect of heat transfer, while for elements in the low-temperature section, there is the problem of corrosion and ash deposition. There are two types of drives used for the rotor: central type and peripheral type. The peripheral type is more commonly used.

7-6 Arrangement of Back-Pass Heating Surfaces

The economizer and the air heater are located at the back end of the flue gas passage. So they are called caudal or rear-end surfaces. There are two types of arrangements: single stage and double stage.

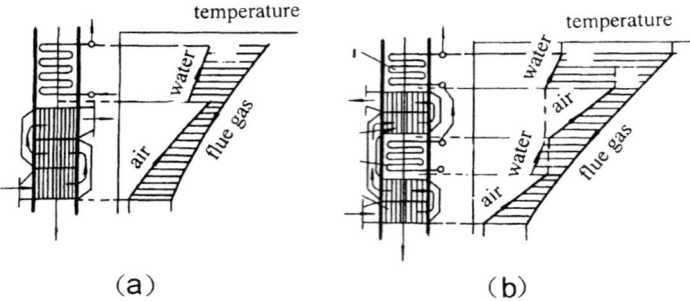

(a)   (b)

FIGURE 7-19. Pinch point or temperature diagram for two arrangements of economizer and air heater: (a) the entire air heater is located downstream of the economizer; (b) the one section of the air heater is placed between two sections of the economizer

7-6-1 Single-Stage Arrangement

The single-stage arrangement (Fig. 7-19a) consists of one stage of economizer and one stage of air heater. The air heater is always located downstream of the economizer. It lowers the exhaust gas temperature and thereby achieves high boiler efficiency. Moreover, this practice helps reduce the corrosion potential of the economizer and thereby avoiding the use of expensive metals for the cold end of the economizer. The single-stage arrangement is simple, but the temperature of hot air can reach only 300°C. Any higher air preheat temperature is impossible with this arrangement because the volume and specific heat of the gas are both higher than those of air. The heat capacity of the gas is larger than that of air. When the gas transfers the heat to air, the drop in gas temperature is less than the corresponding increase in the air temperature. In normal conditions, the air temperature increases 1.25–1.5°C per 1°C decrease in gas temperature. To maintain a stack or exhaust gas temperature of 120°C, the temperature of preheated air should be 320–350°C. The inlet gas temperature of the air heater is so close to the outlet air temperature that it is impossible to raise the air temperature. So if the exhaust gas temperature is fixed, the air exit temperature is automatically restricted to this value. If the air temperature needs to be raised further, the exhaust gas temperature also needs to be raised. In that case, the single-stage design is not economical. To obtain a higher air temperature without increasing the stack loss, the double-stage arrangement should be employed.

7-6-2 Double-Stage Arrangement

The double-stage arrangement consists of two stages of economizers and two stages of air heaters. The first-stage economizer is arranged between the first-stage and the second-stage air heater (along the air-flow direction). The second-stage economizer is located above the second-stage air heater. Because a portion of the air heater surface is arranged in the high gas temperature region, the air can be heated to a temperature higher than that of the single-stage arrangement. Moreover, this type of arrangement increases the log-mean temperature difference in both economizer and air heater, which enhances the heat transfer in the back-pass heating surfaces and therefore reduce the cost of the steel. The temperature profile of gas and the working fluid for the double-stage arrangement are shown in Figure 7-19b. In high-pressure boilers, the reheater is arranged in the caudal, or the rear, duct beside the economizers and air heaters. In some conditions the low-temperature convection superheater is also arranged there. The characteristic of such arrangements follow:

1. As the reheater and sometimes the low-temperature convection superheater are arranged in the caudal duct, the single-stage arrangement is usually employed.
2. The reheater and low-temperature convection superheater are arranged ahead of the economizer.

7-7 Heat Transfer Calculations for Convective Heating Surfaces

7-7-1 Theory and Basic Equations

The convective superheater, convective reheater, economizer, and air heater are all called *convective heating surfaces* as heat transfer in these occurs mainly by convection. However, high-temperature gases passing over tube bundles of convective heat-transfer surface also give out radiation heat. To simplify the calculation, calculation of radiation heat transfer is changed into calculation of convective heat transfer. The heat-exchange surfaces near the furnace, i.e., the rear platen, the screen tubes, and secondary superheater, etc., also absorb part of radiation heat from the furnace.

The heat transfer calculation for convective heating surfaces involves two sets of equations. One is for heat transfer, and the other is for thermal balance. The thermal balance equations determine the entry and exit temperatures of the heating and heated fluids. The heat transfer equation can be used to check whether the quantity of heat determined by the thermal balance can be transferred across the given surface area from the flue gas to the working fluid.

a) Thermal Balance Equation for Convective Heating Surface

The thermal balance works on the premise that the heat given out by the flue gas is equal to the heat absorbed by working fluid.

The heat given out by flue gas is found by

$$Q_{co} = \varphi(H_{gi} - H_{go} + \Delta\alpha H_e) \qquad (7\text{-}18)$$

where φ is the coefficient of heat retention. It takes account of any heat loss to the surroundings; H_{gi}, and H_{go} are enthalpies of the gases at the inlet and outlet of the convective section; and $\Delta\alpha H_e$ is the heat carried by leakage air into the boiler, kJ/kg. H_l is determined based on a temperature enthalpy calculation. The temperature of leakage air is equal to ambient air temperature for all convective heating surfaces except the air heater. For the air heater the temperature of leakage air is equal to $(T_{ai} + T_{ao})/2$ where T_{ai} and T_{ao} are inlet and outlet air temperatures, respectively.

The calculation methods for heat absorption are different for different heat exchanger and working fluids.

(1) Calculation of Heat Absorption for a Platen Superheater

For a platen superheater arranged at the furnace outlet the total heat absorption by steam consists of convection heat and radiation heat coming from furnace.

$$Q_{co} + Q_{ra} = \frac{D}{B}(H_{so} - H_{si}) \tag{7-19}$$

where Q_{co} = convection heat, kJ/kg
 Q_{ra} = radiation heat coming from furnace and absorbed by the
 working fluid of the platen superheater, kJ/kg
 D = steam flow rate, kg/h;
 B = fuel firing rate, kg/h
 H_{so}, H_{si} = outlet and inlet enthalpy, respectively, of steam, kJ/kg

The radiation heat coming from the furnace, Q'_{ra} may not be absorbed entirely by the platen superheater. Some of the heat may pass through the platen superheater and arrive at the next section of the heating surface. This part of heat is marked by Q''_{ra}. In addition, the flue gas inside the platen tube matrix also radiates heat to the next section of heating surface. This part of heat is marked by Q''_{rap}. So radiation heat absorbed by the steam in platen superheater Q_{ra} is

$$Q_{ra} = Q'_{ra} - (Q''_{ra} + Q''_{rap}) \tag{7-20}$$

From furnace heat transfer calculations the radiation heat coming from furnace is given by

$$Q'_{ra} = \eta_h F_c \frac{Q_{fu}}{F} \tag{7-21}$$

where Q_{fu} = total heat absorption in the furnace, kJ/kg
 F = total area of the water wall surface, m^2
 F_c = area of the furnace outlet windows, m^2
 η_h = coefficient of nonuniformity of heat load at the furnace outlet.

The radiation heat Q''_{rap} is found by

$$Q''_{rap} = \frac{Q'_{ra}(1 - \alpha)x_p}{\beta} \tag{7-22}$$

where α is the emissivity of flue gas in platen domain; and β is a factor that takes account of heat exchange between the furnace and the heating surface. It is a function of temperature and the fuel fired. Some typical values are given in

TABLE 7-5. Values of β for different temperatures and fuels.

Temperature (°C) →	900	1000	1100	1200	1300
Gas fired	0.7	0.67	0.62	0.56	0.47
Oil fired	0.9	0.88	0.83	0.76	0.65
Coal fired	1.0	1.0	0.98	0.92	0.81

Table 7-5. The angular coefficient of radiation, x_p, from the platen inlet section to platen outlet section is calculated from Eq. (7-23):

$$x_p = x_p = \sqrt{\left(\frac{b}{s_1}\right)^2 + 1} - \frac{s_1}{b} \tag{7-23}$$

where
b = width of platen, m
s_1 = spacing of platen, m

The radiation Q''_{rap} is calculated by

$$Q''_{rap} = \frac{5.67 \times 10^{-11} \alpha A_{in} T_p^4 \zeta_r}{B} \text{ kW/kg} \tag{7-24}$$

where
α = emissivity of flue gas in platen region,
T_p = average temperature of flue gas in platen region, K
A_{ph} = inlet area of heating surface behind the platen superheater, m^2
ζ_r = correction factor of fuel; for coal and liquid fuel $\zeta_r = 0.5$.

(2) Superheater, Reheater, and Economizer

For the superheater, reheater, and economizer the heat absorption is calculated, as follows:

$$Q_{co} = \frac{D(H_{so} - H_{si})}{B} \tag{7-25}$$

where H_{so}, H_{si} are outlet and inlet enthalpy, respectively, of steam in the reheater, superheater, or water in the economizer, kJ/kg, and D is the flow rate of steam.

(3) Air Heater

For an air heater the absorption of fluid is given by

$$Q_{co} = \left(\beta''_{ai} + \frac{\Delta\alpha_{ai}}{2}\right)(H_{ao} - H_{ai}) \tag{7-26}$$

where
β''_{ai} = excess air coefficient at the outlet of air heater
H_{ai}, H_{ao} = enthalpy of air at the outlet and inlet, respectively, of the air heater
$\Delta\alpha_{ai}$ = leakage air coefficient of the air heater.

b) Convection Heat Transfer Equation for Heating Surface

The convective heat transfer per kilogram of fuel burnt is found by dividing the heat transfer by the fuel firing rate:

$$Q = \frac{K A \Delta t}{B} \tag{7-27}$$

where

Q = amount of convective transfer for 1 kg fuel burnt, kJ/kg
A = area of heat transfer, m^2
K = overall heat transfer coefficient, kW/m^2°C
ΔT = average temperature difference between gases and working fluid, °C
B = fuel consumption

The method for calculation of the overall heat transfer coefficient, K, is discussed below.

7-7-2 Overall Coefficient of Heat Transfer, K

a) Basic Equations

The sensible heat of gases is transferred to the working fluid through several thermal resistances. The overall heat transfer coefficient, K, of this process is given by.

$$K = \frac{1}{\dfrac{1}{h_g} + \dfrac{\delta_{as}}{\lambda_{as}} + \dfrac{\delta_{me}}{\lambda_{me}} + \dfrac{\delta_{sc}}{\lambda_{sc}} + \dfrac{1}{h_s}} \tag{7-28}$$

where h_g is the combined radiation and convective heat transfer of flue gases; λ_{as}, λ_{me}, λ_{sc} are thermal conductivities of layers of ash deposits, tube metal and scale deposits, respectively; δ_{as}, δ_{me}, δ_{sc} are thicknesses of ash deposits, metal wall and scale deposits in side the tubes, respectively; h_s is the convective heat transfer coefficient of water or steam.

The thermal conductivity of the tube metal is very high. So,

$$\frac{\delta_{me}}{\lambda_{me}}$$

is negligible. The scale is not allowed to be deposited inside the tubes of a modern high-capacity boiler. So we can neglect the scale resistance; i.e

$$\frac{\delta_{sc}}{\lambda_{sc}} = 0$$

The heat transfer coefficient equation can thus be simplified as

$$K = \frac{1}{\dfrac{1}{h_g} + \dfrac{\delta_{as}}{\lambda_{as}} + \dfrac{1}{h_s}} \tag{7-29}$$

The outer wall of the tubes is fouled by ash in the gases, which develops a layer, adding a thermal resistance. Since δ_h, λ_h are difficult to find, their influences on heat transfer are taken into account by an *ash deposits coefficient* ε and *effectiveness factor* ψ. Furthermore, the gases sometimes cannot sweep the tubes completely. A *coefficient of utilization* K_u is introduced to take account of the incomplete sweeping of a heating surface. For the platen superheater, the coefficient of utilization is purely the modification of the incomplete sweeping. For the tube-type air heater the coefficient of utilization is used to consider the influences of both ash deposits and incomplete sweeping on heat transfer.

(1) Ash Deposit Coefficient, ε

For the staggered tubes and the platen superheater of the boiler burning solid fuels, the ash deposits coefficient ε used to take account of the influence of ash deposits on heat transfer is determined by

$$\varepsilon = \frac{1}{K} - \frac{1}{K_0} \tag{7-30}$$

where K = heat transfer coefficient of the fouled tubes, kW/(m^2 °C)
          K_0 = heat transfer coefficient of the clean tubes, kW/(m^2 °C)

Based on Eq. (7-28) we obtain

$$K = \frac{1}{\dfrac{1}{h_g} + \varepsilon + \dfrac{1}{h_s}} \tag{7-31}$$

The ash deposit coefficient ε is determined as follows:

For the platen superheater of boilers burning solid fuel, ε can be taken from Figure 7-20 according to the fuel characteristics and the average temperature. For platen superheaters burning heavy oil, $\varepsilon = 0.00516$ m^2 °C/W.

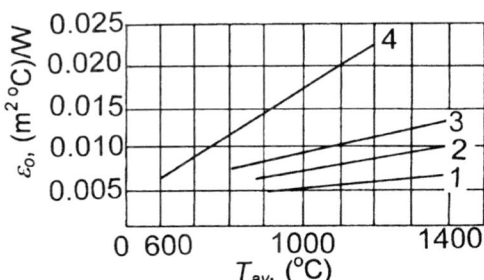

FIGURE 7-20. Ash deposit coefficient of platen superheaters at different temperatures and for different fuels: (1) nonslagging coal, (2) medium slagging coal using soot blower, (3) medium slagging coal without soot blower or serious slagging coal with soot blowing, (4) combustible shale

Staggered tubes of boilers burning solid fuels suffer from a higher degree of fouling. The basic ash deposit coefficient, ε_0, which depends on the flow velocity and the longitudinal pitch (Figure 7-21a), is modified by two correction factors. For economizers operating at gas temperatures below 400°C, an additional correction $\Delta\varepsilon$ is used to account for the higher fouling or deposition potential in the first-stage economizer. Thus ε is found by the equation

$$\varepsilon = C_d C_{R_{30}} \varepsilon_0 + \Delta\varepsilon \tag{7-32}$$

where C_d = correction factor for tube diameter [(Fig. 7-21b)]
$C_{R_{30}}$ = correction factor for ash sizes. It is given by the following equation:

$$C_{R_{30}} = 1 - 1.18 \log \frac{R_{30}}{33.7} \tag{7-33}$$

It is also required for boiler tube banks in a small boiler. For first-stage economizers

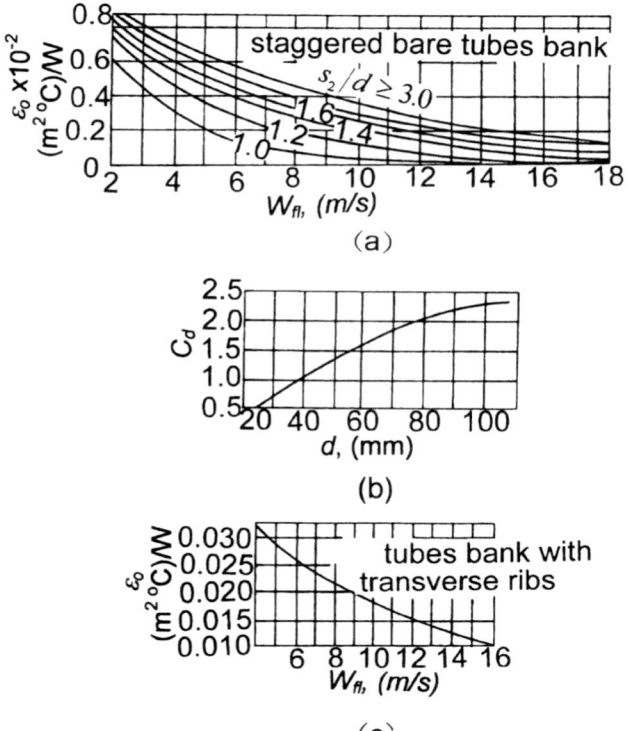

(a)

(b)

(c)

FIGURE 7-21. Ash deposit coefficient of staggered bank of tubes: (a) basic coefficient, (b) correction factor for tube size, (c) deposit coefficient for tube banks with transverse ribs

at gas temperatures below 400°C, for second-stage economizers, for boiler tube banks except as mentioned above, and for superheaters $\Delta\varepsilon = 0.002$. For pulverized fuel burning of anthracite smalls in heating surfaces behind the superheater, the value of $\Delta\varepsilon = 0.002$. (Perkov, 1966).

For tube banks with transverse ribs, the ash deposit coefficient is shown in Figure 7-21c.

(2) Effectiveness Factor, ψ

The *effectiveness factor*, ψ, is the ratio of the heat transfer between fouled and clean tubes. It takes account of the influence of ash deposits on heat transfer tubes. For solid fuel burning boilers it is applied to in-line tubes, while for heavy oil fired boilers it is used for both in-line or staggered tubes, i.e.

$$\psi = \frac{K}{K_0} \qquad (7\text{-}34)$$

So, the heat transfer coefficient, K, for the fouled tube is

$$K = \psi K_0 = \psi \frac{1}{\dfrac{1}{h_g} + \dfrac{1}{h_s}} \qquad (7\text{-}35)$$

For in-line tubes in an anthracite or semi-anthracite burning boiler, $\psi = 0.6$. For bituminous or brown coal, $\psi = 0.65$. For heavy oil fired boilers with staggered and in-line tubes the values of ψ are given in Table 7-6.

(3) Coefficient of Utilization, K_u

The coefficient of utilization, which takes into account the incomplete sweeping of heating surfaces, is shown in Figure 7-22 for platen superheater or screen tubes.

The coefficient of utilization for air heaters is given in Table 7-7.

The following points must be considered when using the Table 7-7:

1. The value of K_u given in the table should be decreased appropriately when there is a division plate in the air path for the tubular air heater. K_u is reduced by 0.1 for one division plate, 0.15 for two division plates.
2. The value of K_u is decreased by 0.1 when the air temperature at the inlet of the tubular air heater is lower than 80°C.

TABLE 7-6. Effectiveness factor, ψ, for staggered tube banks in a heavy oil fired boiler.

Heating surface	Gas velocity	ψ
First and second stage of economizers	4–12 m/s	0.70–0.65
using soot blower	12–20 m/s	0.65–0.6
Superheater in vertical duct using soot blower	4–12 m/s	0.65–0.6
Superheater in horizontal duct and Screen tubes	12–20 m/s	0.6
Economizer in low-capacity boilers with inlet	4–12 m/s	0.55–0.50
temperature less than 100°C		

TABLE 7-7. Coefficient of heat utilization, K_u, for air heaters.

Fuels fired	Tubular air heater, low-temperature section	Without intermediate high-temperature section	Rotary air heater
Anthracite-peat	0.80	0.75	If air leakage ratio $\Delta\alpha =$ 0.15, then $K = 0.9$
Heavy oil wood	0.80	0.85	
Other fuels	0.85	0.85	If $\Delta\alpha = 0.2$–0.25, then $K = 0.8$

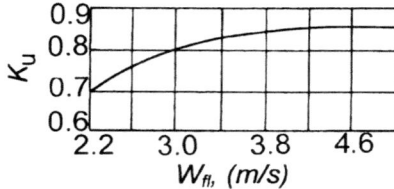

FIGURE 7-22. Coefficients of heat utilization, K_u, for platen superheater and screen tubes

3. The value of K_u is decreased by 0.1 when a regenerative air heater is used, heavy oil is fired, and the input air temperature is lower than 60°C.

b) Overall Heat Transfer Coefficient, K, for Different Convective Heating Surfaces

(1) Coefficient of Heat Transfer, K, for Rear Superheater Platen

The radiant flux from the furnace increases the temperature of outside surface of deposits layer on heating surface increase, which reduces heat transferred by convection. The total coefficient of heat transfer, K, can be obtained as

$$K = \frac{h_a}{1 + \left(1 + \dfrac{Q_{ra}}{Q_{co}}\right)\left(\varepsilon + \dfrac{1}{h_s}\right)h_a} \tag{7-36}$$

where Q_{ra} = heat tranferred by convection kW/m^2
         Q_{ca} = radiative heat coming from the furnace and absorbed by platen, kW/m^2

(2) Heat Transfer Coefficient, K, for Convective Superheaters

In case of solid fuel firing and convective superheaters with staggered tubes, K is calculated based on Eq. (7-31). For convective superheaters with in-line tubes (solid fuel firing) or with staggered or in-line tubes and when liquid fuel is fired, K is calculated based on Eq. (7-36).

(3) Coefficient of Heat Transfer K for Economizer

For economizers with staggered tubes in solid fuel fired systems, K is calculated based on Eq. (7-31). However, since the working fluid in economizer is water and the heat transfer coefficient of water-side is much greater than that of flue gas, the term $1/\alpha_2$ in Eq. (7-31) is very small and can be neglected. Eq. (7-31) is simplified as

$$K = \frac{1}{1 + \varepsilon \alpha_g} \tag{7-37}$$

(4) Coefficient of Heat Transfer, K, for Air Heater

For a tubular air heater, U is given by

$$K = K_u \frac{h_g \cdot h_s}{h_g + h_s} \tag{7-38}$$

where
$\qquad h_g$ = heating coefficient on the gases side
$\qquad h_s$ = heating coefficient on the air side
$\qquad K_u$ = coefficient of heat utilization

For a rotary air heater K is given by

$$K = \frac{K \cdot K_{ic}}{\dfrac{1}{x_1 h_g} + \dfrac{1}{x_2 h_s}} \tag{7-39}$$

where
$\qquad x_1$ = ratio of gas side area and the total area, $x_1 = A_{gas}/A_{tot}$
$\qquad x_2$ = ratio of air side heating surface area and the total area;
$\qquad\qquad x_2 = A_{air}/A_{tot}$
$\qquad A_{gas}$ = heat transfer area on the gas side
$\qquad A_{air}$ = heat transfer area on the air side
$\qquad A_{tot}$ = total flow area
$\qquad K_{ic}$ = influence factor of unsteady heat exchange for the corrugated plate

If the thickness of the heat storage plates is between 0.6 and 1.2 mm, the values of K_{ic} for different rotor speed is given by Table 7-8.

c) Coefficient of Convective Heat Transfer

The total gas-to-wall coefficient of heat transfer h_g consists of the convective heat transfer coefficient, h_{co}, and the radiative heat transfer coefficient, h_{ra}. The flue gas does not necessarily sweep uniformly over all convective tubes. So to take

TABLE 7-8. Relation between N and K_{ic} of the rotary air heater.

N(rpm)	0.5	1.0	≥ 1.5
K_{ic}	0.85	0.97	1.0

account of this nonuniform sweeping of the gas a factor, K, is used to determine the total heat transfer coefficient.

$$h_g = K(h_{co} + h_{ra}) \tag{7-40}$$

For cross-flow $K = 1.0$, but for most mixed-flow conditions $K = 0.95$.

The wall to the fluid heat transfer coefficient, h_g, is only convective. The convective heat transfer coefficient h_{co} is calculated using the following equations.

(1) Coefficient of Convective Heat Transfer for Cross-Flow over In-Line Bank

The convection heat transfer coefficient for the cross-flow in-line bank is determined by the following equation.

$$h_{co} = 0.2C_zC_s\frac{\lambda}{d}\,\mathrm{Re}^{0.65}\,\mathrm{Pr}^{0.33} \tag{7-41}$$

where $\mathrm{Re} = \frac{wd}{\nu}$, $\mathrm{Pr} = \frac{\nu c_p \rho}{\lambda}$

    λ = thermal conductivity at mean temperature of air flow, kW/m°C
    ν = kinematic viscosity at mean temperature of air flow, m^2/s
    d = tube diameter, m
  w_{air} = air flow velocity, m/s
    C_p = specific heat of gases under the certain pressure, kJ/kg
    C_z = correction factor for the tube row number Z_2 along the direction
        of gas flow.

When $Z_2 \geq 10$, $C_z = 1$. When $Z_2 < 10$, C_z is calculated by

$$C_z = 0.91 + 0.125(Z_2 - 2) \tag{7-42}$$

C_s is the correction factor of the geometric arrangement of the tube bank, which is calculated by the following equation:

$$C_s = \left[1 + (2\sigma_1 - 3)\left(1 - \frac{\sigma_2}{2}\right)^3\right]^{-2} \tag{7-43}$$

where $\sigma_1 = S_1/d$ is the transverse pitch and $\sigma_2 = S_2/d$ is the longitudinal pitch. When $\sigma_1 \leq 1.5$ or $\sigma_2 \geq 2$, $C_S = 1.0$.

(2) Convective Heat Transfer Coefficient of for Cross-Flow Staggered Tubes

The coefficient h_{co} for the cross-flow staggered tubes is calculated as follows:

$$h_{co} = C_zC_s\frac{\lambda}{d}\,\mathrm{Re}^{0.6}\,\mathrm{Pr}^{0.33} \tag{7-44}$$

where C_z = correction factor taking account of the number of rows of tube, Z_2 along the direction of gas flow.

    When $Z_2 < 10$ and $S_1/d < 3.0$, $C_z = 3.12Z_2^{0.05} - 2.5$
    When $Z_2 < 10$ and $S_1/d > 3.0$, $C_z = 4Z_2^{0.02} - 3.2$
    When $Z_2 > 10$, $C_z = 1.0$ $(7-45)$

Here C_s is a correction factor that takes into account the geometric arrangement of the tube bank and its spacing of tube bundles,

$$\text{When } 0.1 < \varphi_\sigma < 1.7, \quad C_s = 0.34\varphi_0^{0.1}$$
$$\text{When } 1.7 < \varphi_\sigma < 4.5 \text{ and } S_1/d < 3.0, \quad C_s = 0.275\varphi_\sigma^{0.5}$$
$$\text{When } 1.7 < \varphi_\sigma < 4.5 \text{ and } S_1/d > 3.0, \quad C_s = 0.34\varphi_\sigma^{0.1} \qquad (7\text{-}46)$$

where $\varphi_\sigma = \frac{\sigma_1 - 1}{\sigma_2 - 1}$, $\sigma_2 = \sqrt{\frac{\sigma_1^2}{4} + \sigma_2^2}$

(3) Coefficient of Convective Heat Transfer for Longitudinal Sweeping

In a superheater, reheater, and economizer working fluid flows through tubes. In the tubular air heater, flue gases flow through the vertical tubes. All of them are in longitudinal flow condition. The coefficient h_{co} is calculated as follows:

$$h_{co} = \alpha_{co} = 0.023 C_l C_t \frac{\lambda}{d_{eq}} \, \text{Re}^{0.8} \, \text{Pr}^{0.4} \qquad (7\text{-}47)$$

d_{eq} = equivalent diameter of the longitudinal flow passage. For the noncircular section, d_{eq} is given by

$$d_{eq} = \frac{4A_{flow}}{P_{flow}} \qquad (7\text{-}48)$$

A_{flow} = cross section area of flow passage, m^2
P_{flow} = perimeter of the flow passage, m
C_t = Correction factor for the temperature difference between wall and medium. When gases and air are heated:

$$C_t = \left(\frac{T}{T_{wa}}\right)^{0.5} \qquad (7\text{-}49)$$

When gases are cooled, $C_t = 1$; for water and superheated steam, when the temperature difference between the medium and wall is not great, $C_t = 1$.
C_l is the correction factor of tube length. When $L/d > 50$, $C_l = 1$.

(4) Coefficient of Convective Heat Transfer for Rotary Air Heater

The coefficients of both gases and air sides are calculated by the following equation:

$$h_{co} = \alpha_{co} = C' C_t C_l \frac{\lambda}{d_{eq}} \text{Re}^{0.8} \, \text{Pr}^{0.4} \qquad (7\text{-}50)$$

where C_t, C_l are the same as the factors in Eq. (7-45).
C' is a factor relating to the shape of storage plates. Three types of heat storage plates are shown in Figure 16-4.

For plates in Figure 16-4a:

When $a + b = 2.4$ mm, $C' = 0.027$
when $a + b > 4.8$ mm, $C' = 0.037$
For collector heat plate (Fig. 16-4b) $C' = 0.027$
For collector heat plate (Fig. 16-4c) $C' = 0.021$

(5) Average Flow Velocity of Working Fluid

The average flow velocity used in Reynolds number (Re) calculations is computed as follows

I) Flow velocity

Flue gas flow velocity:

$$w_{fl} = w_{fl} = \frac{BV_{fl}(t + 273)}{A_{fl}273} \text{ m/s} \tag{7-51}$$

Air flow velocity:

$$w_{air} = w_{ai} = \frac{B\beta_{ai}V^0(t + 273)}{A_{ai}273} \text{ m/s} \tag{7-52}$$

where

B = consumption of fuel, kg/s
V_{fl} = specific volume of gases, Nm^3/kg
t = average temperature of flue gas or air in the heating surface, °C
A_{fl}, A_{ai} = flow area of gases or air, m^2
β_{ai} = average excess air coefficient air heater
V^0 = theoretical requirement of air volume, Nm^3/kg

II) Flow passage area

(I). When the flue gas flows transversely over a bank of smooth tubes, and if its cross section area does not change, the flow area is given by

$$A_{fl} = a \cdot b - Z \cdot L \cdot d \tag{7-53}$$

where a, b = length and width of the passage section
           Z, L, d = number, length and outside diameter of tubes respectively, m

When the section area varies continuously, its average area A is

$$A_{fl} = \frac{2A_{in}A_{out}}{A_{in} + A_{out}} \tag{7-54}$$

where A_{in}, A_{out} are inlet and outlet areas of the passage, respectively.
(II). When the working fluid flow inside the tube the area is given by

$$A_{fl} = n\frac{\pi d_{in}^2}{4} \tag{7-55}$$

where n = number of parallel tubes

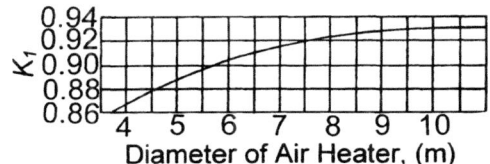

FIGURE 7-23. Flow area factor (K_1), taking account of the area occupied by the rim and division wall, increases with diameter of the air heater

d_{in} = inside diameter of tubes, m

(III) Flow area for rotary air heater

This is given by

$$A_{fl} = 0.785 D_{in}^2 \, x K_1 K_2 \tag{7-56}$$

where D_{in} = inside diameter of the rotor, m; and x = fractional flow area for flue gases or air.

Sectoral area occupied by each section		Fractional flow area, x	
Air passage	Gas passage	Air	Gas
120°	200°	0.278	0.555
120°	180°	0.333	0.500

K_1 = flow factor influenced by the area occupied by the rim and division plate (Fig. 7-23)

K_2 = flow factor for different heat storage plates (=0.89 for Fig. 16-4a; = 0.86 for Fig. 16-4b; = 0.81 for Fig. 16-4c).

d) Radiative Heat Transfer Coefficient

(1) Basic Equations

The radiation heat transfer coefficient is defined as

$$h_{ra} = \frac{Q_{ra}}{T - T_{as}} \tag{7-57}$$

where

Q_{ra} = radiative heat flux between the gas and tube bank, kW/m^2

T = gas temperature, K

T_{as} = temperature of the outer surface (even if it is fouled), K

According to the theory of radiant heat transfer, the sum of radiant heat exchange between gases and the tube bank can be expressed as follows:

$$Q_{ra} = \frac{a_{sh} + 1}{2} \sigma_0 a \left(T^4 - T_{as}^4 \right) \tag{7-58}$$

where

a_{sh} = emissivity of the external ash deposited surface of tubes,

$1 + a_{as}/2$ = average value of a_{sh} and the absolute emissivity, which is used to compensate for the multiple reflection and absorption in the tube bank (for slag screen $a_{sh} = 0.68$, for other tubes $a_{sh} = 0.8$)

σ_0 = Boltzman's constant = 5.670×10^{-11}, kW/m² K⁴

a = emissivity of flue gas at temperature T

From Eqs. (7-57) and (7-58) the coefficient of radiation heat transfer is expressed in following form:

$$h_{ra} = \sigma_0 \frac{a+1}{2} aT^3 \frac{1 - \left(\dfrac{T_{as}}{T}\right)^4}{1 - \left(\dfrac{T_{as}}{T}\right)} \tag{7-59}$$

(2) Calculation of Flue Gas Emissivity

The flue gas emissivity, a, is calculated in the same way as was done for the flame (see Chapter 6). The thickness of the effective radiation layer, S, is still calculated according to the equation $S = 3.6 \ volume/area$. For smooth tube bank and platen superheaters the equation can be used to get the following expressions for the radiation layer thickness:

For smooth tube bank—

$$S = 0.9d_{ou}\left(\frac{4}{\pi}\frac{S_1 S_2}{d_{ou}^2} - 1\right) \tag{7-60}$$

For platen superheater—

$$S = \frac{1.8}{\dfrac{1}{A} + \dfrac{1}{B} + \dfrac{1}{C}} \tag{7-61}$$

For tubular air heater

$$S = 0.9d_{in} \tag{7-62}$$

where

S_1, S_2 = transverse and longitudinal pitch of tube bank, respectively, m

d_{ou} = outer diameter of the tube, m

A, B, C = breadth, height, and depth, respectively, for the platen region, m

d_{in} = inner diameter of the tube, m

(3) Calculation of T_{as}

The temperature of fouled surface T_{as} is calculated as follows: For the radiation of the platen superheater, the convective superheater T_{as} is given by

$$T_{as} = T + \left(\varepsilon + \frac{1}{h_s}\right)(Q_{co} + Q_{ra}) \tag{7-63}$$

where

Q_{co} = convective heat flux, kW/m^2
Q_{ra} = radiative heat flux on platen, kW/m^2
T = temperature of steam, K

For other heating surfaces the following approximate equation may be used:

$$T_{as} = T + \Delta T \tag{7-64}$$

Where ΔT is the temperature difference between the outside surface of the deposits and the working fluid. For the slag screen, $\Delta T = 80°C$; first-stage economizer, $\Delta T = 25°C$; secondary economizer, $\Delta T = 60°C$.

(4) Calculation of Cavity Radiation

The empty space between different heating surfaces forms a gas cavity. The gas cavity increases the thickness of the effective radiation gas layer. As a result the radiation heat exchange between flue gas and the heating surface is increased. So the influence of gas cavity radiation on convective heat transfer is important. In one approach the coefficient of radiant heat transfer, h_{ra}, is modified.

$$h'_{ra} = h_{ra}\left[1 + C\left(\frac{T}{1000}\right)^{0.25}\left(\frac{L_1}{L_2}\right)^{0.07}\right] \tag{7-65}$$

where

h'_{ra} = coefficient of radiant heat transfer corrected, kW/m^2 K
C = empirical constant (for bituminous and anthracite coal $C = 0.4$; for brown coal $C = 0.5$; for heavy oil $C = 0.3$)
L_1 = length or width of the gas cavity between tube banks or wall and tube bank, m
L_2 = length of tube bank along flue gas passage, m
T = average temperature of flue gas in gas cavity, °K

7-7-3 Mean Temperature Difference

In many convective heat transfer situations, the temperature difference between flue gas and working fluid (steam or water) varies continually along the heat surface. So, an average temperature difference is used to simplify the heat transfer calculation. The calculation method of the average temperature difference is determined as follows.

a) Simple Parallel and Counter-Current Flow

The average temperature difference for parallel flow and counter-current flow is expressed in the following equation. It is also called log mean temperature difference:

$$\Delta t = \frac{\Delta t_{max} - \Delta t_{min}}{2.3 \log \dfrac{\Delta t_{max}}{\Delta t_{min}}} \tag{7-66}$$

where

Δt_{max} = greater temperature difference at the inlet or the outlet of the heat surface, °C

Δt_{min} = smaller temperature difference at the inlet or the outlet of the heat surface, °C

When $\Delta t_{max}/\Delta t_{min} < 1.7$, the average temperature may be approximated as

$$\Delta t = \frac{\Delta t_{max} + \Delta t_{min}}{2} \tag{7-67}$$

When the temperature of the working fluid (steam/water) is constant, for examle, in screen tubes and convective evaporating tubes of low-capacity boilers, Eq. (7-67) is also used to calculate average temperature difference.

In convective heat transfer surfaces simple parallel flow and pure counter-current flow are almost nonexistent. Generally, the flow direction of the flue gas is perpendicular to the axis of tubes. However, experience shows that so long as there are four bends of horizontal tubes in the flue gas path, the vertical cross flow of gas can be considered as pure parallel flow or pure counter current flow. This makes the calculation of average temperature difference convenient.

b) Complex Flow

Parallel flow and counter-current flow may simultaneously exist in one section of a convective heating surface. This is called *complex flow* (Fig. 7-24). For a given set of inlet and outlet temperatures of the fluids, the temperature difference of parallel flow is the greatest, while that of counter-current flow is the lowest, and that of complex flow is between that of parallel and counter-current flows.

The average temperature difference can be calculated by the following equation:

$$\Delta t = 0.5(\Delta t_{pa} + \Delta t_{co}) \tag{7-68}$$

when

$$\Delta t_{pa} \geq 0.92 \Delta t_{co} \tag{7-69}$$

where

Δt_{co} = temperature difference of counter-current flow, °C
Δt_{pa} = temperature difference of parallel flow, °C

Alternatively, the temperature difference is determined by

$$\Delta t = K_{td} \Delta t_{co} \tag{7-70}$$

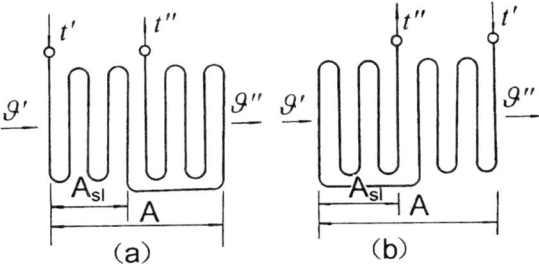

FIGURE 7-24. Complex in-series flow arrangement: (a) first section parallel flow and second section in counter current flow; (b) first section in counter current and second section in parallel flow

where K_{td} is the correction factor of the temperature difference of the complex flow.

The value of K_{td} is determined by the flow type and the thermal parameters.

(1) Multiple Series Flow

These heating surfaces consist of parallel and counter-current flow sections in series. In Figure 7-24a the fluid first flows parallel to the gas; then it turns around to flow counter to the gas flow. In Figure 7-24b the flow is first counter then parallel.

The correction factor, K_{td}, of temperature difference of this type of flow is shown in Figure 7-25. In Figure 7-25, A_{pa} is heating surface area of parallel flow, A is the total heating surface area.

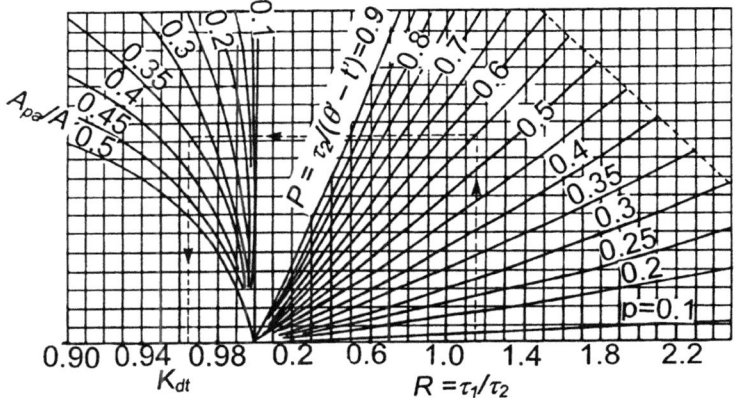

FIGURE 7-25. Correction factor for complex multiple parallel flow arrangement; here $\tau_{max} = \max[t'' - t', \vartheta' - \vartheta'']$, $\tau_{min} = \min[t'' - t', \vartheta' - \vartheta'']$

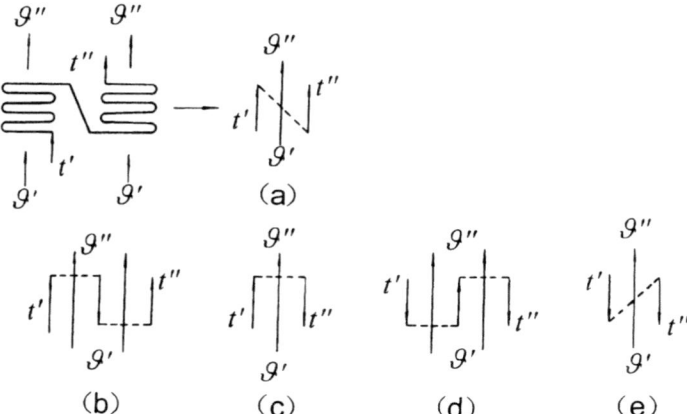

FIGURE 7-26. Multiple parallel-flow arrangement

(2) Multiple Parallel Flow

In a multiple parallel flow (Fig. 7-26) one medium flows through the heating surface in series and the other medium flows through heating surfaces in parallel. The correction factor (K_{td}) of the temperature difference for the multiple parallel flow is shown in Figure 7-25:

$$P = \frac{\tau_{min}}{\vartheta' - t'}, \quad R = \frac{\tau_{max}}{\tau_{min}}$$

(3) Multiple Cross-Current Flow

In a multiple cross-current flow situation, the heating surface consists of one or more sections; and the flow direction of two mediums crosses. There are one, two, three, four flow crossings. When the number of the flow crossings is more than four, it falls into the multiple series flow category.

When the flow is generally counter-current, the correction factor K_{td} is found as shown in Figure 7-27. Here

$$P = \frac{\tau_{min}}{\vartheta' - t'}, \quad R = \frac{\tau_{max}}{\tau_{min}}$$

where $\tau_{max} = \max [t'' - t', \theta' - \theta'']$, $\tau_{min} = \min[t'' - t', \theta' - \theta'']$.

If the nature of the flow is generally parallel, the correction factor K_{td} is still found by

$$K_{td} = \frac{1 - [1 - P(R + 1)]^{\frac{1}{nf}}}{R + 1} \tag{7-71}$$

where nf is the number of the flow crossings.

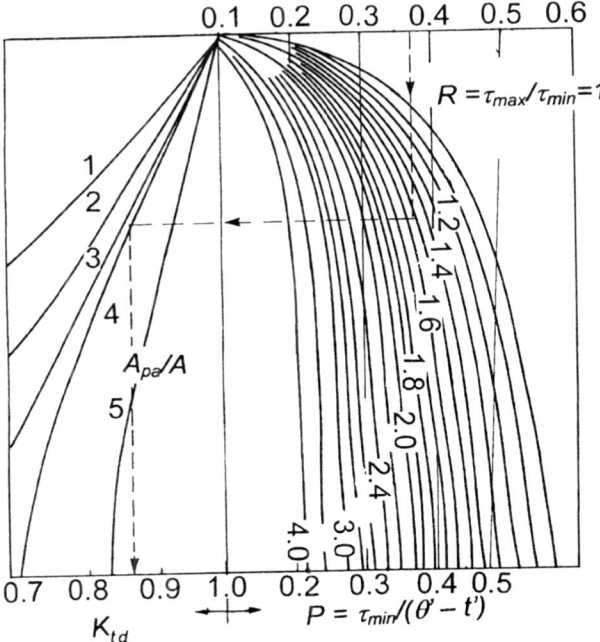

FIGURE 7-27. Correction factor for complex cross-flow arrangement. Here $\tau_{max} = \max[t'' - t', \vartheta' - \vartheta'']$, $\tau_{min} = \min[t'' - t', \vartheta' - \vartheta'']$

7-7-4 Definition of Heating Surface Area

a) Steam and Water as the Working Fluid

The heat transfer coefficient on the gas side is much lower than that on the steam or water side. So heat flux is determined mainly by coefficient on the gas side and the area of outer surface of tube is used as heat transfer area.

$$A = n \cdot l \cdot d_{ou} \tag{7-72}$$

where

n = number of parallel tubes
l = length of the tube, m
d_{ou} = outside diameter of the tube, m

b) Rear Platen Superheater

The rear platen superheater absorbs the radiative heat of the flue gas in the platen region, radiation from the furnace, and the convective heat of the flue gas sweeping over it. However, the radiative exchange is the main form of heat transfer. So, the heat transfer area of the rear platen superheater is calculated as follows:

$$A = 2 \cdot a \cdot b \cdot n \cdot n_p \cdot x_p \qquad (7\text{-}73)$$

where

a = depth of the platen, m
b = number of the platen, m
n_p = number of platens
x_p = angular coefficient of the platen

c) Tubular Air Heater

The heat transfer coefficients on the air and the gas side in the tubular air heater are of similar magnitudes. So, the average between inner and outer diameters of the tube, d_{av}, is used to calculate the heat-transfer area:

$$A = n \cdot l \cdot \pi \cdot d_{av} \qquad (7\text{-}74)$$

where n is the number of tubes of length l.

d) Regenerative Air Heater

The surface area of a regenerative air heater is calculated by the following equation:

$$A = 0.746 D_{in}^2 h \cdot K_1 \cdot C_v \qquad (7\text{-}75)$$

where

D_{in} = inner diameter of the outer shell of the cylinder, m
K_1 = influence coefficient accounting for the volume occupied by the wheelbase, and division plate in the heated surface
C_v = heat transfer area per unit volume of air heater as shown in Fig. 16-4, m^2/m^3
h = cylinder height of the heated surface, m

7-8 Design Methods of Convection Heating Surfaces

If both inlet and outlet gas and steam/water temperatures of a convective heating surface are known, the surface area and the amount of heat absorbed can be calculated by simple heat transfer calculations. However, outlet temperatures of the working fluid or the gas are not always known in advance as they depend on the amount of heat transfer. Thus, the calculation uses an iterative, or trial and error, method. First, the outlet working fluid temperature is assumed. Then the heat absorbed Q_p from the gas and therefore outlet temperature of steam/water can be obtained by a heat balance. After that, the log mean temperature difference between two sides of the heat exchanger is computed. It is used to calculate the heat transferred, Q_d, for the given heat transfer coefficient and the surface area. As the heat transfer should be equal to the heat absorbed, a check should be made as follows. The calculated and chosen values can be taken to be acceptably equal if

$$\left| \frac{Q_d - Q_p}{Q_p} \right| \leq 2\% \quad \text{for major heating surfaces} \tag{7-76}$$

$$\left| \frac{Q_d - Q_p}{Q_p} \right| \leq 10\% \quad \text{for other heating surfaces} \tag{7-77}$$

If the error is larger than the above limits, a new θ'' should be chosen and the calculations repeated until the error is reduced to within this range.

Nomenclature

A	area of heating surface, m^2
A_{flow}	cross section area of longitudinal flow passage, m^2
a	emissivity of flue gas in platen region
a	depth of the platen (Eq. 7-73), m
a	length of the passage section (Eq. 7-53), m
A	flow area of the medium for the rotary air heater (Eq. 7-56), m^2
A	heat transfer area of rear platen superheater (Eq. 7-73), m^2
A	surface area of the regenerative air heater (Eq. 7-75), m^2
A	surface area of the tubular air heater (Eq. 7-74), m^2
A	total heating surface area of parallel flow, m^2
A, B, C	breadth, height and depth, respectively, for the platen region (Eq. 7-61), m
A_{ai}	heat transfer area on the air side, m^2
A_{fl}, A_{ai}	average flow area of gases or air, m^2
A_{ga}	heat transfer area on the gas side, m^2
A_{in}	inlet area of passage, m^2
A_{out}	outlet area of passage, m^2
A_{pa}	heating surface area of parallel flow, m^2
A_{ph}	inlet area of heating surface behind the platen superheater, m^2
b	width of platen, m
b	width of the passage section (Eq. 7-53), m
b	width of the platen (Eq. 7-73), m
B	fuel firing rate, kg/s
C	coefficient relating to the shape of collected heat plate (Eq. 7-49)
C	empirical constant (Eq. 7-65)
C	heat transfer area (Eq. 7-75), m^2
C_d	correction factor for tube diameter (Eq. 7-32)
C_l	correction factor of tube length (Eq. 7-49)
C_p	specific heat of gases at constant pressure, kJ/kg
C_{R30}	correction factor for ash sizes (Eq. 7-33)
C_s	correction factor for geometry (Eq. 7-42)
C_t	correction factor for temperature difference (Eq. 7-49)

C_z	correction factor for tube rows (Eq. 7-42)
D	steam flow rate, kg/s
d	tube diameter, m
d_{av}	average diameter (Eq. 7-74), m
d_n	inner diameter of the tube, m
c_p	specific heat of the steam, kJ/kg $\cdot$ K
c_m	specific heat of the metal, kJ/kg $\cdot$ K
w_f	steam flow velocity at the header inlet, m/s
w_x	steam flow velocity at x position along the header length direction, m/s
H_{ai}'', H_{ai}'	theoretical enthalpy of air at the outlet and inlet of the air heater respectively, kJ/kg
d_{eq}	equivalent diameter, m
D_{in}	inner diameter of the outer shell of the cylinder or rotor, m
d_{in}	inside diameter of the tube, m
d_{ou}	outside diameter of the tube, m
F	total area of the water wall surface
F_c	area of the furnace outlet window
G	steam flow, kg/s
h	height difference between the inlet and the outlet of the header, m
h_{co}	convective heat transfer coefficient, kW/m^2 $^\circ$C
h_{ra}	radiative heat transfer coefficient, kW/m^2 $^\circ$C
h_g	gas side heat transfer coefficient, kW/m^2 $^\circ$C
h_s	steam or water side heat transfer coefficient, kW/m^2 $^\circ$C
H_{so}, H_{si}	outlet and inlet enthalpy of steam, kJ/kg
H_{gi}, H_{go}	enthalpies of gases at inlet and outlet of the convective section, kJ/kg
K	overall heat transfer coefficient kW/m^2 $^\circ$C
K_u	coefficient of utilization
K_l	influence coefficient (Eq. 7-75)
K_1	influence coefficient of flowing area (Eq. 7-56)
K_2	influence coefficient of flowing area (Eq. 7-56)
K_p, K_o	flow resistance factor
K_o	heat transfer coefficient of the clean tubes (Eq. 7-30), kW/m^2 $^\circ$C
l	length of the tube, m
L	length of the tubes, m
L_1, L_2	length of gas cavity and length of tube bank along flue gas passage, m
n	number of parallel tubes, platen crossings
n_p	number of platens
P	parameter nondimensional temperature (Fig. 7-23)
P_{flow}	perimeter of longitudinal flow passage, m
q	heat flux, kW/m^2

Q	quantity of the convective transfer of 1 kg fuel, kJ/kg
Q_{ca}	amount of heat absorbed by convection, kJ/s
Q_{co}	heat transfer by convection, kJ/s
Q_{fu}	total heat absorption in the furnace, kJ/kg
Q'_{ra}	amount of radiation coming from the furnace and falling on the platen, kJ/s
Q''_{ra}	radiation heat passing through platen tubes, kW/m^2
Q_{ra}	radiation heat coming from furnace and absorbed by steam platen superheater, kJ/kg;
Q''_{rap}	radiation heat given out by flue gas in platen section to the next section, kJ/kg
R_{30}	percentage of particles less than 30 μm, %
R	parameter nondimensional temperature (Fig. 7-32)
S	thickness of effective radiation layer, m
s_1	spacing of platen, m
S_1, S_2	transverse and longitudinal pitch of tube bank respectively, m
t	average temperature of flue gas in heating surface, °C
T	temperature, K
T_p	average temperature of flue gas in platen region, K
t''	inlet temperature of the working fluid (steam, water, or air), K
t'	outlet temperature of the working fluid (steam, water, or air), K
T_{ai}	inlet air temperature, K
T_{ao}	outlet air temperature, K
T_{as}	temperature of fouled surface, K
U	medium sweeping perimeter, m
V	volume occupied flue gas, m^3
V_{fl}	specific volume of gases, Nm^3/kg
V^0	theoretical requirement of air volume, nm^3/kg
w	average flow velocity, m/s
w_o	average velocity, m/s
x	fractional flow area for flue gases or air
x_1	flow area fraction of gases (Eq. 7-40)
x_2	flow area fraction of air (Eq. 7-40)
x_p	angular coefficient of radiation (Eqs. 7-23, 7-73)
Z	number of the tubes

Greek Symbols

β_{ai}	average excess air coefficient air heater (Eq. 7-52)
α'_{ra}	coefficient of radiant heat transfer corrected, kW/(m$^2 \cdot$ °C)
$\lambda_{as}, \lambda_{me}, \lambda_{sc}$	conductivity coefficients of ash deposit layer, metallic layer, and scale deposit layer, respectively, kW/mK
α_2	convective heat exchange coefficient of working fluid side (Eq. 7-28), kW/m^{2}K
ρ_m	density of the metal, kg/m^3

$\beta_{ai}^{''}$	excess air coefficient at the outlet of air heater (Eq. 7-26)
α_2	heat transfer coefficient between the inner wall and the steam, kW/m^2K
$\Delta\alpha_{ai}$	leakage air coefficient of the air heater (Eq. 7-26)
β_H	ratio of the internal surface area of the header to that of the tube
β_G	ratio of the mass of metal of the unheated header pipe to that of the tube
β	ratio of the outside diameter to the inner diameter of the tube
λ	flow friction coefficient in the header (Eq. 7-7)
Δt	average temperature difference between gases and working fluid, K
Δp_{ld}	flow pressure drop, Pa
Δp_{lz}	flow resistance, Pa
ρ_f	steam density at the header inlet, kg/m^3
$\delta_{as}, \delta_{me}, \delta_{sc}$	thicknesses of ash deposit layer, metallic layer, and scale deposit layer, respectively (Eq. 7-28), m
δ	thickness of the tube wall, m
$\sigma_1 = S_1/d$	relative transverse pitch (Eq. 7-43)
$\sigma_2 = S_2/d$	relative longitudinal pitch (Eq. 7-43)
Δp_{lz}^{max}	maximum flow resistance, Pa
Δp_{lz}^{max}	maximum rise in static pressure in a header, Pa
η_h	coefficient of nonuniformity of heat load at the furnace outlet
α	emissivity of flue gas (Eq. 7-58)
K_{ic}	influence coefficient of unsteady heat exchange for corrugated plate
λ	thermal conductivity at mean temperature of air flow, $kW/m^{2\circ}C$
σ_0	Boltzman's constant, $= 5.67 \times 10^{11}$ $kW/(m^2 \cdot K^4)$
ζ_r	correction coefficient of fuel (Eq. 7-24)
α_{sh}	emissivity at the outside of the deposits' heated surface
α	emissivity of flue gas in the platen domain (Eq. 7-22)
β	coefficient that takes into account heat exchange between the furnace and the heating surface
$\Delta\alpha H_1$	heat carried by leakage air into boiler, kJ/kg
Δh_o	average enthalpy rise, kJ/kg
Δh_p	enthalpy increase in the specific tube, kJ/kg
Δp_{hl}^{max}	maximum rise in static pressure in the collecting header, Pa
ΔP_o	average pressure drop, Pa
ΔP_p	pressure drop for a specific tube, Pa
Δt	average temperature difference, °C
Δt_{co}	temperature difference of counter-current flow, °C
Δt_{max}	greater temperature difference, °C
Δt_{min}	smaller temperature difference, °C
Δt_{pa}	temperature difference of parallel flow, °C

ε	ash deposit coefficient, $m^2\,°C/kW$
ε_0	inherent coefficient of ash deposite, $m^2\,°C/kW$
η_a	nonuniformity coefficient for heat flux (Eq. 7-4)
η_g	nonuniformity of heating surface areas (Eq. 7-4)
η_q	nonuniformity coefficient for flow rate (Eq. 7-4)
φ	coefficient of heat retention (Eq. 7-18)
Φ	coefficient of nonuniform heat absorption (Eq. 7-1)
λ	coefficient of flow resistance for friction (Eq. 7-12)
ν	kinematic viscosity under the mean temperature of the air flow, m^2/s
ν_o	average specific volume, m^3/kg
θ'	inlet temperature of the gas, K
θ'	outlet temperature of the gas, K
τ_1	temperature difference for the flow of first parallel then counter-current, K
τ_2	temperature difference for the flow of first counter-current then parallel, K
τ_{max}	$max[\tau_1, \tau_2]$, K
τ_{min}	$min[\tau_1, \tau_2]$, K
ξ	coefficient of flow resistance in fittings (Eq. 7-12)
ψ	effectiveness factor (Eq. 7-34)

Dimensionless Numbers

Pr	Prandtl number $Pr = \nu_\rho C_p / \lambda$
Re	Reynolds number $Re = \frac{wd}{\nu}$

References

British Electricity International (BEI) (1991) Modern Power Station Practice, Vol. B, Boilers and Ancillary Plants, London, Pergamon Press.

Lin, Z.H., ed. (1991) "Boilers, Evaporators, and Condensers," S. Kakac, ed. John Wiley & Sons, New York, Chapter 8.

Stultz, S.C., and Kitto, J.B. (1992) "Steam—Its Generation and Use." Babcock & Wilcox, Barberton, Ohio.

8
Swirl Burners

Except for bubbling and circulating fluidized bed boilers, all other boilers use burners as the primary source of energy. The burner plays an important role in boilers as well as furnaces. The ignition of fuel and aerodynamic and combustion conditions in the furnace are all governed primarily by the construction and arrangement of the burners. The performance of the burner determines whether the combustion equipment will operate reliably and economically.

Burners are of many types. Based on their principle of operation they can be broadly classified into two groups

- Swirl types
- Direct or parallel-flow type

A *swirl burner* uses a set of guiding vanes to give a swirling motion to the combustion air. This creates a recirculation zone, which helps ignite the fuel as well as sustain the flame. The recirculation also helps accelerate the process of mixing of fuel and combustion air. Direct burners do not add any swirl to the flow. Here the fuel and air flows are parallel to each other. The direct burners are generally used in corner or tangential fired boilers. So they are also called *tangential burners*. The design method for direct fired burners is described in Chapter 10.

Swirl burners are very different in their construction, aerodynamic conditions, and flame shapes than direct burners. Most front and opposite wall fired boilers use swirl burners. Packaged boilers also use this type of burner. This chapter explains the operating principles and design methods of the swirl burners.

8-1 Design of Swirl Burner

Basic requirements for good performance of a burner may be summarized as follows:

1. Proper aerodynamics at the exit of the burner—required to ensure good ignition, stable, and efficient combustion.
2. Good load control.

TABLE 8-1. Broad classification of swirl burners.

Axial vane burners	Tangential vane burners	Volute burners
Double channel outside-mixing axial vane burners	Direct or parallel flow primary air with adjustable vane	Axial vane-volute burners
Tangential-axial guide vane burners	Swirling primary air and movable stabilizer	Volute and multi-volute burners
Movable axial vane burners		Volute-central cone burners
Tangential guide vane burners		Volute burners

Volute burners are not discussed here as these are less commonly used.

3. Controlled generation of NO_x to meet environmental regulations.
4. Reliable and safe operation.
5. Compatibility with the pulverizing system and the furnace in the case of coal firing systems.

8-1-1 Types of Swirl Burners

Swirl burners are of three major types: volute burners, axial vane burners, and tangential vane burners. Each type is further subdivided as follows (Table 8-1):

a) Axial Vane Swirl Burners

In this design (Fig. 8-1), the secondary air is made to swirl by a set of axial swirling vanes. The primary air, on the other hand, could be either parallel to the burner axis or swirling. The axial vane swirl burner is used mainly to burn lignite and

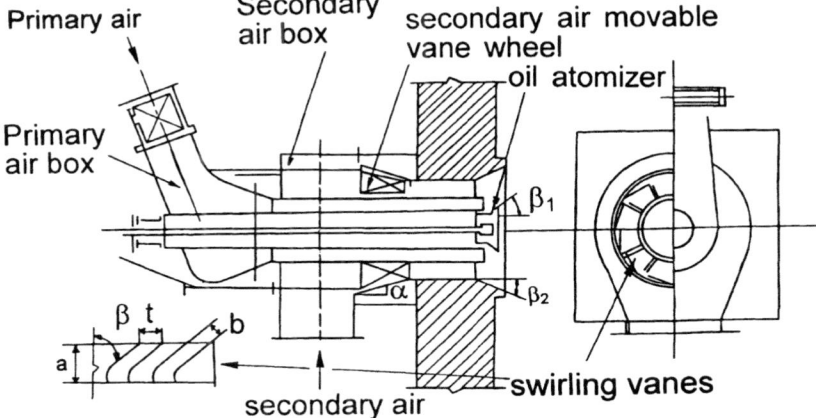

FIGURE 8-1. Axial vane swirl burner without swirling of the primary air

bituminous coal whose volatile content (dry ash-free basis) is more than 25% and whose higher heating value is higher than 16.8 MJ/kg.

The inclination angle β (Fig. 8-1) of the axial vane has a great influence on the burner performance. A large angle will lead to increased resistance of the secondary air. For firing lignite or bituminous coal, the angle β is chosen between 50° and 60°. The swirl intensity of the secondary air is adjusted by moving the axial vane wheel in the conical annulus (Fig. 8-1). This allows a part of the secondary air to flow straight without passing through the axial vane. By adjusting the axial position of the vane, the operator can change the ratio of the straight flow and the swirling flow.

b) Tangential Vane Swirl Burner

In the tangential vane swirl burner (Fig. 8-2) the secondary air enters the furnace through a set of adjustable tangential guide-vanes. Swirl intensity of the secondary air is adjusted by changing the inclination angle of the guide vanes. The number of vanes varies between 8 and 16.

There are two types of tangential vane swirl burners, depending on whether the primary air swirls or not. Figure 8-2 shows a burner without the swirling of the primary air. The primary air flows straight through the central tube. So its flow resistance is much less.

The other model, in which the primary air swirls, is one of the earliest swirl stabilized coal burners used in the industry. The burner input ranges between 15 and 80 MW (Stultz & Kitto, 1992, p. 13-6). The central tube, which carries the primary air and pulverized coal (PC), includes a stabilizer with tangential vanes, as shown in Figure 8-3. A central recirculation zone is created behind the stabilizer, which

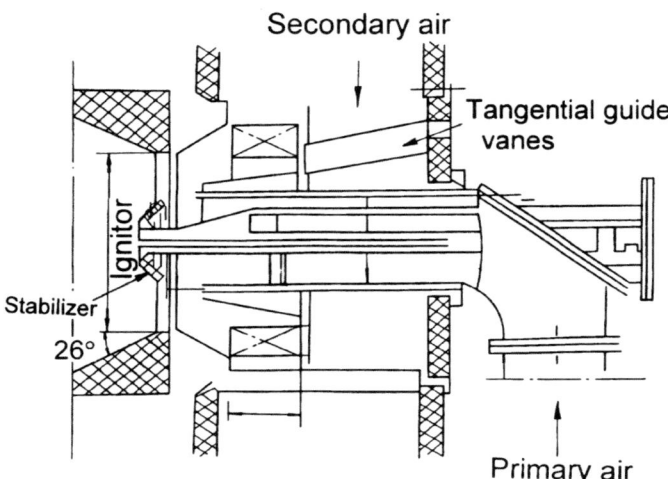

FIGURE 8-2. Tangential vane swirl burner without swirling of the primary air

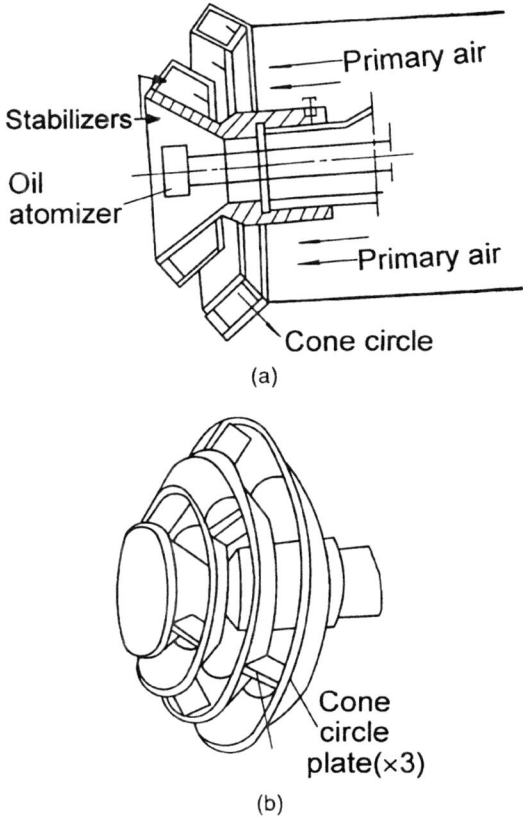

FIGURE 8-3. The central nozzle of a tangential vane swirl burner with conical flame stabilizer

can be moved forward and backward to adjust the dimension of the recirculation zone. The secondary air is admitted through a set of tangential vanes in the annulus around the central nozzle. These vanes of the secondary air register also serve as dampers to control the secondary air flow. The tangential vane swirl burner works well with bituminous coals whose volatile content is above 25%. However, erosion may occur in the stabilizer and at the elbow at the entrance of the primary air and PC. So, the design should allow for easy replacement of these components.

8-1-2 Swirl Intensity

There are three major types of swirlers:

- volute swirler
- tangential vane swirler and
- axial vane swirler

The following section presents a design procedure for tangential and axial vane swirlers.

Swirl intensity, n, represents the intensity of the flow swirl. It is given by

$$n = \frac{M}{KL} \tag{8-1}$$

where M is the swirling momentum (angular momentum) of the air flow ($kg \cdot m^2/s^2$); K represents axial momentum (linear momentum), $kg \cdot m/s^2$; and L is the characteristic size, which is some multiple of the exit diameter, m.

The swirling momentum, M, is given by

$$M = QU_t r \text{ kg} \cdot m^2/s^2 \tag{8-2}$$

where U_t is the tangential velocity of air, m/s; Q is mass flow rate of air, kg/s; r is radius of swirling flow, m.

The axial momentum of the air flow, K, is given by

$$K = QU_x \text{ kg} \cdot m/s^2 \tag{8-3}$$

where U_x is axial velocity of air, m/s.

The swirl intensity, n, is therefore

$$n = \frac{U_t \cdot r}{U_x \cdot L} \tag{8-4}$$

This shows that swirl intensity, which has a great influence on the aerodynamics in the furnace, is expressed by the ratio of the tangential and axial velocities in the swirler. If the swirl intensity is high, the flow swirls strongly, enlarging the expanding angle as well as the recirculation zone. The amount of hot gas recirculating at the root of the flame also increases. This enhances the turbulent diffusion and, therefore, the early mixing. There is also a faster decay of the axial velocity. This helps produce shorter flame lengths. The flow resistance of the burner would, however, increase.

a) Tangential Swirler

In tangential vane burners, tangential vanes are installed (Fig. 8-4) at the entrance of the burner to guide the air flow into the annulus around the central tube carrying the primary air and fuel. This helps form a swirl flow. The swirl intensity of the tangential vane swirler can be obtained as follows:

The tangential velocity, U_t, is found by dividing the air flow rate (Q/ρ) by the flow area ($b \cdot z \cdot t$) between the vanes

$$U_t = \frac{Q}{\rho \cdot z \cdot t \cdot b} \tag{8-5}$$

where t is the minimum space between adjacent vanes at the exit, m

ρ is the density of air flowing, kg/m^3

z is the number of vanes, and

b is the width of vanes, m

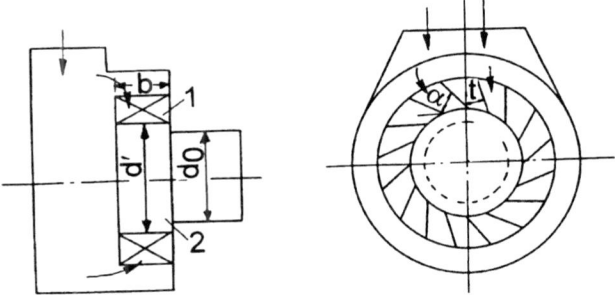

FIGURE 8-4. Tangential vane swirler: (1) tangential vane, (2) cylindrical air exit

The axial flow velocity, U_x, is found by dividing flow rate by the exit area of the burner $(\pi d_0^2/4)$

$$U_x = \frac{4 \cdot Q}{\rho \cdot \pi \cdot d_0^2} \tag{8-6}$$

where d_0 is the exit diameter of the swirler (Fig. 8-4).

If the flow moves along the vane B, the normal distance, r, of this flow stream from the center of the swirler is found from geometry (Fig. 8-5)

$$r = \frac{d'}{2} \cos \alpha r \tag{8-7}$$

where α is the inclination of the vane with the tangent to the central tube and d' is inner diameter of the swirler. Geometric parameters d', d_0, r, and b are shown in Figures 8-4 and 8-5.

An expression for the swirl intensity can be obtained by combining eqs. (8-5), (8-6), (8-7), and (8-4). By taking the characteristic size, L as $(d_0\pi/8)$ we get the

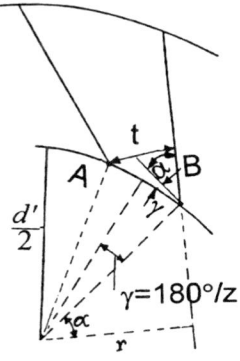

FIGURE 8-5. Diagram showing geometrical parameters of a tangential vane

swirl intensity, n', for a tangential swirler as

$$n' = \frac{d_0 \cdot d' \cdot \cos\alpha}{z \cdot \varepsilon \cdot b} \tag{8-8}$$

The aperture or opening of the moveable tangential vanes is adjustable. To adjust the vane opening, the inclination angle of the vane (α) is changed. This may change the swirl intensity of the flow. From Figure 8-5, the passage area at the exit of the vanes is given by chord AB.

$$AB = d' \sin\left(\frac{180}{z}\right) \tag{8-9}$$

Therefore, from Figure 8-5 the minimum passage opening, t, will be

$$t = d' \sin\left(\frac{180}{z}\right) \sin\left(\alpha + \frac{180}{z}\right) \tag{8-10}$$

Combining Eqs. (8-8) to (8-10),

$$n' = \frac{d_0 \cos\alpha}{z b \sin\left(\dfrac{180}{z}\right) \sin\left(\alpha + \dfrac{180}{z}\right)} \tag{8-11}$$

The atomizer tube at the center of the swirler (Fig. 8-1) occupies a part of the passage area for the axial flow of the primary air. So to account for this we modify Eq. (8-8) and get the revised value of swirl intensity as

$$n' = \frac{\left(d_0^2 - d_1^2\right)\cos\alpha}{z b \sin\left(\dfrac{180}{z}\right) \sin\left(\alpha + \dfrac{180}{z}\right)} \tag{8-12}$$

where d_1 is the diameter of the atomizer tube.

Equation (8-12) suggests that a reduction in the vane inclination angle, α would increase the swirl intensity, but from Figure 8-5 we see that the minimum passage aperture, t, at the vane exit will be reduced. The restricted aperture increases the flow resistance greatly if the inclination angle is very small.

The swirl intensity can be increased by reducing the passage width b and increasing the diameter of the swirler exit, d_0. However, if all other parameters are kept unchanged, there would be a decrease in the axial velocity, and therefore, the flow resistance will decrease.

The ratio of the arc length S_x covered by the blade on its root circle to the arc length S_j between adjacent blades is called *coverage* (Fig. 8-6). This is expressed as k.

$$k = \frac{S_x}{S_j} \tag{8-13}$$

To increase the coverage, the blade root is allowed to extend into the cylindrical air tube as shown in Figure 8-6. However, the extension part should be no more than $(r_0 - r_1)/3$, where r_1 is the radius of center tube, and r_0 is the radius of outer

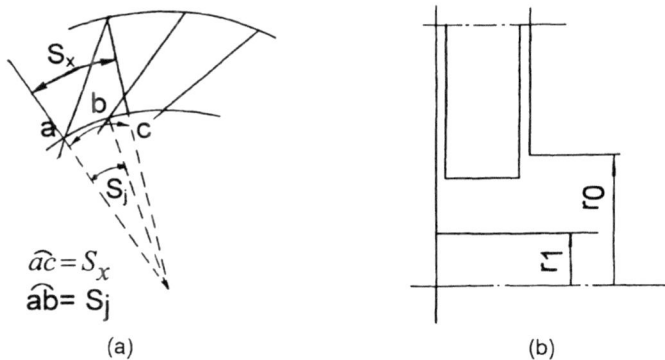

$\overset{\frown}{ac} = S_x$

$\overset{\frown}{ab} = S_j$

(a)   (b)

FIGURE 8-6. (a) Blade coverage is the ratio of arc length, S_x and S_j; (b) extension of the vane into its root

tube. Otherwise, such vanes will lose their flow guiding capability. The width of the vane can be determined by the minimum area of the vane exit, and can be calculated as follows:

$$F_x = z \cdot t \cdot b \tag{8-14}$$

This minimum area should exceed the area of the burner exit.

The width of the vane, b, can be determined by the minimum area of the vane exit, F_x, which should exceed the area of the burner exit:

$$b = \frac{F}{z\varepsilon} \tag{8-15}$$

For a pulverized coal burner, the inclination angle (α) of the vane is usually between 30° and 45°. For high-volatile coal, the value is chosen toward the higher side.

b) Axial Vane Swirler

The construction of a simple axial vane is shown in Figure 8-7. There are several variations of this basic design: screw blade, bent blade, and straight blade. The blade coverage, k ($= S_2/S_1$ in Figure 8-7), which can be defined as in Eq. (8-13), varies in the range of 1.1–1.25 for axial vane swirlers.

In an axial vane swirler (Fig. 8-7), the width of the flow passage at the blade root is smaller than that at the blade tip. This difference is large, especially when the root-to-tip diameters ratio, d_1/d_0 is small. Therefore, the air passage may be narrower at the roots, resulting in a greater resistance. The air would naturally flow preferentially through the tip, where the resistance is small, giving rise to an uneven velocity distribution between the tip and root of the blade. However, for a straight swirl blade (Fig. 8-7), the angle at the root is less than that at the tip, which decreases the resistance at the root. This may partially offset the influence of narrower width at the root.

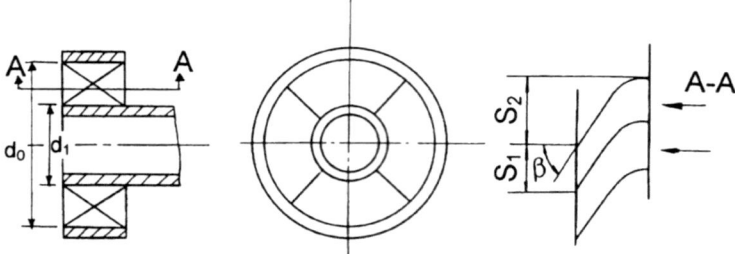

FIGURE 8-7. Geometry of an axial vane swirler

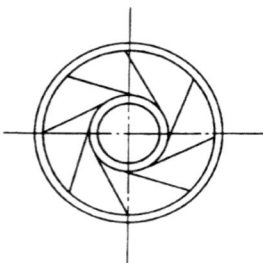

FIGURE 8-8. Cross section of a bent blade axial vane swirler

In case of bent blades (Fig. 8-8), the angles at the root and at the tip are same. So the flow distribution between the tip and the root of the blade would be uneven owing to different flow areas. For the bent blade, if the widths of its tip and root are the same and the entrance edge is in tangential direction, the exit edge of the blade will not be in the tangential direction as shown in Figure 8-8. As a result, eddies are generated at the exit, which disturbs the flow.

To reduce the resistance at the blade root, the blade root is designed to make the blade side projection ladder-shaped, as shown in Figure 8-10. It ensures more even distribution of the exit velocity and lower resistance. For the bent blade, its root should be made shorter, which may make the root coverage closer to that of the tip.

The number of blades, z, can be selected from Table 8-2 using the ratio of the outer and inner diameters. The closer the outer and the inner radius of the blade wheel are, the narrower the annular section of blade wheel will be. So the blade number should be greater. The width of the blade tip can be 0.2–0.4 times that of the outer radius.

TABLE 8-2. Number of blades in axial vane swirler.

Ratio between inner and outer radius, d_1/d_0	0.33	0.5	0.6	0.67
Number of blades, z	12	18	24	30

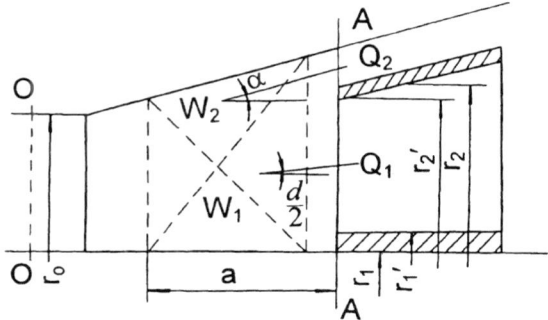

FIGURE 8-9. A cross section view of a moveable axial vane wheel, which moves in a conical enclosure to change the air flow through the swirler; the figure also shows the geometric parameters

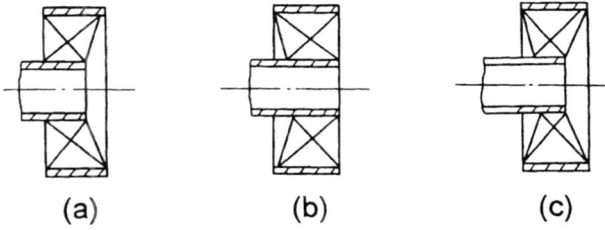

(a)   (b)   (c)

FIGURE 8-10. Three types of axial vane swirler with unequal widths at the root and tip, which helps uniform distribution of air

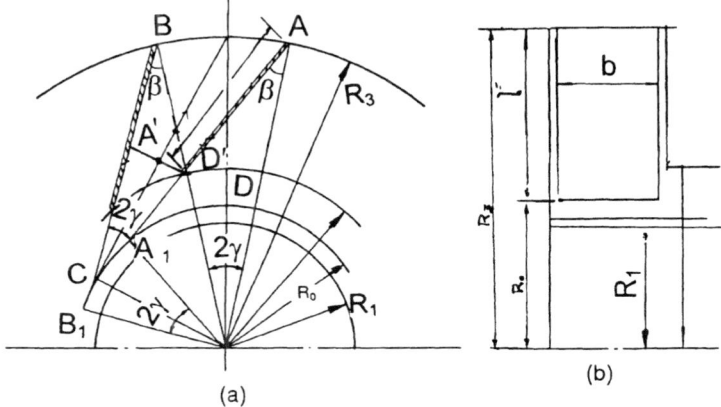

(a)   (b)

FIGURE 8-11. Diagram showing geometric parameters of a tangential vane swirler

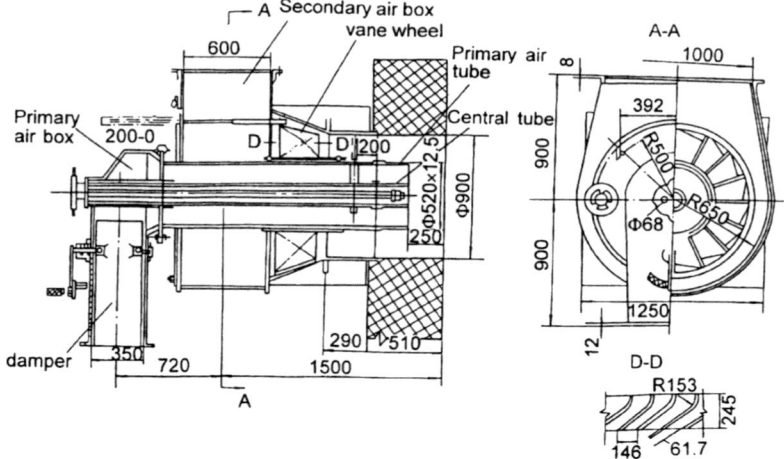

FIGURE 8-12. Details of an axial moveable vane wheel burner

To adjust the swirl intensity in an axial vane swirler, the secondary air vanes are moved axially by means of a vane wheel (shown in Fig. 8-12). The outer enclosure of the vanes is conical (Fig. 8-12). So when the vane wheel is at the extreme right, all the air would flow through the vane, giving the maximum swirl intensity. When the vane wheel is pulled out to the left, an annular passage is formed between the outer periphery of the vane wheel and the conical enclosure. Some of the air bypasses this passage without swirling. So the total swirl intensity decreases. The ratio of bypass flow to swirling flow can be adjusted by changing the position of the vane wheel. This changes the total swirl intensity. If the inclination angle of the vane β (Fig. 8-7) is large, the resistance will be high. So a slight pull will cause a large amount of air flow to bypass the swirler. Consequently, the higher the angle β, the more sensitive the swirl intensity to change in the vane wheel's position. For pulverized coal burners, β is usually taken between $50°$ and $60°$ depending on the type of coal fired. The conical angle of the air channel, α (Fig. 8-9), and that of the moveable vane wheel should be the same. Preferred values of α are in the range of $15°$ to $20°$.

The swirl intensity of an axial vane can be obtained (He & Z 1987) as in Eq. (8-4):

$$n = \frac{4\delta(d_0 + d_1)}{d_0 \varepsilon z} \tag{8-16}$$

where d_0, d_1 = the outer radius and inner radius of exit, respectively, m

ε = average distance of two adjacent blades at exit, m

δ = thickness of blade, m

z = number of blades, m

TABLE 8-3. Swirl intensity of swirl burners (Fen, 1992).

Fuel	Swirl intensity, n		Arrangement of burners
	Primary air	Secondary air	
Wet bottom: anthracite, subbituminous coal	3.5–4.0	4.0–4.5	Opposed jet
Same as above	0–3.5	4.0–4.5	Front wall
Wet bottom: bituminous, lignite	3.0	3.0–3.5	Opposed jet
Dry bottom: bituminous, lignite	2.5	3.0	Opposed jet
Wet bottom: bituminous, lignite	0–2.5	3.0–3.5	Front wall

For the moveable vane wheel, the effective outer and inner radii of the vane exit are different from the exit and the center tube radii. The flow cross section area of the vane exit should not be taken as the annular section area of the rim of the swirler (Fig. 8-9).

It can be seen from Figure 8-12 that when the vane wheel is pulled out proportionally as bypass or the nonswirling flow increases. Therefore, the total swirl intensity decreases.

c) Design of the Swirler

The swirl intensity required for the swirl burner can be selected from Table 8-3 for different fuels. The geometric dimensions of the swirler can be determined based on the swirl intensity, n.

(1) For the single volute burner, a cone is added to the primary air exit; its angle is 90°.
(2) Higher value is for low-volatile coal.

(1) Vane Angle β

For the tangential vane swirler, the inclination angle of the blade, β, can be found (He, 1987) after selecting the vane number z and assuming a vane width b.

$$\beta = \tan^{-1}\left[\frac{\pi \cdot d_{dl} + 2 \cdot \Omega_{dl}}{\pi \cdot d_{dl} \cdot \cot\left(\frac{180°}{z}\right) - 2 \cdot \Omega_{dl} \cdot b \cdot z \cdot \tan\left(\frac{180°}{z}\right)}\right] \qquad (8\text{-}17)$$

where Ω_{dl} = swirl number
d_{dl} = equivalent diameter of the swirler (4 × area/perimeter)

For the moveable vane, the blade number, z, usually varies in the range of 8 and 16, and the blade coverage, k, is 1. For the fixed vane, the blade coverage varies between 1.2 and 1.4. When the blades are arranged radially and the flow does not swirl, the angle covered by two adjacent blades, $2\gamma = 360°/z$ (Fig. 8-5).

(2) Blade Length

When the blades are arranged radially, the blade length, l ($\beta = 0$), (b in Fig. 8-11), we find that

$$l = R_3 - R_0; \quad R_3 > R_0 > R_1$$

where R_3 is the inner radius of secondary air tube, R_1 is the outer radius of primary air tube, and R_0 is the radius of the circle circumscribed by the lower ends of the blades. Under operating condition the blades will be inclined, i.e., $\beta > 0$. So the lower tip the blade AD should lie on the circle of radius, r_2.

From geometry (Fig. 8-11) one can derive the following expression for the blade length, l, and swirl radius, R_3:

$$l = r_2 \frac{\sin 2\gamma}{\sin \beta} \tag{8-18}$$

$$R_3 = r_2 \frac{\sin(2\gamma + \beta)}{\sin \beta} \tag{8-19}$$

In Figure 8-11, the exit section (A'D') of the tangential vane swirler has an width, a, which from geometry twice the length of vane angle bisector at point K. The bisector of the blade angle (CK) represents the average direction of flow in the two blades.

$$a = 2 \, \text{KD}' \tag{8-20}$$

From geometry

$$\text{KD}' = \text{CD}' \sin \gamma \tag{8-22}$$

We also have

$$R_1 = R_3 \sin \beta \tag{8-21}$$

$$\text{CD}' = \text{CA}_1 + (\text{A}_1\text{A} - \text{AD}') = R_1 \tan \gamma + (R_3 \cos \beta - l)$$

Substituting R_3, R_1, and l from Eqs. (8-18), (8-19), and (8-21) we get

$$\text{CD}' = r_2 \frac{\sin(2\gamma + \beta)}{\sin \beta} [\sin \beta \tan \gamma + \cos \beta] - r_2 \tag{8-23}$$

Substituting CD' in Eq. (8-22) we get, after simplification,

$$a = 2r_2 \frac{\sin \gamma}{\sin \beta} [\sin(2\gamma + \beta)\{\sin \beta \tan \gamma + \cos \beta\} - \sin 2\gamma] \tag{8-24}$$

For z number blades the minimum exit section area of blade is

$$F_j = z \, a \, b \tag{8-25}$$

The flow velocity, U_j, in the minimum exit section can be found from the computed value of F_j. This velocity should be less than the secondary air axial velocity U_2 at burner.

8-2 Flow Resistance in Swirl Burners

The aerodynamic resistance of a burner is an important design parameter. This resistance dictates the pressure head required for the combustion air. The following section discusses design methods for the estimation of this resistance.

8-2-1 Resistance Coefficient

The secondary air must overcome a certain amount of aerodynamic resistance while flowing through the swirl vanes in the burner. So the secondary air fan must provide this pressure head. The energy loss owing to this friction is called friction loss. When the air comes out of the burner, its kinetic energy is lost. This energy loss is called exit loss. Thus, the resistance in the swirler comprises both friction loss and exit loss. It is expressed as some multiple of the kinetic energy of flow.

The resistance coefficient for any section is expressed as

$$\xi = \frac{\Delta P}{\frac{\rho U_2^2}{2}} \tag{8-26}$$

where ΔP is the pressure drop across the section and U_2 is the velocity through it.

a) Exit Loss

Assuming a uniform velocity distribution, the exit loss can be written as

$$\Delta H = \frac{U_2^2}{2}\rho = \frac{U_{2x}^2 + U_{2t}^2}{2}\rho \ (\text{Pa}) \tag{8-27}$$

where U_{2x} and U_{2t} are the axial and tangential components of the exit velocity of the swirl vane.

If the exit axial velocity is used as the basis for expressing the loss,

$$\Delta H = \xi \frac{U_{2x}^2}{2}\rho \ (\text{Pa}) \tag{8-28}$$

where $\xi = 1 + (\frac{U_{2t}}{U_{2x}})^2$

As swirl intensity n represents the ratio of tangential velocity to axial velocity of flow, we can write the resistance coefficient as

$$\xi = 1 + n^2 \tag{8-29}$$

b) Resistance Coefficient of the Swirler

The mechanical construction of the swirler is often complex and varied. So it is difficult to estimate its resistance coefficient accurately in advance. It must be obtained experimentally. Some empirical relations based on experimental data are given below for reference:

(1) Tangential Vane Swirler (Cen & Fan, 1991)

When $Re > 2 > 2 \times 10^5$ and $2 < \Omega_{dt} < 5$

$$\xi = 1.1\Omega_{dt} + 1.5 \tag{8-30}$$

where Ω_{dt} is the swirl number.

(2) Axial Bent Blade Swirler (Cen & Fan, 1991)

When $Re > 2 > 2 \times 10^5$ and $2 < \Omega_{dt} < 5$

$$\xi = 0.7\,\Omega_{dt} + 1.0 \tag{8-31}$$

For the axial moveable vane wheel swirler (Cen & Fan, 1991)

$$\xi = (1 + K\beta)(1 + \tan^2 \beta) \tag{8-32}$$

where K is the correction factor for nonuniform distribution of velocity at the exit (Cen, 1991). $K = 0.003$–0.006.

c) Resistance Coefficient of the Primary Air and PC Mixture

The primary air conveys the pulverized coal (PC). When the concentration of PC, μ, is less than 0.75 kg/kg, there is little difference between the resistance coefficient of coal-containing primary air and pure air. However, the influence of PC concentration should be considered if more precision in the calculation is required.

$$\xi = \frac{\Delta P}{(1 + K_\mu \cdot \mu)\dfrac{\rho U}{2g}} + \left(\frac{U_1}{U_2}\right)^2 \tag{8-33}$$

where U is the velocity of the primary air in the tube; U_1 is the velocity of the primary air at the throat; K_μ is the correction factor accounting for the PC concentration. It is generally 0.8.

8-3 Examples of Swirl Burners

8-3-1 Design of a Tangential Vane Burner

A typical moveable tangential vane burner is shown in Figure 8-2. Details of the flame stabilizer placed in the primary air are shown in Figure 8-3. Design details of a typical burner used for firing pulverized coal in a boiler are explained below for illustration.

In a boiler of 380 t/h steam output a total of 24 burners were used to fire bituminous coal with lower heating values in the range of 19.33 to 16.12 MJ/kg. There are eight moveable vanes in the secondary air tube. The angle of the vane can be changed to adjust the swirl intensity of the secondary air. The diameter of the secondary air exit tube is 659 mm, where a refractory cone with an expansion angle of

$52°$ is installed. The primary air tube is 356 mm diameter. It is pulled back into the throat of the burner so as to generate a 400 mm premixing zone required to burn the high-volatile bituminous coal. A cone-shaped flame stabilizer or holder (Fig. 8-3) was installed at the exit of primary air pipe (Fig. 8-2). The flame stabilizer has an outer diameter of 326 mm and a cone angle of $75°$. Immediately downstream of the flame holder, a hot gas recirculation zone is created. The flame holder also provides a mild swirl to the primary air, which helps the mixing of primary with secondary air and the PC. The flame holder can be moved axially by a remote controlled cylinder. This adjusts the shape and size of the recirculation zone. The designed gap between flame-stabilizer and the primary nozzle is 50–125 mm. The degree of openness has a great influence on the size of the recirculation zone and the temperature of the recirculating gas. When the vane opens more than 60%, the coal–air mixture cannot be ignited within the range of the double burner exit diameter, and the temperature decreases rapidly. Therefore, the degree of openness of the vane should be no more than 60% in operation. Typically, the vane angle varies in the range of $30°$–$50°$ while burning bituminous coal. Here, the primary air ratio is 26–27%. The velocity of the primary air is 22 m/s and the secondary air velocity is 26.6 m/s. The thermal efficiency of the boiler can be over 92% at full load.

The turn-down of the burner reaches 50–100%, and the resistance of the burner is small. The resistance coefficient of primary air is about 2.16, while that of the secondary air is about 3.

8-3-2 Design of an Axial Vane Burner

The axial moveable vane wheel burner (Figure 8-12) is a newer development. Here, the swirl intensity of the secondary air can be adjusted by moving the vane wheel. Primary air is nonswirling or direct flow, but by adjusting the position of the tongue damper at the primary air entrance it can also be made to swirl slightly.

Expansion cones (angle β_1) at the exit of the primary air and (angle β_2) at the secondary air exit were used in the burner in Figure 8-1. These can be used in the burner in Figure 8-12. It helps expand the jet and increases the recirculation flow. The angle of expansion is generally not very large. Experience (Cen and Fan, 1991) shows that a large expansion angle leads to disruption in the flow pattern. The greater the inclination angle β of vane (Fig. 8-7) is, the smaller is the maximum extent of expansion.

Cold model tests (Xu, 1990) show that (Fig. 8-13) the expansion angle of the resulting flow cone varies linearly with the variation of the cone angle (α) of the secondary air tube. Tests on a cold model (Cen and Fan, 1991) also show that increasing the cone angle β_2 of the refractory cone at the exit of the burner (Fig. 8-1) decreases the flow expansion angle and shortens the recirculation zone (Figure 8-14). However, beyond $120°$, the flow expanded angle increases sharply (Fig. 8-14) owing to the positive pressure built up behind the cone.

The primary air nozzle is pulled into the burner by a certain distance to create a premixing stage. The primary air and the secondary air begin to mix here. The primary air pressure (0–300 Pa) must be lower than the secondary air pressure (300–600 Pa) as the secondary air entrains the lower-pressure primary air. The

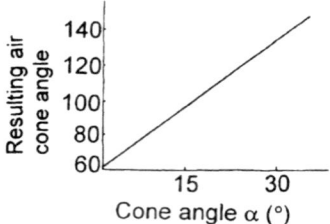

FIGURE 8-13. The influence of secondary air chamber angle (α) on the resulting flow expanded angle for the axial vane swirl burner shown in Fig. 8-1; here the inclination angle of vane is $\beta = 65°$, the cone angle is $90°$, the velocity of the primary air is $U_1 = 25$ m/s, and the velocity of the secondary air is $U_2 = 23.1$ m/s

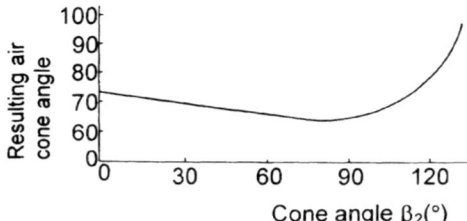

FIGURE 8-14. Influence of cone angle of refractory cone of an axial moveable vane wheel burner on the angle of the air cone produced by the secondary air

greater the pressure of the primary air, the less is the entrainment and therefore the longer the premixing stage. An extended premixing stage reduces the flow expansion angle and the recirculation zone. However, a much too early mixing of the primary with the secondary air adversely affects the ignition as well as the stability of the flame. A pulling back of 100–200 mm is generally considered appropriate. The resistance coefficient of the primary air is about 1.25. This resistance coefficient of the primary air is the smallest amongst several types of swirl burners.

The inclination angle of axial vane, β (Fig. 8-1), has a great influence on the burner. In older designs, where the inclination angle was $65°$, the secondary air resistance was very high ($\xi = 7.4$). This increased the head requirement of the secondary air to 2800 Pa. In later designs, the inclination angle of the vane was reduced to $60°$, and the blade number was decreased from 16 to 12. Then the combustion condition and adjustment range of the vane wheel were improved. The flow guiding capability deteriorates when the number of blades is reduced. For lignite and bituminous coal, β is $50°–60°$ and the distance from the burner to the cooling wall should be 2–2.2 m.

By adjusting the position of the vane wheel one changes the flow expansion angle and the size of the recirculation zone. The adjustment range is related to the primary air flow rate. For example, if the primary flow is 30% of the total, the

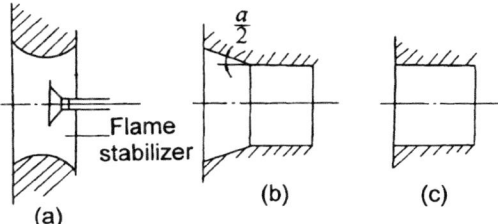

FIGURE 8-15. Shape of air port: (a) contract air port, (b) simple expansion, (c) cylindrical air port

expansion angle and recirculation zone can be varied over a wide range, but when the primary air rate reaches 40% the expansion angle cannot be changed much.

When the vane wheel is pulled all the way back, giving $\alpha = 0$ in Figure 8-9, the vane touches the enclosure. The exit flow touches the wall of the secondary air tube. As all the secondary air passes through the vanes it is guided accurately by the vane.

8-3-3 Design of a Swirl Burner with an Outer Expansion

A swirl burner can have differently shaped exit sections (also called quarl or burner throat section). According to these shapes we can have three types of exit (Fig. 8-15): convergent-divergent air exit, simple expansion exit, and cylindrical air exit. These shapes have a great influence on the exit flow.

1. Convergent-divergent air exit is shown in Figure 8-15a. The air from the swirler is contracted first in the air exit and then it spreads into the furnace. This type of air exit is employed mainly in oil burners.
2. Simple expansion, as shown in Figure 8-15b.
3. Cylindrical air exit as shown in Figure 8-15c.

The type of exit section depends on the fuel fired. For example, with anthracite coal or an inferior bituminous coal, an expansion cone should be employed. The cone should match the expansion angle of the primary and secondary air tubes so as to get a strong recirculation flow. The velocity of the primary air at the exit should be decreased to some degree to improve the ignition. The expansion angle depends on the type of the burner and other conditions. The premix stage in the axial moveable vane swirl burner should be eliminated when burning low-reactive coal, and a proper expansion angle should be added to the quarl.

When an easily ignitable coal such as bituminous coal is burnt, a cylindrical or tube-shaped exit (Fig. 8-12, dia. 900 mm) is used. Such an exit without an expansion may ensure stable combustion. If there is an expansion, the ignition will advance, and the fuel may burn through the burner. In such cases, the primary air pipe has to be withdrawn to form a premix stage so as to delay the ignition.

In the case of anthracite coal the premixing stage is avoided. Chosen values of the cone angle are $\approx 50°$ and that of the quarl length is $0.4D_h$, where D_h is the diameter of the throat of the burner. The central pipe houses the oil gun for ignition. The diameter of central pipe may be increased to get a diameter ratio of the central pipe to the primary pipe of 0.7, and that of the central pipe to the secondary air pipe of about 0.5. Such a combination may enable the swirl burner to burn anthracite coal and inferior coal without expansion quarl.

8-4 Arrangement of Multiple Swirl Burners

So far only the design of individual burners has been considered. A typical boiler would use a combination of several burners. The swirling flows of individual burners interact with each other, affecting the overall aerodynamic condition in the furnace. It also influences the combustion process. So they are to be arranged taking these into account. The following section discusses the arrangements of multiple burners.

8-4-1 Swirl Direction of the Burner

Experiments (He and Zhao, 1987) show that the tangential velocity of two adjacent swirl flows at their junction is nearly equal to their algebraic sum. When the air flow in two burners rotates in opposite directions (Fig. 8-16a), the ratio of tangential and axial velocities (U_t / U_x) at the middle of the two burners increases to almost two times that of a single burner. The velocity of the outside flow is similar to that of an individual burner being in use. When flows of two burners swirl in the same

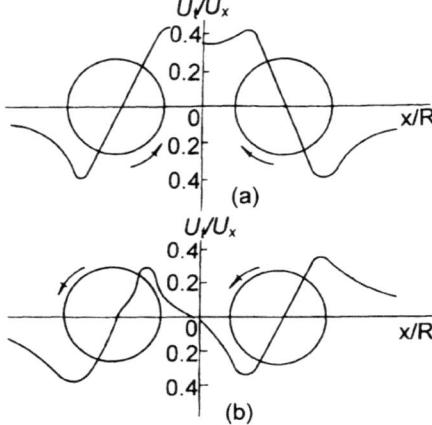

FIGURE 8-16. Distribution of the resultant tangential velocity near two adjacent burners with different swirl directions

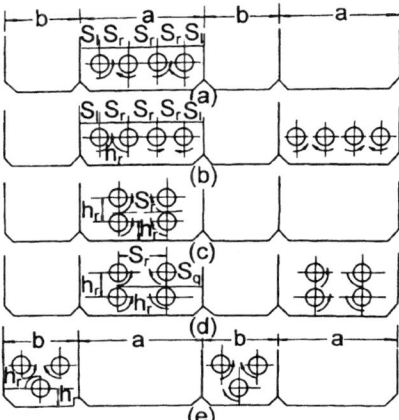

FIGURE 8-17. Arrangement schemes for multiple swirl burners in the furnace showing the distance of burners from the water wall and the swirl direction of flow; the walls of the furnace are folded out to show the arrangement on all four walls: (a) single layer and single side arrangement; (b) single layer, opposed arrangement; (c) single side and double layers staggered arrangement; (d) double layers staggered arrangement in opposite walls; (e) triangular arrangement in opposed walls

direction, the tangential velocity at the center of two burners is small (Fig. 8-16b), but velocity gradient is steep, increasing the heat and mass exchange. Under the former condition, the velocity of the upward flow between two burners is increased, giving an upward tilt to the flame.

So it is important to select the right orientation of the swirl in the burner. Figure 8-17 gives an example of a recommended scheme of a combination of several swirl burners. In any particular burner, the primary air and the secondary air swirl in the same direction.

8-4-2 Arrangement of Swirl Burners

Swirl burners are usually arranged on the front walls or on opposite walls.

a) Front Wall Arrangement

The front wall arrangement has several advantages:

- Pulverizers can be arranged in front of the furnace,
- Pulverized coal pipes leading to the burners can be short and the resistance is small, which contributes to an even distribution of the PC and air flow to each burner.
- The size of the furnace section can be determined to match the size of the back-pass of the boiler without any additional constraints.

• A proper choice of burner capacity and its arrangement can greatly reduce the occurrence temperature nonuniformity at the furnace exit.

One disadvantage of the front wall arrangement is that the flame does not fill the furnace very well, which gives a poor furnace cover. There is a large recirculation zone, which decreases the effective utilization of the furnace capacity. A deflection angle on top of the furnace may increase the utilization of furnace to some degree.

b) Opposed Wall Arrangement

Compared to the front wall arrangement, the arrangement of burners in the front and rear or opposite walls has the following advantages:

• The flame covers the furnace better. It remains at the middle of the furnace giving an uniform distribution of temperature.
• A low-load burner can be used, but the distance between the burner and the furnace wall, as well as the distance between the two burners, should be enough to ensure sufficient time for burn-out of the fuel in the furnace.

To avoid slagging during the operation, minimum distances between the swirl burner and the adjacent furnace wall, and the upper edge of the ash hopper are required. Also, there should be adequate distance between the adjacent burners to ensure that the flame of the burners spreads unhindered without disturbing each other. When the multiple-row arrangement is used, a certain distance between the two rows of burners is required. Table 8-4 gives some values of the distance as

TABLE 8-4. Distance between the swirl burner and the water-cooling wall (d_r is the burner nozzle diameter, S_r is the horizontal spacing, h_r is the vertical spacing).

Parameter	Symbol	Distance in terms of burner nozzle diameter
Horizontal distance between burner central lines		
Single layer arrangement (Fig. 8-17a,b)	S_r	$(2.0-2.5)\, d_r$
Double layer parallel arrangement (Fig. 8-17c,d)	S_r	$(3.5-4.0)\, d_r$
Double layer staggered arrangement (Fig. 8-17e)	S_r	$(2.5-3.0)\, d_r$
Vertical distance to burner central line		
Parallel arrangement (Fig. 8-17c,d)	h_r	$(2.0-3.5)\, d_r$
Staggered arrangement (Fig. 8-17e)	h_r	$(2.5-3.0)\, d_r$
Distance from the outermost burner to adjacent wall (Fig. 8-17d)	S_r	$>(1.8-2.5)\, d_r$
Distance from lowest burner to		
Upper edge of ash hopper (slagging)	h_r	$(1.4-1.6)\, d_r$
Upper edge of slag exit (normal)	h_r	$(1.8-2.2)\, d_r$
Furnace depth		
Distance between front wall and back wall	b	$>(4-5)\, d_r$
Distance between opposing walls	a or b	$>(5-6)\, d_r$

multiples of burner nozzle diameter. Figure 8-17 shows the distance symbols used in Table 8-4 and the correct flow swirl direction of different arrangement.

When the number of swirl burners is large, the axis of the burner should be normal to the furnace wall. When the number of swirl burners is small (only two or three burners), two sides of the burners should be inclined slightly downward. This may ensure an even temperature distribution even if one burner is stopped. Generally, two burners are installed on the front wall. The axis of the burner should be inclined 10°. However, when three burners are used, axes of the burners at the two sides should inclined by 6°.

8-5 Design Procedure of Swirl Burners

The following section discusses the design procedure for swirl burners

8-5-1 Capacity of a Single Burner

With the development of higher-capacity boilers with larger furnace volumes, the energy throughput of a single burner is increasing. Since heavy oil and natural gas are easier to burn than pulverized coal without slagging problems, the capacity of a single oil or gas burner is greater than that of a pulverized coal (PC) burner. For example, the heat input into a high-capacity oil burner now may reach 79–93 MW, while the individual burner input of a pulverized coal burner is in the range of 30–60 MW. The capacity PC burners are limited by two types of factors: one relates to the burner itself, the other relates to the furnace sizes.

a) Burner-related factors

1. In a high-capacity burner the heat input to the furnace is very high even at partial loads.
2. To reduce the NO_x in the exit gas, the maximum temperature of the flame should be low, but this temperature increases with the capacity of individual burners.
3. Large numbers of low thermal capacity burners avoid slagging.
4. If the thermal capacity of individual burners is large, the thickness of both the primary air and the enclosing secondary air streams would also be large. This would adversely affect the mixing of air and coal.
5. During low-load operation, boiler startup and shutdown, some burners must be switched on or off. High-capacity burners make this task difficult.

b) Size-Related Factors

When the furnace cross section area increases, the momentum of the air flow exiting the burner must be increased. Otherwise, the flame will not be able to penetrate through the width and depth of the furnace. This would require a higher-capacity burner. However, beyond a certain value, the furnace depth does not vary noticeably with the capacity of the furnace. Only the furnace width is increased. Thus only

TABLE 8-5. Number of burners for each boiler and its output.

Unit power MWe	Steam capacity t/h	Boiler pressure MPa	Furnace heat input MWth	Number of burners		Coal per burner t/h	Input per burner MWth
				Front wall	Front, rear or side walls		
	35	3.83	27.1	2		3	20.2
	75	3.83	68.1	3–4		3.7–3	25
25	130	3.83	93.1	4	4	3.7	25
50	220	9.81	168.6	4–6	4–6	7.4–3.7	50–25
100	410	9.81	313.9	8–16	8–16	7.4–3.7	50–25
200	670	13.73	608.3	8–24	8–24	11.2–3.7	75–25
300	935	13.73	848.9	8–36	8–36	15–3.7	100–25
500	1600	25	1279		12–48	15–10	100–44
800	2500	25	2000		16–48	18.6–10	126–44

the number and rows of burners are required to be increased. This may restrict the capacity of a individual burner. Some general guidelines on the arrangement, number, and heating power of PC burners are given in Table 8-5.

8-5-2 Design Calculations

a) Determination of Excess Air Coefficient at the Exits of Burner and the Furnace

To determine the size of the burner, air flow at every part of the burner and the furnace should be determined first. Besides the burner excess air, α_x, some additional air enters the furnace as tertiary air and through leakage at different points in the furnace. So the excess air at the exit of the furnace, α_{xf}, is the sum of excess air at the burner, α_x, tertiary air, α_{x3}, and the leakage air coefficient in the furnace, $\Delta\alpha_{xl}$. The tertiary air is needed only when coal is carried by hot air as the drying agent. Some of this coal-conveying air is lost in the coal dust separator. Only a fraction ($\eta_f \approx 0.9$) of it enters the furnace.

$$\alpha_{xf} = \alpha_x + \Delta\alpha_x + \eta_f\alpha_{x3} \qquad (8\text{-}35)$$

If the coal is dried by exhaust gas, then tertiary air is zero. For anthracite and subbituminous coal $\alpha_{xf} = 1.20$–1.25; for bituminous and lignite, $\alpha_x = 1.15$–1.20.

The leakage coefficient of furnace, $\Delta\alpha_{xl}$, represents the air leakage from the screen wall. For tangent tube water wall, $\Delta\alpha_{xl} = 0.1$; for membrane water wall, $\Delta\alpha_{xf} = 0.05$.

The combustion air through the burner comprises both primary and secondary air. So the excess air at the burner exit, α_x, is the sum of primary air coefficient α_{x1} and secondary air coefficient α_{x2}.

$$\alpha_x = \alpha_{x1} + \alpha_{x2} \qquad (8\text{-}36)$$

The primary air conveys the coal. Therefore, the primary air coefficient dictates the coal concentration in the conveying pipe. Another method of expressing the air quantities is by primary air, secondary air rate, and tertiary air rate percentages. They are defined as follows:

$$r_1 = \frac{V_1}{\alpha_{xf} VB} \times 100\% \tag{8-37}$$

$$r_2 = \frac{V_2}{\alpha_{xf} VB} \times 100\% \tag{8-38}$$

$$r_3 = \frac{V_3}{\alpha_{xf} VB} \times 100\% \tag{8-39}$$

where, V_1, V_2, V_3 = flow of primary, secondary, and tertiary air, respectively, (Nm^3/s)

V = theoretical air, Nm^3/kg

B = calculated coal consumption, kg/s

α_{xf} = excess air coefficient at the furnace exit

Similarly, we can define the furnace leakage rate $r_{ef} = (\frac{\Delta\alpha_{xf}}{\alpha_{xf}} \times 100\%)$. Thus we have

$$r_1 + r_2 + r_3 + r_{ef} = 100\% \tag{8-40}$$

The primary air rate and primary air coefficient or leakage rate and leakage coefficient are related as follows:

$$r_1 = \frac{\alpha_{x1}}{\alpha_{xf}} \times 100\% \tag{8-41}$$

$$r_{ef} = \left(\frac{\Delta\alpha_{xf}}{\alpha_{xf}} \times 100\% \right) \tag{8-42}$$

b) Air Rate and Velocity

The primary air rate is usually determined by the drying requirement of the coal, the conveying velocity in pipes, and the ignition conditions. For direct flow or nonswirling burners (where coal is fed directly from the pulverizer) the actual primary air flow rate is often greater than that the air above. When the PC is transported with hot air, the primary air rate percentage, r_1, is allowed to exceed that when the PC is transported with the drying agent like flue gas. Low-grade bituminous coal can also be conveyed by exhaust air, but it should be controlled so that $r_1 \cong 25\%$ and that primary air temperature remain above 90°C. In some high-volatile inferior bituminous coal firing plants, the primary air temperature reaches 200°–250°. The primary air rate can be selected with reference to Table 8-6.

For an exhaust air conveying system, there is no tertiary air. So the secondary air rate is equal to the total air less the primary air rate r_1 and leakage rate r_{ef}.

$$r_2 = 100 - r_1 - r_{ef} \, (\%) \tag{8-43}$$

TABLE 8-6. Primary air rate r_1 (%).

Conveying method	Anthracite	Sub-bituminous	Bituminous		Inferior-grade anthracite		Lignite
			$^a V_{daf} \leq 30\%$	$V_{daf} \geq 30\%$	$V_{daf} \leq 30\%$	$V_{daf} \geq 30\%$	
Exhaust air conveying			20–30	25–35		~25	20–45
Hot air conveying	20–25	20–30	25–40		20–35	25–35	

^{a} where V_{daf} is the volatile matter in dry basis.

TABLE 8-7. Range of air velocities.

Coal group	Anthracite	Sub-bituminous coal	Bituminous	Lignite
Primary air velocity U (m/s)	12–16	16–20	20–26	20–26
Secondary air velocity U_2 (m/s)	15–22	20–25	30–40	25–35
Tertiary air velocity U_3 (m/s)		40–60		

For a hot air conveying system, the tertiary air (exhaust air) rate should also be deducted.

$$r_2 = 100 - r_1 - r_3 - r_{ef}(\%) \qquad (8\text{-}44)$$

The velocity of primary air depends on the ignitability of pulverized coal. When coal is conveyed by exhaust air, the velocity of primary air can be a lower. The velocity of the secondary air depends on jet penetration length, effective mixing, and the need for full combustion. The ranges of velocities of primary, secondary, and tertiary air in swirl burners are given in Table 8-7. The ratio U_2/U_1 is 1.2–1.5 for the swirl burner, while the ratio of secondary air dynamic head $\rho_2 U_2^2/(\rho_1 U_1^2)$ to primary air dynamic head is only 0.6. An optimum dynamic head ratio should be obtained by combustion adjustments.

In hot conveying bin and feeder pulverizing system, tertiary air is determined according to the pulverizer. It occupies 10–18% of the total air in the furnace. For low heating value, high-moisture, and high-ash inferior coal, the tertiary air reaches 25–35%. The velocity of tertiary air is usually selected from Table 8-7. A high velocity ensures deep penetration into the hot furnace gas, contributing to better burn-out of the fuel.

For the front fired furance, the tertiary air nozzle can be arranged on the back wall at the same elevation as the upper row of burners in the front wall. When the burners are arranged in an opposed jet arrangement, the tertiary air nozzle can be arranged tangentially on four corners of the furnace provided its width and depth ratio is below or equal to 1.3. If the ratio is more than 1.3 and the burner is arranged on the front wall or in style of opposed jet, the tertiary air nozzle can be installed on the front wall 2.4–3.0 m higher than the burners of the upper layer. Downward inclination angle of tertiary air is generally 5°–15°. The worse the coal is, the

TABLE 8-8. Allowable mill exit temperature, °C.

Type of coal	Storage	Direct	Semidirect
High-rank, high-volatile bituminous	54	77	77
Low-rank, high-volatile bituminous	54	71	71
High-rank, low-volatile bituminous	57	82	82
Lignite	43	43–60	49–60
Anthracite	93	—	—
Petroleum coke (delayed)	57	82–93	82–93

smaller the inclination angle should be. For swirl burners, the vertical distance from the lower edge of the tertiary air nozzle to the upper edge of the primary burner nozzle can be chosen equal to the diameter of a burner's expansion cone (Fig. 8-1), which is twice the nozzle width of the direct burner. If two pulverizers are in use, each pulverizer should be connected to two diagonal tertiary air nozzles.

The dried coal while being transported by air should avoid self-ignition as well as condensation of moisture. Table 8-8 gives some typical air temperatures required for different types of coal. It also shows the allowable mill outlet temperature for different coals and mills as recommended by Singer (1991).

8-5-3 Design Steps

Tables 8-9 and 8-10 show some essential steps in the procedural design of a swirl burner.

a) Calculation of the Combustion Air Needed

TABLE 8-9. Calculation of the combustion air requirement.

Names	Symbol	Unit	Equations
Design coal			Input parameter
Calculate coal consumption	B	kg/s	By thermal design
Theoretical air	V	Nm^3/kg	By stoichiometric calculations
Excess air coefficient	α_{x1}		See Note 1
Amount of drying agent (kg air/kg fuel)	m_g	kg/kg	Chosen according to pulverizing system (Chapter 3)
As received moisture in coal	W_{ar}	%	Input parameters
Inherent moisture in coal	W_{mf}	%	Input parameter
Type of pulverizing system			Direct conveying by hot air
Evaporated water of per kg dry fuel	ΔW	kg/kg	$(W_{ar} - W_{mf})/(100 - W_{mf})$
Weight of drying gas at exhauster	m_{gw}	kg/kg	$m_g(1 + K_{lf}) + \Delta W$ [K_{lf} — Leakage coefficient in pulverizer
Ratio of drying gas and theoretical air		kg/kg	$m_g(1 + K_{lf})/1.293V$ (for drying gas transport)
Coefficient of primary air	α_{x1}		$0.15 \sim 0.2$ (for hot air transport)
Secondary air coefficient	α_{x2}		$\alpha_x - \alpha_{x1}$
Tertiary air coefficient	α_{x3}		$[m_g(1 + K_{lf}) - m_{gw}]/[1.293V(1 - \eta_f)]$ [See Note 2]

(Contd.)

TABLE 8-9. (Continued).

Names	Symbol	Unit	Equations
PC concentration in primary air	μ	kg/kg	$1/(1.293\alpha_1 V)$
Temperature of hot air	T_h		Selected from Table 3.3 Chapter 3]
Excess air coefficient at furnace exit	α_{xf}	°C	$\alpha_1 + \alpha_2 + \alpha_3 + \Delta\alpha_l$
Specific heat of hot air	C_{rk}	kJ/kg°C	From property table
Specific heat of primary air temperature of T_1	C_{k1}	kJ/kg°C	From property table
Specific heat of PC	C_{mf}	kJ/kg°C	$4.187[0.22(100 - W_{mf})/100 + W_{mf}/100]$
Primary temperature at exit of the mill	T_1	°C	Calculated according to pulverization system (Table 8-8)
Secondary air temperature	T_2	°C	$T_h - 10$ (owing to heat loss in hot air pipe)
Tertiary air temperature	T_3	°C	$T_h - 10$ (owing to heat loss in hot air pipe)
Secondary air flow	V_2	m³/s	$\alpha_2 V B_j (1 + T_2/273) \times 760/P_2$
Primary air flow	V_1	m³/s	$\alpha_1 V B_j (1 + T_1/273) \times 760/P_1$
Tertiary air flow	V_3	m³/s	$\alpha_3 V B_j (1 + T_3/273) \times 760/P_3$

Note 1: The excess air is selected according to the fuel properties and pulverizing system. For example, when conveying coal with hot air, it is 1.05–1.15 for anthracite and subbituminous coal; it is 1.05–1.1 for bituminous and lignite. The lower value is for wet-bottom boilers.
K_{lf} is the leakage coefficient of pulverizing system.
Note 2: It is 0.3–0.5 for a bin feeder system when drying with gas and 0.25–0.45 when drying with air, and 0.18–0.3 for a direct flow system. P_1, P_2, P_3 are pressures of primary, secondary, and tertiary air in mm of mercury. m is number of coal pulverizers used to feed the boiler. B is total coal consumption to feed the boiler, kg/s. η_f is separating efficiency (0.85–0.9). m_{gw} is recirculating amount of drying agent.

b) Calculation of Nozzle of Swirl Burner

TABLE 8-10. Calculation of nozzle of swirl burner.

Name	Symbol	Unit	Equations
Primary air velocity	U	m/s	Select from Table 8-7
Secondary air velocity	U_2	m/s	Select from Table 8-7
Tertiary air velocity	U_3	m/s	Select from Table 8-7
Number of operating burners	Z	—	Select from Table 8-5
Output of a single burner	B_r	kg/s	B_j/Z
Annular area of primary air exit	F_1	m²	$V_1/(Z U_1)$
Annular area of secondary air exit	F_2	m²	$V_2/(Z U_2)$
Outer diameter of central tube	d_0	m	$m\sqrt{\dfrac{1}{1-m^2}\dfrac{4}{\pi}(F_1 + F_2)}$ Where, $m = d_0/d_r$ [*It can be used directly for cylindrical nozzle; for cone nozzle, m = 0.3–0.5*].
Inner diameter of primary pipe	d_1'	m	$\sqrt{d_0^2 + \dfrac{4F_1}{\pi}}$
Wall thickness of primary pipe	δ_1	m	General, $\delta_1 \geq 0.01$
Outer diameter of primary pipe	d_1	m	$d_1' + 2\delta_1$
Inner diameter of secondary pipe	d_2'	m	$\sqrt{d_1^2 + \dfrac{4F_2}{\pi}}$

TABLE 8-10. (Continued).

Name	Symbol	Unit	Equations
Outer diameter of secondary pipe	d_2	m	$d_2' + 2\delta_2$
Cone angle of burner	β_2		Selected from Sections 8-3-2 and 8-3-3
Expanded angle of inner pipe	β_1	m	
Expanded angle of primary pipe	β_2	m	
Angle of burner expansion			
Ratio of burner expansion diameter to outer diameter of primary pipe	d_r/d_1		$\tan^2\beta_2 + \dfrac{1}{\cos^2\beta_1}\sqrt{1 + \dfrac{4F_2\cos\beta_2}{\pi d_1^2}}$
Diameter of expansion	d_r	m	$d_1(d_r/d)$
Length of cone in secondary channel	l		Generally, $l = 0.15$–0.3 m
Inner diameter of secondary annular passage (with a cone exit)	d_{2n}		$d_r - 2x\, lx \tan\beta_2$
Area of secondary annular passage (with a cone exit)	F_2'	m^2	$\pi/4(d_2'^2 - d_1^2)4$
Secondary air velocity in annular passage	U_2'	m/s	$U_2 F_2/F_2'$
Velocity ratio	U_2'/U_2		Select 1 to modify it; let $U_2'/U_2 <= 1.25$

Nomenclature

a	width of a tangential vane swirler, m
b	width of vane, m
B	coal consumption, kg/s
D_h	diameter of the throat of the burner, m
d'	inner diameter of the tube, m
d_{dl}	equivalent diameter of the swirler m
d_0, d_1	outer radius and inner radius of exit respectively, m
d_r	burner nozzle diameter, m
F_1	minimum flow section area of vane wheel exit, m^2
F_2	section area of the direct flow exit, m^2
F_j	minimum exit section area of blade, m^2
F_o	represents the section area of swirler exit, m^2
F_x	minimum area of the vane exit, m^2
g	acceleration owing to gravity, 9.81 m/s^2
h_r	vertical spacing of burners, m
	correction factor in Equation (8-32)
K	axial momentum of the air flow, kg · m/s^2
k	blade coverage
K_μ	correction factor for PC dust concentration (Eq. 8-33)
l	blade length, m
L	characteristic size of the burner, m
m	minimum adjacent space of the vane exit, m
m_g	mass of coal drying gas kg/kg coal

m_{gw}	mass of drying gas at the exhauster kg/kg
M	swirl momentum of the air flow, $kg \cdot m^2/s^2$
n, n'	swirl intensity or number
$\Delta P, \Delta H$	pressure drop across swirler, Pa
Q	mass flow rate of air, kg/s
r	radius of swirling flow, m
R_0	radius of the circle circumscribed by the lower ends of the blades, m
R_1	outer radius of primary air tube, m
R_3	inner radius of secondary air tube, m
r_2	inner radius of secondary air channel, m
r_{pj}	average radius of swirl flow, m
r_1	primary air rate, %
r_2	secondary air rate, %
r_3	tertiary air rate, %
r_{ef}	furnace leakage (Eq. 8-40)
r_o	inner diameter of air exit, m
R_o	radius of the circle shaped from blade extremity when $\beta = 0$, m
Re	Reynold's number
S_j	arc length between two adjacent blades, m
S_x	arc length covered by a blade on blade root circle, m
S_r	horizontal spacing of burners, m
t	minimum distance between two adjacent blades at exit, m
V	theoretical air, nm^3/kg
V_1, V_2, V_3	flow of primary, secondary and tertiary air respectively, nm^3/s
V_{daf}	volatile matter in dry basis, %
U	velocity of the primary air in the tube, m/s
U_1	velocity of the primary air in the throat, m/s
U_2	secondary air velocity, m/s
U_3	tertiary air velocity, m/s
U_{2x}	axial component of the exit velocity of the swirl vane, m/s
U_{2t}	tangential component of the exit velocity of the swirl vane, m/s
U_t	tangential velocity, m/s
U_{t0}	tangential velocity at the vane exit, m/s
U_x	axial velocity at the swirl exit, m/s
z	number of vanes

Greek Symbols

α	angle of the vane
α_{xf}	excess air coefficient at the exit of furnace
α_x	excess air coefficient at exit of burner
$\Delta\alpha_{xl}$	excess air leakage coefficient
α_{x1}	excess air coefficient in primary air
α_{x2}	excess air coefficient in secondary air
α_{x3}	excess air coefficient in tertiary air

η_f	fraction of coal conveying air entering the furnace
γ	angle enclosed by the two adjacent blades (Fig. 8-5)
ξ	flow resistance coefficient
ε	average distance of two adjacent blades at exit, m
δ	thickness of blade, m
β, β_1	angle of vane (Figs. 8-1, 8-11)
β_2	core angle, °
Ω_{dl}	swirl number
ρ	density of air, kg/m^3
$\Delta\alpha$	leakage air coefficient
η_f	fraction (Eq. 8-35)
μ	concentration of pulverized coal in air, kg coal/kg air

References

Cen, K., and Fan, J. (1990) "Theory and Calculation of Engineering-Gas-Solids Multiphase Flow" (in Chinese). Zhejiang University Press, Hangzhou, China, p. 3.

Cen, K., and Fan, J. (1991) "Combustion Fluid Dynamics" (in Chinese). Hydro-Electricity Press, Beijing, p. 9.

Fen, J. (ed.) (1992) "Principles and Calculation of Boilers" (in Chinese), 2nd edition, Science Press, Beijing.

He, P., and Zhao, Z. (1987) "Design and Operation of Pulverized Coal Burners" (in Chinese). Mechanical Industry Press, Beijing.

Singer, J.G., (1981) Combustion-Fossil Power Systems. A Reference Book on Fuel Burning And Steam Generation, 3rd edition, Singer, J.G. Ed., Combustion Engineering, Inc., Windsor, CT, USA.

Stultz, S.C., and Kitto, J.H. (1992) "Steam—Its Generation and Use", 40th edition. Babcock & Wilcox, Barberton, Ohio.

Xu, X. (1990) "Combustion Theory and Equipment" (in Chinese). Mechanical Industry Press, Beijing.

9
Design of Novel Burners

In a pulverized coal (PC) fired boiler, the burner plays an important role. Designs of different types of PC burners have been discussed in earlier chapters. The present chapter discusses improvements made to conventional burners in order to achieve higher levels of performance both in terms of operating efficiency as well as environmental performance.

9-1 Design Requirements and Types of PC Burners

A good burner design aims at increasing the stability of the flame over a wide range of coal flow rates, and it provides a stable ignition source for the coal–air mixture. A proper combination of designs of the coal preparation system, the burner, and the furnace design helps achieve the above. Before going into improvements let us revisit the essential requirements of a good burner. The main requirements of the design are listed below (Cen, 1991).

(1) To create a favorable ignition condition, the local aerodynamic condition must ensure a good heat and mass exchange between the cold primary air and the hot gas at the exit of the nozzle or the burner.
(2) The arrangement of nozzle and air register must ensure the availability of a sufficient amount of ignition energy, the correct distance between the ignition region and the nozzle, the required heating rate of primary air, good oxygen supply, adequate mixing time for the primary, and the secondary air, and the required flame length.
(3) The burner must provide a stable source of heat energy for ignition and for flame stability.

In the last decade, several new types of pulverized coal burners and combustion techniques were developed in different countries to improve the combustion and to save fuels. Some of those new burners include

1. Blunt body burner
2. PC precombustion chamber burner

3. Flame stabilization boat burner
4. Co-flow jet burner with high velocity difference
5. Counter-flow jet burner
6. Dense–lean phase burner
7. Down-shot burner
8. Low NO_x burner

The down-shot burner is developed specifically to burn low-volatile, low-grade coal. Detailed descriptions and design approaches to these burners are given in the following sections.

9-2 PC Burner with Blunt Body

The blunt or bluff body burner (Fig. 9-1) is used to stabilize flame and to improve the combustion when burning low-grade bituminous coal, semi-anthracite, and anthracite. A wedge-shaped blunt body (Fig. 9-2) is installed at the exit of the

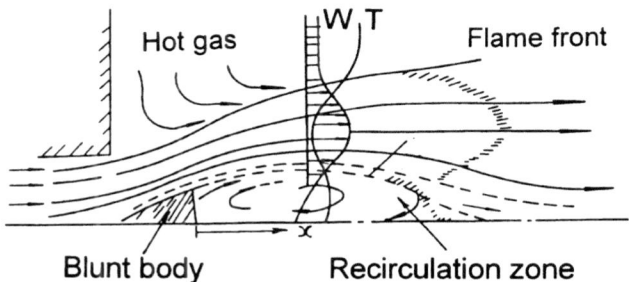

FIGURE 9-1. The principle of flame stabilization using a blunt or bluff body

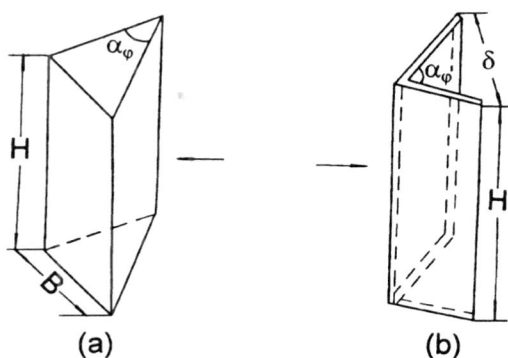

FIGURE 9-2. Wedge-shaped blunt body

primary air nozzle of a PC burner to deflect the air carrying the pulverized coal (PC).

Behind the blunt body a low-velocity recirculation zone is created (Fig. 9-1). The heat of the high-temperature gas is retained by this recirculation zone. Thus it can act as an ignition source. Pulverized coal can easily reach its ignition temperature in the recirculation zone and burn behind it. The flame is formed in the wake of the blunt body where the flow recovers (Qian and Ma, 1986).

9-2-1 Design of the Blunt Body

The size and flow rate of the recirculation zone are very sensitive to the cone angle α_φ of the blunt body (Fig. 9-2). The blunt body may be installed either outside or inside the nozzle. It covers a part of the air flow passage. The fraction of the flow area covered by the blunt body is called the blockage ratio, ϕ. It is defined as the ratio of cross section area occupied by the blunt body, f, to the total cross section area of primary air nozzle, F_n.

$$\phi = \frac{f}{F_n} \tag{9-1}$$

In a tangential burner the nozzle exit is rectangular and so is the blunt body. When the blunt body covers the entire height of the nozzle exit (Fig. 9-3), the blockage ratio is

$$\phi = \frac{\delta}{D} \tag{9-2}$$

where D is the width of primary air nozzle, δ is the width of the cross section occupied by blunt body at the exit of nozzle.

Experimental results (Zen & Ma, 1986) in cold models (Fig. 9-4) show that the ratio of the mass flow rate of the recirculating gas and the jet mass flow rate reaches a peak at a certain nondimensional distance (x/δ) from the exit of the nozzle, where

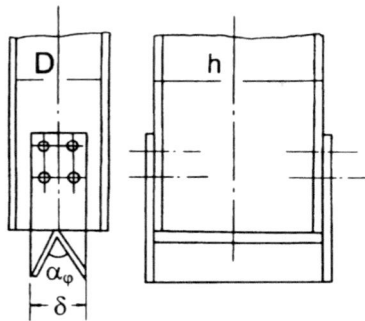

FIGURE 9-3. Cross section view of the exit of a primary air nozzle with a blunt body

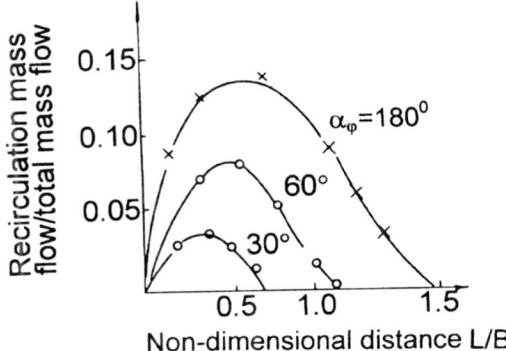

FIGURE 9-4. Effect of blunt body cone angle upon the mass flow of recirculation region

TABLE 9-1. Suggested geometrical parameters for blunt body burners.

Cone angle of blunt body	$\alpha_\varphi = 50°-65°$
Shape of blunt body	Isosceles triangle
Nozzle to blunt body width ratio	$\delta/D = 0.40-0.65$
Height of blunt body, H	Equal to the height of nozzle

δ is the width of the nozzle and x is the distance from the nozzle. This peak mass flow ratio increases with the angle of the wedge, α_φ. So the increase in wedge angle strengthens the ignition stability of the flame. This would, however, increase the flow resistance. For this reason, the blunt body cone angle should be chosen according to the furnace configuration and the type of coal. Some recommended values of these parameters are given below (Table 9-1):

9-2-2 Design of the Nozzle

Primary air nozzles, which carry the PC and the primary air, are rectangular in cross section. The ratio of its height to its width is greater than unity, i.e., $h/D > 1$ (Fig. 9-3). When the ratio of the height to the width increases, the vertical surface area of the jet increases. So the jet can draw in more hot gas from the furnace. However, the length of the potential flow core will reduce. Another consequence of increasing the h/D ratio is that the contact area between the primary air and the secondary air above will reduce. This leads to a delayed mixing of the primary with the secondary air.

The height-to-width ratio of the primary air nozzle should not be too high, or else the PC distribution would be uneven. Owing to gravity, the concentration of PC in the lower part of the jet will be higher. The separation of PC from the conveying air could be worse after it hits the blunt body. This would increase the unburned

carbon in the bottom ash. An optimum height-to-width ratio of the primary air
nozzle, h/D, is 1.5–2. However, the ratio can be lower if the distance between two
adjacent primary air nozzles is very small.

The presence of a blunt body in the primary air jet increases the half-angle of
the jet cone. For example, the expansion angle of a free jet is about 34°, while that
after the installation of a blunt body increases to 52–56°. This increase in the jet
angle enlarges the diameter of a firing circle. Therefore, the diameter of the ideal
or theoretical tangential firing circle (d_{jb}) should decrease when a blunt body is
installed. The following approximate relation gives an estimate of the change in
the diameter of the theoretical firing circle after the installation of a blunt body in
a tangential burner:

$$d_{jb} = (0.5\text{–}0.8)\,dj \qquad (9\text{-}3)$$

where d_{jb} is the ideal tangential circle diameter with blunt body, and dj is that
without a blunt body.

9-3 Precombustion Chamber Burner

The precombustion chamber (Fig. 9-5) of a burner helps ignite and stabilize the
flame in the main burner. It is made of refractory material or lined with refractory.
During startup fuel oil or natural gas is burnt in the pre-combustion chamber to heat
its refractory walls. Since the volume of the chamber is small, the wall heats up

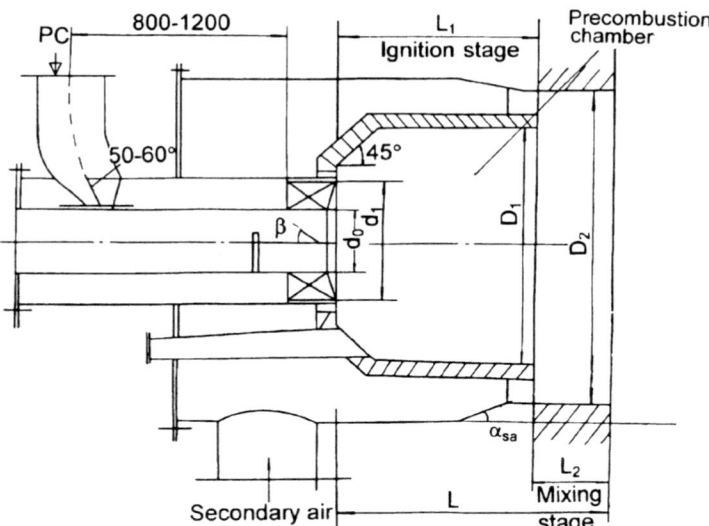

FIGURE 9-5. Precombustion chamber of a swirl burner with a mixing stage to fire bituminous
coal, the refractory lined precombustion chamber is shown by hatched lines

fast; and then the pulverized coal (PC) and primary air are allowed to pass through it. The mixture could be either in the form of a swirling jet or parallel flow, which mixes with the burning oil (gas) flame. The radiative heat from refractory wall ignites the PC, and helps sustain its combustion. The primary air flow is adjusted to anchors the PC flame to the root of the primary air jet. Thus, the PC–air mixture can ignite within a short time and continues to burn without any oil support.

9-3-1 Design of Precombustion Chamber

The design of the precombustion chamber varies with the type of coal, boiler configuration, and the system (Xu et al., 1987). A good design of the precombustion burner needs (a) adequate recirculation of gas, (b) favorable aerodynamic conditions, (c) high concentration of the PC mixture, and (d) high temperature and a high oxygen zone. All these create favorable conditions for stable ignition. Some basic design steps are given below with reference to Figure 9-5.

The concept of burner cross section heat release rate, q_F, is used for the determination of the precombustion chamber diameter D_1

$$q_F = \frac{Q}{F} = \frac{Q}{\frac{\pi}{4} D_1^2}, \text{MW/m}^2 \qquad (9\text{-}4)$$

$$D_1 = \sqrt{\frac{4}{\pi} \frac{Q}{q_F}}, \text{m} \qquad (9\text{-}5)$$

where F is the cross section area of the precombustion chamber, m^2, and Q is the rate of heat release in the precombustion chamber, MW.

The burner heat release rate, q_F, is a chosen design parameter. If it is very high, the chamber diameter will be too small, making it difficult to have adequate recirculation required for the ignition. If the value of q_F is too low, the design of the precombustion chamber will be too complex, and the PC particles will accumulate in the precombustion chamber. Assuming that the entire heat of the fuel is released in the precombustion chamber.

$$Q = B \, LHV \text{ MW} \qquad (9\text{-}6)$$

where B is the coal flow rate in a precombustion chamber in kg/s and LHV is the lower heating value of the coal in MJ/kg. The cross section heat release rate of most precombustion chambers firing bituminous coal and semi-anthracite is in the range of 15–25 MW/m^2.

9-3-2 Other Parameters of the Precombustion Chamber

Recommended values for other major dimensions of a precombustion chamber are as follows:

1. Angle β of the axial vanes for swirling the primary and secondary air (Fig. 9-5) is chosen as 20°–25° for bituminous coal with high heating value and high volatiles, and 30°–35° for bituminous coal with low heating value.

2. The chamber length L_1 and the ratio of length to diameter L_1/D_1 are taken as 500–600 mm and 0.65–0.8, respectively, for bituminous coal (Fig. 9-5)

With these geometrical proportions, the precombustion chamber can operate from cold with air temperature just above 45°C. A higher value of the ratio of diameter of the chamber and the outer diameter of the primary air swirler, D_1/d_1 is needed if the ignition temperature of the coal is high.

Secondary air is injected into burning primary air flow through a wedged flow passage having an angle α_{sa} (Fig. 9-5). This passage is in the annulus around the primary air tube at the exit of the precombustion chamber. The constriction angle, α_{sa}, should be reduced or eliminated when β is increased to match the coal type.

The swirling strength, S, of the primary air is given by

$$S = \frac{2}{3} \left[\frac{1 - \left(\dfrac{d_0}{d_1}\right)^3}{1 - \left(\dfrac{d_0}{d_1}\right)^3} \right] \tan \beta \qquad (9\text{-}7)$$

where d_0 and d_1 are the diameters of primary nozzle and the annular passage around it (Fig. 9-5). The above shows the influence of the geometry of the precombustion chamber on the swirl aerodynamic character. A weak swirl jet is adopted in the PC precombustion chamber. Generally, $S = 0.3$–0.45, while axial vanes of the primary air are at angle $\beta = 20°$–$35°$.

9-4 Boat Burner

Here, a boat-shaped flame stabilizer is installed in the primary air nozzle of a normal parallel flow PC burner (Fig. 9-6). Details of the boat-shaped flame stabilizer are given by Xu (1988). The boat is made of heat- and abrasion-resistant steel. Its service life can be up to six years in a 200 MW unit. The stabilizer can be moved back and forth with the center pipe in the primary air nozzle to adjust the combustion condition.

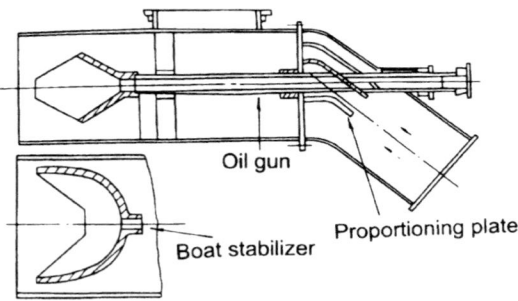

FIGURE 9-6. Cross section of a burner with a boat-type flame stabilizer

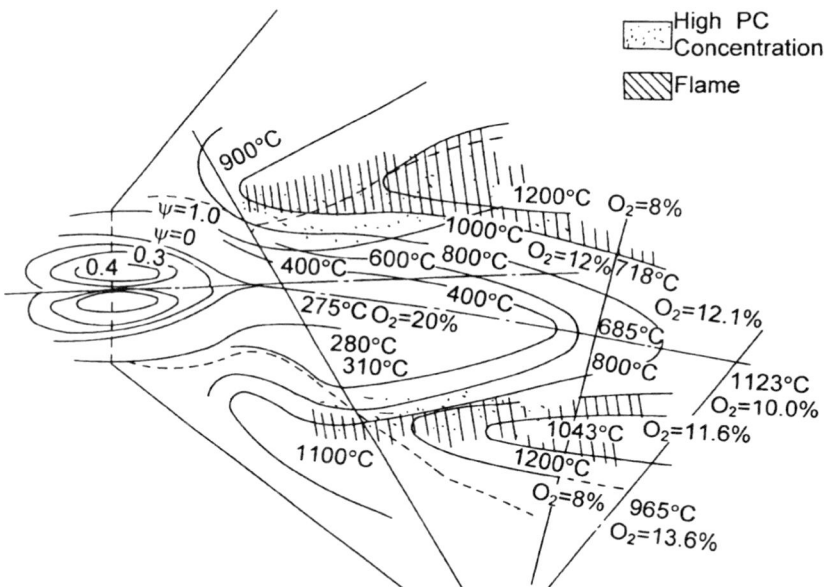

FIGURE 9-7. The distribution of temperature, oxygen concentration, and PC concentration (calculated) near the primary air nozzle of a boat burner fired in parallel with another one

The recirculating zone in a boat burner is small, and it is formed immediately behind the boat stabilizer. The gas temperature in this region is as low as 100–300°C. The walls of the boat body would be in the same temperature range. Thus, it cannot directly help ignite the PC. However, it promotes a mixing downstream, which creates a favorable condition to help the ignite and stabilize the PC flame.

Temperature measurements in a boat burner flame firing Datong bituminous coal are presented in Figure 9-7. The outer profile of the primary air flow stream is shown by the dotted line. Outside the dotted line is a high-temperature region where the temperature is above 900°C. This is formed by high-temperature gas from an adjacent burner. This gas is induced by the nozzle jet in a tangential firing furnace. The oxygen concentration measured here is above 10%. The region shown by scattered dots is the region of high PC concentration. It is also located at the high-temperature region. Therefore, high-temperature, high PC concentration, and high oxygen concentration appear at the same time: i.e., three favorable regions combine. It is easy to heat the air–fuel mixture in this region to its ignition temperature. Thus, this burner provides a stable heat source for PC ignition.

9-5 Co-Flow Jet Burner with High Differential Velocity

In ordinary burners the highest PC concentration does not necessarily occur where the gas temperature is highest. The co-flow jet burner attempts to make the PC

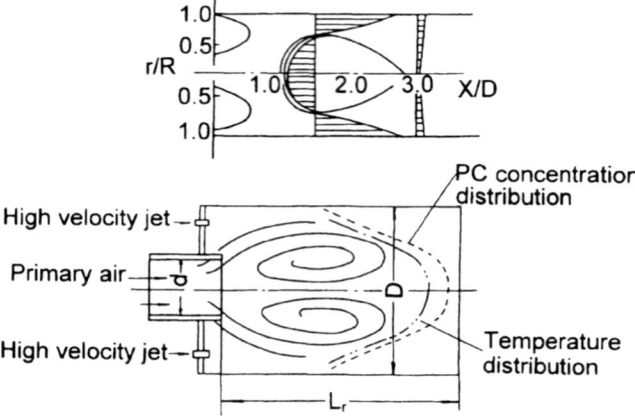

FIGURE 9-8. Cross section view of a co-flow jet burner: (a) the temperature and velocity profile overlaps each other, (b) velocity profiles across the cross section of the burner

concentration distribution coincide with the temperature profile Figure 9-8a. The primary air flows through a central tube of large diameter, at a speed of 15–20 m/s. Several small holes are made around the primary air tube. These holes can be arranged with or without axial symmetry. The shape of the holes can be round or 1–2 mm wide and slit shaped. Steam or compressed air is injected from the small holes, forming a high-velocity jet whose velocity is about 300 m/s.

In the confined space of the precombustion chamber of diameter D, the high-velocity jet entrains the gas from the primary air flow. Thus it forms a recirculating region at the back of the primary air (Fig. 9-8b). The length of this zone is about three times the length of recirculation region formed by a swirler. Because of this high temperature gas can be easily entrained and the recirculating region of the burner can be above 1000°C, which is much higher than that of the normal burner. This recirculation region is favorable to flame stability and combustion enhancement (Fu, 1987).

Based on cold flow tests (Fu, 1987) the following design norms can be used.

- Primary air velocity = 15–20 m/s
- Pressure of co-flow jets = 350–400 kPa
- Ratio of momentum of high-velocity jet and primary air momentum >5

9-6 Counter-Flow Jet Burner

The opposed jet, or counter-jet, burner (Fig. 9-9), as developed in China at Zhejiang University (Cen & Fan, 1991; Wang et al., 1988), uses the principle of opposed jet flame stabilization. Primary air enters through a central tube. One or more

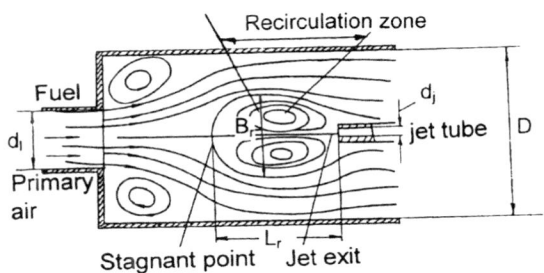

FIGURE 9-9. Diagram of an opposed jet burner showing its flow field fuel, primary air, critical region, stagnant point, jet outlet, nozzle

jet nozzles are installed on the burner axis opposing the primary air (Fig. 9-9). The counter-flow jet nozzle injects a high-velocity jet axially opposed to the primary air. The distance of this nozzle, L_j, from the primary air nozzle can be adjusted as needed within $(2–4)xD$, where D is inner diameter of flame tube (Fig. 9-9). The diameter of the counter-flow nozzle, d_j, is in the range of $1/30$–$1/20$ D. The secondary air nozzles are installed tangentially at a certain axial distance. The primary air flow is parallel to the axis, while the secondary air is injected tangentially.

The opposed jet forms a recirculation zone on the flow axis with a stagnant zone, as shown in Figure 9-9. This zone slows down the primary air locally and the flame is anchored to this low-velocity zone. Large coal particles, owing to their higher momentum, may directly enter the stagnation zone from the PC flame. Since the gas temperature is very high in the recirculation zone, large coal particles are easily ignited in the stagnation zone.

The position of the stagnation point or the length of the recirculation, L_r, can be adjusted by controlling the velocity of the counter-flow jet, U_j. The velocity of this jet is generally in the range of 50–70 m/s.

A dimensional analysis identified the following parameters to affect the recirculation zone. The length of the recirculation zone, L_r, can be computed from the following empirical equations based on experimental results (Cen & Fan, 1991)

$$L_r = D[A_1\phi(Re)^2 + B_1\phi(Re) + C_1] \qquad (9\text{-}8)$$

where $\phi(Re) = Re_1/(Re_1 + Re_j)$; D is the diameter of the flame tube; Re_1 and Re_j are Reynolds numbers for the primary air and jet flow, respectively. The coefficients A_1 B_1 and C_1 are as follows.

$$A_1 = -(d_1/d_j)^5\{0.01945[\ln(L_j/D)]^2 - 0.03595\ln(L_j/D) + 0.01868\}$$
$$B_1 = -10^{-5}(d_1/d_j)^{-5}\{0.05934[\ln(L_j/D)]^{1/3} + 0.1168\ln(L_j/D)^{3/2} - 0.05676\}$$
$$C_1 = -(d_1/d_j)^{-0.85}[0.5186(x/D)]^2 - 2.7189(x/D) + 10.872$$

Similarly, the width of the recirculation zone, B_r is found as

$$B_r = D[A_2\phi(Re)^2 + B_2\phi(Re) + C_2] \qquad (9\text{-}9)$$

where $A_2 = 0.3020(L_j/D)^2 + 2.4795(L_j/D) - 5.5258$
          $B_2 = 0.2353(L_j/D)^2 - 1.9395(L_j/D) + 4.1401$
          $C_2 = 0.03922(L_j/D)^2 + 0.3258(L_j/D) - 0.0968$

The maximum recirculation flow rate and the total flow rate are found from the above regression as

$$\frac{q_{mr,\max}}{q_{ml} + q_{mj}} = A_3\phi(Re)^2 + B_3\phi(Re) + C_3 \qquad (9\text{-}10)$$

where $A_3 = 0.09974(L_j/D)^2 + 0.6289(L_j/D) + 0.7564$
          $B_3 = -(d_1/d_2)[-0.07632(L_j/D)^2 + 1.0168(L_j/D) + 0.006664]$
          $C_3 = 0.8438(L_j/D)^2 - 4.589(L_j/D) + 9.7618$

These empirical relations are helpful for a quantitative assessment of the characteristics of the recirculation zone.

EXAMPLE

> Find the length of the recirculation zone and its width for an opposed jet burner. Given (referring to Fig. 9-9) $L_j/D = 2.667$, $d_1/d_j = 3.064$, and $d_l = 20$ mm; $U_1 = 15$ m/s; $U_j = 50$ m/s; $\rho_1 = \rho_j = 1.2$ kg/m^3. Also find the increase in recirculation flow if the opposed jet velocity U_j is increased from 50 to 60 m/s.

SOLUTION:

Mass flow of jet,

$$q_{mj} = \frac{\pi}{4}d_j^2\rho_j U_j = 0.785 \times (.02)^2 \times 50 \times 1.2 = 0.01885 \text{ kg/s}$$

similarly mass flow of primary air, $q_{ml} = 0.785 \times (.02 \times 3.064)^2 \times 15 \times 1.2 = 0.05309$ kg/s.

For maximum recirculation rate the empirical Eq. (9-10) is used. Since all parameters except the velocity remains unchanged the ratio of Reynolds number is taken as the ratio of velocities: $\phi(Re) = Re_1/(Re_1 + Re_j) = 15/(15 + 50) = 0.479$
    Substituting $L_j/D = 2.667$ in Eq. (9-10) we get

$A_3 = 0.09974 \times 2.667^2 + 0.6289 \times 2.667 + 0.7564 = 3.145$
$B_3 = -3.064 \times [-0.07632 \times 2.667^2 + 1.0168 \times 2.667 + 0.006664]$
$\quad = -6.681$
$C_3 = 0.8438 \times 2.667^2 - 4.589 \times 2.667 + 9.7618 = 3.4231$

$$\frac{q_{mr,max}}{q_{ml} + q_{mj}} = 3.145(0.479)^2 - 6.681 \times 0.479 + 3.423 = 0.944$$

$$q_{m,rmax} = 0.944 \times (0.01885 + 0.05309) = 0.06810 \text{ kg/s}$$

$$q_{m,rmax}/q_{ml} = 0.06810/0.05109 \times 100\% = 128.3\%$$

If the counter-flow jet velocity is increased to 60 m/s, keeping other parameters unchanged, we get

$$q_{mj} = \frac{\pi}{4}d_j^2\rho_j = 0.785 \times (.02)^2 \times 60 \times 1.2 = 0.02262 \text{ kg/s}$$

$$Re_1/(Re_1 + Re_j) = 0.434$$

Thus from Eq. (9-10)

$$q_{m,rmax} = 0.08462 \text{ kg/s}$$

$$q_{m,rmax}/q_{ml} = 0.08462/0.05309 \times 100\% = 159.4\%$$

From the above, we see that when U_j is increased from 50 to 60 m/s, the recirculation flow increases by (0.08462−0.06810/.06810) or 24%.

9-7 Dense and Lean Phase PC Burner

A dense mixture or highly concentrated pulverized coal (PC) air stream is easier to ignite than a leaner mixture. However, the mass ratio of coal and air is fixed by the stoichiometry. So it is difficult to increase the amount of combustibles in a mixture. Lower grade coals are difficult to ignite at normal concentration. Dense and lean phase burner gets around the problem by splitting the primary air–PC mixture into a fuel rich (high PC concentration) stream and a fuel lean (low PC concentration) stream. This splitting is done by various forces like centrifugal force, inertial force, or other methods. Two streams are injected separately into the furnace through different nozzles. Thus, a fuel rich flow can be obtained without reducing the total quantity of primary air flow. These types of burners can burn a wide range of coal, from high-ash bituminous coal to high-moisture brown coal, and anthracite to semi-anthracite.

9-7-1 Design of a PC Concentrator Burner (Cen & Fan, 1991)

Several types of dense and lean phase PC burners with different separation methods are briefly described below:

a) Elbow Concentrator (PAX Burner)

When the PC air mixture passes through an elbow of a pipe, denser PC particles tend to concentrate toward the outer edge of the elbow owing to the centrifugal force (Fig. 9-10a). The downstream tube is split to carry the lean and rich mixture

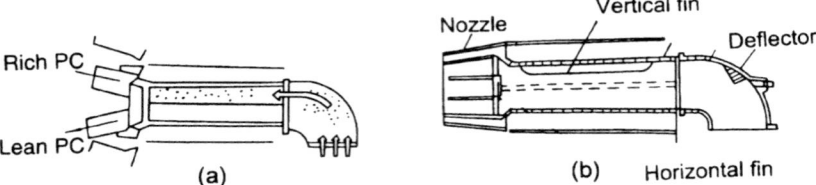

FIGURE 9-10. A bend in the fuel line is used to concentrate the PC mixture: (a) simple bend, (b) bend with a deflector

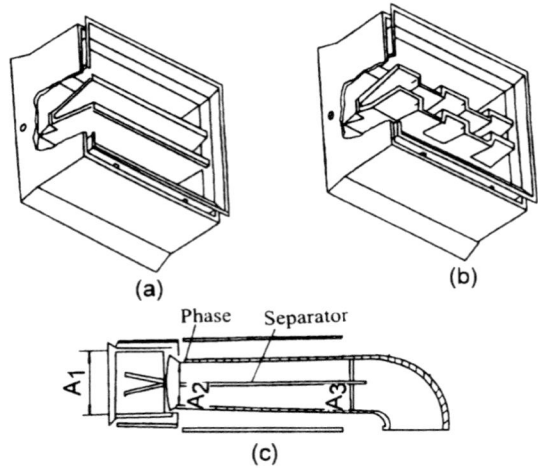

FIGURE 9-11. PC-concentrated burner with outlet expand cone: (a) corrugated cone, (b) simple triangle cone, (c) cross section of the nozzle

separately. In another variation, a deflector located on the outer edge of the elbow (Fig. 9-10b) deflects the PC toward the inner edge, separating the flow in rich and lean streams. In one variation of this burner, the fuel rich mixture is mixed with hot secondary air. Such a mixing prompts quick ignition and flame stability.

b) Expanding Section

The design of the burner shown in Figure 9-11 is similar to the burner described above except that it uses two different designs for the burner exit. The one shown in Figure 9-11a uses a waveform expanding section at the outlet of nozzle. It can increase the contact surface between the primary air flow and the recirculating gas. The other one, shown in Figure 9-11b, uses a simple triangular expanding section. It has, however, a small edge at the tip of the expanding section. This edge

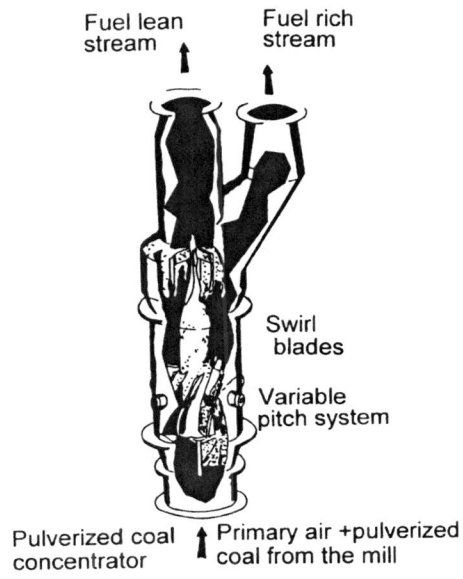

Fuel lean stream

Fuel rich stream

Swirl blades

Variable pitch system

Pulverized coal concentrator

Primary air +pulverized coal from the mill

FIGURE 9-12. PC-concentrated burner with swirl vane of GEC Alsthom Stein Industries setup in a 600 MW plant of the Electrocute de France

has a great effect on the recirculation of high temperature gas. The angle of the expanding section is 20°. The cross section of the nozzle as shown in Figure 9-11c.

c) PC Concentrator Burner with Rotary Vanes

Figure 9-12 shows the working principle of the PC concentration burner with rotary vanes. The PC air flow is separated into dense and lean mixtures by the centrifugal force in a vortex generator. The dense PC flow is injected into the furnace from the two lower layers of nozzles (Lani et al., 1995). At the same time, the upper and lower secondary air is injected and the lean phase is injected above the ignition zone. Thus the flame stability is intensified.

d) Cyclone PC Concentrator Burner

In this design the PC air mixture is split into rich and lean mixtures by using the centrifugal action of reverse flow cyclones. The design of the cyclone PC concentrator burner is shown in Figure 9-13. The PC air mixture is first divided into two streams by a flow divider. The mixture then enters the cyclone, which separates it into dense coal flow and lean coal flow by centrifugal force. The lean PC flow is injected into the furnace through the air exhaust pipe at the upper part of the cyclone. The dense coal flow is injected into the furnace through the nozzles at the bottom of cyclone.

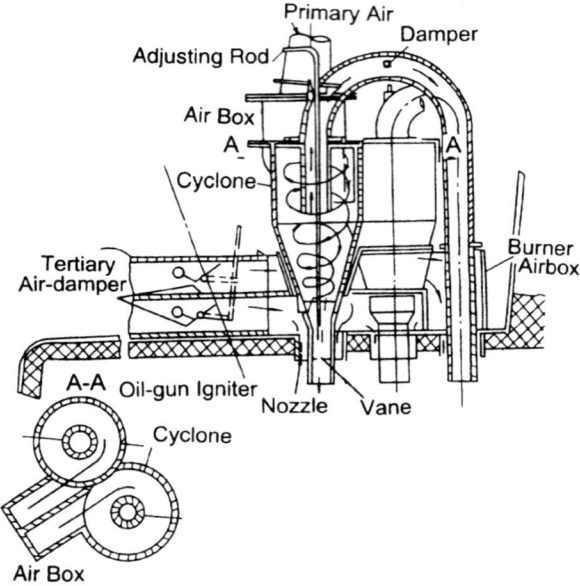

FIGURE 9-13. PC-concentrated burner with a reverse-flow cyclone

9-7-2 Operation of Dense-Lean Phase Burners

In a tangentially fired boiler, the dense phase moves toward the flame but away from the wall, while the lean phase moves away from the flame but toward the wall. The concentration of PC in the lean phase is low (0.1–0.2 kg/kg). So it reduces the number of PC particles hitting the wall. The concentration of dense phase is in the range of 0.8–1.2 kg/kg. When the dense suspension hits the flame it is easily ignited improving the flame stability.

In a tangentially fired boiler, when parallel bends are used to separate PC, the dense phase of the primary air flow at two corners faces the flame and the lean phase of the primary air flow at the other two corners faces the wall. PC begins to ignite first from the region where the temperature and the PC concentration is high and the oxygen is sufficient. The dense phase of the primary air flow can be ignited immediately when facing the flame because of its the high temperature. The slagging at the wall can be prevented when the lean phase of the primary air flow faces the wall. The dense and lean phases are often reversed to balance the effect of concentration. This means the dense phase is made to shift toward the wall and the lean phase toward the flame. Usually a spiral plate is used to make the flow reversal. But the resistance of spiral plate is high, and there exists the problem of abrasion, so a special type of streamlined guiding equipment (Chi, 1995) can be used.

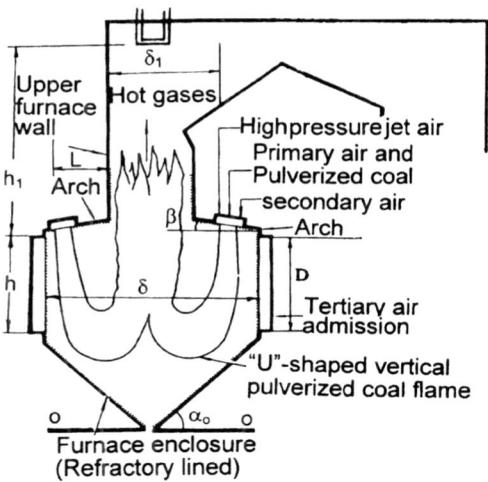

FIGURE 9-14. Down-shot flame furnace

9-8 Down-Shot Flame Combustion Technique

This type of burner makes the flame turn $180°$ (Fig. 9-14) to utilize the radiation from the tip of the flame igniting the coal at its base. It is especially effective in burning low-volatile coal. The schematic of the furnace of a down-shot flame boiler is shown in Fig. 9-14. Recommended fuels for down-shot furnace are given in Table 9-2, which shows that the down-shot boiler is suitable for both good- and poor-quality coal.

The primary air carries the PC with about 5%–15% of total air flow. When the volatile is less than 11%–14%, the concentration of PC in the primary air should be increased, and the velocity of primary air should be decreased to ensure that the PC air flow ignites at proper time. When the volatile is higher than 12%–14%, the primary air need not be concentrated.

The secondary air flows around the primary air. This air flow can make the burning gas recirculate to form a vortex to increase the ignition region temperature.

TABLE 9-2. Applicable range of coal quality in typical down-shot flame furnaces (Cen and Fan, 1990, 1991).

Proximate analysis	Typical values	Range
Moisture (%)	8	5–14
Ash (%)	60	30–70
Volatile (%)	15	13–20
Fixed Carbon (%)	40	20–55
Heat value (kJ/kg)	12,500	8800–25,500

The gas in the ignition region and the furnace arch wall are heated by the side radiation of the flame. When a low-volatile coal is burned, it is necessary to add tertiary air to help complete the combustion after ignition. When burning with high-volatile coal, the tertiary air can be reduced. If the volatile matter is above 14%–16%, the tertiary air can be decreased to zero.

The main features of the down-shot firing are as follows:

1. It is easy to achieve staged combustion. So the excess air and NO_x emission can be low.
2. Since the center of the down-shot flame is near the PC nozzle, low-grade coal is easy to ignite (especially low volatile coal).
3. The residence time of the PC in the furnace is long. This helps improve the combustion efficiency, especially for difficult to burn coal.
4. The fly ash can be separated out because the down-shot flame turns 180° near the bottom. However, excessive separation at the bottom may increase unburned carbon losses.
5. It is easy to use a PC-concentrated burner, which may facilitate ignition.
6. The down-shot flame boiler can adapt to a wide range of coal types because of its staged air and the adjustable PC concentration.
7. The furnace top of the front and back wall provides a part of heat source of ignition because they reflect the radiant heat.

Depending upon the type of PC concentrators used down-shot burners can be of three types:

1. With cyclone separator
2. With parallel flow slitted burner
3. With primary air exchange burner (PAX burner)

Following are some typical values of a down-shot furnace given with reference to Figure 9-14.

1. The height of lower furnace, h:
   (a) Large boilers, $h > 8$–10 m; (b) small boilers, $h = 6$–7 m, (c) $h_1/h = 1.4$–3.0 m.
2. The ratio of upper to lower furnace depth, $b_1/b \approx 0.5$.
3. The vertical-type cyclone chamber burner of standard design requires an opening diameter of 0.508–4.52 m on the furnace arch. The furnace arch depth L cannot be less than 3 m; for boilers with capacities of 250–600 MW, $L = 3$–6 m.
4. The inclination angle of hopper $\alpha_h = 49.5°$–58°. For low-volatile coals it is 58° or higher. The width of slag-drain opening can be 1.4 m.
5. The inclination angle of furnace arch β is 10°–15°. If β is reduced, the reverse flow will increase, which will in turn help the ignition though it would increase slagging.

TABLE 9-3. Air velocity and percentage air rate for down-shot firing.

Manufacturer	Foster Wheeler	B&W	Stein	Steinmuller
Burner Type	Cyclone separator	Slitted	Slitted	Slitted
Coal type	Anthracite	Low-rank bituminous	Anthracite	Semi-anthracite
V_{daf}, %	4–10	18	9.3	13.89
Ash, %		25	30.5	19.94
Primary air rate, %	18	29	10-15	18
Secondary air rate, %	15	71		52
Tertiary air rate, %	67	0	0–10	30
Primary air velocity, m/s	25	10	7.5(0–12)	11.3
Secondary air velocity, m/s	10–20	30	35–40	33.8
Tertiary air velocity, m/s	8–10	0		24.8

6. Volume heat release rate of the furnace (lower and upper furnace) = 116–140 kW/m^3.
7. Gas velocity at the exit of the burn-out zone is 9–10 m/s.
8. To create proper aerodynamic conditions in the furnace, an appropriate choice of velocity and air flow rate of the burner must be made. The data chosen by several boiler manufacturers are listed in Table 9-3.
9. The temperature of the preheated air can be 380°–430°C for anthracite.
10. Required fineness of the pulverized coal is $R_{90} \leq 5\%$. Generally 85% PC should pass through 200 mesh screen ($R_{75} = 15\%$). For anthracite whose volatile as received $V_{daf} = 4\%$–5%. The fineness required $R_{75} < 10\%$.

9-9 Low NO$_x$ Burner

Reduction of nitric oxide emission is a major concern for most industrialized nations. Thus special attention is given to the reduction of NO$_x$ produced in burners. Nitric oxide is produced by the high-temperature oxidation of the nitrogen in the combustion air (thermal NO$_x$) as well as by the oxidation of the nitrogen in the fuel (fuel NO$_x$). The first reaction is favored by high temperature, while the second one is favored by high oxygen availability (Talukdar & Basu, 1995). Many measures can be taken to reduce the NO$_x$. Burners are designed to use one or a combination of these measures (Cen & Fan, 1991, Fen, 1992). Some of those measures are discussed below.

9-9-1 Reduction in Flame Temperature

One effective method to reduce the NO$_x$ formation is to lower the flame temperature which reduces the thermal NO$_x$ formation. There are many ways to decrease the flame temperature. For example:

• Decrease the temperature of preheated air

- Divide one large flame into several smaller flames
- Use a lower burner zone heat release rate
- Enlarge the radiative heat absorbing area of the flame
- Adopt a gas recirculation technique

When burning heavy oil, if the air preheating drops by 300°C at an excess air coefficient = 1.2, the flame temperature would be reduced by only 50°C. The NO_x would be reduced by 10% after this reduction (Cen & Fan, 1991), owing to the decrease in thermal NO_x production. However, the stack gas temperature will rise owing to the reduction in air preheating with consequent decrease in the boiler efficiency.

When heavy oil is burned, the injected oil spray drops are made to collide with another object near the nozzles such that each drop is divided into smaller droplets. Alternately, several small oil nozzles are also used to form several small but finer oil sprays, which are then mixed with air. Thus several small flames are formed instead of one large flame, and the total radiation area of the flames is larger than that of a single one. So, radiation to the wall increases decreasing the flame temperature, which in turn reduces the thermal NO_x formation.

When other conditions are the same, a reduction in the heat input to the burner zone per unit effective projected heat transfer surface reduces the flame temperature and therefore the thermal NO_x. Following are some alternative means for NO_x reduction.

a) Water Injection

The temperature of a heavy oil flame may also be reduced by injecting steam into the combustion air or injecting steam near the oil spray. To decrease thermal NO_x formation, a large amount of steam is needed, which adversely affects the boiler efficiency and increases the low-temperature corrosion of the cooler section of the boiler. Sometimes water is mixed with heavy oil to form an oil–water emulsion. The burner produces water 2 and 10 μm droplets. When the oil droplet is burnt, the micro-fine water droplets will vaporize and expand, bursting the oil droplet. Thus this "micro-explosion" helps atomize the oil droplet, which is helpful for the burn-out of oil and reduction of the carbon black in the gas. Another effect of this is the reduction of NO_x formation.

The above-described two methods for reduction of NO_x formation by decreasing the flame temperature rely on the reduction of oxidation of thermal nitrogen in the air. They are very effective when burning low nitrogen ($N_{daf} < 0.4\%$) fuel.

b) Flue Gas Recirculation

Flue gas recirculation dilutes the air and helps decrease the flame temperature. In one type, exhaust gas recirculation, a part of the flue gas is bled before the air heater, and mixed with combustion air before being sent into the furnace (Fig. 9-15). Normally, the recirculation rate (ratio of recirculating gas and the total flue gas) is about 15%–20%. NO_x emission will decrease by 35% when burning natural gas; 13%–30% when burning oil; and by less than 25% when burning pulverized

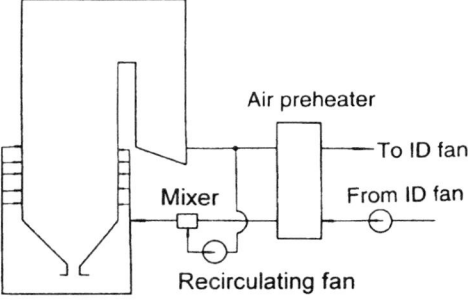

FIGURE 9-15. Flue gas recirculation system

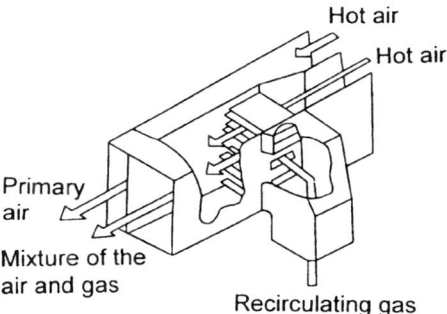

FIGURE 9-16. Mixing of the recirculation flue gas with the secondary air hot air, hot air, primary air, gas and air, recirculation gas

coal. Because the cold gas is mixed in the combustion region, the flame temperature will decrease, and this helps reduce the thermal NO$_x$ formation. Moreover, because the oxygen concentration in the gas and air mixture will decrease from 21% to 17%–18% when the recirculation rate is 15%–20%, it would also help reduce the fuel NO$_x$ formation. So this method can be used when burning residual oil as well as pulverized coal with low nitrogen. The low-temperature inert gas in primary air may hinder good ignition, which is acceptable for the easily ignitable brown coal, but it will affect the flame stability. In order to reduce its impact on the flame stability, instead of mixing the recirculation gas with the primary air, it is mixed with the secondary air as shown in Figure 9-16. Sometimes, the recirculation gas is not mixed with preheated air and is sent to the furnace through separate nozzles or from the furnace bottom.

9-9-2 Dense and Lean Phase Method

This type of burner, described earlier in Section 9-7, is also a low NO$_x$ burner. Here, a part of the fuel will burn under an excess air coefficient, while the other part will burn under deficient air conditions. The combined excess air coefficient

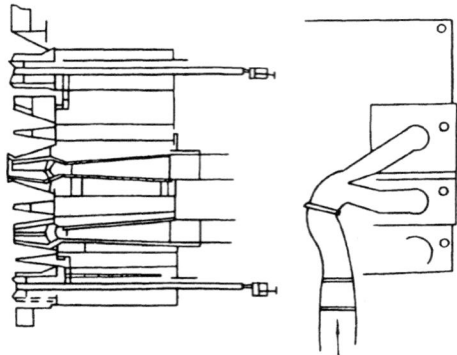

FIGURE 9-17. PM low NO$_x$ parallel-flow burner firing oil

is, however, still in the range of 1.03–1.15. A pollution minimum (PM) low NO$_x$ parallel flow burner developed by Mitsubishi Co. is shown in Figure 9-17. It is used in a tangential fired boiler. The fuel delivered by some nozzles is low, giving a low fuel concentration. This results in a lower flame temperature, and hence lower thermal NO$_x$ formation. Other nozzles inject larger amounts of fuel, and therefore, the oxygen concentration is decreased. This helps reduce the fuel NO$_x$ formation. At a certain distance from the nozzles, the air flows mix completely, the excess air coefficient is kept within the required overall range, and the fuel burns completely. Sometimes, a recirculation gas nozzle is inserted between the dense phase nozzle and the lean phase flame nozzle. This is called separate gas recirculation (SGR). This helps further reduce the NO$_x$ formation.

The characteristics of a low NO$_x$ PM oil burner are such that when the fuel burns under different excess air coefficients, α, without gas recirculation, the NO$_x$ emission is very high at $\alpha = 1$. If the nozzle is divided into dense and lean phases, the NO$_x$ formation in both streams is lower than that at $\alpha = 1.0$. Thus the average NO$_x$ formation of this type of burner is low.

If flue gas recirculation is used, further reduction is achieved. When α is less than 1, the NO$_x$ decreases owing to insufficient air supply near the flame surface, but the resulting diffusion flame produces carbon black owing to the shortage of oxygen. Such a soot-laden flame cannot be used in the boiler. One needs to ensure that enough air is supplied to the base of the fuel nozzles; and the fuel spray is mixed with air as quickly as possible to reduce the soot.

The arrangement of PM low NO$_x$ pulverized coal burner for a tangential fired furnace is shown in Figure 9-18. The fuel is divided into dense and lean phases at the bend and then is sent to the upper and lower primary air nozzles. The upper nozzles are for the dense phase, the lower ones are for the lean phase, and a separate gas recirculation nozzle is installed between them. The PC concentration in the original primary air pipe is about 0.5 kg PC/kg air. After separation, the concentration of pulverized coal in the upper primary air nozzle is more than 1.5 kg/kg(air) where

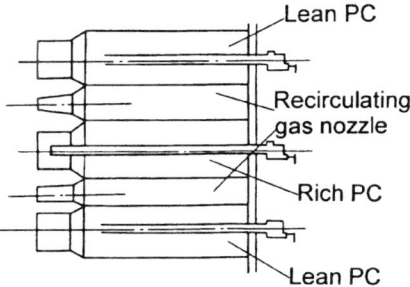

FIGURE 9-18. PM low NO$_x$ parallel-flow PC burner

the primary air rate is much smaller. Although the PC concentration in the lower primary air nozzle is small, the total air quantity is still less than the theoretical air. The total NO$_x$ formation of this kind of burner is low. A similar burner developed by GEC Alsthom Stein Industrie and by EVT Energie (Whitworth et al., 1994).

9-9-3 Staged Combustion Method

Staged combustion relies on the reduction of the total air at the burners in one of the following three ways:

a) Over-Fire Air

The excess air coefficient in the burner is 0.7 for gaseous fuel, 0.8 for oil, and 0.8– 0.9 for coal. This region is called the primary combustion region. The remaining combustion air is injected into the furnace between 1.5 and 3 m above the upper edge of the top nozzles. This part of the air is called secondary air or *over-fire air* (OFA). The ratio of the OFA and the total air is usually in the range of 0.1– 0.2 in the PC boiler. In this method, because of the insufficient air in the primary combustion region, the fuel cannot burn completely and the temperature is too low. The temperature being low, the thermal NO$_x$ is low. As air is insufficient, fuel NO$_x$ formation decreases too. The method is able to reduce not only high-temperature NO$_x$ but also fuel NO$_x$. So it is effective for both high- and low-nitrogen fuel. Dense phase–lean phase techniques and flue gas recirculation can be combined with this method for further reduction in NO$_x$.

b) Air Staging in Burners

Babcock & Wilcox of Barberton, Ohio (Stultz & Kitto, 1992) have used the stage combustion principle in their model DRB-XCL low NO$_x$ swirl burners (Fig. 9-19). The primary air is only 15%–20% of the total air. The secondary air is divided into two parts. The inner air is 15%–45%, while the outer one is 55%–65%. The secondary air is designed to mix into the flame gradually. To separate the primary air from the secondary air and to delay their mixing, the recirculation air or cool

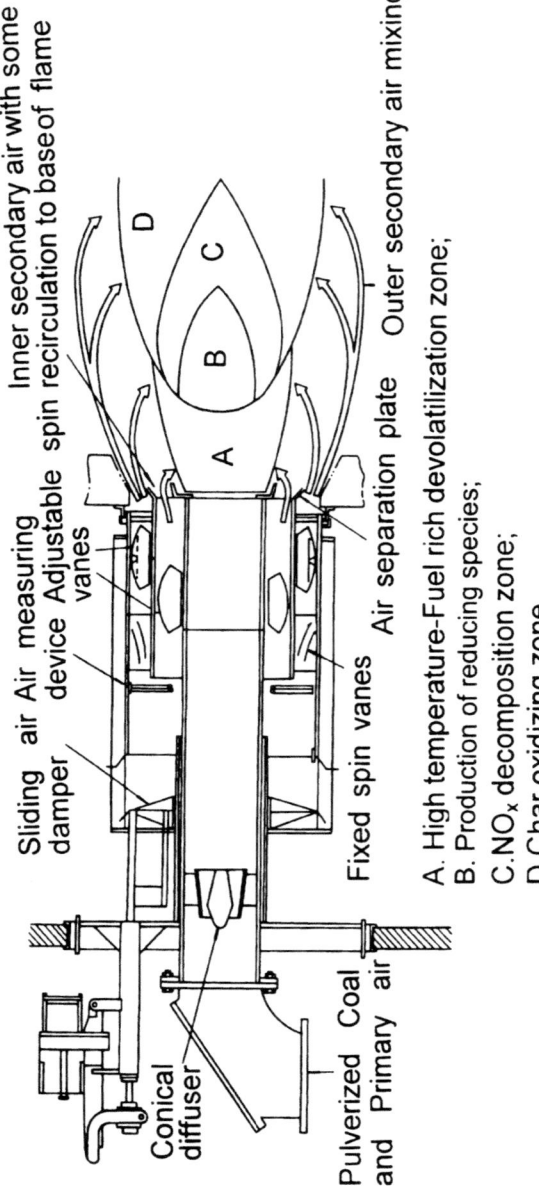

Sliding air Air measuring Inner secondary air with some
damper device Adjustable spin recirculation to baseof flame
          vanes

Fixed spin vanes   Air separation plate   Outer secondary air mixing

Conical
diffuser

Pulverized Coal
and Primary air

A. High temperature-Fuel rich devolatilization zone;
B. Production of reducing species;
C.NO_x decomposition zone;
D.Char oxidizing zone.

FIGURE 9-19. Low NO_x swirl burner model DRB-XCL of Babcock & Wilcox

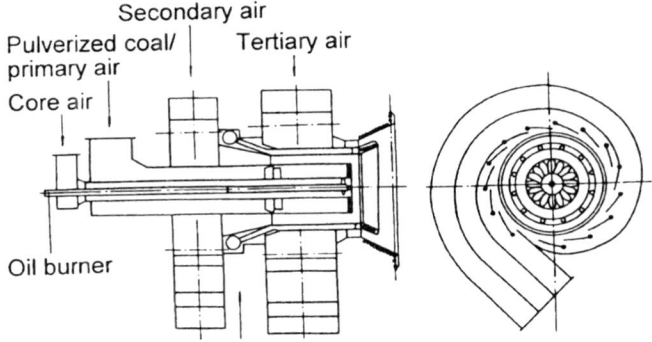

Secondary air

Pulverized coal/ | Tertiary air
primary air

Core air

Oil burner

FIGURE 9-20. Model DS low NO$_x$ burner of Deutsch Babcock

air is sent around the primary air nozzle. As there are two air registers on each burner, this type is called a twin air registers low NO$_x$ burner. In some designs, the quantity of the secondary air is reduced, and four nozzles of the tertiary air are added around one swirl burner at 90° intervals. Because of the insufficient air and the problem of increased combustibles in fly ash during operation, there is greater slagging potential. Deutsch Babcock added a spiral to the secondary and tertiary air in addition to the swirler inside the primary PC-air tube (Fig. 9-20). This gave better reduction of NO$_x$ (Brinkmann & Schuster, 1993).

c) Reburning Technique

Reburning is another staged combustion method to control NO$_x$. It sends a part of the fuel into the furnace, bypassing the main combustion region. This is done in several stages. About 80%–90% of the fuel is sent underneath the combustion chamber, which is called main *primary combustion stage*. This fuel is burnt at an excess air coefficient exceeding 1.5 and at a normal primary to secondary air ratio. The remaining 15%–20% fuel is injected into the second stage, which is also called the *reburning stage*. This part of the fuel is called *reburn* fuel. In the reburning stage hydrocarbon-based radicals strip oxygen from NO molecules. So the remaining nitrogen ions combines to produce molecular nitrogen.

$$2NO + 2C_nH_m + (2n \pm 1)O_2 \Leftrightarrow N_2 + 2nCO_2 + mH_2O \qquad (9\text{-}11)$$

The reaction requires a minimum residence time of about 0.1 s and a temperature of about 1200–1350°C. The reburn fuel (PC) is carried by inert recirculated flue gas to maintain the required low oxygen atmosphere in the reburning zone.

The efficiency of NO$_x$ reduction increases with decreasing ratio of fixed carbon to volatile matter content of the reburn fuel. Natural gas, if available, can be used as an effective reburn fuel. Additional air is added above the reburning zone to complete the combustion. This zone also requires 0.1 s minimum for complete combustion.

While it is a simple technique and is especially suited for retrofitting an existing furnace, it suffers from the disadvantages like higher tube wastage, extra furnace height requirement, and increased carbon in ash.

9-9-4 Partial PC Concentration at the Burner Nozzle

This method is different from the dense and lean phase method of separating the fuel before it enters the burner. Here the flow pattern in the burner is arranged such that pulverized coal is separated from the conveying air partially in the chamber after the nozzle, which makes the PC pass through the high concentration stage during the combustion. So the method can decrease fuel NO_x. If the method is adopted, secondary air is purposely reduced, and the excess air coefficient decreases in the burner region. The tertiary air is injected into the furnace at a distance of 1.5–3 m from the burner so that the NO_x emission decreases more effectively. Tests (Xu et al., 1987) in tests on swirl PC precombustion chambers have shown that the NO_x can be reduced below 200 ppm when firing anthracite.

Nomenclature

$A_1 B_1 C_1$	coefficients in Eq. (9-9)
$A_2 B_2 C_2$	coefficients in Eq. (9-10)
$A_3 B_3 C_3$	coefficients in Eq. (9-11)
B	coal flow rate in a pre-combustion chamber, kg/s
b	width of the blunt body section, m
D	width of primary air nozzle, m
d_1	outer diameter of the primary air swirler, m
D_1	pre-combustion chamber diameter, m
d_0	diameters of primary nozzle, m
d_1	diameter of the annular passage around the primary nozzle, m
d_j	ideal tangential circle diameter without blunt body, m
d_{jd}	ideal tangential circle diameter with blunt body, m
f	cross section area occupied by the blunt body of primary nozzle, m^2
F	cross section area of the precombustion chamber, m^2
F_n	total cross section area of primary nozzle (Eq. 9-1), m^2
H	height of lower furnace, m
h	height of primary air nozzle, m
L	furnace arch depth, m
L_1	chamber length, m
L_j	length of jet (Eq. 9-11), m
L_r	length of recirculation zone, m
Q	heat release of pre-combustion chamber, MW
q_F	heat release rate of burner based on its cross section, MW/m^2
q_{mj}	flow rate of primary air, kg/s
q_{ml}	jet flow rate, kg/s
$q_{mr,max}$	maximum recirculation flow, kg/s

LHV	lower heating value of the coal, MJ/kg
q_v	average volume heat release rate, kW/m^3
S	strength of the primary air

Greek Symbols

α_b	cone angle of blunt body
α	excess air coefficient
α_χ	inclination angle of hopper
α_{sa}	angle of constriction of the secondary air exit
β_1	angle of the axial vanes for swirling (Eq. 9-7)
β	inclination angle of furnace arch
ϕ	blockage ratio (Eq. 9-1)
$\phi(Re)$	Function of Reynolds number (Eq. 9-9)

Dimensionless Numbers

| Re_l | Reynolds number for the primary air |
| Re_j | Reynolds number for the jet flow |

References

Brinkmann, C., and Schuster, H. (1993) Advanced new combustion technologies for coal fired steam generators with high steam capacity. Power-Gen '93, Europe, 7–8:387–419.

Cen, K., and Fan, J. (1991) "Combustion Fluid Dynamics" (in Chinese). Hydroelectricity Press, Beijing, p. 9.

Cen, K., and Fan, J. (1990) *Theory and Calculation of Engineering Gas-Solids Multi-Phase Flow* (in Chinese). Zhejiang University Press, Hangzhou, China, p. 3.

Chi, Z., Yao, O., and Cen, K. (1995) "Development of a new technology—Rich/lean combustion of the pulverized coal." *The 20th International Conference on Coal Utilization and Fuel Systems, USA.*

Fen, J. (ed.) (1992) *"Principles and Calculations of Boilers"* (in Chinese), 2nd edition. Science Press, Beijing.

Fu, W. (1987) Co-flow jets with large velocity difference—principle of a new type of flame stabilizer. Journal of Science of China (in Chinese), Vol. 8.

Lani, P., Maissa, P., Degousee, P., and Caron, V. (1995) SO$_2$ and NO$_x$ emission reduction by in-furnace techniques—Main results from EDF development program. Power Gen Conference, Amsterdam.

Qian, R., and Ma, Y. (1986) Investigation into the mechanism of pulverized coal combustion flame with a blunt body stabilizer. Journal of Engineering (in Chinese), Vol. 7.

Talukdar, J., and Basu, P. (1995) A simplified model of nitric oxide emission from a circulating fluidized bed combustor. Canadian Journal of Chemical Engineering 73(October):635–643.

Stultz, S.C., and Kitto, J. B. (1992) "Steam—Its generation and Use "Babcock & Wilcox, Barberton, Ohio.

Wang, Y., Cao, X., Cen, K., and Yao, Q. (1988) Theory and experimental research on combustion behavior of CWS PC. Journal of Engineering Thermophysics (in Chinese), Vol. 9.

Whitworth, C.J., Meixner, R.F., and Gehrke, D. (1994) German-Australian cooperation for building two 500 MW units Loy Yang B. Power Gen '94, Europe, 6–7:387–412.

Xu, X. (1988) The function of the boat-shape flame stabilizer on the pulverized coal flame. Journal of Engineering Thermophysics (in Chinese), Vol. 9.

Xu, X., Jing, M., and Yao, Q. (1987) "The Principle of Pulverized Coal Combustion in the Pre-Combustion Chamber" (in Chinese), Vol. 8.

Zen, C., and Ma, Y. (1986) Research on numerical simulation and flame stabilization mechanism of the pulverized coal blunt body burner. Journal of Huazhong University of Science and Technology, Vol. 14.

10
Tangentially Fired Burners

There are three main methods of firing pulverized coal in a furnace. Based on their orientation in the furnace these are classified into horizontal (opposed jet or front firing), down-shot, and corner firing. The horizontal and down-shot firing burners are described in Chapter 9. The present chapter describes operating principles and design procedures of corner-type burners. These are also called tangential-type burners because here pulverized coal and air are injected tangentially into an imaginary circle in the center from four corners of the furnace. This type was introduced and used by International Combustion in the UK, and ABB Combustion Engineering in the USA. Presently, it is widely used around the world, especially in the boilers designed by ABB Combustion Engineering and their associates.

10-1 General Descriptions

The arrangement of tangential burners is simple but critical. Four sets of burners are located at four corners of a furnace (Fig. 10-1a). These burners fire tangentially at an imaginary circle at the center of the furnace, forming a whirling fireball as shown in Figure 10-1b. These burners are thus called *tangentially fired* or *corner fired*. Alternate layers of coal and air injectors (burners) are stacked at each of the four corners of the furnace. Generally a stack of burners at a particular corner receives coal from all coal mills. So if for some reason one mill is out of order, the stability of the flame is not affected. Oil burners, used for startup or low-load operations, are located such that the pulverized coal burners have good access to the oil flame for ignition. Similarly, secondary air injectors are also stacked in the corners to provide the secondary air for combustion (Figure 10-9).

10-1-1 Principle of Corner Firing

Corner firing, as shown in Figure 10-1, is widely used in large and medium-size coal fired and oil fired boilers. The primary air jet carries the pulverized coal (PC). This jet entrains hot gases from its vicinity, and absorbs radiation heat of the flame in the furnace. Thus the temperature of the mixture increases, leading to ignition

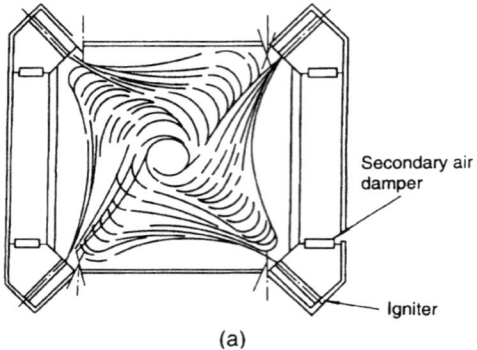

(a)

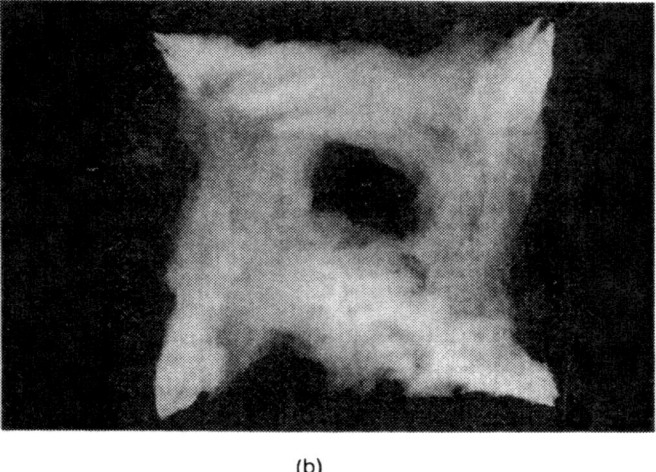

(b)

FIGURE 10-1. Plane view of a tangential fired furnace showing the (a) scheme of burner arrangement and (b) photograph of the resulting flame

of the PC. The intensity of the flame is apparent from the photograph (Fig. 10-1b) of a corner fired furnace taken from the top. The secondary air also enters the furnace as a jet. The geometric axis of the injected air flows is tangential to an imaginary circle in the furnace center. This intensifies the mixing in the furnace. Figure 10-2 shows the flow pattern established by this tangential injection of air into the furnace. A strong air spiral ascends along the axis of the furnace. The axis of the jet injected from the nozzle should be designed to be tangential to the imaginary circle whose diameter is d_0 (Fig. 10-2). More than one burner is stacked in each corner of the furnace. So, all the air jets combine to form a strong vortex in the center. In practice, owing to the interaction of neighboring jets, each jet is

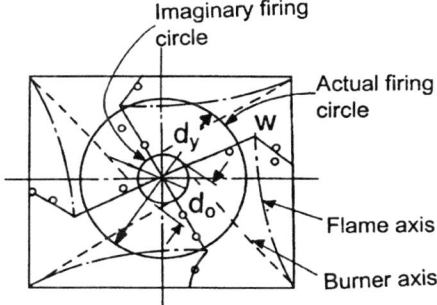

FIGURE 10-2. Aerodynamics in a tangentially fired furnace: (a) axial velocity distribution at different heights of the furnace, (b) tangential velocity distribution on the horizontal plane at section I

deflected from its burner axis. So the diameter of the actual vortex formed, d_y, is larger than the diameter of the imaginary circle (d_0) formed by burner axes. The actual flame circle will be referred to as *real circle*. The diameter of the real circle, d_y, is generally defined by a circle circumscribed by the flow lines representing the maximum tangential velocity the furnace. The strength of all four jet stacks and their angle of entry are not identical, as shown in Figure 10-2. So the shape of actual tangential circle is elliptical at times.

The velocity profile, shown in Figure 10-2, may tentatively separate the rotational flow into two parts. The part of the vortex core within the real circle is called quasi-solid zone. Here the rotary motion of the gas–solid suspension is similar to the rotation of a rigid body. The other part, outside the real circle, is called the quasi-equivalence zone. Here the tangential velocity of the gas decreases along the radial distance from the furnace center. Therefore, the tangential velocity in the furnace can be tentatively expressed by the following equation:

$$\omega = k \left(\frac{d}{2} \right)^{-n} \qquad (10\text{-}1)$$

where k is the experimentally determined constant. The exponent n is such that $0 < n \le 1$ in the quasi-equivalence zone and $0 > n \ge -1$ in the quasi-solid zone; d is the rotation diameter.

Some major features of tangential firing are the following:

1. Ignition stability of tangential firing is very good because of the favorable aerodynamic conditions in the furnace. However, the flow pattern in the furnace is affected strongly by the combustion condition of the burners because the combustion modifies the flow streams.
2. The strong vortex, formed by the four impinging strands of jets, greatly increases the turbulence exchange, which in turn enhances the heat, mass, and momentum transfer. Thus the burning of fuel particles accelerates after their ignition.

3. The turbulence dissipation in the furnace is good. Consequently the heat release rate is uniform in the furnace. This helps the flame to spread all over the furnace, giving a better furnace coverage.
4. Each burner strand includes nozzles for primary, secondary, and tertiary air. These can be controlled and adjusted easily when the load varies.
5. The furnace configuration of a corner fired boiler is simple. So large capacity boilers can easily use this type of firing.
6. When tilting burners are used, the gas temperature at the furnace exit can be adjusted by varying their tilting angle. This controls the superheat temperature of the steam.
7. It is easier to adapt these burners to staged injection of air, which is necessary for control of NO_x emission, etc.

10-1-2 Furnace Shape and Burner Arrangement

The burners can be arranged in different configurations depending on the chosen shape of the furnace. The different shapes of the furnace cross section and the corresponding burner arrangements are schematically shown in Figure 10-3.

a) Types of Furnace Configuration and Arrangement of Burners

Following is a list of possible burner arrangements and shapes of the furnace for tangential firing. Some configurations are used more commonly than others.

   (i) *Square pattern (Fig. 10-3a):* This is the most commonly used arrangement. Here, the air and pulverized coal pipes can be arranged symmetrically. Sometimes this arrangement cannot be used owing to the location of boiler columns at the four corners.
  (ii) *Rectangular furnace with burners on the wider side (Fig. 10-3b):* The ratio of the length and the width of the rectangular furnace is about 1.25–1.35. Owing to the absence of total symmetry of jets the actual shape of the fireball is elliptical.
 (iii) *Rectangular furnace with burners on shorter sides (Fig. 10-3c):* Its characteristics are similar to those of the rectangular wide-side pattern of Figure 10-3b.
 (iv) *Two side-wall pattern (Fig. 10-3d):* If the furnace columns are located at its four corners, the burners cannot be arranged exactly at the corners. They may have to be arranged on two opposite sides. The angle between adjacent air jets would, therefore, differ. This results in a higher imbalance of gas flow along the height or width of the furnace.
  (v) *Tangential and opposed firing pattern (Fig. 10-3e):* This system is used in a square furnace. Here two burners from opposite corners are directed at each other, while the other two are tangential to the imaginary circle. Since the flames of two burners are directed away from the wall, the slagging potential is low compared to that of pure corner firing (Fig. 10-3a), but the intensity of the vortex is low here.

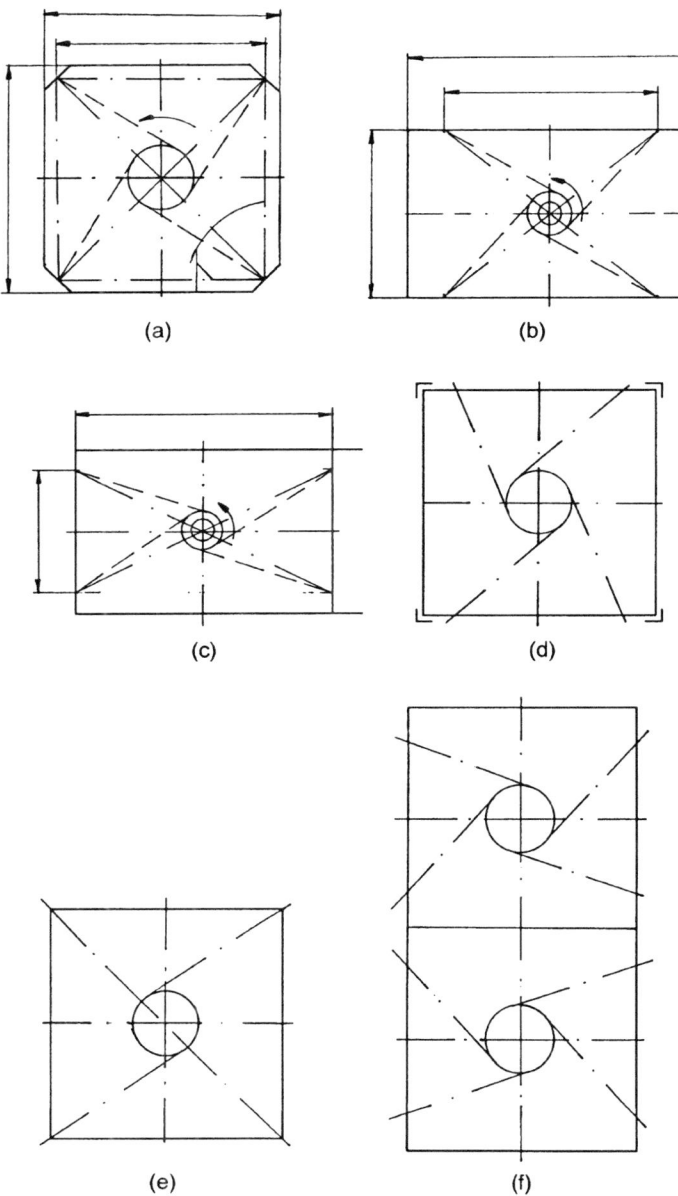

FIGURE 10-3. Shapes and burner arrangements in different tangentially fired furnaces:
(a) square furnace with burners on corners, (b) rectangular furnace with burners on wider
sides, (c) rectangular furnace with burners on shorter sides, (d) two side-wall pattern,
(e) combined tangential and opposed jet firing, (f) twin furnace tangential firing pattern

(vi) *Twin-furnace tangential firing pattern (Fig. 10-3f):* It is suitable for large-capacity (>500 MW) units. The furnace is split in two by a heat-absorbing water wall. All burners are on two opposite sides. So the angles between the burner axes and the side wall are not the same. Consequently, the air supply conditions differ substantially. An appropriate distance between two burners in the middle is allowed to facilitate cleaning of slag on the two water walls, removal of adjacent burners for examination and repairs, and arrangement of air and PC pipes and pillars. The direction of rotation of the air flow in the two furnaces would affect the temperature variation in the gas exiting from the two furnaces. The temperature variation can be adjusted through varying the tilting angles of the burners in the two furnaces.

In general, various tangential arrangements of burners lead to different aerodynamic conditions in the furnace. We can select the one most suitable for the specific fuel and for the capacity of the boiler.

b) Types of Air Registers

Tangential firing can also be classified according to the mode of distribution of primary, secondary, and tertiary air and registration patterns in the burners. These are described below:

i) Unequal tangential circle arrangement: The inclination angles of primary and secondary air nozzles are different. The bottom-most secondary air is arranged tangential to the firing circle, but all other air jets are kept opposed to one from the opposite corner. In this way the air flow in the furnace is stable and the turbulence level of the opposed air flow is strong.

ii) Inverse-tangential arrangement: Here both primary and secondary air are injected tangentially at the firing circle, but they are in reverse tangential direction. The primary air nozzle and the secondary air nozzle are directed such that two different imaginary circles rotating in opposite direction are formed. As a result, the residual rotation of gas at the furnace exit is decreased. This helps control the temperature imbalance in the superheater.

iii) Opposed tangential arrangement: The secondary air is injected tangentially to the firing circle, but the primary air supplies are opposed to each other. This reduces the diameter of the real primary air firing circle. This gives a better controlled PC flame, which does not to scour the walls. So, slagging is also prevented.

iv) Burners with side-secondary air (Fig. 10-4a): Secondary air is added from ports located alongside of the burners. These burners are used for bituminous or subbituminous coal fired boilers. Here the secondary air enters just after the coal is ignited. Thus the coal is prevented from sticking to the wall.

v) Burners with peripheral air (Fig. 10-4b): A high-velocity secondary air curtain is added around primary air to enhance PC combustion. It also helps flame stabilization.

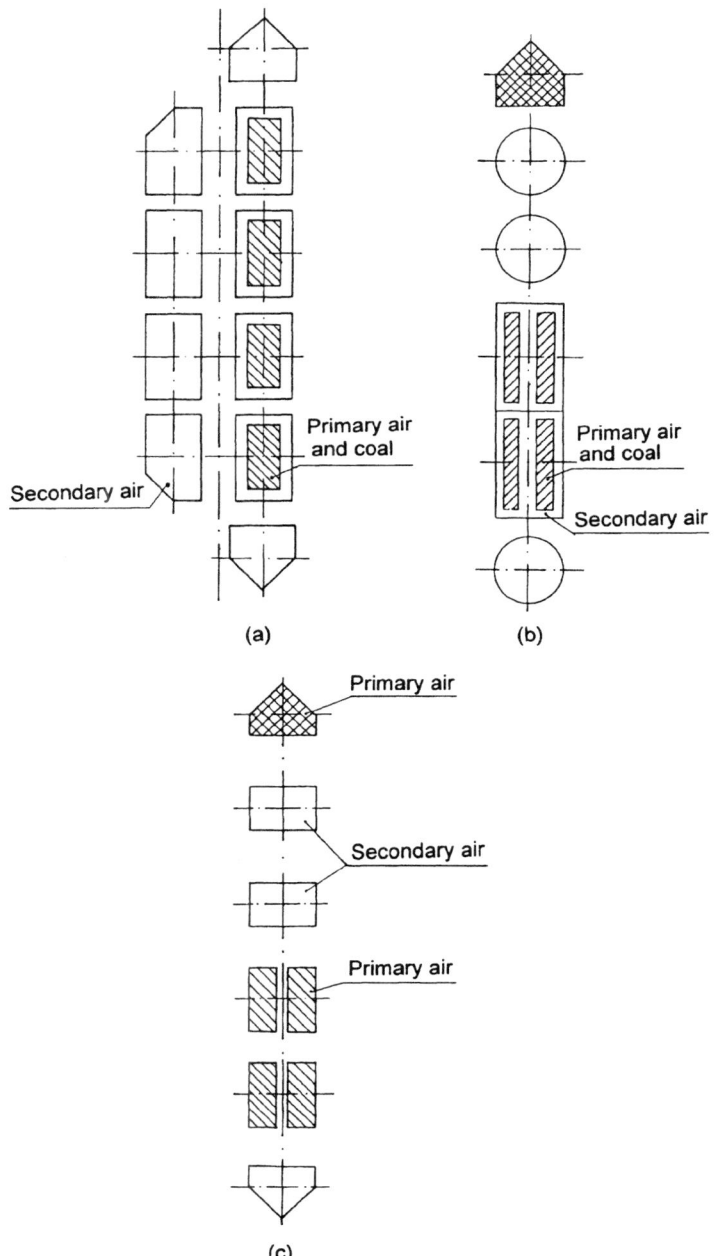

FIGURE 10-4. Examples of some air registration patterns of tangential firing burners

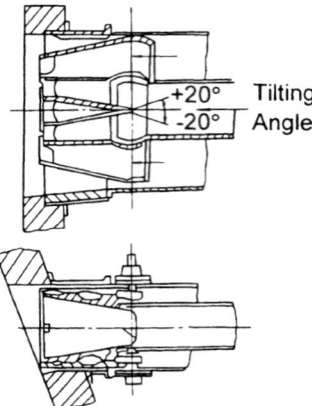

FIGURE 10-5. A view of a short total tilting nozzle tangential burner

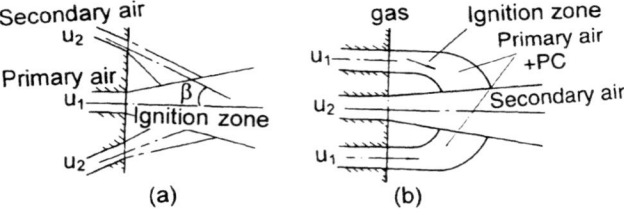

FIGURE 10-6. Ignition schemes of the tangential firing burners: (a) ignition triangle scheme, (b) cluster attracting ignition

vi) Burners with sandwich air (Fig. 10-4c): Here two high-velocity secondary air jets are arranged in between two primary air jets such that the secondary air jet is sandwiched between primary air jets, hence, *sandwich air*.

vii) Tilting-type burners (Fig. 10-5): Here the angle of the burner is changed to adjust the position of the flame axis. An important feature of this is that it can control the heat flux profile in the furnace and the gas temperature at the furnace exit. It is very effective in steam temperature control.

c) Types of Ignition Systems

A comprehensive understanding of the process of ignition and combustion in tangential firing is still wanting. The present understanding, though incomplete, is based on the model of the ignition triangle scheme and the cluster attracting ignition scheme.

i) Ignition triangle scheme. As shown in Figure 10-6a, the secondary and primary air intersect at an angle (β). The primary air carrying PC travels faster with a higher momentum. This intersection takes place after primary air and PC travel a certain distance (500–800 mm) into the furnace. Within this distance the primary air jet draws by entrainment hot gases from its periphery and absorbs radiation heat from high-temperature flame. This helps raise the jet temperature steadily to the ignition point and the ignition occurs on the periphery of the primary air jet. The intersection of the ignited primary air with the secondary air at this stage provides the oxygen needed for PC combustion and strengthens jet turbulence and mixing. So the combustion process is stabilized and intensified. Therefore, the intersection of primary and secondary air must occur after the air–PC jet is ignited. A much too early intersection of primary air or a large quantity of the secondary air will delay ignition, which may lead to local extinction. The position of the intersection point is closely related to the combustion characteristics of the coal.

ii) Cluster attracting ignition scheme: This situation (Fig. 10-6b) arises when a large amount of secondary air travels with a high velocity and high momentum. Although the secondary and primary air are injected in parallel, the primary air is drawn into secondary air soon after its injection because of the higher momentum of the secondary air jet. The primary jet bends and forms a turning fan-shaped surface, which is good to draw in the peripheral hot gas and to form a peripheral ignition area. Since the ignited primary air is drawn into the secondary air of high momentum, the mixing is good and the flame jet strong. So the flame cannot be disturbed so easily, and the coal can be fired steadily.

The above two ignition schemes are used, depending on the air registration method employed in the burners. Two types, an equal air-registration pattern and a step air-registration pattern, are shown in Figure 10-7.

(1) Equal air registration arrangement (Fig. 10-7a): The nozzles of primary and secondary air are spaced alternately a small distance apart. According to cluster attracting ignition theory, the stronger secondary air entrains primary air and thereby gives a good mixing of primary and secondary air. This mixing helps the burning of bituminous and brown coal whose volatile contents are high.

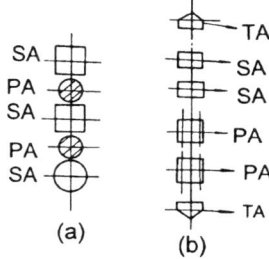

(a)   (b)

FIGURE 10-7. Typical nozzle arrangements of the tangential firing burners: (a) equal air registration, (b) step air registration (PA = primary air; SA = secondary air; TA = tertiary air)

(2) Step air registration arrangement (Fig. 10-7b): Primary air nozzles are arranged adjacent to each other. The secondary air jets are also arranged together but placed at some distance away from the primary air nozzles. The central air jet comes from the primary air nozzles. The ignition triangle scheme (Fig. 10-6a), described above, is used here to improve the ignition of fuel. Thus this system meets the demand of firing low volatile fuels like anthracite and sub-bituminous coal.

10-1-3 Design Requirements of Tangential Firing

a) Deviation of the Jet

Figures 10-1 and 10-2 show typical deviations of the air jets from the geometrical axis of the burner, which is tangent to the imaginary firing circle. There are two main reasons for this deviation. First, the spiraling air flow collides with the jet, laterally deviating the jet. Second, the jet injected from the burner into the confined furnace space entrains peripheral gas. Higher entrainment means higher velocity and hence lower pressure. So a static pressure difference would occur between the inner and the outer sides of the jet. Thus the jet swings to the lower-pressure side. Results of cold tests show that the diameter of the actual air flow circle, d_y, is about 2.5 to 4 times that of the designed firing circle, d_0. If the diameter of the firing circle is designed properly with a suitable vortex of air flow, the flame from the adjacent corner will be closer to the jet root. This is favorable for ignition, and the flame better covers the furnace. But if the diameter of the imaginary circle and the height-width ratio of the burners are too large, serious slagging may occur because the air flow deviates excessively toward the furnace wall, scouring it.

b) Flame Stability

The stabilization of the flame under all operating conditions without an external support is an essential condition for well-designed burners. The following measures can be taken to enhance flame stabilization.

1) Increase the number of primary air nozzles to enlarge the contact area of the primary air with the hot gas in the furnace. An excessive number of nozzles will, however, make the jet weak, and the flame will scour the wall, causing slagging.
2) Introduce a low volume of secondary air but at a high velocity around the primary air. The high-speed jet curtain of secondary air can efficiently entrain hot gas in the furnace and mix it with the primary air (Fig. 10-4b). This part of the secondary air is called peripheral secondary air.
3) Increase the distance between the nozzles for the primary and secondary air. The fuel with the primary air will ignite well before it mixes with secondary air further downstream.

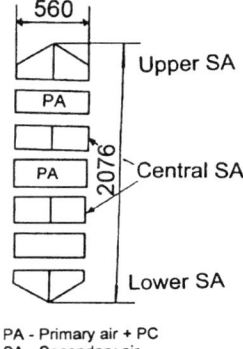

PA - Primary air + PC
SA - Secondary air

FIGURE 10-8. The nozzle arrangement in a brown coal fired 200 t/h boiler

c) Secondary Air

The secondary air of tangential firing PC burners may be roughly divided into upper, central, and bottom air (Fig. 10-8). The upper secondary air depresses the flame such that it does not drift upward excessively. In case of staged air registration the percentage amount of this secondary air is the highest, and it is the main air source for PC combustion. The upper secondary air jet will be inclined 0–12° downward. In all other systems, the percentage of central secondary air is highest, and it is the main source of air for PC combustion. When an equal air registration pattern is used, the air is inclined at about 5–15° downward. The bottom secondary air can prevent PC from separation and hold the flame such that it does not drop downward too much. It is also necessary to prevent excessive slagging in the hopper. The percentage of bottom secondary air is relatively low, about 15–26% of the total secondary air.

10-2 Design of Burners with Peripheral Air

In many pulverized coal corner fired boilers, a high-velocity curtain of secondary air envelops the primary air jet carrying the coal from the burner. The following section elaborates further on the working of this type of burner and the basic design considerations. Although peripheral air is useful to intensify the combustion and prevent the separation of pulverized coal powder from primary air flow, it is not always favorable to PC ignition. Therefore, the need for peripheral air shold be evaluated according to the specific situation.

10-2-1 Peripheral Air

The thickness of the curtain of peripheral secondary air varies with the characteristics of the fuel. It is generally in the range of 15–25 mm. The velocity of the peripheral air is very high (30–44 m/s), and it constitutes 10% or more of the total amount of the secondary air. The main functions of peripheral air are as follows:

1. To cool primary air nozzles to prevent them from burning out or deforming.
2. To supplement oxygen. Because PC is ignited at the periphery of primary air jets, the oxygen is not adequate around the flame. The peripheral air serves the function of supplementing oxygen.
3. To hold down the PC jet and thereby avoid separation.
4. The injection of peripheral air helps entrain hot gas and thereby promote ignition. But inappropriate amounts of peripheral air (i.e., too low air speed or too high air rate) will separate primary air and high-temperature gas, preventing their mixing.
5. To serve as a means of adjustment of combustion when the coal rank or load varies.
6. To enhance the mixing process of primary and secondary air.

10-2-2 Effect of Peripheral Air on Combustion

Detailed research on mixing of jets and the peripheral air burner (Cen & Fan, 1991) reveals the following effects of peripheral air.

a) Effect of Peripheral Air on the Mixing Pattern of Primary and Secondary Air

Radial profiles of velocity were measured at different axial distances at two peripheral air velocities (Cen & Fan, 1991). They showed that the velocity gradient at the jet outlet increases and turbulence and mixing are intensified when the peripheral air velocity increases from 12.8 to 31.7 m/s. At this higher peripheral air rate, primary air, secondary air, and peripheral air mixed well and formed a well-developed jet at a distance three times the width from the burner. At the lower peripheral air rate the mixing is delayed and the velocity profile is not even at the largest axial distance. Therefore, the peripheral air can enhance the mixing of primary air and secondary air and advance the intersection point of primary air and secondary air. For example, experimental data (Cen & Fan, 1991) showed that for peripheral air speeds of 20, 40, and 60 m/s, strongest mixing points are located at $x/b = 3.2$, 2.1, and 1.6, respectively.

b) Effect of Peripheral Air Rate on Combustion Conditions

The peripheral air rate should not exceed 10%–15% of secondary air in the case of coals with poor combustion characteristics. In some poorly designed and operated boilers firing such coals, the peripheral air rate was 20%–40% because of air leaks. This adversely affected the combustion (Cen & Fan, 1991). However, when burning better fuels like bituminous coal, a larger amount of peripheral air rate (>30%) can be used with beneficial effects on the combustion performance. In some burners, the peripheral air rate is designed to be 35%–40% that of the secondary air.

10-3 Design of Tilting Burners

The tilting burner introduced earlier in Section 10-1-2 is widely used in Europe and North America. Tilting direct-flow burners are arranged in corner or tangentially fired furnaces to adjust the position of the flame, heat flux profile in the furnace, the gas temperature at the furnace exit, and the temperature of steam in the superheater and reheater. This can be done through tilting the burners up or down.

10-3-1 Types of Tilting Burners

There are two main types of tilting burners:

1. Total tilting type
2. Partial tilting type

The main features of a total tilting burner are that the air nozzles tilt up and down synchronously and the usual tilting angle is less than 30°. The position of the flame center in the furnace can be changed to adjust the superheat and reheat steam temperature. In a twin furnace boiler, the temperature variation between the two furnaces can be adjusted efficiently by varying the angle of the upper nozzles. Moreover, nozzle tilting is favorable for adjusting the combustion during startup as well as during burning. It is also useful in changing combustion conditions and heat release rate in the burner area, and in decreasing slagging and overheat temperature. However, the design of the full tilting burner is complex and it is more difficult to design, manufacture, operate, and repair than the partial tilting type.

The design of a partial tilting burner, on the other hand, is simpler. These are used primarily to vary the mixing condition of primary and secondary air. The mixing point is changed to vary the ignition and combustion rates and to adjust the combustion conditions through varying the intersection angle of primary and secondary air flow. With changes in the coal type, adjustments can be made within a certain range.

According to the geometry and design of the exit nozzle, tilting burners may also be classified as

a) Long nozzle
b) Short nozzle
c) Tilting nozzle with peripheral air type, etc.

Figure 10-5 shows a short tilting nozzle burner. Here, the length of the nozzle is about 1–1.2 times its height. During operation the nozzle can tilt up and down within 20°. This type of design is in extensive use.

The design of a tilting burner with long nozzle is shown in Figure 10-9. The length of the nozzle is 6–7 times its equivalent diameter. Its tilting angle is within ±20°. This design is widely used in the former Soviet Republics and East European countries. The performance of the long nozzle is relatively good, but its operational

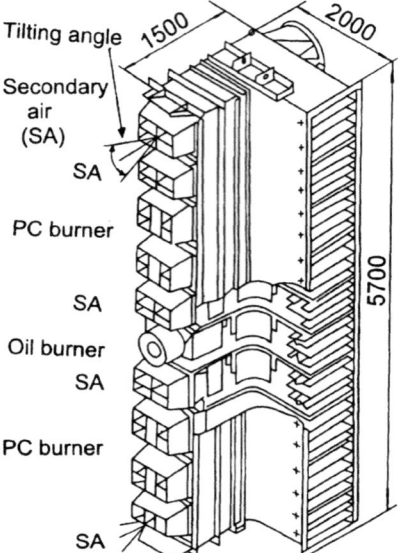

FIGURE 10-9. Design of a long tilting nozzle tangential burners

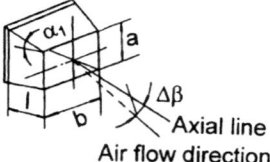

FIGURE 10-10. Geometric parameters of a short tilting nozzle

reliability is not very good. The moment needed for tilting is large and it often gets locked in position.

10-3-2 Aerodynamic Characteristics of Tilting Burners

The effect of various design and operating parameters on aerodynamic characteristics of tilting burners is far from being fully understood. So only some preliminary test results of short nozzles are presented here to illustrate how the above parameters might affect the tilting burners. The arrangement of geometric parameters of the short nozzle is shown in Figure 10-10.

a) Effect of Nozzle Length-to-Height Ratio on the Angle of Jet

The nozzle guides the air flowing through it. When the nozzle tilts, the air that flows parallel to the nozzle body will tilt and the jet coming out of the nozzle will incline by an angle close to the tilt angle. If the nozzle is too short, i.e., its length, l, is smaller than its height, a, the guiding function of the nozzle is poor. So, the inclined angle of the jet will be less than the tilt angle of the nozzle. This difference between the tilt and inclined angle, $\Delta\beta$, increases the decrease in l/a ratio (Fig. 10-10). When $l/a > 1$, the difference in angle is negligible, but increases rapidly as $l/a < 1$.

b) Effect of Nozzle Cone Angle α, on the Injection Direction of the Jet

The nozzle cone angle, α and the angle difference $\Delta\beta$ of jet is shown in Figure 10-10. The difference between the tilt and jet angle, $\Delta\beta$, attains a minimum value when the cone angle is $10°-15°$. This is, therefore, the optimum cone angle. The jet velocity decreases away from the nozzle exit. This attenuation of the jet shows minimal influence of the nozzle angle and the nozzle length.

c) Effect of Nozzle Design on the Flow-Resistance Coefficient, ξ

Test results show that the flow resistance coefficient, ξ, increases with the length-to-height ratio of the nozzle. The nozzle tilt (β) has a marginal effect on the resistance coefficient, while its cone angle (α) has a major influence.

10-3-3 Some Design Norms

The following suggestions for designing tilting burners are based on experience.

a) Tilting Nozzle

Good results can be obtained for $l/a = 1-1.2$, $\alpha \approx 10°-20°$. Also, the flow divider inside the burner nozzle should be strong enough to give the required rigidity to the burner and serve the function of flow guiding. The problem of thermal expansion of the nozzle should also be considered during its design.

b) Air Pipe (Air Chamber)

The air pipe should have a good rigidity. A convergent shape can be adapted for the outlet segment to meet the space demand of the tilting nozzle and to increase the uniformity of air flow. The flow-guiding plate should be fixed at the bend.

c) Fitting Clearance

The fitting lateral clearance and vertical clearance with the nozzle should be 3–4 mm and 5–8 mm, respectively.

d) Rotary Axis

The rotary axis between the nozzle and the air pipe is made to decrease blockage potential due to rusting. The fitting clearance is less than 0.5 mm and should not be too large or too small.

e) Tilting Mechanics

The tilting mechanism is made up of a set of connecting rods, pins, and axis. It should have enough rigidity and strength. A heat-resisting stainless steel, 18 Cr-8Ni may be used. The axis diameter is usually 20–30 mm. The fitting clearance among axis pores is about 0.1 mm. Safety devices, including safe pins, spring pins, a caging device, and a signal transmitter must be used with the tilting mechanism.

f) Operating Mechanism

The nozzle-tilting mechanism is driven by a driving mechanism capable of remote operation and automatic control. The driving mechanism can be pneumatic, fluid driven, or electric driven with corresponding auxiliary equipment and apparatus.

g) Cooling and Ash-Blowing

Under ideal conditions, a little secondary air may be used to cool the nozzle and tilting mechanism. However, it helps to blow ash at the same time in order to prevnt the tilt mechanism from blockage caused by overheating.

10-4 Burners for Bituminous Coal

This section briefly describes the operating principles and design considerations for a burner for tangential firing of normal bituminous coal (not including low-rank bituminous coal).

Normal bituminous coal has high volatile content and high heating value, but its ash content is low. This makes bituminous coal easy to fire. So the problem of ignition requires less attention. The coal powder (PC) generally ignites within a distance of 300–400 mm of the nozzle, and it mixes with secondary air promptly after ignition. The required oxygen for PC combustion comes from this secondary air. Therefore, this type of burner uses an equal air registration pattern, where the primary and secondary air are divided into several layers and are arranged alternately (Fig. 10-7a) at a certain span for vigorous early mixing of the primary and secondary air.

Owing to the presence of two high-velocity (45 m/s) secondary air jets above and below the low-velocity (25 m/s) primary air jet (Fig. 10-11), the latter is drawn into the secondary air jets promptly after ignition. Thus the needed oxygen for PC combustion is obtained from the secondary air. At the same time, higher levels of turbulence and mixing help the combustion. Figure 10-9 shows a typical design of this type of burner for burning bituminous coal. In the center there is an oil burner

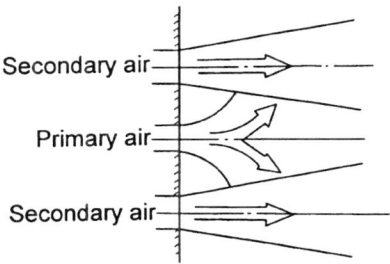

FIGURE 10-11. The low-velocity primary air is entrained by high-velocity secondary air jets from adjacent nozzles in a burner firing system of bituminous coal

to help ignition and sustain the flame under adverse operating conditions. The burner can tilt up and down within ±20°. Most bituminous coal fired direct-flow burners belong to tilting type.

Except for the top secondary nozzles, all the other nozzles can be adjusted together to tilt up and down within ±30°. The top secondary air nozzle is designed to be adjusted independently. So the inclined angle adjustment of the other burners will not be affected when the adjustment of top secondary air burners is out of service because of a blockage caused by slagging.

In order to adapt to many kinds of fuel, especially a low heating value fuel like blast furnace gas, and to meet the demand of relatively high burner throughput, the the furnace should be high. The volume heat release rate is, therefore, relatively low (about 0.1 MW/m³). The grate or cross section heat release rate of the furnace is 4.9 MW/m². The wall heat release rate in the PC burner zone is 2.3 MW/m². But although the volume heat release rate is low, the PC flame length cannot increase much because PC burners are arranged at the top part and the axis of the top layer PC burner is about a 17 m below the bottom of the screen superheater.

The peripheral air rate of the burner is related to the coal combustion performance. In order to decrease the NO_x emission during combustion, over-fire air nozzles have been arranged at the top of the PC burners. Owing to the over-fire air the oxygen content is low in the lower zone of the burner, which is favorable to NO_x control. The required air for PC burning is fed from over-fire air nozzles at the top of the burner.

Two-stage firing is another characteristic of the combustion engineering direct-flow burner (Table 10-1).

TABLE 10-1. Design characteristics of direct-flow burner.

Diameter of the imaginary firing circle	(0.13–0.14)(Furnace depth + width)/2
Primary air velocity w_1	25 m/s
Primary air	18–20% of total air
Over-fire air	15% of total air
Secondary air velocity w_2	45–50 m/s

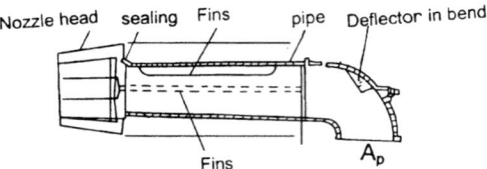

FIGURE 10-12. Primary air nozzle for direct flow bituminous coal burners; the improvement of the primary air nozzle is shown here; the deflector in the bend is replaced with horizontal fins to encourage nonuniformity of the PC mixture

A moderate-speed pulverizer system with one or two standby pulverizers is widely used in this type of units. Correspondingly there should be one or two standby nozzles for the direct burner at each corner.

The design of primary air nozzles is modified (Cen & Fan, 1991) (Fig. 10-12) in order to improve combustion stability at low load. In old designs, a deflector is fixed at the inlet bend of the primary air nozzle to prevent nonuniform PC concentration caused by centrifugal separation in the bend. Vertical fins are fixed in the nozzle to eliminate vortex in the air flow after turning. The air velocity is made to increase gradually in the nozzle to prevent PC deposit. The cross section area of the inlet of the nozzle was increased from 0.77–0.83 A_p in the old design to 0.96 A_p in the new design; the outlet area was increased from 0.90–0.95 A_p to 1–1.2 A_p. (A_p is the area of the primary air pipe where the velocity is in the range of 22–25 m/s). After modification, the deflector is removed. Additional horizontal fins are added to prevent mixing of rich and lean in the nozzle. The nozzle head is arranged such that the cross section area at its outlet is equal to that at the inlet. The main goal of the modification is to use PC separation at the bend, which gives a relatively high PC concentration of the outlet air flow in the upper nozzle. It is favorable to ignition, and the primary air outlet velocity decreases suitably.

The load adjustment range of the burner can be increased by making the nozzle head of two parts that can tilt separately, and the difference of their tilting angles attains 24°. Similar to the design of a blunt body, it can be arranged at the nozzle outlet, creating a recirculation zone in the primary air flow. This further improves ignition stability. In addition, the nozzle outlet section area increases to 130% of the inlet section area. Additional improvements suggested by Cen & Fan (1991) involve the use of either square wave (Fig. 10-13a) or triangular exit (Figure 10-13b) at the outlet of the nozzle. These designs greatly increases the contact area between the primary air and hot recirculation gas, improving the flame stability. The wedge angle is 20°.

10-5 Anthracite and Lean Coal Fired PC Burner

The volatile content of anthracite, the lowest in coal rank, is under 10%, but its fixed carbon content is highest. So it is difficult to ignite and to sustain the flame of anthracite fuels. With 10%–20% volatile content, the ignition performance of a

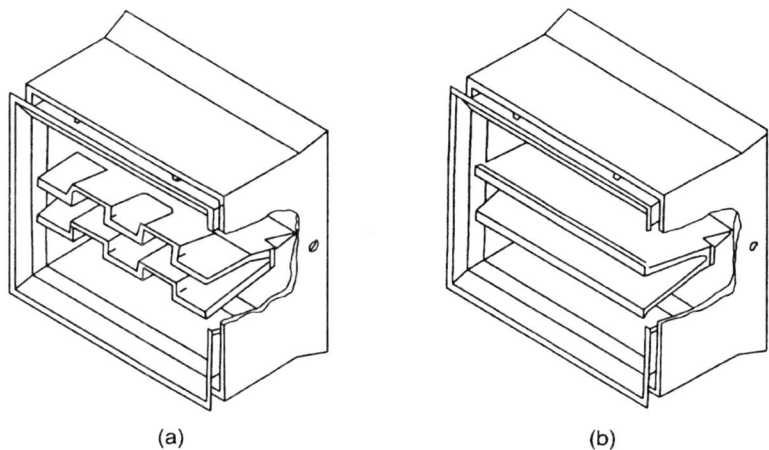

(a)   (b)

FIGURE 10-13. Use of special section at the nozzle exit to improve flame stability: (a) sawtooth wave form, (b) triangle wedge form

lean coal is better than that of the anthracite coal. The heating value of anthracite and lean coal is generally high (>20.9 MJ/kg). Only a few types with a relatively high (30%) ash content have low heating values (~18.8 MJ/kg). In many anthracite and lean coals, the ash fusion temperatures are low. The initial ash deformation temperature is lower than 1200°C, and the ash fusion temperature is under 1300°C. When anthracite or lean coal is fired, the furnace temperature must be high to ensure good ignition. If the fuel has low ash fusion temperatures, this high furnace temperature increases slagging potential. Therefore, the main problems faced by anthracite and lean coal fired burners are as follows:

a) Unsteady ignition
b) Slagging
c) Poor burning rate and high unburned carbon content in ash

In most anthracite and lean coal burners, the primary air is fed in one step, but the secondary air is fed progressively in several stages. The injection of all primary air in one stage delays the mixing of primary and secondary air. Thus it increases the PC concentration in the ignition area, which favors steady ignition. Secondary air is fed progressively in stages as the need of combustion develops. The ignition properties of lean coal are better than those of anthracite coal. So in some lean coal fired burners, an equal air registration pattern or side secondary air pattern is used.

10-5-1 Primary Air

The primary air is heated by radiation from the flame and furnace walls and by mixing with the high-temperature gas in the furnace. The larger the contact area between the primary air jet and the high-temperature gas, the larger the entraining

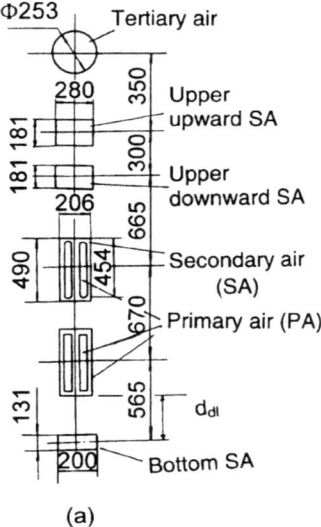

(a)

FIGURE 10-14. Burner arrangement for an anthracite fired boiler; tertiary air is fed from the top nozzle

capacity of the primary-air flow and, consequently, the steadier the ignition. So, the primary air nozzle of the anthracite coal-fired burner is designed to be straight, narrow, and tall to increase the peripheral area of the primary air flow jet.

An example of burner arrangement for firing an anthracite coal (volatile = 7.5%, LHV = 18.7 MJ/kg) is shown in Figure 10-14. Here the peripheral air flows around the narrow and tall primary air nozzle. A steady flame can be maintained down to 55% of the rated load without support from the oil firing. Though the rectangular nozzle, used in Figure 10-14, is good for ignition, the strength of the air jet is poor. The jet may deviate from the designed direction of flow and shift toward the wall, causing slagging. Therefore, the height-width ratio of the nozzle section should be maintained within 2–2.5. Even when the primary air nozzles are arranged together and are surrounded by a common peripheral air, the height-width ratio should not exceed 4.0. When several primary air nozzles are used, appropriate spaces should be kept between two adjacent nozzles. A large spacing between the adjacent nozzles is good for drawing hot gas into the primary air jet, but too large a spacing will increase the total height of the burners. Therefore, the spacing is usually less than the nozzle width.

To stabilize the ignition of a low volatile coal like anthracite, several measures can be taken. One is to reduce the heat required to heat the PC-primary air flow mixture to the ignition temperature. This is done by decreasing the amount of the primary air. Moreover, a very high primary air velocity will cause the ignition point to move far from the nozzle. Then the temperature of gas in the neighborhood of the burners will decrease, adversely affecting ignition. For anthracite coal, therefore, the primary air rate is recommended to be 18–22% and primary air velocity within the range of 20–22 m/s.

10-5-2 Secondary Air

When anthracite or sub-bituminous coal is fired, secondary air should be provided in stages with the development of the combustion process. Early mixing with the primary air should be avoided as it affects the flame stability. The secondary air nozzles of an anthracite burner are, in general, arranged above or below the primary air nozzles. When the secondary air nozzles are located above the primary air nozzles, they are usually divided into two parts (Fig. 10-14): upper upward secondary air and upper downward secondary air.

The function of the bottom secondary air is to hold the pulverized coal so that it does not drop into ash hopper. The bottom secondary air flow, which constitutes about 15%–25% of the total secondary air, is less than that of upper upward secondary air and upper downward secondary air. The bottom secondary air should mix with the pulverized coal flow from adjacent primary air nozzle only after ignition. So a certain distance should be kept between the nozzles of the bottom secondary air and the primary air. The ratio of such a distance to the equivalent diameter of nozzle is about 1.8–3.5. The equivalent diameter of a rectangular nozzle can be calculated by the following equation:

$$d_{dl} = \frac{2ab}{a + b} \tag{10-2}$$

where a and b are the length of the two sides of the primary air nozzle. The nozzle of the bottom secondary air is generally arranged in parallel.

The function of the upper downward secondary air, which constitutes about 27–37% of the total secondary air, is to supply the oxygen needed for combustion in proper time after the ignition. It has a great influence on the flame stability, especially when a low volatile anthracite coal is fired. The distance between the nozzles of the secondary air and the primary air is increased such that the ignition is not affected. The net distance can be more than 400 mm. Besides this, the downward inclination angle of the upper secondary air nozzle is another important factor that affects the mixing of upper downward secondary air and primary air. A lower volatile content coal is more difficult to ignite. So a smaller inclination angle is used.

The functions of the upper upward secondary air, which constitutes no more than 36%–40% of the total secondary air, are (a) to supply the air for downstream combustion, (b) to enhance the mixing and (c) to facilitate the burn-out of fuel particles. The inclination angles of the upper upward and the upper downward secondary air nozzles have great influence on combustion conditions, and they should be determined by trials. A tilting design should be used for the upper upward, upper downward secondary nozzle, and the tertiary air nozzle such that the inclined angle can be adjusted according to the coal rank.

The velocity of the secondary air should be sufficiently high to ensure that it has enough momentum to penetrate the primary air jet. To increase the strength or momentum of the secondary air flow, the cross section of the secondary air nozzle can be rectangular, with its width larger than its height. During low-load operations, the primary air velocity is kept nearly unchanged to avoid blockage of PC in the pipe, but the secondary air velocity has to be decreased at low load. This requires that a certain minimum secondary air velocity at low load be kept when designing the burner. In boilers of less than 50 MW capacities, the velocity of the upper upward and the upper downward air should be in the range of 40–45 m/s. For higher-capacity boilers, it is in the range of 45–55 m/s.

10-5-3 Tertiary Air

The tertiary air constitutes about 20% of the total air. For anthracite-fired boilers, a hot air PC supply system is used. The tertiary air is used as a drying agent in the pulverizer in the pulverized coal preparation system (see chapter on coal preparation). It is injected into the furnace separately. It carries about 10%–15% by weight of the coal at a low temperature of less than 100°C, but with a high moisture content. The tertiary air is injected separately into the furnace (Fig. 10-14).

The amount of the tertiary air is difficult to control, as its requirement is dictated by the moisture content in the raw coal. If the amount is high, the tertiary air may lead to a decrease in the temperature of the combustion zone, which will delay combustion and affect combustion conditions. As a result, tertiary air should be arranged at a certain distance from upper secondary air nozzles and at a downward angle that will not affect the combustion in the primary air flow, but the disturbance and later mixing will be strengthened. The design of the furnace height and the tertiary air nozzle location should, however, ensure that the small particles in the tertiary air can be burnt out completely. Some typical distances between tertiary air nozzle and upper secondary air nozzle are tabulated in Table 10-2 and shown in Figure 10-14.

Tertiary air has a great influence on anthracite combustion. Excess tertiary air will lead to increased unburned carbon in ash, decreasing the boiler efficiency and causing delayed or downstream combustion of carbon particles to increase. Such burning may lead to an excessive superheated steam temperature threatening the safety of the boiler.

Besides the distance between the tertiary air nozzle and the upper secondary air nozzle, the downward inclination of the tertiary air also has a great influence on the flame stability. In the past, the inclination angle of the tertiary air was usually

TABLE 10-2. Spacing of tertiary air nozzles.

The ratio of distance between upper tertiary air nozzle and upper secondary air nozzle to width of upper secondary air nozzle	The distance from upper tertiary air nozzle to edge of upper secondary air nozzle
1.1–1.7	210–395 mm

between 8° and 12°, but this trend has decreased in recent years. The velocity of tertiary air should not be so low that it will not be able to penetrate high-temperature gas to enter the furnace center. Velocity in the range of 50–55 m/s is desirable. In higher-capacity boilers, a value near the upper limit should be selected.

10-6 Brown Coal Fired Direct Burner

Brown coal is characterized by high ash and high volatile contents. If the ash has a low fusion temperature it easily fuses at the furnace temperature and sticky slag deposits are formed on the water wall. However, the high volatile content of the fuel is favorable for PC ignition and flame stability. Considering the above two characteristics, a low-temperature combustion technique is most suitable for brown coal.

10-6-1 Design Parameters of Brown Coal Burners

Combustion stability, minimization of slagging, and economical operation are three guiding principles for the design of a good PC burner. These are especially important for brown coals. These features can be achieved by low-temperature combustion, where a stable but low-temperature profile is maintained in the furnace. The average flame temperature is around 1100–1200°C. This prevents a localized high-temperature zone that would lead to slagging. Therefore, the distance between the primary air nozzles should be increased. High-capacity burners should be arranged so as to avoid very high concentrations of PC in the furnace. In addition, the volume heat release rate, grate heat release rate, as well as burner zone heat release rate should all be limited. The combustion rate should also be controlled. These can be achieved by choosing the right air flow rate and air velocity and by controlling the mixing of primary and secondary air. The mixing of burnt gas with the air is also a means for controlling the burning rate of PC. The temperature level in the furnace depends on the theoretical (adiabatic) flame temperature and the exit gas temperature. Mixing of the low-temperature gas with the air in the combustion zone is an effective means to control peak furnace temperature.

   With increasing boiler capacity, the height of burners and the height-width ratio increase correspondingly. To ensure the strength of the flame (or the air jet), the burners must be divided into groups according to their heights with a certain space between them. This may reduce the flow deviation and help the wall heat release rate in the burner zone. To avoid slagging in boilers of different capacities, the wall heat release rate is chosen to be about 1.16 MW/m². Furthermore, with increasing furnace cross section (owing to higher capacities), a higher secondary air velocity is required to retain its penetrability.

   If the furnace is very large, the double furnace configuration is preferable for proper organization of the aerodynamic field for tangential combustion. However, several problems are associated with it. For example, arrangement and thermal expansion of the water-cooled wall are major issues. Slagging on the division furnace walls is not easily removed, and the arrangement for pulverizers is also difficult. Experience suggests that the double furnace is unnecessary for the 200–300 MW capacity range. The single furnace is used now even on 600 MW boilers, but beyond this size one must use a double furnace.

In high-capacity brown coal fired boilers (He & Zhao, 1987; Cen & Fan, 1991) single furnace corner firing is used. The burners are arranged at four corners, six corners, or eight corners, which form a tangential circle in furnace. Each group of burners is distributed well within a certain distance. Several overlapping swirl flames form, filling the furnace and thereby ensuring a moderate local heat release rate in the burner zone. The absence of a high local heat release rate decreases the slagging tendency.

10-6-2 Examples of Some Brown Coal Burners

Some examples of brown coal burners are given in Table 10-3.

a) 220 T/H Boiler

A high value of volumetric heat release rate has been chosen deliberately for economic reasons. The high volumetric heat release rate allows for a smaller furnace, which requires less materials. The arrangement of burner nozzles is shown in Figure 10-15. The alternate positioning of primary air and secondary air nozzles is not only good for mixing, but it also avoids an excessively concentrated flame.

b) 600 MW Boiler

In the 600 MW brown coal boiler (He, 1987), eight pulverizers are arranged around the furnace. Each pulverizer supplies PC for all the burners at one corner. Burners at each corner are divided into three groups, and are arranged with a certain distance between them. The space between each group of burners is large. It may help equalize the flow and decrease the pressure difference of both sides of outlet flow, making flow difficult to deflect, thereby reducing slagging.

TABLE 10-3. Design details of some PC burners for brown coal firing.

Capacity of boiler	220 T/h	600 MW
Volume heat release rate, MW/m^3	0.146	0.099
Grate heat release rate, MW/m^2	2.95	
Furnace cross section area, m^2	7.73×7.73	16×19.6
Firing circle diameter, m	0.8	
Height to width ratio	4	5 for each of 3 groups
Primary air flow rate and velocity	35%, 20 m/s	12–16 m/s
Secondary air flow rate and velocity	61%, 45 m/s	30–35 m/s
Tertiary air		16–18 m/s

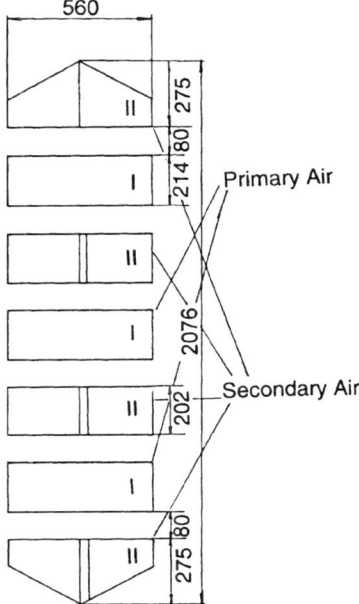

FIGURE 10-15. Burner stack arrangement for a 220 T/h boiler

When total water and ash content of brown coal is below 60%–65%, such a combustion system with an exhausted air separator is not recommended because the temperature in the burner zone will be high, and if the softening temperature of the ash is low, serious slagging may occur.

10-7 Multifuel Burner

Figure 10-16 shows a sketch of a burner where designed to burn pulverized coal, blast furnace gas (BFG), coke oven gas (COG), and heavy oil either in isolation or in combination. The burner output would, however, be reduced by 25% when gas and oil are fired alone.

The burner shown in Figure 10-16 is designed to fire Datong coal of China in a corner fired 1160 t/h boiler (He, 1987). Here, eleven layers of burners are arranged on each corner. The blast furnace gas burners are at the lowest layers A, B, C, D; coke oven gas and heavy oil burners are at layers E and F layers; pulverized coal burners are arranged in layers G–K. The whole burner stack is about 14.1 m high. The ratio of the total height to nozzle width of PC burner is 24.3. In practice not all nozzles operate at the same time. When the PC burners are put into operation alone, the ratio of burner height to width is 12. The cross section

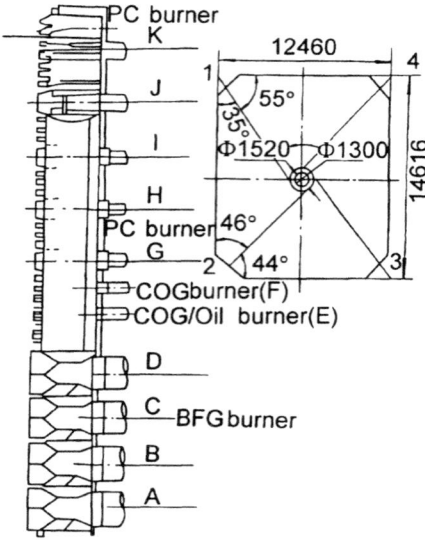

FIGURE 10-16. A multifuel burner to burn PC, blast furnace gas, and oil

of the furnace is nearly square (width : depth = 1.08). The twin tangential circle arrangement is used. The diameters of the imaginary firing circles are 1500 and 1300. Details of the primary and secondary nozzles are shown in Figure 10-16. Longitudinal-lateral plates are fixed on all primary and secondary air nozzles to increase rigidity, to enhance nozzle cooling, and to improve the outlet air flow distribution. The section area at the outlet of the primary air nozzle is 0.13 m^2. Peripheral air passages of relatively large areas are arranged around the primary air with the outlet section area 0.127 m^2. Secondary air is arranged between two primary air jets.

Equal air registration is required for sub-bituminous coal burners. Here, the secondary and primary air are arranged alternately. To improve ignition conditions, the nozzle is designed with larger height/width ratio. A V-shape bluff body flame holder is adopted in the primary air nozzle, which splits the nozzle into upper and lower parts. A recirculation zone is formed after this bluff body and the primary air can mix with the high temperature gas to enhance ignition of the pulverized coal.

10-8 Design Methods for Tangentially Fired Boilers

The following section presents a typical design calculation method for tangentially fired burners for corner fired boilers.

TABLE 10-4. Number and capacity of burner for different boilers.

Power plant capacity	Boiler capacity		Details of tangential burners	
Electrical generation (MWe)	Steam generation (t/h)	Furnace cross section m × m	Number of burner sets × layers of primary air nozzles	Thermal capacity of each primary air nozzle or burner (MW)
12	65, 75		4 × 2	7–9.3
25	120, 130		4 × 2	9.3–14
50	220(230)		4 × (2–3)	14–23.3
100–125	410(400)		4 × (3–4)	18.6–29
200	670	10.9 × 10.9	4 × (4–5)	23.3–41
300	≈1000		Double furnace 8 × 4 single furnace 4 × (5–7)	23.3–52
550		15 × 15	16–28 in 4–7 layers	40–75
600	≈2000	16 × 19.6	Single furnace 8 × 6 or 4 × 6	41–67.5

10-8-1 Selection of Burner Capacity

The number of burners, in the present context, refers to the number of primary air nozzles. It is chosen from experience, taking into consideration the fuel fired, ash fusion temperature, capacity of the unit, and the furnace configuration. Several examples are given in Table 10-4 to illustrate this choice. As the boiler capacity increases the number of layers of burners is increased and the capacity of each burner decreases. Boilers larger than 600 MW capacity need to use double-furnace configuration.

The maximum capacity of a tilting burner increases with the furnace cross section area but reduces with increasing ash fusion temperature (Fig. 10-17). For example, for a 200 MW boiler, firing coal with a fusion temperature of 1173°C, the furnace cross section area is 10.9×10.9 m^2, and the permitted maximum capacity of each burner is 29 MW. Table 10-4 presents the number and the burner capacities of several boilers.

10-8-2 Design Method

a) Air Rate and Air Velocity

The criteria of selection of flow rate and velocity of air for pulverized coal fired burners are given in Section 8-5-2 of Chapter 8. Some special characteristics of the tangential burners are discussed below.

The secondary air serves the upper burner, the middle burner, and the lower burner. The upper secondary air, which constitutes the highest amount in the step registration system (Fig. 10-7b), depresses the flame to prevent it from moving up. Its downward inclination angle is $0°-12°$. The middle secondary air constitutes the

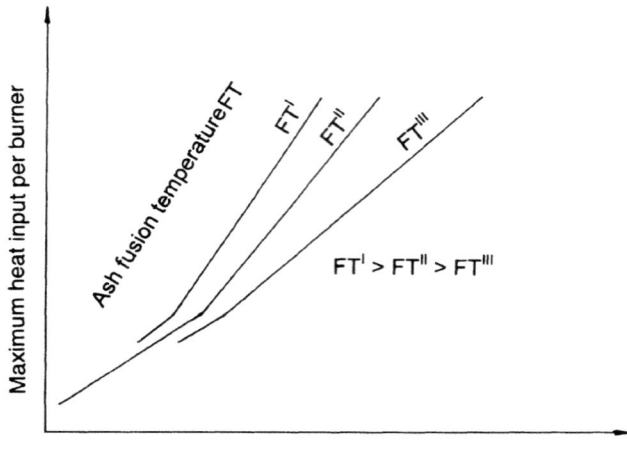

Furnace section area

FIGURE 10-17. The maximum allowed heat input to each burner increases with the ash fusion temperature (FT) and the furnace section area

largest amount in the equal air registration burners. It is the main oxygen supplier for coal combustion. In large-capacity boilers, two over-fire air nozzles, carrying 15% of the total air, can be provided above the upper secondary air nozzles to reduce the NO_x emission.

The recommended velocities of primary and secondary air in the burners are given in Table 10-5. In larger furnaces one should chose higher secondary air velocities to ensure an adequate swirl momentum of the air flow.

b) Design of Burner Geometry

The height (h) to width (b) ratio of a burner is an important parameter. For example, in boilers larger than 410 t/h steam capacities, the ratio h/b is greater than 6–6.5. Bigger boilers use higher ratios of height to width. Here, the packing density of burners can be reduced or the nozzles can be divided into several groups. The h/b ratio of each group is about 4–5 and the space between adjacent groups should be no less than the nozzle width. The space is taken advantage of to equalize the pressure difference between both sides and decrease deflection of flow. For

TABLE 10-5. Recommended primary (W_1) and secondary (W_2) air velocities.

Thermal capacity of one burner (MW)	Anthracite & lean coal			Bituminous & brown coal		
	W_1(m/s)	W_2(m/s)	W_2/W_1	W_1(m/s)	W_2(m/s)	W_2/W_1
23.3	18–20	28–30	1.5–1.6	24–26	36–42	1.5–1.6
34.9	18–20	29–32	1.6–1.7	26–28	42–48	1.6–1.7
52.3	20–22	34–37	1.6–1.7	28–30	48–50	1.6–1.7

TABLE 10-6. Selection of inclination angle of air nozzle.

Air nozzles	Tertiary air nozzle	Upper upward secondary	Upper downward secondary
Inclination angle	5°–15°	0°–12°	5°–15°

anthracite and lean coal one needs to use a higher ratio than that used for bituminous coal.

For a low-volatile coal, the distance between burners is increased to facilitate ignition. For example, when a short nozzle burner is used to fire anthracite, the direct pulverizing system is employed, and one or two layers of spare primary air burners are provided. These burners are generally not operated. They serve the function of separating the burner into groups.

The diameter of the imaginary firing circle, d_0, of a corner fired boiler should be selected carefully to ensure good furnace coverage. The *coverage* is the fraction of the furnace volume filled by the flame. The higher the coverage, better is the heat distribution. The choice of firing circle should avoid flame deflection, adherence to the wall, and excessive local heat release rate. The diameter of the firing circle is given by

$$d_0 = (0.05-0.13)a_1$$
$$a_1 = \frac{Per}{4} \qquad (10\text{-}3)$$

where a_1 is average width of the furnace and *Per* is the periphery of the furnace cross section.

The diameter of the imaginary tangential circle is usually in the range of 600–1600 mm. If burners are arranged in groups, then the imaginary circle of the upper group is greater than that of the lower group. The selected diameter of the imaginary circle should also meet the requirement that the included angle α between the tangent to the imaginary circle and the diagonal line of the furnace should be within 4°–6°.

The range for downward inclination angle may be found from Table 10-6. Except for tilting burners, tertiary air nozzles, upper upward, and second-upward secondary air nozzles are designed to tilt for low-volatile coal and inferior coal, while other nozzles are not tilted.

c) Calculation of Burner Nozzle Size

The calculation procedure for burner nozzles is shown in Table 10-7.

10-9 Example of Burner Design

An example of design calculations for a bituminous coal fired burner for a 200 MW boiler follows. The bin and feeder and hot air conveying systems are equipped with a steel ball pulverizer.

TABLE 10-7. Calculation of burner size.

Name	Symbol	Unit	Calculation
Primary air velocity	W_1	m/s	Selected from Table 10-5
Secondary air velocity	W_2	m/s	Selected from Table 10-5
Tertiary air velocity	W_3	m/s	Select within 50–55 m/s range
Total primary air nozzle area	ΣF_1	m²	V_1 / W_1
Total secondary air nozzle area	ΣF_2	m²	V_2 / W_2
Total tertiary air nozzle area	ΣF_3	m²	V_3 / W_3
Primary air density	ρ_1	kg/m³	
Secondary air density	ρ_2	kg/m³	
Tertiary air density	ρ_3	kg/m³	
Number of primary air nozzle layers	Z_1	—	Selected, generally chosen from Table 10-4
Nozzle width	b_1	mm	Selected 200–800 mm
Primary air nozzle height	h_1	mm	$\Sigma F_1/(4b_1 z_1)$ [4 nozzles in 4 corners]
Primary air nozzle section area per layer	F_1	m²	$\dfrac{V_1}{4W_1 Z_1}$
Secondary air nozzle section area per layer	F_2	m²	Same as above
Tertiary air nozzle section area per layer	F_3	m²	Same as above
Coal burned per set of burners	B_r	t/h	$B_j/4$
Coal burned per primary air nozzle	B_1	t/h	$\dfrac{B_j}{4Z_1}$
Clearance between nozzles	H_j	m	Selected from the arrangement of nozzles
Total height of the burner	H_r	m	
Relative clearance between nozzles	$\dfrac{H_j}{H_r}$	—	
Total height and width ratio of the burner	$\dfrac{H}{b}$	—	H/b
Ignitor heat power requirement		MJ	2% of the adjacent primary air power

TABLE 10-8. Example of burner design.

Parameters	Symbol	Unit	Calculation	Results
Coal burned	B_j	kg/h	From the boiler design	97,000
Heating value of coal	Q	MJ/kg	Given	21.2
Excess air coefficient	α_1		Chosen	1.2
Combustion air temperature	T_a	°C	Selected from the boiler thermal design	340
Theoretical air needed	V^0	nm³/kg	From the boiler thermal design	5.673
Total air flow to the furnace	V	nm³/s	$\alpha_1 \times V^0 \times B_j/3600 =$ $1.2 \times 5.673 \times 97000/3600$	183.42
Air leakage coefficient	$\Delta\alpha_1$	—	From the boiler thermal design	0.05
Air leakage rate of the furnace	α_e	%	$\Delta\alpha_1/\alpha_1 \times 100 = 0.05/1.2 \times 100$	4.16
Primary air	r_1	%	Chosen (Section 10-5-1)	20
Tertiary air	r_3	%	Calculated in a later step	18.3
Secondary air	r_2	%	$100 - \alpha_e - r_1 - r_3$	57.48

TABLE 10-8. (*Continued*)

Parameters	Symbol	Unit	Calculation	Results
Number of the primary air nozzles	n		4×4 (Four in each corner)	16
Energy per primary air nozzle	Q_1	MJ	$Bj \times Q/n = 97,000 \times 21.2/16$	128525
Furnace section area	F_f	m^2	$A \times B = 11.92 \times 10.88$ (from furnace area given)	129.68
Primary air velocity	W_1	m/s	Chosen (Table 1-5)	25
Secondary air velocity	W_2	m/s	Chosen (Table 1-5)	45
Velocity ratio	W_2/W_1		45/25	1.8
Secondary air temperature	T_2	C	$T_a - 5 = 340 - 5$ (owing to the heat loss of the heat pipe)	335
Secondary air flow	V_2	m^3/s	$V \times (273 + T_2)/273 \times (r_2/100)$	235
Secondary air outlet area	F_2	m^2	V_2/W_2	5.21
Moisture content in PC	W_{mf}	%	From the pulverizing system	10
Temperature of PC	T_{mf}	C	From the pulverizing system	60
Specific heat of PC	C_{mf}	kJ/kg C	From the pulverizing system	0.921
Specific heat of air before mixing	C_{pa}	kJ/kg°C	at $T_1 = 335°C$	1.03
Specific heat of air after mixing	Cam	kJ/kg C	at $T = 250°C$	1.026
Temperature of primary air mixture	t_1	C	First guess value $\Sigma q = \Sigma q'$ below	250
Heat needed to heat PC	q_1	kJ/kg	$(100 - W_{mf})/100 \times C_{mf} \times (T'_1 - T_{mf})$	$0.829T'_1 - 49.7$
Heat needed to heat moisture in PC	q_2	kJ/kg PC	$W_{mf}/100 \times [2491 + 1.88 \times (T'_1 - T_{mf})]$	$238 - 0.188T'_1$
Heat given by primary air	q_3	kJ/kg PC	$1.285 \times \alpha_1 \times V^0 \times r_1 \times (Ca \times T_2 - C_{am} \times T'_1)$	$603 - 1.8T'_1$
Temperature of primary air mixture	T_1	°C	From $q_1 + q_2 = q_3$ one finds	170
Moisture vaporized per kg fuel	ΔW	kg/kg PC	From the pulverizing system	0.0514
Number of mills	z_m		From the pulverizing system	2
Output of the mills	B_m	kg/h	From the pulverizing system	57200
Total PC in primary air	B_{mf}	kg/h	$(1 - \Delta W) \times (Bj - 0.15 \times z_m \times B_m)$	75736.2
PC total heat consumed	Σq	kJ/h	$B_{mf} \times (q_1 + q_2)$	77047.6
Primary air temperature at outlet	T_1	C	$T'_1 - 5$	165
Primary air flow	V_1	m^3/s	$V \times r_1 \times (273 + T_1)/273$	58.7
Primary air oulet area	F_1	m^2	$V_1/w_1 = 58.7/25$	2.35
Calculation of tertiary air				
Dry agent per kg of fuel	g_1	kg/kg	From the pulverizing system	1.85
Hot air rate in dry agent	r_{rk}		From the pulverizing system	0.485
Air leakage in the pulverizing system	K_{lf}		From the pulverizing system	0.25
Tertiary air flow	g_3	kg/kg	$g_1 \times (r_{rk} + K_{lf})$	1.36
Tertiary air rate	r_3	%	$g_3 \times z_m \times B_m/(1.285 \times \alpha_1 \times V \times Bj)$	18.3
Exit temperature of the mill	T''_m	°C	From the pulverizing system	100

Contd.

TABLE 10-8. (*Continued*)

Parameters	Symbol	Unit	Calculation	Results
Tertiary air temperature	T_3	°C	$T_m'' - 10$ owing to loss in pipe	90
Tertiary air volume	v_3	m³/kg	$(g_3/1.285 + \Delta W/0.804) \times (273 + T_3)/273$	1.49
Tertiary air flow	V_3	m³/s	$v_3 \times B_m \times z_m/3600$	47.4
Tertiary air velocity	W_3	m/s	Selected	50
Tertiary air nozzle section area	F_3	m²	V_3/W_3	0.95

TABLE 10-9. Final results of the burner designed.

Name	Air split (%)	Air temperature °C	Air velocity (m/s)	Outlet area (m²)
Primary air	20	165	25	2.35
Secondary air	57.48	335	45	5.23
Tertiary air	18.35	90	50	0.945

Nomenclature

A_p	area of the primary air pipe, m²
a	tilting burner nozzle height, m
a	length (Eq. 10-2), m
a_1	average width of the furnace, m
b	length (Eq. 10-2), m
b	width, m
d_0	diameter of the ideal firing circle, m
d_y	diameter of the actual firing circle, m
d	rotation diameter, m
d_{dl}	equivalent diameter of a rectangular nozzle
h	height, m
k	experimental constant (Eq. 10-1)
l	tilting burner nozzle length, m
n	exponent (Eq. 10-1)
Per	perimeter of furnace, m
q_t	section heat release rate, Mw/m³
q_v	furnace volume heat release rate, Mw/m³
u	periphery of the furnace cross section, m
w	tangential velocity, m/s
V_1, V_2, V_3	volume flow rate of primary, secondary and tertiary air flow respectively, m³/s
W_1	primary air velocity, m/s
W_2	secondary air velocity, m/s
W_3	tertiary air vlocity, m/s

Greek Symbols

α	nozzle cone angle (Fig. 10-10)
β	angle of intersection of primary and secondary air jet (Fig. 10-6a)
β	nozzle tilt of a tilting burner
$\Delta\beta$	difference between the tilt and inclined angle (Fig. 10-10)
ω	tangential velocity
ξ	flow-resistance coefficient of a tilting burner nozzle

References

Cen, K., and Fan, J. (1990) "Theory and Calculation of Engineering Gas-Solids Multiphase Flow" (in Chinese). Zhejiang University Press, Hangzhou, China, p. 3.

Cen, K., and Fan, J. (1991) "Combustion Fluid Dynamics" (in Chinese). Hydro-Electricity Press, Beijing, p. 1.

Fen, J. (ed). (1992) "Principles and Calculations of Boilers" (in Chinese), 2nd edition. Science Press, Beijing.

He, P., and Zhao, Z. (1987) "Design and Operation of Pulverized Coal Burners" (in Chinese). Mechanical Industry Press, Beijing.

Xu, X. (1990) "Combustion Theory and Equipment." Mechanical Industry Press, Beijing p. 11.

11
Fluidized Bed Boilers

Fluidized bed boilers use a firing technique where the fuel is burnt in a bed or suspension of hot, noncombustible granular solids. This type of firing process has proved successful in addressing some of the long-standing problems of fossil fuel boilers. As a result, the traditional market for conventional solid fuel firing techniques like stoker and pulverized fuel firing is being progressively taken over by fluidized bed boilers.

The earliest application of fluidized beds can be traced to the coal gasifier of Fritz Winkler of Germany (1921). However, serious use of this novel gas–solid contacting process did not begin till 1950s, when petroleum industries started using it for cracking heavy oil. The effort to use fluidized beds for steam generation began in the midsixties. Prof. Douglas Elliott of CEGB, UK, is credited as the "father of fluidized bed boilers" for promoting the use of the fluidized bed process as practical means for generation of steam. Today, fluidized beds are used in a wide variety of processes from hospital beds for patients (Basu, 1995) to 250–600 MWe power plants. The advent of fluidized combustion techniques removed impediments to fossil fuel utilization. For example, it is now possible to generate energy from coal in an environmentally acceptable manner and to burn even the worst grade of fuel in a moderate-size of furnace. Furthermore, the introduction of the fluidized bed based partial gasification-combustion combined cycle and supercritical boilers may soon make it possible to push the overall efficiency of a coal fired power plant beyond the 50% mark.

11-1 Fluidized Bed Boiler

A fluidized bed boiler is a type of steam generator in which fuels burn in a special hydrodynamic condition called the *fluidized state* and transfer the heat to boiler surfaces via some noncombustible solid particles. There are two main types of fluidized bed boilers:

1. Bubbling fluidized bed boiler
2. Circulating fluidized bed boiler

FIGURE 11-1. A view of a bubbling fluidized bed boiler [Reprinted with permission from J.T. Louhimo (1993) Combustion of pulp & papermill sludges and biomass in BFB. "12th International Conference on Fluidized Bed Combustion," L.N. Rubow ed., ASME, New York, p. 259, Fig. 2]

Figure 11-1 shows a cut-view of a bubbling fluidized bed boiler, where fuel is burnt in a bed of hot (800–900°C) noncombustible solids (ash, sand, or limestone). The bed, which is typically 0.5 to 1.5 m deep, is created by blowing air through the grid (called the *distributor* at certain velocities. The fuel particles burn in this bed and transfer the combustion heat to inert solids. The bed solids, in turn, transmit the heat to boiler surfaces inserted directly in this bed and to the flue gas leaving the bed.

In a *circulating fluidized bed* boiler (Fig. 11-2) a much higher velocity of air through the grid is employed. This produces a bed much less dense than that in the bubbling fluidized bed boiler but it extends to the top of the furnace. As a result solids continuously leave the furnace. These solids are captured in a gas–solid separator and returned to the base of the bed at a sufficiently high rate. This recycling of solids is done by a device called loop seal. The solids are vigorously back-mixed. As a result the temperature of the furnace is fairly uniform, and is kept in the range of 800–900°C for optimum combustion and emission control. Heat transfer surfaces are located primarily on the enclosing walls of the furnace as in a PC boiler.

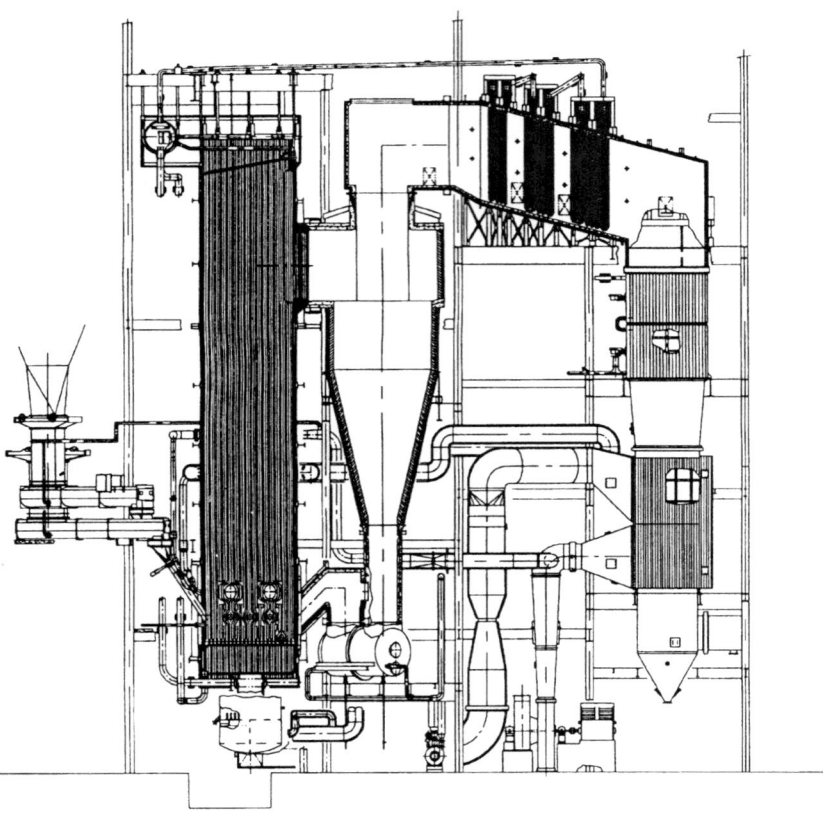

FIGURE 11-2. Cross section view of a typical CFB boiler [Reprinted with permission from P.A. Jones et al. (1995) Case history of the scrubgrass circulating fluidized bed project, "13th International Conference on Fluidized Bed Combustion," K.J. Heinschel, ed., p. 717, Fig. 1]

11-2 Major Features of Fluidized Bed Boilers

The following section presents a number of features common to both main types of fludized bed boilers. There are some specific to particular types, which are described later.

11-2-1 Advantages

Fluidized bed boilers have a number of unique characteristics that make them more attractive than other types of boilers. Some of these attractions are discussed below.

a) Fuel Flexibility

This is one of the major attractions of fluidized bed boilers, especially in the present free market of fuel. A fluidized bed boiler can accept a wider variety of fuel with lesser loss in performance than any other boilers. Thus, it enables the plant owner to diversify the procurement source of fuel, obviating the need to be tied to one source of fuel. The plant owner can make the choice based on price or other advantage of the fuels. This flexibility is a great boon to some countries where the quality of coal is deteriorating progressively owing to the depletion of deposits of better-quality fuels. Over the 20–30 year life span of some boilers the quality of the available fuel in these countries dropped so much that many of the conventional boilers had to accept either a severe loss in output and or use expensive fuel oil to maintain it. The fuel flexibility feature of fluidized bed boilers provides better protection to a new plant against any future uncertainty in the fuel supply.

Combustibles constitute less than 1%–3% by weight of all solids in the furnace of a typical fluidized bed boiler. The rest of the solids are noncombustibles: ash, sand, or sorbents. The special hydrodynamic conditions in the furnace of a fluidized bed boiler facilitate an excellent gas–solid and solid–solid mixing. Thus fuel particles fed to the furnace are quickly dispersed into the large mass of hot bed solids, which rapidly heat the fuel above its ignition temperature. Since the thermal capacity of the noncombustible particles is a few orders of magnitude higher than that of the fuel particles, there is no significant drop in the temperature of these solids for preheating even the worst quality of fuel.

This feature of a fluidized bed boiler would ideally allow it to burn any fuel without the support of an auxiliary fuel, provided its heating value is sufficient to raise the combustion air and the fuel itself above its ignition temperature. However, in an existing plant, practical considerations like capacity of auxiliary equipment, heat absorption, metal temperature, etc., limit the fuel flexibility. In any case, the penalty for operation on off-design fuel is minimal in fluidized bed boilers.

b) Efficient Control of Sulfur Dioxide Emission

Unlike other types of boilers, the processes of combustion and heat extraction continue simultaneously in a fluidized bed boiler. This feature along with its good solid mixing capability help the fluidized bed boiler maintain a nearly uniform temperature in the entire furnace. As a result it is possible to maintain the combustion temperature within the range of 800–900°C. This narrow temperature window is ideal for the following reaction for absorption of SO_2 gas by limestone ($CaCO_3$).

$$CaCO_3 = CaO + CO_2$$
$$SO_2 + CaO + 1/2O_2 = CaSO_4$$

The reaction product is a relatively benign solid calcium sulfate, $CaSO_4$. So it can be disposed of in land fill or used as gypsum. Limestone, when used in a fluidized bed boiler furnace, can easily help retain more than 90% of the sulfur in the fuel as solid residues. These dry granular solids are easy to handle and have less disposal problems than sludge from a wet scrubber.

This sulfur capture capability is a result of uniform and low-temperature properties of the fluidized bed firing technique. A fluidized bed boiler does not need a separate scrubber to control the SO_2 emission.

c) Low NO_x Emission

The typical level of emission of nitric oxide from a fluidized bed boiler is in the range of 100–300 ppm (dry volume). Such a low level of nitric oxide emission is a direct result of the low combustion temperature, devolatilization of coal in a reducing atmosphere, and air staging in fluidized bed firing. Thus, most fluidized bed boilers can meet the local NO_x emission limit without unburnt losses or any extra provision. Some local areas stipulate stricter limits for the emission of nitric oxide. In those cases, additional suppression can be achieved by injection of ammonia downstream in the hot cyclone (Chelian & Hyvrinen, 1995).

The low emission is an intrinsic property of the fluidized bed combustion. In conventional boilers most of the nitrogen oxides comes from the oxidation of the nitrogen in the combustion air. This reaction is significant above 1480°C. At typical combustion temperatures (800–900°C) of fluidized bed boilers the above reaction is negligible. However, the fuel nitrogen is oxidized to NO_x, which can be subsequently reduced to nitrogen by means of the staged addition of air in a circulating fluidized bed boiler.

d) Operational Convenience

Fluidized bed boilers enjoy a number of practical operating conveniences that make their operation much easier, for example:

(1) No Flame-Out

Burner-type boilers (pulverized coal, oil, or gas fired) have to have a sophisticated flame supervisory system. If for some reason the flame is extinguished momentarily, the furnace must follow an elaborate purging procedure before re-igniting the flame in the boiler again. Thus, even a brief fuel stoppage takes the unit out of service with a series of consequences on the turbine generator system. A fluidized bed boiler does not have a flame. It has, instead, a large mass of inert hot solids. Thus, even after a temporary fuel stoppage, neither the furnace temperature nor the steam temperatures drop immediately. If the fuel feed is resumed within several minutes, no restart is required on the boiler.

In pulverized coal fired (PC) boilers, the ignition of low-volatile, less reactive fuels is a major problem, but in the case of CFB boilers, ignition does not pose any problem.

(2) Excessive Oil Consumption

The operating cost of a PC boiler often increases owing to the use of large amounts of expensive fuel during startup and during low-load operation. The PC furnace is required to be heated slowly to avoid thermal stress. This low-intensity heating in a PC boiler can only be done by firing oil. In fluidized bed boilers, the slow heating of boiler components can be done by gradually increasing the amount of coal fired in the furnace. Alternative startup procedures (Muthukrishnan et al., 1995) can even eliminate the use of fuel oil entirely.

PC boilers require partial support of oil flame at very low load. This support may also be needed when there is some ignition problem. The fluidized bed boilers can be designed to burn these fuels or operate at low load without the oil support.

(3) Short Hot Start-Up Time

Some industrial boilers operating on two shifts need to be lighted after 8 h. A fluidized bed boiler can be started up in a relatively short time without going through the elaborate light-up procedure. A furnace, left under defluidized condition, loses very little heat. Even after several hours, the bed solids retain an adequate amount of heat for ignition. So, when the fuel is injected into the bed during the next light-up, it ignites immediately and the boiler comes on line in a relatively short time.

(4) Stability

Fluidized bed boilers, CFB boilers in particular, can take much more abuse or operational lapses than can PC fired boilers.

e) Reduced Fouling

The alkali metals in the fuel ash cannot evaporate at the low operating temperatures of a fluidized bed furnace. As a result the probability of their condensing in a cooler section of the boiler to cause fouling is much less. Thus, a fluidized bed boiler can fire even high-slagging or high-fouling fuels without a major slagging problem. This greatly reduces the need for the use of soot blowers. Soot blowers may be used only to keep the tubes in the back-pass clean such that the heat absorption is not affected.

f) Reduced Erosion

The ash produced in fluidized bed boilers is relatively soft because it does not melt at the low (800–900°C) temperature of the furnace. This reduces erosion of tubes in the convective section or the blades of the induced draft fan further downstream.

g) Simpler Fuel Preparation

The size of coal fed to a fluidized bed boiler is typically 70% below 6000 μm (6 mm), whereas it is 70% below 75 μm for pulverized coal (PC) fired boilers. As a result a PC boiler has to have both a crusher and pulverizer. The pulverizer is not only a costly and sophisticated, it requires a relatively high level of maintenance. A majority of the boiler outage results from the failure of its pulverizers. A fluidized bed boiler, on the other hand, does not need a pulverizer. Thus its fuel preparation is considerably more maintenance free and simple than that of a PC boiler.

11-2-2 Specific Features of Circulating Fluidized Bed Boilers

The above discussion lists the advantages common to both the bubbling and circulating fluidized bed (CFB) boilers. CFB boilers enjoy some additional advantages. They are as follows:

a) Lower Sorbent Consumption for Sulfur Retention

Circulating fluidized bed boilers require 1.5 to 2.5 times the stoichiometric amount of limestone to reduce the SO_2 emission by 90% compared to 2.5 to 3.5 for bubbling fluidized bed boilers. This is attributed to finer particle sizes and longer gas particle residence time used in a CFB boiler. The average residence time of gas in the combustion zone of a bubbling fluidized bed is on the order of 1–2 s, while it is 3–6 s for CFB. The average particle size of limestone in a bubbling fluidized bed is 1000 μm, while that in a CFB is 150 μm. The latter has a much larger specific surface area for reaction than the sizes used in bubbling bed boilers.

b) Staged Combustion

In a CFB boiler the primary air, which passes through the grid or around it, is slightly less than the stoichiometric amount. The rest of the combustion air, constituting 20% excess, is added further up in the furnace. Staged combustion further reduces the NO_x emission from the CFB boiler to about 50–100 ppm.

c) Good Turndown and Load Following Capability

In a CFB boiler the heat absorption in the furnace can be easily controlled by varying the suspension density in the upper furnace. This allows faster response to varying load. In a PC boiler firing high-ash coal it is difficult to maintain a very load without the support of expensive fuel oil. In a CFB unit the suspension density can be reduced to nearly negligible amount dropping the furnace heat absorption to the level of freeboard of bubbling bed boilers. Thus it is possible to operate a CFB boiler in very low load conditions without the assistance of auxiliary fuels.

11-2-3 Limitations

There are certain shortcomings of fluidized bed boilers. They are explained below.

a) Higher Fan Power Requirement

Both bubbling and CFB fired boilers require powerful forced draft fans because the primary combustion air has to overcome the pressure of a grid plate and the mass of bed solids in the furnace. This higher level of power consumption is offset to some extent by the elimination of the pulverizer.

b) Large Grate Area (Bubbling Fluidized Bed)

Owing to its lower grate heat release rate, a bubbling fluidized bed boiler uses a larger grate area than a PC boiler. However, this area is comparable to that used by a stoker fired boiler. A CFB boiler uses a grate heat release rate similar to that of a PC boiler, but has the additional requirement of a cyclone and external heat exchanger. So the total boiler plant footprint may be larger than that for a PC boiler.

c) Higher Surface Heat Loss (CFB Boiler)

A cyclone, solid recycling system, and external heat exchanger are additional components of CFB boilers. Some of these surfaces are uncooled. As a result the convective and radiation heat loss from a CFB boiler is higher than that from a bubbling or PC boiler.

d) Lower Combustion Efficiency

The combustion efficiency of a normal PC boiler is higher than that of a CFB boiler because of its higher combustion temperature, finer particle size, and longer residence time in the combustion zone. However, the PC boiler using over-firing air and low excess air for NO_x control would have a higher combustible loss. The combustion efficiency of a CFB boiler is higher than that of bubbling fluidized bed boilers because the combustion zone of the former extends up to 15–25 m height of the furnace and bulk of the unburnt carbon particles leaving the furnace are recycled back into the furnace. In a bubbling fluidized bed the main combustion zone is the 1–1.5 m deep bed. Unburnt carbon fines, escaping the furnace, are not always recycled.

11-3 Basics of Fluidized Beds

Fluidization may be defined as *the operation through which fine solids are transformed into a fluid-like state through contact with either a gas or a liquid.*

In the fluidized state, the gravitational pull on granular solid particles is offset by the fluid drag on them. Thus the particles remain in a semi-suspended condition . A fluidized bed displays characteristics similar to those of a liquid. For example:

- The static pressure at any height is approximately equal to the weight of the fluidized particles per unit cross section above that level.
- The bed surface remains horizontal irrespective of how the bed is tilted. Also, the bed assumes the shape of the vessel.
- The solids from the bed may be drained like a liquid through an orifice at the bottom or on the side.
- An object denser than the bed will sink, while one lighter than it will float.
- Particles are well mixed, and the bed maintains a nearly uniform temperature throughout its body when heated.

TABLE 11-1. Physical processes used in different types of boilers.

Process	When does it occur	Where is it used
Fixed bed	Velocity through the bed is so low that solids sits on the grate	Stoker fired grate
Bubbling bed	Gas velocity is high enough to lift them and keep them suspended	Bubbling fluidized bed boiler & in parts of CFB boiler
Circulating fluidized bed	Gas velocity is high enough to blow the solids out of the furnace, but they are recycled back at a sufficiently high velocity; this makes solids to have a refluxing motion in the furnace	Main furnace of CFB boiler
Transport bed	Gas velocity is very high; solids are carried through by the air	The burner and furnace of PC boiler

Like a stoker or pulverized coal burner, a fluidized bed is also a gas–solid contacting process. One moves from one type to the other by changing the way gas passes, through the solids. As the gas velocity increases, the solid density in the bed decreases changing the nature of gas–solid contacting process (Table 11-1). Figure 11-3 shows a bed density vs. gas velocity graph for several common types of boiler firing systems.

11-3-1 Fixed Bed

In a fixed bed the particles do not move relative to each other. As the gas flows through the solids, it exerts a drag force on the particles, causing a pressure drop across the bed. The pressure drop, ΔP, across a height, L, of a packed bed of uniformly sized particles is correlated as (Ergun, 1952)

$$\frac{\Delta P}{L} = 150\frac{\mu U(1-\varepsilon)^2}{(\phi d_p)^2 \varepsilon^3} + 1.75\frac{\rho_g U^2(1-\varepsilon)}{\phi d_p \varepsilon^3} \tag{11-1}$$

where U is the superficial velocity (volume flow rate of gas per unit area of the grid), ε is the void fraction in the bed, and d_p and ϕ are the diameter and sphericity of bed solids, respectively. Also μ and ρ_g are the viscosity and density of the gas, respectively. The velocity, U, must be smaller than a characteristic velocity, U_{mf}, described below.

11-3-2 Bubbling Fluidized Beds

If the gas flow rate through a fixed bed is increased, the pressure drop continues to rise (Fig. 11-4) according to Eq. (11-1), until the superficial gas velocity reaches a critical value, U_{mf}, which is known as the *minimum fluidization velocity*. It is the velocity at which the fluid drag is equal to the gravitational pull on the particle less the buoyancy force. At this stage the particles feel "weightless," and become more mobile. Then the fixed bed transforms into fluidized bed. Further increase in

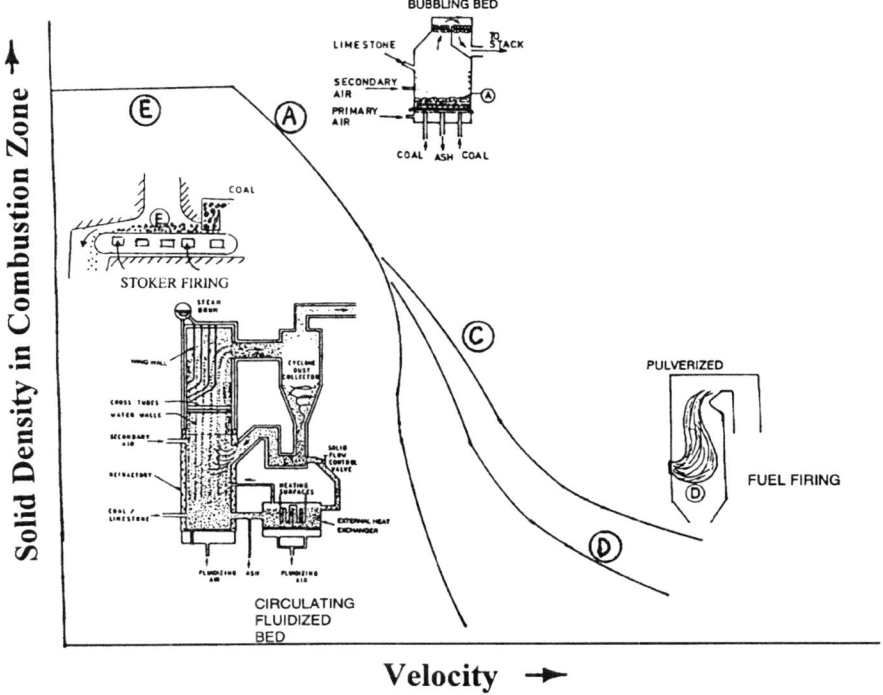

Velocity ➤

FIGURE 11-3. Regime of operation of different types of boiler furnaces shown on a bed density vs. gas velocity diagram

gas flow rate through the bed does not result in any increase in the pressure drop (Fig. 11-4). By equating $\Delta P/L$ with the bed weight less the buoyancy force, we can get an expression for the minimum fluidization velocity:

$$U_{mf} = \frac{\mu}{d_p \rho_g}\left[\left(C_1^2 + C_2\frac{\rho_g(\rho_p - \rho_g)gd_p^3}{\mu^2}\right)^{0.5} - C_1\right] \qquad (11\text{-}2)$$

where d_p is the surface-volume mean diameter of particles, C_1 and C_2 are functions of ε, μ, ϕ, etc. Values of the empirical constants are $C_1 = 27.2$; $C_2 = 0.0408$ (Grace, 1982).

At minimum fluidization, the bed behaves as a pseudoliquid. For group B and D particles (typically larger than 100 μm sand, ash, or sorbents), a further increase in gas flow can cause the extra gas to flow through the bed in the form of *bubbles*, as shown in Figure 11-4. These bubbles are voids with very little solids in it. Owing to the buoyancy force, bubbles rise through the emulsion phase, bypassing the particles. The bubble size increases with particle diameter, d_p; excess gas velocity $(U - U_{mf})$; and its position above the distributor or grid.

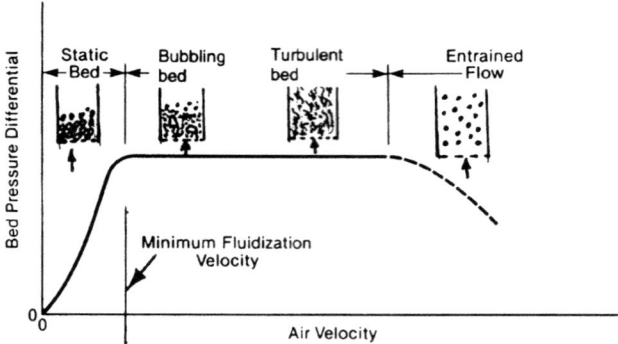

FIGURE 11-4. Pressure drop vs. gas velocity curve in different types of beds

11-3-3 Fast Beds (Circulating Fluidized Beds)

When the fluidization velocity is increased in a bubbling bed, it expands. Eventually it may lead to collapse of bubbles into a state of continuous coalescence or deformed bubble. This state is called *turbulent bed*. The exact definition of this state and transition point to this stage is still being debated. Some (Rhodes, 1997) prefers to call this regime a transition to transport regime.

a) Terminal Velocity

In any case, if the gas velocity through the bed is increased continually another critical value, U_t, is reached. It is called *terminal velocity*. At this velocity the drag on individual particles is equal to their weight less the buoyant force. Using the drag coefficient of particles in different flow regimes the expression for terminal velocity of spherical particles can be obtained as (Basu & Fraser, 1991)

$$U_t = \frac{\mu}{d_p \rho_g} \left[\frac{Ar}{18} \right] \qquad 0 < Re < 0.4$$

$$U_t = \frac{\mu}{d_p \rho_g} \left[\frac{Ar}{7.5} \right]^{0.666} \qquad 0.4 < Re < 500$$

$$U_t = \frac{\mu}{d_p \rho_g} \left[\frac{Ar}{0.33} \right]^{0.5} \qquad Re > 500 \qquad (11-4)$$

where $Ar = \frac{\rho_g (\rho_p - \rho_g) g d_p^3}{\mu^2}$; and $Re = \frac{U d_p \rho_g}{\mu}$

b) Fast Fluidized Bed

A gas–solid bed, operated at a velocity exceeding the terminal velocity, may create a special hydrodynamic condition called *fast fluidization* if solids can be recycled back to it at a sufficiently high rate. The upper furnace of a CFB boiler is believed to operate in this regime. So we first try to define it in the context of CFB boilers. *A fast fluidized bed is a high velocity gas–solid suspension where particles, elutriated by the fluidizing gas above the terminal velocity of individual particles, are recovered and returned to the base of the furnace at a rate sufficiently high as to cause a degree of solid refluxing that will ensure a minimum level of temperature uniformity in the furnace.*

A clear understanding of the transition to or from fast fluidization is lacking at the moment. However, it is generally recognized that to achieve fast fluidization the superficial velocity must exceed a certain minimum value in excess of the terminal velocity of average particles and at the same time the solids must recycle to the bed at a rate exceeding a minimum value. At the time of writing no reliable expression for estimation of these critical values of fast fluidization was available. So, the boiler designer may be guided by the values used in typical CFB boilers which are 4–6 m/s at recirculation rates in the range of 1–30 kg/m^{2}s cross section of the furnace. The average particle size is in the range of 100–300 μm. Fast fluidized beds are also used in petrochemical industries where much finer particles (<100 μm) and higher circulation rates (100–400 kg/m^{2}s) are used. This regime is sometimes called '*High density transport*' regime.

Some interesting properties of a fast fluidized bed are the following:

1. The bulk density of the bed is a function of gas velocity, circulation rate, and the solid inventory. By changing any or combination of the above parameters, one could change the suspension density.
2. The heat transfer coefficient is proportional to the square root of the suspension density. So one could change the heat absorption in the furnace by changing any of the three parameters indicated in one.
3. Solids flow upward in the center of the furnace in a relatively dilute suspension and flow downward along the wall in a denser suspension with a very good exchange of solids between these two streams. As a result of this internal recirculation of solids the temperature is fairly uniform throughout except for about 0.5 m near a heavily cooled wall, where a thermal boundary is developed (Basu & Nag, 1995).

The furnace is less dense in the center than near the wall. However, these are directly related to the cross section average suspension density, which decreases exponentially from the lower to the higher section of the furnace. A typical axial distribution of suspension density of a CFB boiler is shown in Figure 11-5. This distribution of suspension density may be expressed as (Johnsson & Leckner, 1995).

$$\rho_{sus} = (\rho_x - \rho_{xb})\exp[-a(x - x_b)] + \rho_{exit}\exp[K(X_{exit} - x)]$$

$$\rho_{xb} = \rho_{exit}\exp[K(X_{exit} - x_b)] \tag{11-5}$$

where empirical values are $K = 0.23/(U - U_t)$; $a = 4U_t/U_o$; U_o is the velocity in the lower and U is the velocity in the upper section of the furnace.

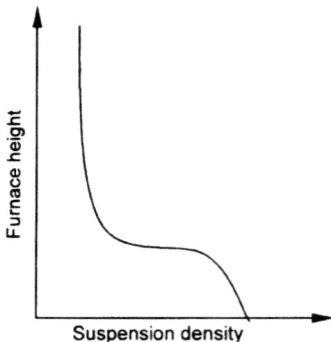

FIGURE 11-5. Suspension density vs furnace height profile of the furnace of a 125 MWe commercial CFB boiler

The density near the wall may be approximated from the equation given below

$$\rho_{wall} = \rho_p \left[1 - \left(1 - \frac{\rho_{sus}}{\rho_p} \right)^2 \right] \qquad (11\text{-}6)$$

11-4 Bubbling Fluidized Bed Boilers

Although several large bubbling fluidized bed boilers are in operation in up to 350 MWe capacity, their use is generally limited to smaller units and retrofitting of older plants. Low furnace height requirements and the absence of a cyclone make them ideal for converting old stoker fired boilers into environment-friendly fluidized bed firing systems. The following section describes the typical arrangement of a bubbling fluidized bed boiler.

11-4-1 General Arrangement

Figure 11-6 shows schematically the different components of a bubbling fluidized bed boiler and the disposition of its heating surfaces. The primary combustion zone is a bubbling fluidized bed of particles of average size of about $1000 \mu m$ (1 mm). It is supported on a grate called distributor. Coal is fed into this bed through coal feeders. The primary combustion air enters the bed through the grate. Ash generated in the bed accumulates in the bed raising its height. Fluidized solids behave like water. So solids built above the level of overflow pipe drop into an ash hopper. In practice, larger ash particles tend to accumulate at the bottom of the bed. So, an ash drain pipe (Fig. 11-6) extracts these solids on an occasional basis. The secondary air, not shown in the figure, is injected just above the bed level to burn the volatile and unburnt combustibles leaving the bed. The furnace space above the bed is called *freeboard*. The flue gas travels upward through this

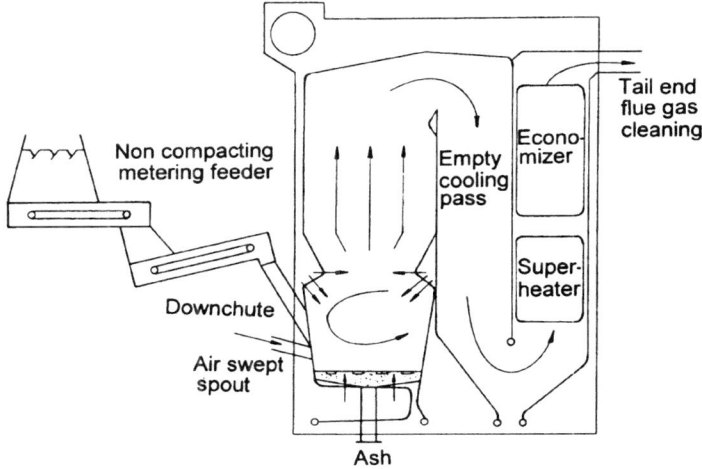

FIGURE 11-6. Over-fire arrangement for a bubbling fluidized bed boiler firing waste products

space along with some fine solids. A limited amount of combustion takes place here.

The hot flue gas, leaving the furnace, passes through the superheater and evaporative bank of tubes. As the dust-laden flue gas turns around the convective pass of tubes, some of the dust is separated. These can be collected in a carbon burnout bed, where the less reactive unburnt carbon could burn at a higher temperature of 900–950°C. The hot gas finally passes over the economizer tubes and then to the dust collector.

The feed water is first heated in the economizer. Then it goes to the boiler drum. This water flows down to the lower header through the vertical tubes in the cooler section, from where water flows to a set of inclined evaporator tubes immersed in the fluidized bed where combustion is occurring. This water absorbs heat directly from the bed generating steam. The mixture of steam and water then flows to the drum through evaporator tubes on the furnace walls. These tubes also absorb some heat (mostly radiate) from the furnace. The superheater tubes are located just behind the rear wall of the furnace to protect it from direct radiative heat and also to facilitate better control of superheater temperature.

11-4-2 Design of Bubbling Fluidized Bed Boilers

The basic design steps are common to all boilers. These involve stoichiometric and heat balance calculations, as shown in Chapter 3. So these are not discussed here. In the heat balance calculations one uses a carbon loss similar to that for a spreader stoker fired boiler. The convective and radiative heat loss should lie somewhere between those of spreader stoker and PC boilers.

a) Grate Area

This is determined from the following basis. A thumb rule suggests that grate heat release rate for a wide range of fuels can be found from

$$q = \frac{3.3U_{27}}{\alpha} \text{ MW/m}^2 \qquad (11\text{-}7)$$

where U_{27} is the superficial velocity referred to 27°C and α is the excess air coefficient. The superficial velocity (referred to the bed temperature) used in bubbling fluidized bed boilers is generally in the range of 1–3 m/s (Howard, 1989). A grate heat release rate of 2 MW/m² can be used in a boiler with cooling tubes in the bed. It would be about 1 MW/m² in furnaces, which are without cooling tubes in the bed.

The depth of the bed is determined such that (a) it accommodates the cooling tubes, (b) provides for minimum gas residence time, and (c) minimizes the pressure drop through the bed.

The amount of heat to be extracted from the bed, Q_{ext}, is found from a heat balance around the bed.

$$Q_{ext} = Q_{comb} + Q_{in} - H_{rad} - Q_{gas} - Q_{drain} \qquad (11\text{-}8)$$

where Q_{comb} is the combustion heat released in the bed, Q_{in} is the heat carried by air and feed-stocks into the bed, H_{rad} is the heat radiated from the upper surface of the bed, Q_{gas} is the enthalpy of the flue gas and solids leaving the bed, and Q_{drain} is the enthalpy of solids drained directly from the bed.

$$Q_{ext} = h \cdot LMTD \cdot S \cdot F \qquad (11\text{-}9)$$

where S is the external surface area of the tubes and $LMTD$ is the log-mean temperature difference in individual compartment. The correction factor F is close to one. The overall heat transfer coefficient h for a bubbling fluidized bed boiler is in the range of 220–340 W/m²K. It is found as

$$h = \frac{1}{\dfrac{1}{h_c + h_r} + \dfrac{R}{K_m} \ln\left(\dfrac{R}{r}\right) + \dfrac{R}{h_i \cdot r}} \qquad (11\text{-}10)$$

where R and r are the external and internal radii, respectively, of the tube having thermal conductivity K_m. The internal heat transfer coefficient, h_i, can be calculated using any reliable forced convection correlation like one by Colburn (Holman, 1992). The external radiative heat transfer coefficient, h_r, can be calculated from the following equation:

$$h_r = \frac{\sigma \cdot \left(T_{ehe}^4 - T_{tube}^4\right)}{[1/e_b + 1/e_s - 1](T_{ehe} - T_{tube})} \qquad (11\text{-}11)$$

where are T_{ehe} and T_{tube} are temperatures of the external fluidized bed heat exchanger and that of the tube surface, respectively, in °K. The bed emissivity, $e_b \sim 0.5(1 + e_p)$, where e_p and e_s are emissivities of particles and the heat transfer surface, respectively.

The convective heat transfer coefficient, h_c, can be calculated using either of the following two equations, depending on the size of average bed particles (Andeen & Glicksman, 1976):

$$h_c = 900(1 - \varepsilon)\frac{K_g}{d_{tube}}\left[\frac{2RU_{ehe}}{d_p^3\rho_p g}\right]^{0.326} \text{Pr}^{0.3} \quad d_p < 800\,\mu m; \quad U_{ehe}\rho_g d_p/\mu_g < 10$$

$$(11\text{-}12)$$

$$h_c = \frac{k(1 - \varepsilon)}{d_p}\left[\frac{3600\rho_s d_p C_p U_{ehe}}{K_g}C_5 + C_6\right] \quad d_p > 800\,\mu m; \qquad (11\text{-}13)$$

where C_5 and C_6 are experimental constants (Glicksman & Decker, 1980).

To use the above data for a bank of tubes the convective heat transfer coefficient may be multiplied by an arrangement factor F_{ar} (Gelperin et al., 1969):

$$F_{ar} = \left[1 - \frac{R}{2.S_n}\left(\frac{4R + S_p}{2R + S_p}\right)\right]^{0.25} \qquad (11\text{-}14)$$

where S_n and S_p are normal and parallel tube spacing in the tube bank.

The heat transfer to the lean zone or freeboard, h_{fb}, above the bed is given by (Stulz & Kitto, 1992):

$$h_{fb} = C_4[1 - C_3(1 - \varepsilon)] + 0.023\frac{K_g}{D_{bed}}\text{Pr}^{0.3}\left(\frac{U_f D_{bed}\rho_g}{\mu\varepsilon}\right)^{0.8} \qquad (11\text{-}15)$$

where C_4 and C_3 are constants and ε is the voidage of the dense bed. D_{bed} is the equivalent diameter of the bed and U_f is the gas velocity in the freeboard.

The tube bank is often exposed to the splash zone of the bed, where the bed density decreases exponentially. The heat transfer coefficient, h_{sp}, in this zone may be estimated from following interpolation between dense and the lean bed zones (Tang et al., 1983).

$$h_{sp} = h\exp\left\{-\left[\frac{10 + 38.7Z}{25.8}\right]^{2.2}\right\} \qquad (11\text{-}16)$$

where Z is the height in the splash zone above the bed surface in m, and h is the overall heat transfer coefficient on tubes fully submerged in bubbling fluidized bed.

b) Fan Power Requirement

The power required to pump air through a bubbling fluidized bed is a major component of the auxiliary power consumption of the boiler. So an estimation of this parameter is required at an early stage of the design. The pumping power requirement, P, is given by

$$P = \frac{\Delta P \cdot m_a}{\rho_g \eta} \qquad (11\text{-}17)$$

where m_a is the mass of air handled by the fan, ρ_g is the density of the air at average temperature in the fan impeller, and η is the efficiency of the fan. The resistance through the system is generally dominated by that through the bed, ΔP_b, and that through the distributor, ΔP_{dist}. For preliminary estimation we can use the suggested convention for the design of distributor plate (Section 11-6-2) that $\Delta P_{dist} = 0.3\ \Delta P_b$. So the total resistance can be written as

$$\Delta P = (1 + 0.3)\Delta P_b = 1.3 \times \rho_p(1 - \varepsilon_{mf})H_{mf} \cdot g \qquad (11\text{-}18)$$

EXAMPLE

Find the power required by the forced draft fan of a bubbling fluidized bed to deliver 20,000 kg/h air. Minimum fluidization velocity of average particles ($\rho_p = 2300\ kg/m^3$) is 0.15 m/s. Large-scale elutriation occurs at 1.8 m/s. The minimum gas residence time required for sulfur capture reaction is 0.8 s.

1. The choice of fluidizing velocity is guided by the minimum fluidization velocity, 0.15 m/s, and the velocity for excessive elutriation 1.8 m/s. For a combustion bed, the risk of defluidization is more dangerous than elutriation because the former will lead to complete shutdown while the latter may only increase the loss in combustibles. Based on this we chose a value 1.5 m/s.

2. The bed depth can be determined using relationships for bed expansion (Kunii & Levenspiel, 1969), but they depend on the bubble size, whose value is uncertain. The best option is to measure this in pilot plants. In absence of that an average voidage of 0.6 may yield results for a first level approximation. The bed depth can be determined from the gas residence time.

$$H = \text{velocity} \times \text{residence time} = 1.5 \times 0.8 = 1.2\ m$$

3. The pressure drop through the bed is equal to the weight of the solids, and using Eq. 11-18 we can calculate the combined resistance of bed and distributor.

$$\Delta P_b = (1 - \varepsilon)H \cdot \rho_p g = (1 - 0.6) \times 1.2 \times 2300 \times 9.81 = 10,830\ N/m^2$$

$$\Delta P_{dist} = 0.3 \times 10830 = 3249\ N/m^2$$

For the rest of the ducting, bends, dampers we add another 250 mm WG.

$$\text{Total resistance} = 3249 + 250 \times 9.81 + 10,830 = 16,531\ N/m^2$$
$$= 16.531\ kN/m^2$$

We assume that the air temperature inside the fan is about 60°C. So $\rho_g = 1.0\ kg/m^3$.

4. For a forced draft fan we chose a backward curved blade centrifugal fan, whose efficiency is in the range of 75–85%.

$$\text{Power required} = \frac{\Delta P \cdot m_a}{\rho_g \eta} = \frac{16.531 \times 20000}{3600 \times 1.0 \times 0.80} = 114\ kW$$

So the electrical power requirement of the fan motor is 114 kW.

c) Design for Incineration or Waste Fuel Firing

Some bubbling fluidized bed boilers are used for burning waste fuels or biomass. These fuels contain a large amount of volatiles or moisture. For a better thermal balance and reduced emission of NO, a small amount of air is passed through the bed. The bed will typically have no heating surface in it. A strong over-fire secondary air is applied just above the splash bed to help complete the combustion. A boiler shown in Figure 11-6. Here the bed is tapered to avoid defluidization of heavier tramp materials which tend to settle on the grid. This particular design uses directional air nozzles similar to those shown in Figure 11-9 to push these solids toward the bed drain.

11-5 Circulating Fluidized Bed Boilers

The circulating fluidized bed (CFB) boiler is becoming the solid fuel boiler of choice, especially in process and utility industries. The following section presents a general description of the boiler and some design considerations.

11-5-1 General Arrangement

A circulating fluidized bed boiler has a water-cooled furnace (Fig. 11-2). The lower part of the furnace is refractory lined and tapered in most designs. Coal is fed into this section of the furnace. Depending on the moisture content, the fuel may flow into the furnace by gravity. Limestone, used for sulfur capture, is much finer than the coal. So it is injected into the furnace by air. The coarse fraction of the ash is drained from the bed and cooled by a small fluidized bed cooler or by a water-cooled screw. The fly ash exits through the cyclone and is collected in a bag-house or electrostatic precipitator. The split of ash between coarse and fly ash depends on the ash characteristics of the fuel, cyclone efficiency, and the fluidization regime of operation. The primary air is heated in the air heater, and then passes through the grid. In some designs a direct-fired air heater is used to preheat the primary air during start up. The secondary air is injected from the sides of the furnace at 3–6 m above the grid.

The bed solids are carried up the furnace and are separated from the flue gas in hot cyclones attached to the furnace. Some CFB boilers use either a square cyclone (Fig. 11-14) or other types of separators, as shown in Figure (11-13). Most cyclones are uncooled refractory lined to prevent erosion. Some cyclones are, however, steam or water cooled. Still the heat absorption in the cyclone is small as it is covered with a thin layer of refractory. The collected solids drop into a solid recycling device, which could be either a loop seal or an L-valve. In some designs a part of the recirculating hot solids passes through an external bubbling fluidized bed. Here, heat exchanger tubes of economizer, superheater, or reheater absorb heat from the hot circulating solids. The rest of the heating surfaces are located in the convective section of the boiler, which is similar to that of a pulverized coal fired boiler.

11-5-2 Design of Circulating Fluidized Bed Boilers

As in the case of bubbling fluidized bed boilers we skip the design steps of stoi-chiometric calculations and heat balance. These are described in Chapter 3. More details specific to the circulating fluidized bed (CFB) boilers are available else-where (Basu and Fraser, 1991). So we start with the design of the furnace:

11-5-3 Furnace Design

The design of a CFB furnace has three major aspects:

• Furnace cross section
• Furnace height
• Furnace openings

The furnace cross section is chosen primarily from a combustion consideration. Conventions followed vary from one manufacturer to the other. The furnace height, on the other hand, is determined from considerations of heat transfer, as well as the gas–solid residence time. The cyclone height may have some bearing on the furnace height. The moisture content of the fuel has a major bearing on the design of the furnace. A higher moisture content (or hydrogen content as found in the ultimate analysis) gives a larger flue gas volume per unit heat release. To maintain the same superficial velocity through the furnace and through cyclone, a larger flow cross section is needed. This also affects the design of the convective section if a rigid limit on the gas velocity in this section is imposed by the boiler buyer.

The furnace is divided into two regions: upper and lower. The lower section generally operates in a sub-stoichiometric condition. Thus to avoid corrosion po-tential this part is free from boiler heating surfaces. The absence of heat absorption also helps the lower section to act a thermal flywheel for the furnace. The lower section also accommodates most of the furnace openings needed for entry of feed stock and recycle solids. The upper section constitutes bulk of the furnace. It operates in usual excess air conditions. Heating surfaces are located in this region.

The cross section of the upper furnace may be designed on the basis of superficial gas velocity, which is in the range of 4.5–6.0 m/s. The upper limit of velocity is set by erosion potential, and a lower limit of 4 m/s is set at full load to ensure fluidization under low loads (Lafanechere & Jestin, 1995). The furnace area may also be designed on the basis of grate heat release rate. The grate heat release rates based on the upper furnace are given in Table 11-2 for a number of commercial CFB boilers. As one can see from Table 11-2, the grate heat release rate for CFB boiler is in the range of 2.5 to 4.0 MW/m^2.

a) Furnace Aspect Ratio

For a given cross section area of the furnace, one may have a different width-to-depth, or aspect, ratio of the furnace. Major factors that influence this ratio include

1. Heating surface necessary in the furnace

TABLE 11-2. Heat release rates on the basis of grate area (q) and furnace volume (q_r) of some commercial circulating fluidized bed boilers.*

Fuel type	FC %	M %	Ash %	VM %	HHV kJ/kg	W m	D m	H m	Q_{comb} MW	q MW/m^2	q_v MW/m^3
ANT	47.2	20.1	27.6	7	17,453	9.42	5.84	27.4	230	4.18	0.15
BIT	43.5	9.5	26.3	28.4	22,460	13.8	7.37	31	327	3.21	0.10
CUL	25.7	7.11	61	7	8907	5.46	5.46	22.8	97	3.25	0.14
BIT	63.6	6.5	15.6	32	27,376	10.9	5.45	30	211	3.55	0.12
BIT	40.5	17	7.5	35	29,000	2.28	2.28	30.4	18.5	3.56	0.12
BIT	47	11	28.8	33	26,820	14.5	7.9	22	464	4.05	0.18
LIG	36.6	30.7	15.5	27	15,658	12.45	10.2	30	422	3.3	0.11
BIT	47	11	18.8	33	24,167	18	7.52	32.6	474	2.91	1993
BIT	52	10	16	22	24,820	7.24	7.24	33	208	3.97	0.12
SUB	40	11	30		15,565	11.5	2 × 7.4	37	600	3.53	0.095

*Computed on the basis of an overall efficiency of 35% on the gross power output where the furnace heat input is not available. FC, fixed carbon; M, moisture; VM, volatile matter, HHV, higher heating value; W, width; D, depth; H, height. Q_{comb}, combustion heat released in furnace.

2. Secondary air penetration into the furnace
3. Lateral dispersion of solids fed into the furnace

The furnace should not be so deep as to result in a poor penetration of the secondary air fed from one side. This may also give a nonuniform dispersal of the volatile matter from the coal if fed from one side. From operating experience (Table 11-2) we note that the depth does not usually exceed 8 m. As can be seen from the table, a shorter furnace uses lower furnace depth. This is necessary to ensure good mixing of solid and gas.

b) Design of Lower Furnace Section

The lower section of the furnace can be of several designs, as shown below:

1. Straight vertical wall
2. Tapered from smaller width to the full width at the top
3. Egg-curtain shaped or pant-leg design

Only a part of the total air is passed through the furnace grid. The rest is passed above the lower section. So in the lower section with vertical walls the superficial gas velocity will be low especially under low load. Since the superficial gas velocity in the lower section is low, the bed is close to bubbling fluidized stage here. Coarser and heavier particles selectively gather in this section. The tapered lower furnace has the advantage of being able to use nearly the same superficial velocity as the upper section. Under low load when the secondary air is greatly reduced, the tapered bed still remains vigorously fluidized and avoids any risk of agglomeration.

In very high capacity boilers, the bed becomes too deep for effective secondary air penetration. To avoid this problem the *pant-leg* design is used. These egg-crate-type beds allow addition of air from both sides of the lower section, while the upper section is still the common furnace.

The lower furnace is also a junction point for a number of gas and solid flow streams. These streams enter the furnace through the lower section. For each of the following openings the designer must specify the number of openings, size of openings, and the location of openings.

- Coal feed port
- Sorbent feed port
- Bed drain and air vent from bed cooler.
- Solid recycle point
- Solid transfer ports to the external heat exchanger

In addition to these, some designs may need openings for bypass primary air, secondary air.

c) Fuel Feed Ports

A circulating fluidized bed boiler requires feed points fewer than those used by a bubbling bed boiler. No theoretical method for calculating the number of feed points is yet available in the published literature. However, qualitatively one can say that a less reactive fuel would require a smaller number of feed points than a more reactive fuel especially those with high volatile content. In early designs it was found that a larger number of feed points gave higher combustion efficiency owing to better mixing of fuels. Table 11-3 lists feed points allocation in some commercial boilers. This shows a wide variation of bed area served by one feed point. One needs to be careful about using these data, because the table does not take into account the spare feed points provided. The smaller the boiler, the higher the percentage of spare capacity provided.

Coal (<6000 μm) is generally fed by gravity. So, the feed line should have a pressure higher than that in the furnace to prevent blow-back of hot gases from the furnace. More details about feeding systems are described in Chapter 3. The fuel feed port is generally located in the lower furnace as far below the secondary air entry point as possible. This would allow longer residence times, especially for fine fuel particles. In some designs coal is fed into the loop seal. This helps mixing with hot bed solids. It is particularly advantageous for high moisture or sticky fuels.

TABLE 11-3. Feed point allocation of some CFB boilers.

Heat input MW	Feed points	MW/point	Upper bed area (m^2) per feed point	Year of commissioning
123	2	62	13	1985
67	2	33	9.3	1985
124	2	62	17.7	1986
177	2	88.5	27.5	1986
327	4	82	24.8	1987
97	2	48.5	14.9	1987
18.5	1	18.5	5.2	1990
208	4	52	13.1	1993
611	4	152	42.5	1995
474	8	60	16.8	1994

Limestone is fine (100–400 μm). So it is pneumatically injected into the bed. Since limestone reacts much slower than coal, the number of feed points chosen is not as critical as it is for coal.

d) Combustion Air

In a CFB boiler furnace, a significant part of the combustion air (35%–60%) is injected into the bed as secondary air to reduce the NO_x emission. The lower furnace is usually operated in a sub-stoichiometric condition. This sets the upper limit of the quantity of the primary air. The lower limit is set by the minimum fluidized condition under the lowest possible load. The primary air generally passes through the bottom of the air distributor. So this air has to overcome the resistance of the grid, the dense lower section of the bed. It thus calls for a fan with a high head. The grid offers a high flow resistance. So to reduce the auxiliary power consumption, sometimes a minimum amount of air as indicated above is passed through the grid. The remaining primary air is injected from the side wall just above the grid.

e) Height of Upper Furnace

The height of the upper furnace is determined from the following considerations:

- Furnace heat absorption
- Pressure balance around CFB loop
- Control of NO_x emissions
- Control of SO_x emissions

Of these the furnace heat absorption is most important. It is discussed below in Section 11-5-3. The pressure balance is discussed in Section 11-7. Many mathematical models of varying degrees of sophistication and accuracy are available in the literature (Basu et al., 1999). However, for proposal design simpler closed form equations may be used. Basu and Fraser (1991) gave an expression to determine the minimum furnace height to attain a certain level of sulfur capture for a given set of design parameters and given reactivity of the sorbent. A similar engineering model (Talukdar & Basu, 1995) can be used to check whether the designed furnace height is adequate to control the NO emission to the desired level.

11-5-4 Design of Heating Surfaces

In a CFB boiler solids recirculate around the CFB loop, which comprises the lower furnace, upper furnace, cyclone, return leg, loop-seal, and the external heat exchanger. Boiler heat absorbing surfaces are arranged within this loop such that they absorb enough heat to cool the flue gas to about 850–900°C before leaving the loop. The rest of the surfaces are arranged in the convective or back-pass section. Typical arrangements of the heating surfaces for boilers with and without external heat exchangers are shown in Figure 11-7. The furnace enclosure is made of water wall evaporative surfaces. Additional evaporative surfaces may be used in the form of tube panels projecting into the furnace, running the full furnace height, or bank tubes before the back-pass or on the walls of the cyclone. A part of the superheater

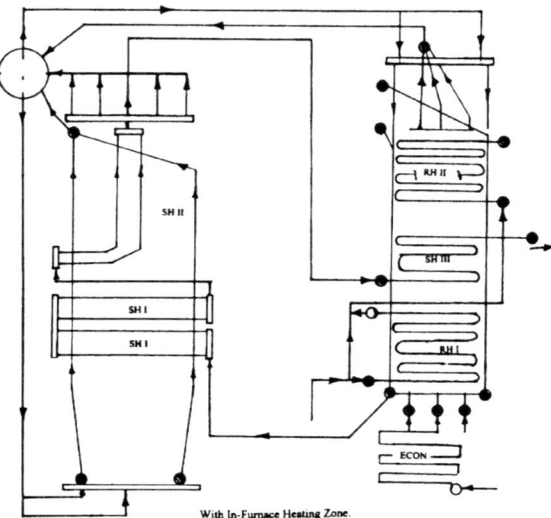

With In-Furnace Heating Zone.

FIGURE 11-7. Steam and water circuit in a CFB boiler with external heat exchangers

or reheater may be in the back-pass. The other part is located either in the external heat exchanger (Fig. 11-7b) or in the furnace as tube banks (Fig. 11-7a). Suitable protection can be provided on these surfaces to prevent erosion.

In order to maintain the furnace temperature within the desired 800–900°C range a certain percentage of combustion heat must be extracted by the CFB loop. It can vary from 68% to 35% for fuels having LHV from 32,000 to 5,000 kJ/kg (Lafanechere & Jestin, 1995). The heat duty of the CFB loop affects the overall design of the boiler. A detail study on the effect of steam parameters (Lafanechere et al., 1995a) and of fuel parameters (Lafanechere et al., 1995b) shows these effects. A simple one-dimensional mass and energy balance of the furnace can be carried out to ensure that the design meets following important criteria under all loads:

a) The temperature at any point of the furnace remains within the 800–930°C range
b) The furnace height does not exceed 40 m, as it may increase the cost excessively.

A mass and energy balance equation is written for a horizontal slice of the furnace using cross section average properties of all parameters.

$$\Delta x \cdot \rho_{sus}(x) \cdot q_v = (m_g \cdot C_{pg} + G_s \cdot C_{ps}) \cdot \Delta T + \frac{Per \cdot \Delta x \cdot h(x)}{A_{fur}} \cdot (T - T_{sat})$$

$$+ m_a \cdot C_p \cdot (T - T_{ai}) + m_{si} \cdot C_{ps}(T - T_{si}) \qquad (11\text{-}19)$$

where $G_u(x) = \rho_{sus}(x) \cdot (U - U_t)$; $\rho_{sus}(x)$ is found from Eq. (11-5). Per is the perimeter of the evaporative surfaces and A_{fur} is the cross section area of the bed.

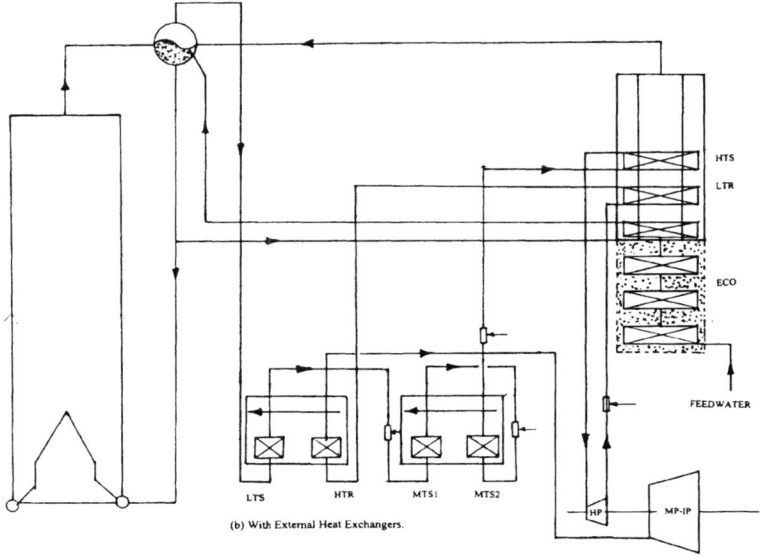

(b) With External Heat Exchangers.

FIGURE 11-7. *continue*

The furnace may be divided into at least three major zones, as shown in Figure 11-2.

Zone I:

The first zone is the lowest zone with no heat transfer to the wall. Here, primary air, m_a, enters with temperature T_{ai}. Recycled solids, m_{si}, enter with temperature T_{si}; and feed stock, m_f, enters with temperature T_o. Only the volatiles are assumed to burn here, releasing q_v, the amount of heat per unit volume of solids in the lower section. Since this section is refractory lined [i.e., $h(x) = 0$] one would note a rise in temperature.

Zone II:

The second zone is the zone of mixing of secondary air only. So the temperature will decrease.

Zone III:

The third zone is the rest of the upper furnace where no more solid or gas is added ($m_{si}, m_{gi} = 0$). All the char is assumed to burn in the upper furnace. So we take q as being equal to the total heat release rate from fixed carbon per unit volume of solids in the upper furnace. The suspension density, $\rho_{sus}(x)$ in the upper or lower furnace may be taken from the expression given in Eq. (11-5). If the boiler uses external heat exchangers (EHE), separate heat balance equations (Eq. 11-21) are used to estimate the temperature of recycled solids ($T_{si} = T_{po}$). In the absence of EHE, one can take T_{si} as equal to the furnace temperature at the exit of the furnace for a first approximation.

Heat transfer coefficient, $h(x)$, in the upper furnace may be calculated using the empirical correlation in the form (Basu & Nag, 1996):

$$h(x) = k \cdot (T - T_w)^m [\rho_{sus}(x)]^n \qquad (11\text{-}20)$$

where k, m, and n are empirical constants found from experiments; T_w is the tube wall temperature.

The heat transfer coefficient is a strong function of the suspension density, which in turn depends on the particle size distribution and cyclone design. If a boiler is operated with coal having larger amounts of intrinsic ash, then the ratio of fly ash/bed ash will increase. A similar reduction in particle size in the bed may be experienced with an improperly operating limestone crusher. An increase in fines in the bed may affect suspension density, furnace heat absorption, and therefore the steam output. Thus the particle size control is very important in a CFB boiler.

11-5-5 External Heat Exchanger

The external heat exchanger (EHE) bed of a CFB is a bubbling fluidized bed, where heat transfer tubes are generally horizontal. So it is difficult to use natural circulation here. Forced flow sections like economizer, superheater, and reheater surfaces (Fig. 11-7b) are located in the EHE.

One part of the solids recirculating in the CFB loop is diverted through the EHEs. The rate of solid flow through the EHE is controlled by means of a mechanical valve. Thus by controlling the diversion rate of solids one can control the heat absorption in these beds. One EHE may be split into several compartments in series, each heating separate sets of heat exchanger tubes, as shown in Figure 11-7b. The heat lost by the circulating solids is equal to the sum of heat absorbed by the mass of steam/water and the fluidizing air less the combustion heat generated, Q_c, in individual compartments

$$m_p C_{p_s}(T_{pi} - T_{po}) = \sum [m_w(H_{so} - H_{si}) + m_{ae} \cdot C_p(T_{ao} - T_{ai}) - Q_c] \qquad (11\text{-}21)$$

where m_p is the mass flow rate of circulating solids, T_{pi} and T_{po} are the entry and exit temperatures, respectively, of these solids. The combustion heat generated, Q_c, is very low. The heat absorbed by the steam or water in tubes, $m_w(H_{so} - H_{si})$, of one particular compartment of the EHE can be found from overall heat transfer coefficient of EHE as shown below:

$$m_w(H_{so} - H_{si}) = h \cdot LMTD \cdot S \cdot F \qquad (11\text{-}22)$$

where S is the external surface area of the tubes and $LMTD$ is the log-mean temperature difference in individual compartment. The correction factor F is close to 1. The overall heat transfer coefficient h depends on both inside and outside heat transfer coefficients of heat exchanger tubes.

$$\frac{1}{h} = \frac{1}{h_{ehe}} + \frac{R}{K_m} \ln\left(\frac{R}{r}\right) + \frac{R}{h_i r} \tag{11-23}$$

where R and r are the external and internal radii, respectively, of the tube, having thermal conductivity K_m. The internal heat transfer coefficient, h_i, can be calculated using any reliable forced convection correlation like the one by Colburn (Holman, 1992). The external heat transfer coefficient, h_{ehe}, can be calculated from the following equation

$$h_{ehe} = h_c + h_r \tag{11-24}$$

where h_c and h_r are convective and radiative heat transfer coefficients, which can be calculated from Eqs. (11-11) and (11-12).

DiMaggio et al. (1995) used the correlation of Borodulya et al. (1985) for h_{ehe}, which overpredicts experimental data from commercial units gave values by 2-3 times. This disagreement led them to speculate that a part of the hot solids pass through the EHE without exchanging heat with the immersed heat exchange surfaces. DiMaggio et al. (1995) suggested that this bypass amount increased with solid flow rate, but it decreased with fluidizing velocity and the length of the EHE.

11-5-6 Back-pass

This part of fluidized bed boilers (both bubbling and CFB) is similar to that of a conventional boiler. The high-temperature superheater is located on the top of the back-pass. Below this there are low-temperature reheater and economizer sections. The air preheater comes last in the flue gas path. The roof and the side walls of the back-pass (also called convective section) may be made of evaporative tubes, whose primary function is to reduce the heat loss to the boiler house and to absorb heat. Unlike the pulverized coal fired boilers, ash does not stick to the tubes in a fluidized bed boiler. However, light dusts are deposited on the tubes reducing the heat absorption on these tubes. Heat transfer coefficients measured (DiMaggio et al., 1995) on the high-temperature superheater tubes of a 125 MWe CFB boiler before and after cleaning of the dusts show that although soot blowing consumed equivalent to 1.3% thermal input to the back-pass, it increased the heat absorption by 12.8%.

11-6 Distributor Plates

An air distributor or the grid plate distributes the air through the base of a fluidized bed. It also supports the weight of the bed solids in most designs. In a CFB boiler the distributor plates are required in three different locations: the lower furnace, the loop seal, and the external heat exchanger.

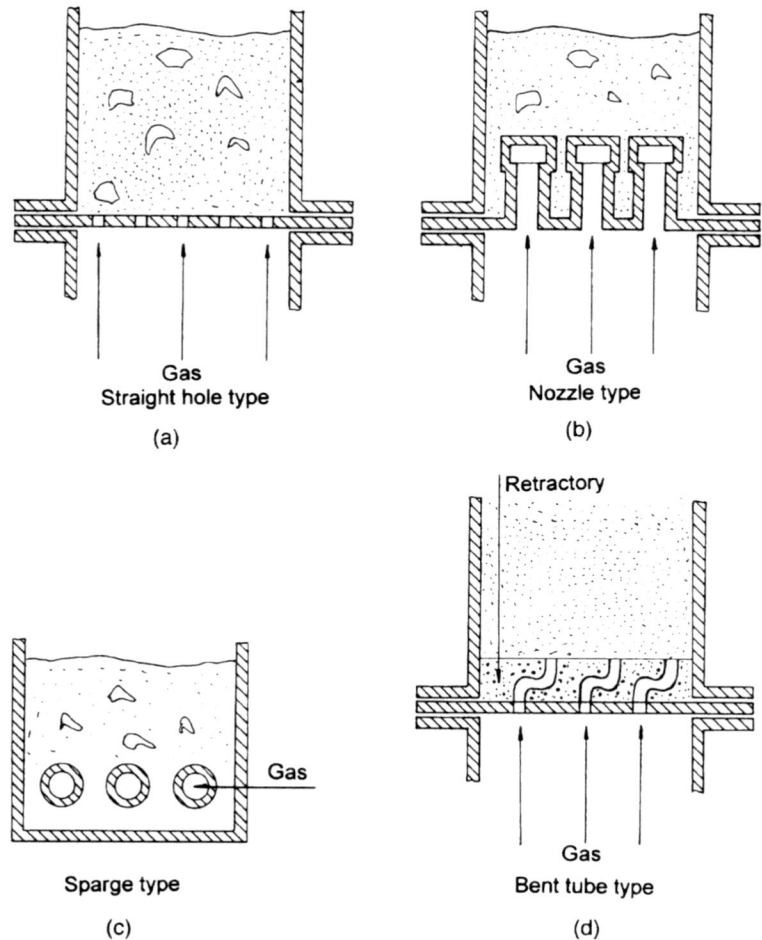

FIGURE 11-8. Different types of nozzles and distributor plates

11-6-1 Types of Distributors

There are three main types of distributor plates (Fig. 11-8):

1. Porous and straight hole orifice plate (Fig. 11-8a)
2. Nozzle type (Fig. 11-8b)
3. Sparge tube type (Fig. 11-8c)

In a sparge tube, air enters through a tube inserted into the fluidized bed. Commercial fluidized bed boilers usually use the nozzle-type distributor. The nozzles

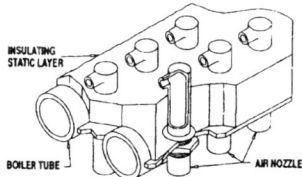

FIGURE 11-9. Nozzles are mounted on membrane wall and directed in one direction to facilitate solid movement [Reprinted with permission from R. Voyles et al. (1995) Design considerations for a 250 MWe CFB, "13th International Conference on Fluidized Bed Combustion," K.J. Heinschel ed., p.709, Fig. 5]

are supported on a grid plate, made of either a water-cooled membrane wall or a refractory coated steel plate. The water-cooled membrane type grid is effective in handling very hot air sometimes used for startup. Below are some desired properties of a distributor plate. These are arranged in the sequence of importance in fluidized bed boilers:

• Uniform and stable fluidization over the entire range of operation
• Minimum back-flow of solids into the air box
• Minimum plugging over extended periods of operation
• Minimum dead zones on the distributor
• Minimum erosion of heat exchanger tubes and attrition of bed materials

Stable fluidization at the lowest boiler load is a critical requirement of the design of a fluidized bed distributor. At low load the air flow through the distributor is reduced, but the mass of solid in bed remains the same or may even increase. Such a situation often leads to defluidization in parts of the bed. This leads to heat accumulation in defluidized areas. Eventually the temperature of those zones may increase to form a fused bed mass. Similarly there is another concern at the maximum load when the air flow through the distributor is maximum. The pressure drop through the distributor plate may be too high, which may force the fan to reduce its volume discharge according to the performance characteristic of the fan.

Back-flow of solids may occur when the boiler is shut down. Fine solids may drop into the air plenum. Next time, when the boiler is started, these particles are picked up by the air and are blown through the nozzles at velocities exceeding 30 m/s. At such high velocities the nozzle holes are rapidly eroded by back-flow solids. Severe damage to nozzles has been noticed in some plants. In a special design, the nozzle is a tube bent in the form of a goose neck (Fig. 11-8d). Here, the solids drop on the floor of the trough of the nozzle during shut down. When the bed is fluidized the high-velocity air easily blows it straight up without damaging the walls of the nozzle. Nozzles directed in a specific direction is another special design (Fig. 11-9). This helps force coarser ash particles toward the drain exit.

11-6-2 Design Procedure

To ensure uniform fluidization, designers often recommend a high-pressure drop across the distributor. A high-resistance distributor with its high orifice velocity easily cleans a temporary blockage in an orifice. It also overrides any local difference in pressure drop through the bed. For bubbling bed Zenz (1981) recommends that the ratio of distributor to bed pressure drops be 0.3.

To achieve uniform and stable fluidization, the resistance offered by the distributor to the air should be significant compared to that offered by the bed solids. Thus should there be a small difference in the resistance between one section of the bed and the other, the combined resistance of bed and distributor to the air passing would not vary appreciably from one section to the other. Thus uniform air flow through all sections of the bed is ensured. Experience suggests that for bubbling fluidized beds the ratio of distributor to bed pressure drops should be around 0.1–0.3.

The pressure drop across a bubbling fluidized bed is equal to the weight of bed solids above the grid. So

$$\Delta P_b = \rho_p(1 - \varepsilon)H \cdot g = \rho_p(1 - \varepsilon_{mf})H_{mf} \cdot g \qquad (11\text{-}25)$$

where ε and H are the voidage and bed heights, respectively. The subscript mf refers to their values under minimum fluidizing conditions. For fine group A solids (<100 micron) the bed voidage ε may be estimated as $(U + 1)/(U + 2)$ (King, 1989). The pressure drop through the distributor is taken as a multiple of that across the bed, i.e., $\Delta P_{dist} = k \cdot \Delta P_b$.

The value of k lies between 0.3 and 0.1. The pressure drop through the distributor is related to the orifice velocity by the following equation:

$$U_0 = C_d \left[\frac{2 \cdot \Delta P_{dist}}{\rho_{gor}} \right]^{0.5} \qquad (11\text{-}26)$$

where the value of orifice coefficient C_d may be taken as 0.8 (Zenz, 1981).

An orifice velocity, U_0 less than 30 m/s is considered safe (Pell, 1990), while one exceeding 90 m/s is risky (Geldart & Baeyens, 1985). It is related to the superficial gas velocity, U, by the number of orifices per unit area of the grid, N.

$$\text{Fractional opening of the grid} = N \left(\frac{\pi}{4} \right) d_{0r}^2 = \frac{U \cdot \rho_g}{U_0 \rho_{gor}} \qquad (11\text{-}27)$$

where ρ_g and ρ_{gor} are the densities of the fluidizing gas at average temperatures in the gas and inside the orifice, respectively.

In case of a CFB boiler the bed depth is much higher than that of a bubbling bed boiler. Owing to its high fluidizing velocity and highly expanded bed, the probability of nonuniform fluidization is less. So the choice of distributor pressure drop 30% that of the bed pressure drop, which would lead to a very high pressure drop distributor plate, is not necessary here. The design is guided by a net pressure drop ΔP_{dist} across the plate, which is usually in the range of 350 mm WG to prevent the orifices from being plugged.

EXAMPLE

Design a distributor grate for the loop seal of a CFB boiler. The loop seal operates at 0.5 m/s at 850°C. The mean size of solids is 220 μm. The bed depth is 1.0 m and the air is entering the plenum at 27°C. Estimated bed voidage at fluidizing condition as 0.6.

1. The pressure drop through the bed is found from Eq. (11-25).

$$\Delta P_b = \rho_p(1 - \varepsilon)H \cdot g = 2500(1 - 0.6)1.00 \times 9.81 = 9810 \, \text{Pa}$$

2. Take the distributor pressure drop to be 30% that of the bed.

$$\Delta P_{dist} = 0.3\Delta P_b = 0.3 \times 9810 = 2963 \, \text{Pa}$$

3. From Eq. (11-26) we find the orifice velocity. Here we take the orifice coefficient $C_d = 0.8$. The cold air will be heated while passing through the orifice owing to back radiation from the bed. If we take a temperature rise of 300°C, the average air temperature inside the orifice is $27 + 300/2 = 177°C$. From air table, $\rho_{gor} = 0.774$ kg/m^3.

$$U_0 = C_d \left[\frac{2 \cdot \Delta P_{dist}}{\rho_{gor}}\right]^{0.5} = 0.8 \left[\frac{2 \times 2963}{0.774}\right]^{0.5} = 70 \, \text{m/s}$$

4. We chose the diameter of the orifice as 2 mm. Now we use Eq. (11-27) to calculate the fractional opening and the number of orifices per unit area of the grid.

$$\text{Area fraction} = \frac{U \cdot \rho_g}{U_0 \rho_{gor}} = N\left(\frac{\pi}{4}\right) \cdot d_{0r}^2 = \frac{1.0 \times 0.31}{70 \times 0.774} = 0.0057$$

From here we calculate N as 1822 orifices per m^2.

NOTE: An orifice velocity of 70 m/s is on the higher side, though it is less than the upper limit of 90 m/s. This can be reduced by taking the ΔP_{dis} to be a smaller fraction, e.g., 0.186 of the ΔP_b. After repeating the above calculation we get a more reasonable orifice velocity of 55 m/s and the number of orifices as 2065 per m^2.

11-7 Loop Seals

The loop seal, an important component of a CFB boiler, is one type of non-mechanical valve for transfer of solids from the cyclone to the furnace. Proper operation of the loop seal is critical for the operation of a CFB boiler. If it fails to transfer solid at the required high rate, the furnace moves from fast to entrained bed conditions. This results in a large temperature gradient along the furnace, low heat absorption, and high back-pass temperature. The boiler suffers from loss in steam output, excessive steam temperature, and high stack temperature. The loop seal also prevents leakage of air from the lower furnace to the cyclone. When the loop seal functions properly no air bypasses the furnace. If for some reason a loop

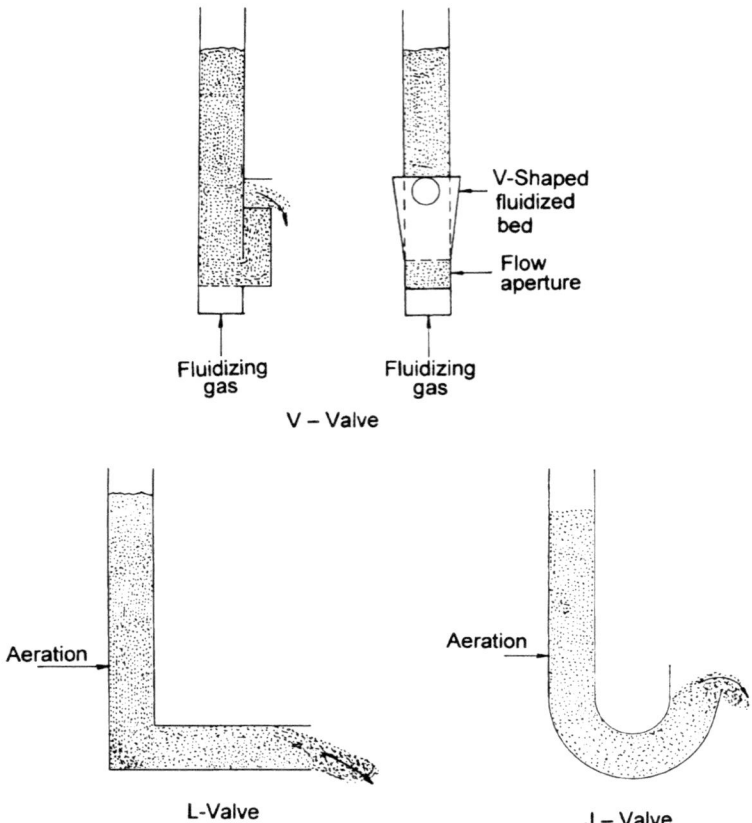

FIGURE 11-10. Schematic diagram of some types of nonmechanical valves

seal malfunctions, there is a large-scale bypassing of furnace air through it. It is one indication of malfunction of the loop seal.

11-7-1 Design and Working Principles of Nonmechanical Valves

Besides the loop seal there are other types of nonmechanical valves, e.g., L-, J-, and V-valves. Schematics of these are shown in Figure 11-10. Some CFB manufacturers also use L- and J-valves. For brevity we will confine the present discussion to the loop seal alone.

Like a U-tube manometer, a loop seal works on the principle of pressure balance does. Under steady-state conditions the static pressure on both sides of the seal is balanced. The loop seal is fluidized by air coming through its distributor plate.

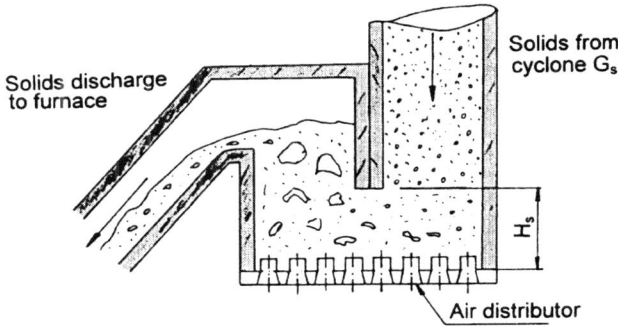

FIGURE 11-11. Working principle of a loop seal

Under normal operating conditions the bulk of this air goes up through the left arm (delivering solids to the furnace) of the loop seal (Fig. 11-11) and then to the boiler furnace. The amount of air passing through the right arm (receiving solids from the cyclone) of the loop seal must be sufficiently low so as not to fluidize the solids in the cyclone return leg. For air to flow along the left arm of the loop seal the following pressure balance needs to be satisfied:

$$P_s - \Delta P_{dist,s} - \Delta P_{sb} - \Delta P_{dis} \geq P_0 - \Delta P_{dist} - \Delta P_{tb} \qquad (11\text{-}28)$$

where P_s and P_b are pressures of the air entering the air box of the loop seal and the main bed, respectively. $\Delta P_{dist,s}$, ΔP_{sb}, ΔP_{tb}, ΔP_{dis}, and $\Delta P_{dist,b}$ are pressure drops across the loop seal distributor, loop seal bed, section of the furnace between the grid and the entry point of loop-discharge pipe, discharge pipe of the loop seal, and across the grid plate of the furnace respectively.

In order to minimize the power requirement of the loop seal blower, the superficial velocity through the loop seal may be kept just higher than the minimum fluidization velocity of larger particles circulating. So the pressure balance around the CFB loop gives

$$\Delta P_{sb} = \Delta P_{sp} - \Delta P_{hs} - \Delta P_{dis} - \Delta P_{fur} + \Delta P_{tb} \qquad (11\text{-}29)$$

where ΔP_{sp}, ΔP_{fur}, ΔP_{hs} are pressure drops across the standpipe, the entire furnace, and the horizontal section of the loop seal, respectively.

The pressure drop in the main bed below the discharge pipe, ΔP_{tb}, is given as

$$\Delta P_{tb} = (1 - \varepsilon_t)\rho_p H_t g \qquad (11\text{-}30)$$

where H_t is the height of lower edge of the discharge pipe above the distributor plate of the furnace. The voidage of this section, ε_t, may be taken as 0.8.

11-7-2 Design Procedure

The loop seal receives solids from the cyclone through the standpipe. Solids flow down this pipe as a moving packed bed with a velocity V_s. In a CFB boiler, the descending velocity of the solids, V_s, in the standpipe is typically 0.15 m/s or less, although in CFB fluid catalytic cracker units, V_s can go above 1 m/s. This velocity would not exceed the minimum fluidizing velocity of solids in a CFB boiler. The diameter of this pipe, d_s, then can be calculated from its relation with the solid circulation rate, W_s.

$$G_s A_{fur} = (\pi/4)d_s^2(1 - \varepsilon)V_s \qquad (11\text{-}31)$$

where G_s is the external recirculation flux through the furnace of area, A_{fur}. The external circulation rate in CFB boilers varies in the range of 1–10 kg/m$^2 \cdot$ s.

This solid flows from right-hand side of the loop seal (Fig. 11-11) to its left-hand side with a linear velocity, V_h, which is in the range of 0.15–0.01 m/s. The aperture area, A_{ls}, between two sides of the loop seal can be calculated from

$$A_{ls} = \frac{G_s A_{fur}}{(1 - \varepsilon_s)\rho_p V_h} \qquad (11\text{-}32)$$

The height of the aperture should not exceed the depth of loop seal fluidized bed, which is set by the lower end of the discharge pipe on the loop seal. It should at the same time be sufficiently deep to allow free flow of solids. Further details of the design are available in Basu and Cheng (1999).

11-8 Gas–Solid Separators

In a circulating fluidized bed boiler furnace, solids, leaving the furnace, are continuously separated from the flue gas and recycled back to the base of it. A high degree of separation at the furnace exit is required so that the solids escaping do not exceed those fed through the feed stock. Large-diameter reverse cyclones are most commonly used, but other devices like impact separators are also being introduced. Table 11-4 shows different types of gas solids separators that can be used in CFB boilers and their characteristics.

11-8-1 Cylindrical Cyclones

The cyclone works on the principle of centrifugal separation. A particle entering a cyclone tangentially travels in a vortex (Fig. 11-12). It is subjected to a centrifugal force, F_c, toward the wall and a corresponding viscous drag force, F_d. The centrifugal force attempts to throw the particle toward the wall from where it slides down to the standpipe for collection. The drag force opposes this motion and attempts to take it along the gas stream out of the cyclone. Under equilibrium conditions, these two forces are equal.

$$F_c = F_d$$
$$\frac{\pi d_p^3 \rho_p}{6} \frac{V_i^2}{r} = 3\pi \mu V_r d_p \qquad (11\text{-}33)$$

TABLE 11-4. Performance of different separator.

Types of Separators	U-shape	Louver	Swirling plate	S-shape	Channel tubes	U-beam	Finned tubes	Square separator	Square uniflow	Cyclone	Uniflow cyclone	Two-stage separator
Separation efficiency η (%)	50–70	55–70	60–99	80–90	80–90	90–94	94–95	97–99	96–99	98–100	97–99	>99
Pressure loss P (mmH$_2$O)	10–50	15–30	10–40	20–40	10–30	20–30	10–25	80–120	40–90	150–200	80–120	90–150
Particle diameter d_p (mm)	0.286	0.177	0.05–0.3		0.2–0.3	0.26	0.26	0.1–0.26	0.16	0.1–0.3	0.1–0.26	0.26
Circulation rate	bubbling fluidized beds (BFB)	BFB	BFB	1–2	1–2	2–3	3–4	6–20	5–20	>10	6–20	>20

335

Solving this, one gets an expression of radial velocity, V_r, toward the wall.

$$V_r = \frac{d_p^2 \rho_p V_i^2}{18 \mu r} \qquad (11\text{-}34)$$

where d_p and ρ_p are the diameter and density, respectively, of the particle. Here, V_i is the tangential velocity of the particle, r is the radial distance of the particle, and μ is the viscosity of the gas. Under ideal conditions, solids reaching the wall are fully captured. To be captured a particle must reach the wall before being caught in the reverse vortex leaving the cyclone. So, V_r may be taken as an index of the cyclone efficiency.

From Eq. (11-34) one can infer that the cyclone efficiency will increase for

- Higher entry velocity, V_i
- Larger particle size, d_p
- Higher particle density, ρ_p
- Smaller radius of the cyclone wall or cyclone exit
- Lower viscosity (μ) of the gas, which means lower gas temperature

The pressure drop increases, with a decrease in the size of mean size of particle captured. An analysis of data (Lafanechere et al., 1995c) on industrial cyclones shows the following interdependence of above parameters.

$$(\Delta P_{cy})^2 \cdot d_{50}^2 = \text{constant} \times U_b^3 \cdot D_c \qquad (11\text{-}35)$$

where U_b is a characteristic velocity (= gas volume flow rate divided by the cyclone body cross section), D_c is the cyclone body diameter, and $d_{p50\%}$ is the size of particles collected with 50% efficiency.

In a CFB boiler the situation is more complex owing to particle-to-particle interactions, which lead to formation of particle strands. However, it is apparent from Table 11-4 that the reverse cyclone gives the best collection efficiency, but has the highest pressure drop.

a) Design Procedure

Industrial cyclones generally follow a geometric proportion on the basis of its barrel diameter. The proportion varies from manufacturer to manufacturer as well as on the field of application. Figure 11-12 shows the proportion of a typical cyclone used in a CFB boiler. CFB cyclones use a shorter exit pipe, which is about 40% of the height of the inlet section ($\beta_2 D_c$). The height of the barrel section of a CFB cyclone is shorter than that in other cyclones. The inlet Reynolds number is a good index for the performance of a CFB cyclone (Lewand et al., 1993). However, Lewnard et al. (1993) noted a sharp decrease in the collection efficiency above a Reynolds number of 500,000 for 80 μm particles. This is due to the re-entrainment of solids. For 140 μm particles in a 125 MWe CFB boiler, Lafanechere et al. (1995c) did not observe any appreciable drop in efficiency even up to a inlet Reynolds number of 650,000. Based on these the following steps may be suggested for the design of a cyclone for a CFB boiler.

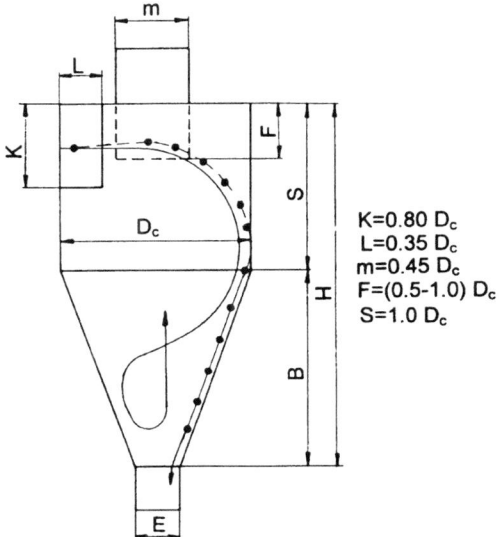

K=0.80 D$_c$
L=0.35 D$_c$
m=0.45 D$_c$
F=(0.5-1.0) D$_c$
S=1.0 D$_c$

FIGURE 11-12. Principle of operation of a reverse-flow cyclone

1. If the gas flow rate through each cyclone is V_{gas}, the cyclone diameter, D_c, for the chosen Reynolds number, Re_{max}, is found as

$$D_c = \frac{2V_{gas}}{Re_{max}(\beta_1 + \beta_2)v} \qquad (11\text{-}36)$$

2. Where $\beta_1 D_c$, $\beta_2 D_c$ are inlet width and height, respectively, and v is the kinematics viscosity of the gas.
3. If the diameter is greater than 7 m, more than one cyclone is used and the value of V_{gas} is adjusted accordingly.
4. Calculate the other dimensions of the cyclone using the proportions given in Figure 11-12. The cone angle should be greater than the angle of repose of the solids.
5. The diameter of the return leg or standpipe, d_s, may be calculated on the basis of bulk density of solids, ρ_{ret}, and the solid flow rate, M_s (Zenz, 1962). It may also be based on solid velocity, as shown in Eq. (11-31)

$$M_s = 2.29 \, \rho_{ret}(d_s)^{0.5} \qquad (11\text{-}37)$$

11-8-2 Impact Separators

Impact separators are less efficient but simpler and more compact than cyclones. These gas–solid separation devices are used by some manufacturers. A major advantage with this system is that it can be built inside the existing boiler shell and therefore avoid the need for extra space and surface areas required by an external

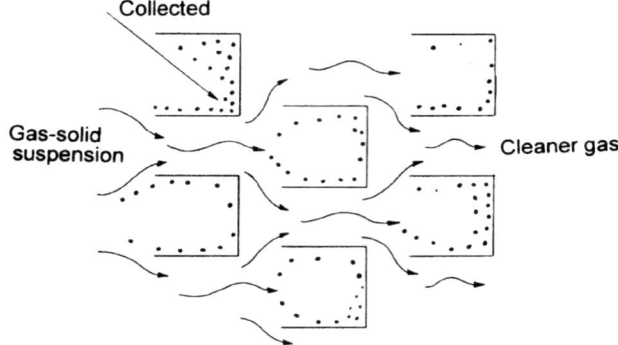

FIGURE 11-13. A simple impact separator

hot cyclone. Figure 11-13 shows the flow through a simple impact separator. Here the gas–solid mixture is subjected to a series of sharp turns. Solids, by virtue of their larger inertia, are separated and drop along the target wall as the gas flows through. The collection efficiency is a function of the Stokes' number ($\rho_p d_p^2 \cdot V_0/9\mu W$), where W is the width of the target and V_0 is the approach velocity.

11-8-3 Newer Designs

The designs of CFB boilers discussed so far are generally more expensive than PC boilers without flue gas scrubbing. Also, thermal efficiency is also slightly lower than that of a PC boiler of comparable size. This difference in performance and cost are major impediments to wide-scale use of CFB boilers where enviornmental stipulations are less rigid and availability of good fuel is not a problem.

Extra components like a hot cyclone, external heat exchanger, and loop seal are responsible for the extra cost and higher level of surface heat loss of a conventional CFB boiler. To address the need for less costly and better-performing CFB boilers several new designs have been developed. Some of them are

1. Square cyclone with bubbling fluidized bed heat exchanger in the return leg (Fig. 11-14)
2. In-furnace U-beam separator obviating the need for hot cyclone (Fig. 11-15)
3. Internal multi-inlet cyclone CFB boiler (Fig. 11-16)

a) Square Cyclone

The square cyclone comprises a square chamber with a cylindrical gas exit (Weitzke and Ganesh, 1997). Hot gas from the furnace enters this chamber through a tangential inlet. This chamber is a part of the boiler heating surface. The spiral generated in the chamber separates the solids, which flows into the fluidized bed loop seal. Solids from here are returned to the furnace at a controlled rate. The desired amount of heat is extracted from this fluidized bed heat exchanger.

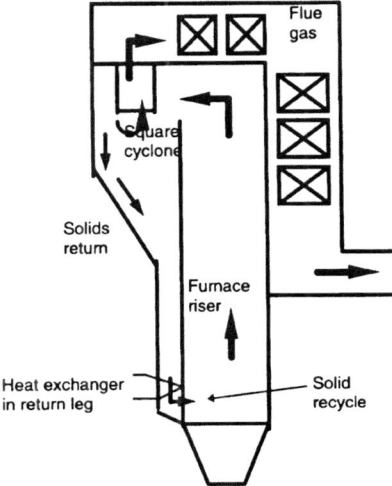

FIGURE 11-14. Schematic of a CFB boiler with square cyclone

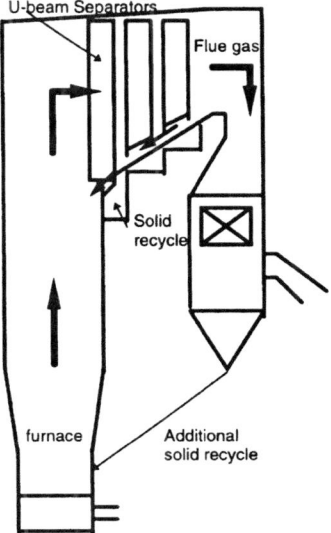

FIGURE 11-15. A CFB boiler with internal separator

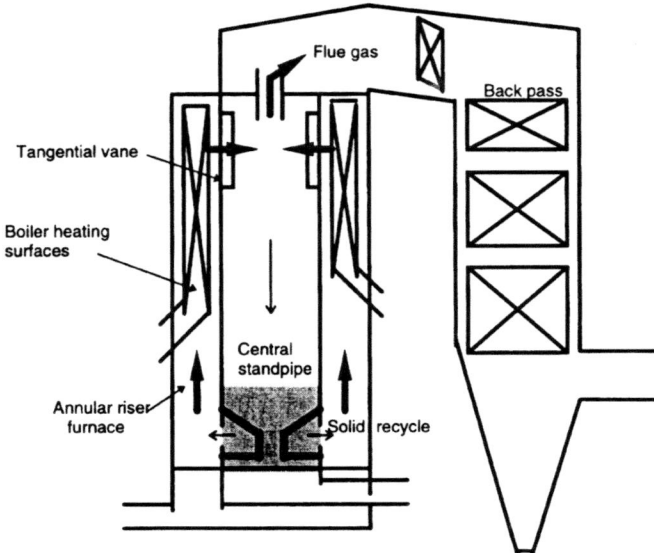

FIGURE 11-16. CFB boiler with multiple-entry internal separator

b) In-Furnace Separator

Here the hot cyclone is entirely eliminated by separating the solids by a set of inertial separator at the exit of the furnace (Belin et al., 1995). The separated solids drop into the furnace, thus obviating the need for a loop seal or return leg. A cold end separator is added downstream of the back-pass to maintain the solid balance.

c) Multiple Entry Cyclone

This boilers uses a cylindrical shaft in the middle with multiple tangential entry at its top. The main fast fluidized furnace is outside the central shaft. Gas solids leaving the annulus enter the top of the central shaft through tangential entry vanes. This creates a vortex that separates the solids in the shaft and flue gas exit from the shaft exit (Fig. 11-16). The separated solids collect on the bubbling fluidized at the bottom of the central shaft. These solids flow into the outer fast bed through an opening around it. Larger units can have several separating shafts (Koko et al., 1995). Both central shaft and the furnace enclosures are made of heating surfaces.

Nomenclature

A_{ls} aperture area between two sides of the loop seal, m^2
A_{fur} furnace area, m^2
C_1, C_2 empirical constants in Eq. (11-2)
C_5, C_6 experimental constants in Eq. (11-13)
C_4, C_3 constants in Eq. (11-15)

C_d	orifice coefficient
$C_p\ C_{pg}$	specific heat of gas, kJ/kg · K
C_{ps}	specific heat of circulating solids, kJ/kg°C
d_{50}	diameter of the particle collected with 50% efficiency, m
D_{bed}	equivalent diameter of the bed, m
D_c	cyclone body diameter, m
d_p	surface volume mean diameter of particles, m
D_{rat}	diameter of the return leg or standpipe, m
d_{tube}	diameter of bed tubes, m
d_s	diameter of standpipe, m
d_{or}	diameter of orifice, m
e_b	bed emissivity
e_p	emissivity of particle
e_s	emissivity of gas
F	correction factor in Eq. (11-9)
F_{ar}	arrangement factor (Eq. 11-14)
F_c	centrifugal force, N
F_d	viscous drag force, N
G_s	external recirculation flux through the furnace, kg/m² s
G_u	downflow solid flux through the furnace, kg/m² s
g	acceleration owing to gravity, m/s²
h	overall heat transfer in fluidized bed, kW/m² K
H	depth of bed, m
$h(x)$	heat transfer coefficient at height x, kW/m² K
h_c	convective heat transfer coefficient, kW/m² K
h_{ehe}	outside heat transfer coefficient of the tubes in EHE, kW/m² K
h_{fb}	heat transfer to the lean zone of freeboard, kW/m² K
h_i	internal heat transfer coefficient, kW/m² K
h_o	outside heat transfer coefficient, kW/m² K
H_{mf}	height bed for minimum fluidizing bed, m
h_r	radiative heat transfer coefficient in EHE, kW/m² K
H_{rad}	heat radiated from the upper surface of the bed, kW/m² K
H_s	height of aperture, m
h_{spo}	constant in equation, kW/m² K
h_{sp}	heat transfer coefficient in splash zone, kW/m² K
H_{si}	enthalpy of steam/water entering EHE, kW/kg
H_t	height of lower edge of the discharge pipe of loop seal above the distributor, m
K	correlated constant (Eq. 11-5)
k	empirical constant (Eq. 11-11)
K_g	thermal conductivity of gas, kW/m K
K_m	thermal conductivity of metal, kW/m K
L	bed height, m
m	exponent in Eq. (11-11)
m_a	mass of primary air per unit bed area, kg/m²

m_{ae}	mass flow rate of air through EHE, kg/s
m_g	gas flow rate per unit bed area, kg/m^2
m_p	mass flow rate of solids through EHE, kg/s
m_w	mass flow rate of steam/water through EHE, kg/s
M_s	solid flow rate through one cyclone, kg/s
m_{si}	solid feed rate per unit bed area, kg/m^2
n	exponent in Eq. (11-11)
N	number of orifices per unit grate area, 1/m^2
N'	characteristic number of cyclone
P	pumping power, kW
Per	perimeter of the furnace wall, m
P_s	pressure in loop seal, Pa
P_o	pressure of primary air below the distributor of the furnace, Pa
q	grate heat release rate, MW/m^2
q_v	combustion heat release per unit volume of furnace, kW/m^3
Q_{comb}	combustion heat released in the bed, kW
Q_{drain}	enthalpy of solids drained directly from the bed, kW
Q_{ext}	heat extracted from the bed, kW
Q_{gas}	enthalpy of the flue gas and solids leaving the bed, kW
Q_{in}	heat carried by air and feed stocks into the bed, kW
R	external radius, m
r	internal radius, m
S	external surface area, m^2
S_n	normal tube spacing in tube bank, m
S_p	parallel tube spacing in tube bank, m
T	average temperature of the furnace at location x, °C
T_{ehe}	temperature of the external heat exchanger, °K
T_f	temperature of the furnace, °C
T_{ai}	inlet temperature of the primary air, °C
T_{ao}	temperature of the air leaving an EHE, °C
T_o	initial temperature of the fuel or ambient temperature of the ash, °C
T_{pi}	temperature of solids entering an EHE, °C
T_{po}	temperature of solids leaving an EHE, °C
T_{sat}	saturation temperature, °C
T_{si}	inlet temperature of the recirculating solid, °C
T_{tube}	surface temperature tubes in EHE, °K
T_w	tube wall temperature, °C
U	fluidizing velocity, m/s
U_b	characteristic velocity of a cyclone, m/s
U_f	velocity in the upper section of the furnace, m/s
U_{27}	superficial velocity referred to 27°C or 300 K, m/s
U_{ehe}	fluidizing velocity in the external heat exchanger, m/s
U_{mf}	minimum fluidization velocity, m/s
U_o	orifice velocity, m/s
U_t	terminal velocity of spherical particles, m/s

V_h	horizontal velocity of solids in the loop seal, m/s
V_i	tangential velocity of particles in cyclone, m/s
V_r	radial velocity of solids in the cyclone, m/s
V_s	descending velocity of solids in stand pipe, m/s
V_0	approach velocity of solids to impact separators, m/s
V_{gas}	volume of gas flow rate through a cyclone, m^3/s
W	width of impact separators, m
X_{exit}	height of the exit from the furnace, m
x_b	depth of the lower dense bed, m
x	height above the distributor, m
Z	height of tube in splash zone measured above the bed surface, m

Greek Symbols

α	excess air coefficient
β_1, β_2	geometric proportion for the width and breadth of the inlet of a cyclone
Δx	elemental height of the furnace, m
ΔP	pressure drop, Pa
ΔP_b	pressure drop across a bubbling fluidized bed, Pa
ΔP_{cy}	pressure drop across cyclone, Pa
ΔP_{dis}	pressure drop across the discharge pipe of the loop seal, Pa
ΔP_{dist}	pressure drop across the distributor plate of the furnace, Pa
$\Delta P_{dist,s}$	pressure drop across the loop seal distributor, Pa
ΔP_{fur}	pressure drop across the entire furnace, Pa
ΔP_{hs}	pressure drop across the horizontal section of the loop seal, Pa
ΔP_{sb}	pressure drop across the loop seal bed, Pa
ΔP_{sp}	pressure drop across the standpipe, Pa
ΔP_{tb}	pressure drop in the furnace between the distributor and the loop seal discharge point, Pa
ΔT	rise in temperature of gas solid suspension across a height Δx, °C
ε	void fraction in the bed
ε_{mf}	void fraction in the bed for minimum fluidizing bed
ε_t	voidage of the section (Eq. 11-19)
σ	Stefan Boltzman constant
ϕ	sphericity of bed solids
μ	dynamic viscosity of gas, kg · m/s
η	efficiency of fan, fraction
ρ_g	density of air or gas, kg/m^3
ρ_p	density of solids, kg/m^3
ρ_{gor}	density of fluidizing gas at average temperature inside the orifice, kg/m^3
ρ_{ret}	bulk density of solids kg/m^3
ρ_{sus}	suspension density, kg/m^3
ρ_{wall}	density near the wall, kg/m^3
ρ_x	suspension density at height x, kg/m^3
ρ_{xb}	suspension density at the top of the lower dense bed, kg/m^3

Dimensionless Numbers

Ar	Archimedes number, $\rho_g(\rho_p - \rho_g)gd_p^3/\mu^2$
Pr	Prandtl number, $C_p\mu/K$
Re	Reynolds number, $Ud_p\rho_g/\mu$

References

Andeen, B.R., and Glicksman, L. (1976) Heat Transfer Conference. ASME paper No. 76-HT-67.

Basu, P., and Cheng, L. (1999). Effect of pressure on loop seal operation for a pressurized circulating fluidized bed, Powder Technology 103:203–211.

Basu, P., Luo, Z., Boyd, M., Cheng, L., and Cen, K. (1999) An experimental investigation into a loop seal in a circulating fluidized bed. 6th International Conference on Circulating Fluidized Bed. J. Werther, ed. Wurzburg, Germany.

Basu, P., and Fraser, S.A., (1991) "Circulating Fluidized Bed Boilers—Design and Operations." pp. 146–153. Butterworth-Heineman, Stoneham, USA.

Basu, P., and Nag, P.K. (1996) A review of heat transfer to the walls of a circulating fluidized bed furnace. Chemical Engineering Science 51:1.

Belin, F., Maryachik, M., Fuller, T., and Perna, M.A. (1995) CFB combustor internal solids recirculation—Pilot testing and design applications. Proceedings of 13th International Conference on Fluidized Bed Combustion. K.J. Heinschel, ed. ASME, New York, pp. 201–209.

Borodulya, V.A., Epanov, Y., Ganzha, V.L., and Teplitsky, V.L. (1985) Heat transfer in fluidized beds. Journal of Engineering Physics, pp. 621–626.

Chelian, P.K. and Hyvarinen, K. (1995) Operating experience of pyroflow boilers in a 250 MWe unit. In Proceedings of 13th International Conference on Fluidized Bed Combustion. K.J. Heinschel, ed. ASME, New York, pp. 1095–1103.

DiMaggio T., Piedfer, O., Jestin, L., and Lafanechere, L. (1995), Heat transfer and hydrodynamic analyses in an industrial circulating fluidized bed boiler. Proceedings of 13th International Conference on Fluidized Bed Combustion. K.J. Heinschel, ed. ASME, New York., pp. 577–584.

Ergun, S. (1952) Fluid flow through packed columns. Chemical Engineering Progress, 48:89–94.

Gelperin, N.I., Einstein, V.G., and Korotyanskaya, L.A., (1969) International Chemical Engineering, 9 (1):137–142.

Geldart, D., and Baeyens, J. (1985) The design of a distributor for gas fluidized beds. Powder Technology 42:67–78.

Grace, J.R. (1982) "Fluidized Bed Hydrodynamics. Handbook of Multiphase Systems." G. Hestroni, ed. Hemisphere, Washington, ch. 8.1.

Grimson, E.D. (1937) Correlation and utilization of new factors on flow resistance and heat transfer for cross flow of gases over tube banks. Transactions of ASME 59:583–594.

Holman, J.P. (1992) "Heat Transfer." McGraw Hill, New York. 7th edition.

Howard, J.R. (1989) "Fluidized Bed Technology—Principles and Applications." Adam Higler, Bristol.

Johnsson, F., and Leckner, B. (1995) Vertical distribution of solids in a CFB furnace. Proceedings of 13th International Conference on Fluidized Bed Combustion. K.J. Heinschel, ed. ASME, New York, pp. 671–679.

Koko, A., Krvinen, R., and Ahlstedt, H. (1995) CYMIC© Boiler scale up and full scale demonstration experience. Proceedings of 13th International Conference on Fluidized Bed Combustion. K.J. Heinschel, ed. ASME, New York, pp. 211–223.

Kuñii, D., and Levenspiel, O. (1969) Fluidization Engineering, New York, Wiley.

King, D.F. (1986) in Grace, J. Bergongnou, M. and Schemitt, S. (eds.) Fluidization VI AIChE, New York. p. 1.

Lafanchere, L., and Jestin, L. (1995) Study of a circulating fluidized bed furnace behavior in order to scale it up to 600 MWe. Proceedings of 13th International Conference on Fluidized Bed Combustion. K.J. Heinschel, ed. ASME, New York, pp. 971–980.

Lafanchere, L., Basu, P., and Jestin, L. (1995a) Effects of steam parameters on the size and configuration of circulating fluidized bed boilers. Proceedings of 13th International Conference on Fluidized Bed Combustion. K.J. Heinschel, ed. ASME, New York, pp. 1–8.

Lafanchere, L., Basu, P., and Jestin, L. (1995b) The effects of fuel parameters on the size and configuration of circulating fluidized bed boilers. Journal of the Institute of Energy, December 68:184–192.

Lafanechere, L., Jestin, L., and DiMaggio, T. (1995c) Aerodynamics and heat transfer analyses in order to scale up and design of industrial circulating fluidized bed boilers. 3rd International Symposium on Multiphase Flow and Heat Transfer, Sept 19–21, Xian, China. ISBN 7-5605-0668-3/TK.51, pp. 1–8.

Lafanechere, L., Jestin, L., Bursi, J.M., Roulet, V., and DiMaggio, T. (1996) Circulating fluidized bed boiler-numerical modeling—Parts 1 & 2. "Circulating Fluidized Bed Technology," V. M. Kwauk and J. Li, eds. Science Press, Beijing, pp. 466–480.

Lewnard, J.J., Herb, B.E., Tsao, T.R. and Zenz, F.A. (1993) Effect of design and operating parameters on cyclone performance for CFB boilers. "Circulating Fluidized Bed Technology IV." A. A. Avidan, ed. AICHE, New York, pp. 525–531.

Muthukrishnan, M., Sundarajan, S., Viswanathan, G., Rajaram, S., Kamalanathan, N., and Ramakrishnan, P. (1995) Salient features and operating experience with world's first rice straw fired fluidized bed boiler in a 10 MW power plant. Proceedings of 13th International Conference on Fluidized Bed Combustion. K.J. Heinschel, ed. ASME, New York, pp. 609–614.

Pell, M. (1990) "Gas Fluidization." Elsevier, New York, pp. 21–30.

Rhodes, M. (1997) "Circulating Fluidized Bed Technology, IV." M. Kwauk and J. Li, eds. Science Press, Beijing, pp. 307–312.

Stultz, S.C., and Kitto, J.B. (1992) "Steam—Its Generation and Use." Babcock & Wilcox, Barberton, Ohio, pp. 16-8–16-11.

Tang, J.T., et al. (1983) AIChE Spring National Meeting Houston, Texas, March 27–31.

Talukdar, J., and Basu, P. (1997) Effect of fuel parameters on the performance of a circulating fluidized bed boiler. "Circulating Fluidized Bed Technology, IV." M. Kwauk and J. Li, eds. Science Press, Beijing, pp. 307–312.

Talukdar, J., and Basu, P. (1997a) A simplified model of nitric oxide emission from a circulating fluidized bed combustor. Canadian Journal of Chemical Engineering 73(October) 635–643.

Weitzke, P.E., and Ganesh, A. (1997) Design aspects of supercritical steam conditions in a large CFB boiler. "Circulating Fluidized Bed Technology, IV." M. Kwauk & J. Li, eds. Science Press, Beijing, pp. 289–294.

Zenz, F.A. (1962) Petroleum Refiner 41 (2):159–168.

Zenz, F.A. (1981) Elements of grid design. Presented at the Gas Particle Industrial Symposium, Engineering Society, Western Pennsylvania, Pittsburgh.

12
Steam–Water Circulation in Boilers

Circulation refers here to the flow of water, steam, or their mixture around the steam–water circuit in a boiler. In early days of boilers, steam was generated in a large vessel. The vessel was heated from outside, which would create a convective current within the large body of water. This process did not pose any threat to the safety of the boiler. So a detailed knowledge of steam–water circulation or flow pattern was not important. Modern water tube boilers, on the other hand, are subjected to increasingly higher levels of heat flux, temperature, and pressure. So the designers can no longer ignore the process of circulation of water and steam in the boilers. Inadequate circulation of water would fail to remove heat from the tube surface at a sufficient rate, which may lead to increased tube wall temperature. The allowable stress of the metal reduces with increasing metal temperature. In an extreme case the tube may rupture. An adequate circulation of steam–water is also essential for proper separation of steam from water. Without a good circulation the quality of steam would suffer.

Three main types circulation are used in commercial boilers. Most commonly used is *natural circulation*. Here, water and the steam–water mixture move as a result of their density difference. Once-through forced circulation is the second type. Here, a pump feeds the water at one end and forces the steam through the other end of the boiler. Unlike natural circulation, it does not circulate around a boiler circuit. Assisted circulation is the third type. Here, a circulation pump helps the steam–water mixture to overcome the resistance through the circuit. Besides these, there is also a combination system which will be discussed later.

12-1 Natural Circulation System

In natural circulation, the steam and water move around the boiler under the action of buoyancy force. Figure 11.7 of Chapter 7 shows the steam–water circulation circuit of a circulating fluidized bed boiler. Here we see that a feed water pump forces water into the economizer (1). The water, leaving the economizer, goes up the side wall of the back-pass (2). The subcooled water enters the steam drum at (3). The water from the drum flows down under gravity through large-diameter pipes called *downcomers* (5). It then passes through bottom headers (6) of the water wall to enter vertical water wall riser tubes (7) of the furnace. Steam is generated in the riser tubes. The mixture of steam and water from the water wall enters the steam drum at (8). A set of separators inside the drum separates the steam from the mixture. The remaining water flows back to the bottom headers through the downcomer (5). This flow occurs under natural circulation. The steam leaves the drum at (9) to enter the back-pass enclosure at (10). The steam leaves the back-pass enclosure at (12) and then enters the primary superheater (12) exiting the enclosure at (11). The superheated steam leaves a set of primary superheaters (SH-I, SH-II) at (13). It is heated further in the secondary superheater at (14) and leaves the boiler at (15). The superheated steam expands in the steam turbine to drive the electric generator. The partially expanded steam, sometimes returns to the boiler at (16) for reheating. This heating is also done in several stages of reheaters (16–18). The pressure of steam forces it to flow through the superheater and reheaters. It is, therefore, said to be in *forced circulation*.

12-1-1 Flow of Steam–Water Mixture

In boiler riser tubes relatively cold water enters at the bottom. Since the temperature of the water is below its saturation temperature, it is called *subcooled*. In the lower section heat received from the furnace may be high enough to evaporate the water in the immediate neighborhood of the tube wall, but the bubbles condense owing to the subcooled water around it. This process is known as *subcooled boiling* (Fig. 12-1e). There is no net steam generation in this process.

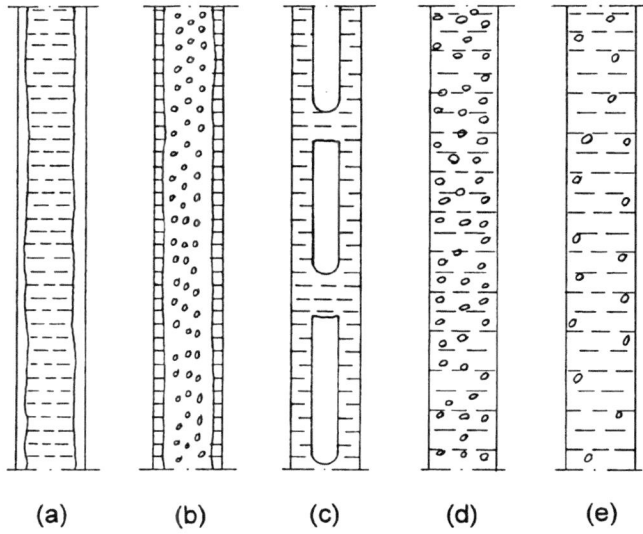

(a)   (b)   (c)   (d)   (e)

FIGURE 12-1. Steam–water mixture flow modes in vertical tubes

As the water is heated close to its saturation temperature the steam bubbles formed do not collapse. They move up through the water as bubbles dispersed in it. This is called *nucleate boiling* (Fig. 12-1d). The heat transfer coefficient in this case is very high. This bubbly type of flow continues till the steam fraction in the steam–water mixture is low. At a low flow rate but higher heat flux the steam fraction in the mixture increases. Then bubbles coalescence into larger bubbles, which may be big enough to form slugs in the tube (Fig. 12-1c). With a further increase in the steam fraction the water body between consecutive slugs merge into a continuous vapor core with atomized water droplets in it (Fig. 12-1b).

If the heat flux exceeds a certain critical value, some steam bubbles on the wall may join together to form a vapor blanket (Fig. 12-1a). This steam film may occur on the wall, leaving the liquid water to fill the core (Fig. 12-1a). This is called *film boiling condition*, which leads to a boiling crisis and is referred to as *departure from nucleate boiling* (DNB). The wall temperature immediately shoots up because of the drastic drop in heat transfer rate from the wall. The boiling heat transfer is replaced by thermal conduction and radiation through the vapor blanket alone. This leads to deterioration in heat transfer or boiling crisis, called *dry out*. It is associated with terms like critical heat flux, dry out, critical vapor mass fraction, etc.

12-1-2 Principle of Natural Circulation

In a subcritical boiler, the density of the steam–water mixture is lower than that of the water in the unheated downcomers outside the furnace. Owing to the lower density, the hydrostatic pressure in the riser tubes at the bottom of the furnace is lower than that in the downcomers filled with water. This creates a motive head, P_m, which drives the mixture around the circuit

$$P_m = H(\gamma_w - \gamma_m) \text{ N/m}^2 \tag{12-1}$$

where H is the height of the circuit, and γ_m and γ_w are the specific weights (density × g) of water and mixture, respectively.

Friction in pipes, eddies in fittings, and flow acceleration offer resistance to the flow through the riser. The flow resistance, ΔP_r, consumes a part of the motive head. Thus the head available, P_{av}, to force the mixture through the riser is

$$P_{av} = P_m - \Delta P_r \text{ N/m}^2 \tag{12-2}$$

This head must be provided by the downcomer. The downcomer will lose some head owing to its own flow resistance, ΔP_{dc}. So for a given steady flow rate the following condition must be satisfied:

$$P_{av} = \Delta P_{dc} \text{ N/m}^2 \tag{12-3}$$

The resistance through the downcomer, ΔP_{dc}, increases with the flow through the downcomer (G) (Figure 12-2). The resistance of the riser, ΔP_r, also increases with the increase of flow, G, through individual tubes. The available head across these tubes, P_{av}, therefore decreases with increasing flow rate (Eq. 12-2) (Fig. 12-2). Equation 12-3, which is the condition of steady circulation, is satisfied at the point of intersection of the two curves. It should be noted that depending on the resistance

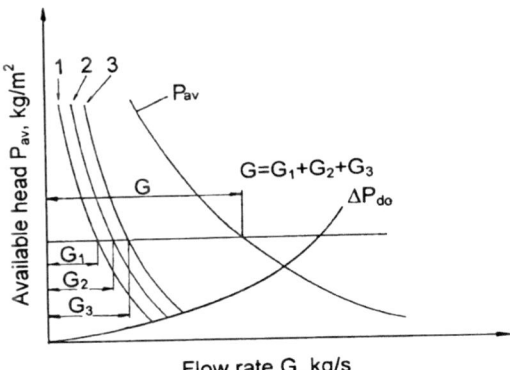

FIGURE 12-2. Variation of downcomer resistance (ΔP_{dc}) and available head (P_{av}) with circulation flow rate (G); such graphs are used for graphical solution of circulation

coefficients, the flow through individual tubes may be different; but the resultant available head, P_{av}, must be same for all of them, as it has to match the ΔP_{dc}, which is governed by the total flow rate.

12-1-3 Circulation Ratio

The *circulation ratio* (K) is defined as the ratio of water passing through the riser (or evaporator tube in natural circulation boiler) and the steam generated in it.

$$K = \frac{G}{G''} \qquad (12\text{-}4)$$

where G is total flow rate of fluid in the circuit, and G'' is the steam leaving the circuit.

The circulation rate of a circuit is not known in advance. So the calculation is carried out for a number of assumed values of fluid flow rate, G, and the corresponding water velocity in the circuit, W_o, is assumed. For each assumed flow rate the resistance in the downcomer (ΔP_{dc}), riser (ΔP_r), and the corresponding motive head are calculated. The head available calculated from Eq. (12-2) is plotted against the assumed flow rates G (Fig. 12-2). The downcomer resistance, calculated in the same way, is also plotted against the assumed flow rate, G. The intersection of these two graphs gives the circulation rate that is likely to occur for the given thermal and mechanical conditions.

This procedure is illustrated by a worked-out example given at the end of this chapter. The equations for computation of system resistance and the motive head are given below.

12-1-4 Design Equations

The heat flux in a PC boiler being nonuniform, the heat absorbed by all riser tubes in the furnace is not the same. The standard method of calculation of boilers

TABLE 12-1. Coefficient of nonuniformity of heat flux distribution (CKTI).

Type of furnace	Location of furnace	Coefficient of nonuniformity
Opposed jet & tangential firing	All walls along perimeter	1.0
Front firing	Side walls	1.0
Front firing	Front wall	0.8–1.0
Front firing	Rear wall	1.2
Upper quarter	All walls: PC firing	0.75
Upper quarter	All walls: Oil firing	0.6
Upper third	All walls: PC firing	0.8
Ceiling	PC firing	0.6
Ceiling	Oil firing	0.5

used in the former Soviet Union (CKTI) recommends the use of a coefficient of nonuniformity of heat absorption, Y, to express the heat absorption, q, by each part of the water wall in terms of the average heat flux on riser tubes, q_m.

$$q = Y q_m \tag{12-5}$$

The value of Y varies around the furnace. Table 12-1 gives typical values for a PC boiler. The heat flux on the water wall tubes is assumed to be uniform for initial design.

12-1-5 Specific Weight of Steam–Water Mixture

The total water flow rate, G, at the entrance section (also known as the economizer section) of a riser eventually transforms into a mixture of steam flow, G'' and water flow G' in the downstream evaporative section of the riser. So, from mass balance we can write

$$G = G' + G'' \text{ kg/s}$$
$$\text{or} \quad W_o A \gamma' = W_o' A \gamma' + W_o'' A \gamma'' \text{ m}^3/\text{s}$$
$$W_o = W_o' + W_o''(\gamma''/\gamma') \text{ m/s} \tag{12-6}$$

where W_o, W_o', and W_o'' are the reduced velocities of water in the economizer section, water in the evaporator section, and steam in the evaporator section of a tube, respectively. A is the flow cross section of the riser tubes. The specific weights of water and steam are γ' and γ'', respectively.

From Eq. (12-6) we can write β, the volume fraction of steam in the mixture, as

$$\beta = \frac{W_o'' \cdot A}{A \cdot W_o'' + A \cdot W_o'} = \frac{W_o''}{W_o'' + W_o'} = \frac{1}{\dfrac{W_o}{W_o''} + 1 - \dfrac{\gamma''}{\gamma'}} \tag{12-7}$$

The total flow cross section of the tube is A, but only a fraction (ϕ) of it is occupied by steam. The rest is occupied by saturated water. So the actual velocity of steam and water will be:

For steam

$$W_s = \frac{W_o''}{\phi} \text{ m/s;} \qquad \text{For water, } W_w = \frac{W_o'}{1 - \phi} \text{ m/s}$$

It is very difficult to determine ϕ, the portion of the flow area occupied by the steam. CKTI (Perkov, 1965) carried out extensive measurements on ϕ. On the basis of their measurement they produced a set of graphs (Figs. 12-3a,b, 12-4, and 12-5) that shows the dependence of ϕ on the ratio of $(W_{o,av}''/W_o)$, (W_o^2/d), pressure and the angle of inclination of tubes. The reduced steam velocity

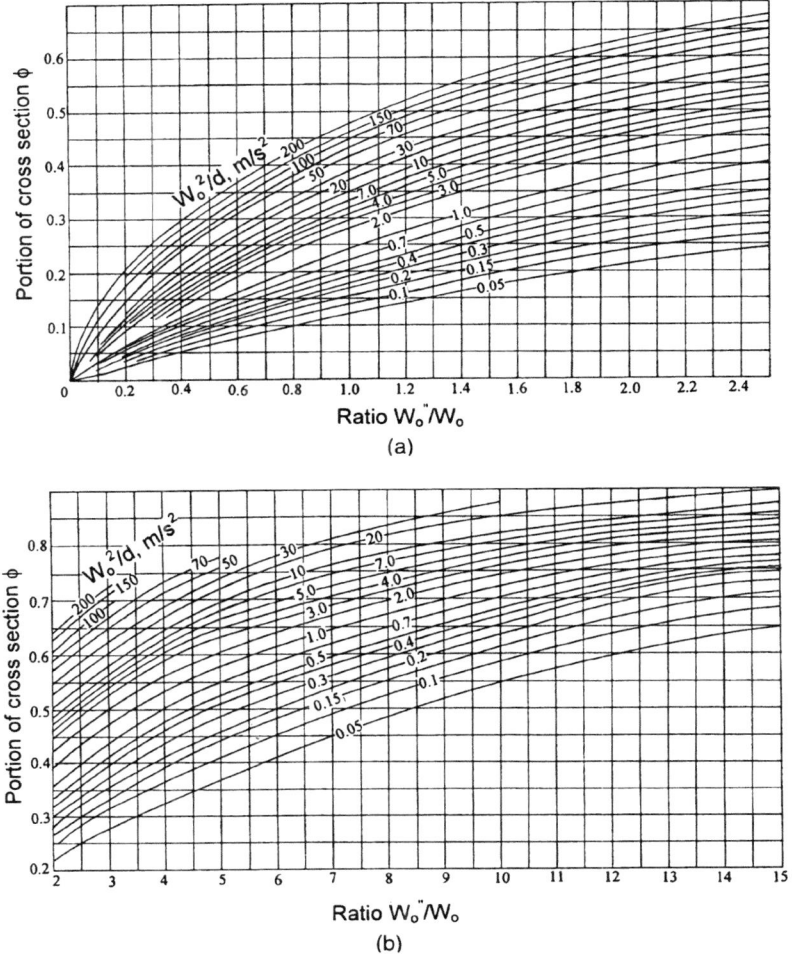

(a)

(b)

FIGURE 12-3. Chart by CKTI for determination of the portion of cross section occupied by steam in vertical tubes, $(P = 32$ atm abs. and W_o''/W_o from 0 to 15) (Perkov, 1965)

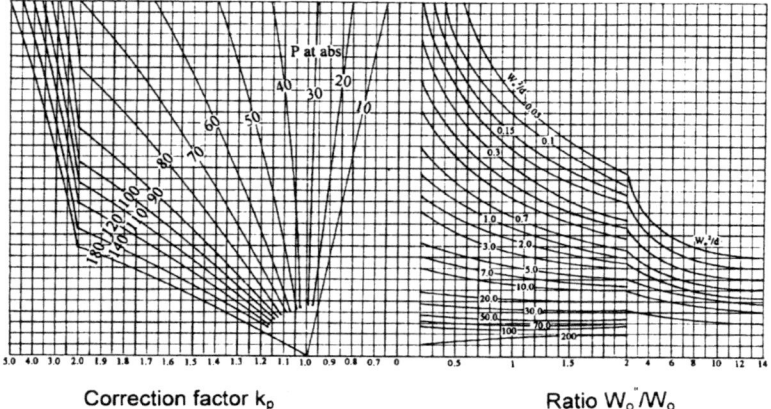

Correction factor k_p Ratio W_o''/W_o

FIGURE 12-4. Chart by CKTI for determination of correction factor for pressure (Perkov, 1965)

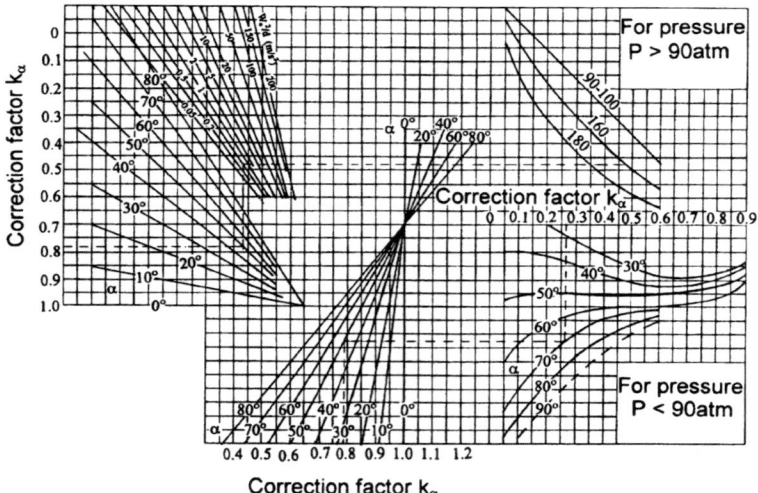

Correction factor k_α

FIGURE 12-5. Chart by CKTI for definition of correction factor for pressure (Perkov, 1965)

$(W_{o,av}'')$ averaged over a length of evaporator section of the tube may be approximated as

$$W_{o,av}'' = \frac{G_1'' + G_2''}{2 \cdot A \cdot \gamma''} \tag{12-8}$$

where G_1'' and G_2'' are the flow rates of steam at the entrance and exit of the section in question.

If γ_m is the specific weight of the mixture, then for a small length of the tube we can write

$$A \, \Delta L \cdot \gamma_m = A \, \Delta L \phi \cdot \gamma'' + A \, \Delta L (1 - \phi) \cdot \gamma'$$
$$\gamma_m = \phi \cdot \gamma'' + (1 - \phi) \cdot \gamma' \tag{12-9}$$

The steam velocity is higher than water velocity. So, comparing it with Eqs. (12-6) and (12-7), we note that ϕ is generally less than β. For simplification, if we take $\phi = \beta$, Eq. (12-9) can be simplified after substituting Eq. (12-7) for ϕ in Eq. (12-9) as follows:

$$\gamma_m = \frac{\gamma'}{1 + \dfrac{W_o''}{W_o'}\left(1 - \dfrac{\gamma''}{\gamma'}\right)} \tag{12-10}$$

Since mass flow rate through any section is constant, the mean velocity of the mixture may be found as

$$W_m = \frac{W_o'' \cdot \gamma'}{\gamma_m} = W_o + W_o''\left(1 - \frac{\gamma''}{\gamma'}\right) \tag{12-11}$$

The buoyancy is the weight of displaced liquid. So, the motive head, which is essentially a buoyancy force can be determined by using the volume fraction of the tube filled steam. Thus the motive head of an evaporator section of vertical height H_s is

$$P_m = H_s \phi (\gamma' - \gamma'') \tag{12-12}$$

The actual volume fraction of steam needs to be corrected for pressure (K_p) and velocity ratio (K_α).

In a steady state the sum of all resistance in the circuit would be equal to the difference in the hydrostatic pressure between the downcomer and the riser. The pressure balance around the downcomer-riser-drum loop can be written as

Hydrostatic pressure − Flow resistance = Hydrostatic + Flow resistance
+ Resistance in downcomer in
downcomer + Pressure in riser
+ Pressure in drum

$$\rho_l g Z - \Delta P_d = \sum \rho_m(z) g \, dz + \Delta P_r + \Delta P_{drum}$$

where Z is the total height of the water wall and $\rho_m(z)$ is the local density of steam–water mixture at a height z.

The water–steam mixture flows through a large number of parallel tubes. These tubes are not necessarily of identical geometry. Since all of them are subjected to the same pressure difference, the flow through individual tubes adjusts themselves to match the imposed pressure difference. This feature is used to estimate the flow through the tubes.

12-1-6 Calculation of Pressure Drop in Tubes

The pressure drop, ΔP, through a tube comprises several components: friction (ΔP_{fric}), entrance loss (ΔP_{en}), exit losses (ΔP_{ex}), fitting losses (ΔP_{fit}), and hydrostatic (ΔP_{st}).

$$\Delta P = \Delta P_{fric} + \Delta P_{en} + \Delta P_{ex} + \Delta P_{fit} + \Delta P_{st} \qquad (12\text{-}13)$$

The total pressure drop through a section carrying water is expressed as

$$\Delta P_w = k \frac{W_o^2}{2 \cdot g} \gamma' \ \text{N/m}^2 \qquad (12\text{-}14)$$

For steam–water mixture, after using Eq. (12-10), the total pressure drop is written as

$$\Delta P_{mix} = k \frac{W_{m,av^2}}{2g} \cdot \gamma_m = \frac{W_o^2}{2g} \cdot \gamma' \left[1 + \frac{W_o''}{W_o} \left(1 - \frac{\gamma''}{\gamma'} \right) \right] \text{N/m}^2 \qquad (12\text{-}15)$$

If the pressure drop is due to friction alone the loss coefficient k is

$$k = \frac{f \cdot l}{d} \qquad (12\text{-}16)$$

where the friction factor, f, is a function of the flow Reynolds number and relative surface roughness. In most boiler tubes the flow Reynolds number is in a fully turbulent regime where the friction factor, f, is defined by the ratio of roughness and tuber diameter. For roughness 0.1 mm, the following values of friction coefficient may be used (Table 12-2). For fittings, entrances, and exit sections the following values of k may be used (Table 12.3).

TABLE 12-2. Friction coefficient, f, of Eq. (12-14) for fully developed rough zone flow for surface projection 0.1 mm.

Diameter of tube (d) mm	40	50	64	76.3	90	100
Friction coefficient (f)	0.025	0.0235	0.0224	0.0212	0.0198	0.019

TABLE 12-3. Loss coefficient (k) for fittings and flow sections.

Types of fittings or flow situation	Loss factor, k
Entrance to tubes from drum	0.5
Exit from header to tubes	1.0
Exit from tubes into drum	1.0
Exit from tubes into header	3.0
Elbow of curve less than 30°	0.0
Elbow of curve 30–70°	0.1
Elbow of curve more than 70°	0.2

Sometimes the steam and water mixture rises above the drum level. This offers an additional resistance, ΔP_{elev}, for a height of H_{el}:

$$\Delta P_{elev} = H_{el}(\gamma' - \gamma'')(1 - \phi) \text{ N/m}^2 \qquad (12\text{-}17)$$

12-1-7 Heat Transfer in Steam–Water Mixture in Riser

One major concern of circulation is the protection of tube walls from failures owing to excessive heating. To understand this one needs to examine the basic boiling process and the heat transfer process.

Boiling Heat Transfer

In subcooled boiling, the heat transfer is sufficiently high. So it maintains the wall temperature (T_w) close to that of the water. The temperature difference between the wall and saturation temperature of water depends on the heat flux, q_f.

$$T_w - T_{sat} = 25(q_f)^{0.25} e^{-P/62} \text{ °C} \qquad (12\text{-}18)$$

where q_f is the heat flux in MW/m^2 and P is pressure in bar absolute. (Jens & Lottes, 1951).

Eq. (12-18) may be used in the nucleate boiling regime within a permissible margin of error (Kakac, 1991).

The boiling crisis, which occurs in film boiling, may lead to tube failure through either of two mechanisms:

1. Local metal temperature rises to levels where its creep life is rapidly exceeded.
2. Local concentration of dissolved solids at the steam/water interface leads to rapid corrosion.

If the steam fraction is high, the water travels as a thin annular layer on the wall and the steam flows in the tube core with fine water droplets suspended in it (Fig. 12-1b). The boiling crisis occurs here as soon as the heat flux is sufficient for the annular layer to tear off or dry out. The metal temperature increases immediately, but the drop in heat transfer coefficient is not as severe as in the previous case as convective cooling of the wall is higher than conduction cooling in that case. The critical heat flux is lower in the latter case of dry out.

12-1-8 Critical Heat Flux and Critical Steam Fraction

Critical heat flux depends on the nature of the liquid, flow conditions, subcooling and the steam mass or void fraction (x), geometry of the heated surface, material of tube, and tube surface finish. Figure 12-6 shows that in the range of film boiling the critical heat flux (q_{cr}) drops slowly with increasing steam mass fraction (x), but it drops sharply in the dry-out regime. Finally it flattens out in the burn-out region. One of many correlations for the critical heat flux is discussed below.

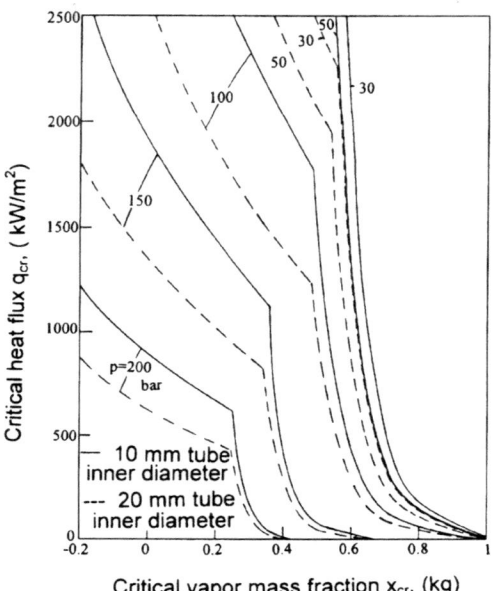

Critical vapor mass fraction x_{cr}, (kg)

FIGURE 12-6. Critical heat flux as a function of the critical steam fraction for a mass velocity 1000 kg/m² · s (VDI Heat Atlas, 1993)

a) Vertical Tubes

(1) Film Boiling

Within the following range of film boiling (low vapor fraction) Doroshuchuk et al. (1975) derived the following equations for the critical heat flux, q_{cr}:

$$q_{cr} = A_1 \cdot B_1 \qquad (12\text{-}19)$$

where

$$A_1 = 10^3 \left[10.3 - 17.5 \left(\frac{P}{P_c} \right) + 8 \left(\frac{P}{P_c} \right)^2 \right] \left(\frac{0.008}{d} \right)^{0.5}$$

and

$$B_1 = \left[\left(\frac{m}{1000} \right)^{[0.68.(P/P_c)-1.2x-0.3]} \right] \cdot e^{(-1.5x)}$$

It is valid in the range: 29 < pressure, (P) < 196 bar; 500 < mass flux (m) < 5000 kg/m²s; 0 < subcooling at inlet (ΔT_{sub}) < 75 K; 4 < tube diameter (d) < 25 mm. P is the pressure, P_c is the critical pressure in bar, m is the mass velocity in kg/m² · s; and x is the mass fraction of steam.

The critical vapor mass fraction, x_{cr}, was found as

$$x_{cr} = \frac{\ln\left(\dfrac{m}{1000}\right)\left(0.68\dfrac{P}{P_c} - 0.3\right) - \ln(q_{cr}) + \ln(A_1)}{1.2\ln\left(\dfrac{m}{1000}\right) + 1.5} \tag{12-20}$$

These were checked with more than 3000 data points; standard deviation was 16%.

(2) Dry-Out Range

For high steam mass fraction, where dry out occurs, the following equation of Kon'kov (1965) is recommended by VDI Atlas (1993).

For $4.9 < P < 29.4$ bar:

critical heat flux, $q_{cr} = 1.8447 \cdot 10^8 \cdot \exp(0.1372\ P)$
$$\cdot [x^8 \cdot m^{2.664} \cdot (d \cdot 1000)^{0.56}]^{-1}$$

critical steam fraction, $x_{cr} = 10.795\exp(0.01715\ P)$
$$\cdot [q^{0.125}m^{0.333} \cdot (d \cdot 1000)^{0.07}]^{-1}$$

For $29.4 < P < 98.0$ bar:

$$q_{cr} = 2.0048 \cdot 10^{10}[\exp(0.0204\ P) \cdot x^8 \cdot m^{2.664} \cdot (d \cdot 1000)^{0.56}]^{-1}$$
$$x_{cr} = 19.398[\exp(0.00255\ P) \cdot q^{0.125}m^{0.333} \cdot (d \cdot 1000)^{0.07}]^{-1}$$

For $98.0 < P < 196$ bar:

$$q_{cr} = 1.1853 \cdot 10^{12}[\exp(0.0636\ P) \cdot x^8 \cdot m^{2.664} \cdot (d \cdot 1000)^{0.56}]^{-1}$$
$$x_{cr} = 32.302[\exp(0.00795\ P) \cdot q^{0.125}m^{0.333} \cdot (d \cdot 1000)^{0.07}]^{-1} \tag{12-21}$$

The range of validity is $200 < m < 5000$ kg/m^{2}s and $4 < d < 32$ mm and stable flow.

For lower pressure, $P < 5$ bar, lower mass flow, < 300 kg/m^{2}s, and steam mass fraction $< -0.2 < x_{in} < 0$, the equation by Alad'yev et al. (1969) is recommended by VDI Heat Atlas (1993). The steam fraction, x, can be negative for subcooled water as it is given as $[(H_{inlet} - H_{sat})/\Delta H_{vaporisation}]$.

$$q_{cr} = 460\left[\frac{m \cdot d}{l}\right]^{0.8} \cdot (1 - 2 \cdot x_{in}) \tag{12-22}$$

where x_{in} is the mass fraction of steam at the inlet of the tube, l is the length of tube, and d is the inside tube diameter.

Figure (12-6) shows the variation of critical heat flux with critical mass fraction as computed from the above equation. For design purposes one should take the lower value of critical heat flux or steam fraction from either Eq. (12-19), (12-20), or (12-21).

b) Horizontal or Inclined Tubes

The evaporator tubes are at times inclined at an angle. Then the lighter steam vapor collects preferentially in the upper part of the tube circumference. So the steam fraction in the upper side is higher than that on the underside of the tube. Thus, in an inclined tube the boiling crisis occurs in the upper and underside of the tube at a steam fraction lower or higher, respectively, than that for vertical position. They are found as

$$x_{cr,up} = x_{cr} - (\Delta x_{cr}/2) \quad \text{for the crest of tube}$$

$$x_{cr,low} = x_{cr} + (\Delta x_{cr}/2) \quad \text{for the underside of tube} \qquad (12\text{-}23)$$

where x_{cr} is the critical steam fraction for vertical tubes as found from Eqs. (12-20) and (12-21).

The difference in critical steam fraction between the upper and lower part of a tube inclined at an angle, θ, is a function of inertia of the fluid and the buoyancy force. It may be estimated from the following equation (Kefer, 1989).

$$\Delta x_{cr} = \frac{16}{(2 + Fr')^2} \qquad (12\text{-}24)$$

where the modified Froud number, Fr', is

$$\frac{x_{cr} \cdot m}{\sqrt{g \cdot d \cdot \rho_s (\rho_l - \rho_s) \cos \theta}}$$

Here, ρ_s and ρ_l are the saturated density of the steam and water, respectively. This equation is valid for $25 < P < 200\,\text{bar}$; $500 < m < 2500\,\text{kg/m}^2\text{s}$; and $200 < q < 600\,\text{kW/m}^2$.

EXAMPLE:

Estimate the critical steam mass fraction at the upper and underside of a 20 mm diameter tube inclined at 45° through which 1000 kg/m²s of water flows at a pressure 150 bar. It receives a uniform heat flux of 300 kW/m².

SOLUTION:

We can find the critical steam fraction, x_{cr}, for the given heat flux q_f (300 kW/m²) either from the eqs. (12-21) or from the graph (Fig. 12-6) as 0.4. The densities of saturated steam and water at 150 bar are found from the steam table as $\rho_s = 96.7\,\text{kg/m}^3$ and $\rho_l = 603.1\,\text{kg/m}^3$.

$$Fr' = \frac{x_{cr} \cdot m}{\sqrt{g \cdot d \cdot \rho_s (\rho_l - \rho_s) \cos \theta}} = \frac{0.4 \cdot 1000}{\sqrt{9.81 \cdot 0.02 \cdot 96.7(603.1 - 96.7) \cos 45}}$$
$$= 4.85$$

From Eq. (12-24),

$$\Delta x_{cr} = \frac{16}{(2 + Fr')^2} = \frac{16}{(2 + 4.85)^2} = 0.341$$

Now the critical steam fraction in the upper and lower side of the tube is found from Eq. (12-23) as

$$x_{cr,up} = x_{cr} - (\Delta x_{cr}/2) = 0.4 - (0.341/2) = 0.230 \text{ for the crest of tube}$$
$$x_{cr,low} = x_{cr} + (\Delta x_{cr}/2) = 0.4 + (0.341/2) = 0.57 \text{ for the underside of tube}$$

It is instructive to note that if the same calculation is carried out for half the heat flux and for a horizontal tube, the critical steam fraction at the upper and lower side of the tube will be at 0.239 and 0.627, respectively.

c) Nonuniform Heating

In most boilers evaporator tubes receive heat on only one side. This makes the heat flux nonuniform around the tube circumference. The maximum heat flux for the onset of boiling crisis is greater than that for the case of uniform heating described earlier. Butterworth (1971) postulated that while for the nonuniformity of heat flux ($q_{crest}/q_{circum\ average}$) increases from 1.0 to 2.0, the ratio of critical heat flux for the crest of the tube and circumference found for a uniform heated tube varies from 1 to 1.6.

12-2 Calculations for Simple and Complex Tube Circuits

There are different methods for circulation calculations. These methods vary in complexity and therefore precision. The foregoing section on heat flux and pressure drop can be used to carry out a fairly comprehensive computation of circulation. However, the calculation could be too large to show here. An alternative procedure developed by the CKTI in the former Soviet Union is presented here firstly because this spreadsheet type method is relatively simple, and also because this method has been tested in a large number of boilers in the former Soviet Union.

This method starts out by assuming several velocities for the circulating water in the riser tubes. For each of these flow rates the available pressure head in the riser circuit is obtained by subtracting hydraulic resistance from the motive head. The available head in the downcomer is also found in the same way by subtracting the flow resistance through it. Under natural circulation the available head in the downcomer and the riser must match each other. Thus the circulation rate at the prevalent condition is found graphically. This procedure is shown in an worked-out sample calculation later in this chapter.

12-2-1 Checking the Reliability of Natural Circulation

In a natural boiler large number of tubes run in parallel. Nonuniform heat flux on the tubes may lead to circulation crisis. For example, if a tube receives less heat than average tubes or if its hydraulic resistance is much greater than average tubes, the flow through the tube will reduce. In an extreme situation the amount of water flow at the bottom of the tube may be less than the steam generated in the tube. Thus, instead of steam–water mixture only steam will exit from this tube. If the tube ends in the steam space in the drum, a free water surface forms. This is the case of *stagnation circulation*. If the tube is connected to the water space of the drum, water from the drum flows into the tube, slowly filling it. Thus the tube acts as a downcomer, which is not as bad as it sounds. However, owing to the change in heat flux in the furnace, it may move from downcomer to riser operation with a higher probability of a stagnation circulation in between.

To avoid the stagnation circulation the hydraulic resistance of the downcomer should be less than P_{st}.

$$P_{av} < 0.9 \Delta P_{st} \text{ (tubes are connected to the steam space)} \qquad (12\text{-}25)$$

$$P_{av} < 0.9[(P_{st} - h(\gamma_w - \gamma_s)] \text{ (tubes are connected to the water space)} \qquad (12\text{-}26)$$

12-2-2 Example of Natural Circulation Calculation

The calculation procedure for natural circulation is very complex and involves iteration. A worked-out example shown in Section 12-5 explains the basic procedure of circulation calculation.

12-3 Two-Phase Flow Resistance

A more comprehensive method of calculation of pressure drops through the tube circuit is presented below. Symbols in the Appendix carry the same meaning as that given in the Nomenclature except those defined here.

The pressure drop through any section may be written as

$$\Delta P = \Delta P_{fric} + \Delta P_{st} + \Delta P_{acl} \qquad (12\text{-}27)$$

In the riser evaporator tube a part of the tube carries pure subcooled water, and the rest of the tube carries steam–water mixture in which the steam mass fraction, x, varies along its length. VDI (VDI Heat Atlas, 1993) recommends the following equations for computations of different components of the pressure drop.

12-3-1 Friction

The hydrodynamic resistance in a tube may be due to friction in a straight length of tube or eddy losses in fittings like a bend, elbow, etc. The losses through the fittings may be expressed in equivalent straight length of the tube. Values of equivalent tube length are found in textbooks on fluid dynamics.

In the case of subcooled water the frictional pressure drop is given as

$$\Delta P_{fric}^{subcooled} = f \frac{m^2}{2\rho_w d} \Delta l \qquad (12\text{-}28)$$

The friction coefficient, f, should be taken from the Moody friction diagram for the given tube diameter, d, its surface roughness, e, and the flow Reynolds number,

Re. For numerical calculation one could use an appropriate empirical relation valid for the given flow regime. For example, the following equation is valid for Re > 2300.

$$\frac{1}{\sqrt{f}} = -2 \log \left[\frac{e}{3.7d} + \frac{2.51}{Re\sqrt{f}} \right] \tag{12-29}$$

For steam–water mixture appropriate values of the density and correction need to be used. Here, two situations may arise. If the mixture has a core-annulus structure as in Figure 12-2b, the friction of each phase needs to be found separately. Details are given in VDI Heat Atlas (1993). If the steam is dispersed uniformly in water (Fig. 12-1e), we use the following equation:

$$\Delta P_{fric}^{2\text{-}phase} = f \frac{m^2}{2\rho_w d} \Delta l \left[1 + x \left(\frac{\rho_w}{\rho_s} - 1 \right) \right] \left[1 - x \left(\frac{\rho_w}{\rho_s} - 1 \right) (K - 1) \right] \tag{12-30}$$

The friction factor f may be calculated from Eq. (12-29) by substituting an average value of the kinematic viscosity, $v_w[1 - x(1 - v_s/v_w)]$ for the kinematic viscosity, v_w, of water alone. The coefficient K depends on the volume ratio of steam and water, β, which is given as

$$\beta = [(1 - x)\rho_s/(x \cdot \rho_w)]$$
$$K = 1 + 0.09\beta \quad \beta < 0.4$$
$$\frac{1}{K} = 1 - \frac{2.97\beta^{-0.66} + 1}{6.(1.83\beta^{-0.66} + 1)(3.43\beta^{-0.66} + 1)} \quad \beta > 0.4$$

If the steam flows at very high velocity in the core and water creeps along the wall (Fig. 12-1b), we use a separate equation.

$$\Delta P_{fric}^{2\text{-}phase} = f \frac{m^2 \cdot x^2}{2\rho_s d} \Delta l \left[\frac{1}{1 - (1 - E)\eta_F - E\eta_E} \right]^2 \tag{12-31}$$

The friction factor may be found from a Moody diagram as above, or one could use the following correlation of Prandtl:

$$\frac{1}{\sqrt{f}} = 2 \log \left[\frac{m \cdot x \cdot d}{v_s} \sqrt{f} \right] - 0.8 \tag{12-32}$$

The value of E in Eq. (12-31) lies between 0 and 1. The other parameters of this equation are found as follows:

$$\eta_F = 1 - \left(1 + \frac{(1 - x)\rho_s}{x\varepsilon\rho_w} \right)^{-1.19}$$

and

$$\eta_E = \left(1 + \frac{6.67}{\left(\dfrac{1-x}{x}\right)^{0.45}(1+3x^4)\left(\dfrac{v_w}{v_s}-1\right)^{0.25}}\right)^{-1}$$

$$\varepsilon^{-3} = \varepsilon_1^{-3} + \varepsilon_2^{-3}$$

$$\varepsilon_2 = 9.1\varphi$$

$$\varphi = [(1-x)/x](Re_w Fr_w)^{-1/6}(\rho_w/\rho_s)^{-0.9}(v_w/v_s)^{-0.5}$$

where $Re_w = m.d.(1-x)/v_w$ and $Fr_w = m^2(1-x)^2/(\rho_w^2 g \cdot d)$.

$\varepsilon_1 = 1.71\,\varphi^{0.2}(1/x-1)^{0.15}(\rho_s/\rho_w)^{0.5}(v_s/v_w)^{0.1}$ for $e/d < 5.10^{-4}$

$\varepsilon_2 = \varepsilon_1(5 \cdot 10^{-4}d/e)^{0.13}$ for $e/d > 5.10^{-4}$

It should be noted that since the steam fraction x varies along the height of the riser, Eq. (12-31) should be integrated over the height to get the overall pressure drop.

12-3-2 Acceleration

The pressure drop due to acceleration is found by integrating over the section over which the steam fraction changes from $x1$ to $x2$.

$$\Delta P_{accl} = m^2 \left| \frac{x^2}{\alpha \rho_s} + \frac{(1-x)^2}{(1-\alpha)\rho_w} \right|_{x1}^{x2} \qquad (12\text{-}33)$$

12-3-3 Hydrostatic Head

The hydrostatic head is given as

$$\Delta P_{static} = [\rho_w(1-\alpha) + \rho_s\alpha]g \sin\theta \cdot \Delta l \qquad (12\text{-}34)$$

where α is the local volume fraction of steam and θ is the angle of inclination of the tube.

12-4 Height of Economizer Section in the Riser

The water enters the riser tubes at a temperature below the saturation temperature of water at that point. So the steam generation does not necessarily start right at the lowest heated point of the riser. Water is heated up to its saturation temperature as it climbs through the riser before the steam generation starts. Since this part of the riser tube provides only the sensible heat to the water like the economizer does, it may be called the "economizer section of the riser."

The pressure, at the top of the riser where boiling starts P_{ec}, is higher than that in the drum owing to the hydrostatic pressure. This pressure less the frictional resistance in the downcomer is added to the drum pressure to get the local pressure in the downcomer. Thus the rise in pressure of water above the drum pressure, ΔP:

$$\Delta P = P_{ec} - P_{drum} = H\gamma'_{dc} - H_{ec}\gamma'_{ec} - \Delta P_{dc} \qquad (12\text{-}35)$$

where H = height of water level in the drum above the point from which
                 the heating of the riser tube begins, m

γ'_{dc} and γ'_{ec} = specific weights of water in the downcomer and in the lower
                 section of the heated riser tube, kg/m^3

H_{ec} = height of the section of the riser from the starting point of
                 heating of the tube up to the point at which boiling starts, m

ΔP_{dc} = hydraulic resistance of the downcomer and headers up to the
                 point where boiling starts, N/m^2

The pressure drop in downcomer, ΔP_{dc}, is due to friction and drum entrance loss.

$$\Delta P_{dc} = [f(H + H_{ec})/d + \zeta_{en}]W^2\gamma/(2g) \qquad (12\text{-}36)$$

where w is the velocity of water in downcomer and ζ_{en} is the entrance loss coefficient.

If $[dh/dp]$ is the rate of change of saturated enthalpy with change in pressure, the rise in saturated enthalpy owing to the increase in pressure at that pressure is $[dh/dp] \cdot \Delta P$.

The saturation enthalpy at the point of boiling, h', is therefore

$$h' = H_{sat} + dh/dp \cdot \Delta P_{dc} \qquad (12\text{-}37)$$

Feed water enters the drum below the saturation temperature with a total enthalpy h_1. The drum receives boiling water $(G - G')$ with an enthalpy $(G - G')h'$. After the feed water has been mixed with boiler water, the mixture (G) leaves the drum to enter the downcomer with an enthalpy $h_m G$. So, neglecting losses, one could write the heat balance as

$$h_1 \cdot G' + h' \cdot (G - G') = h_m \cdot G$$

which gives the enthalpy, h_m, of the water leaving the drum as

$$h_m = \frac{(K - 1) \cdot h' + h_1}{K} \qquad (12\text{-}38)$$

where h_1 = enthalpy of water entering drum, kJ/kg
                 K = circulation ratio (G/G')

The amount of heat the water must absorb to reach the boiling or saturation condition is found from Eq. (12-38) as

$$\Delta h_{ud} = h' - h_m = (h' - h_1)/K \qquad (12\text{-}39)$$

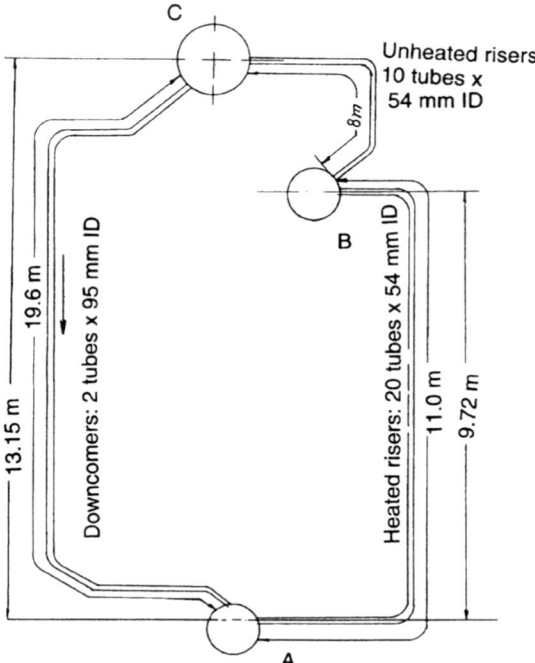

FIGURE 12-7. Diagram of a steam–water circuit in a natural circulation boiler

This term is called lack of heating or saturation shortfall. Since the saturation enthalpy depends on pressure, one must find the static pressure at the point in the riser where water is to reach saturation

If the heat flow to the lower section of height H_h is Q, and the flow rate of recirculating water is G, we have heat absorbed in the economizer section of riser = saturation short fall.

$$Q = \Delta h_{ud} \cdot G \qquad (12\text{-}40)$$

The height, H_h, can be found by substituting the above equations into Eq. (12-40).

12-5 Worked-Out Example

Determine the circulation velocity and check the reliability of circulation in the front water wall of a steam generator whose details are given in the table below.

General dimensions of the steam generator is shown in Figure 12-7. Subcooled water from the drum (C) descends through the downcomers (CA). During this flow it overcomes hydraulic resistance in the exit from the drum, bends, and elbows and then while entering the bottom header (A). From the header the water is split into several riser tubes. Although this part of the tube receives furnace heat steam is not immediately formed in it. Water travels up to a certain height (E) before it reaches saturation temperature at the pressure existing at that point. So, this part of the riser tube (A–E) should be considered a part of the downcomer, and when calculating the motive head only the vertical distance of the drum water above this level should be considered. As the water travels up the riser tubes (E–B), progressive amounts are converted into steam. As a result the specific weight of the mixture and therefore the velocity continuously changes along the height. To make the matter more complex the furnace heat flux which is responsible for evaporation of water also changes along the height. In the present example the furnace is divided into two zones with two heat fluxes. After leaving the furnace the riser tubes join together to form a smaller number of tubes and they pass through unheated space (B–C) terminating into the drum.

SOLUTION

The problem is solved using the following spreadsheet-type calculations shown below:

An example of circulation calculation in a natural circulation boiler (Fig. 12-8)

Parameter	Symbol	Units	Equation	Calculation	Value
Diameter of tubes (inside)	dhr	mm		Given	54
Number of tubes	n			Given	20
Pitch		mm		Given	80
Width of the section		mm		Given	1520
Total heated height of water wall	Hr	m		Given	9.72
Total area of heating surface	At	m²	$(n-1)(p/1000)Hr$	$(20-1)*80/1000*9.72$	14.8
Total area of cross section of tubes	Ar	m²	$n(3.14/4)(d/1000)^2$	$20*3.14*.054^2/4$	0.0458
Total length of risers	lr	m		Given	11.0
Length of unheated riser	luh			Given	1.3
Number of 90 bed				Given	1.0
Total height of circuit from bottom header to the water level in the drum	H	m		Given	13.15
Unheated part of riser					
Unheated tubes from risers to drum			(Risers are connected with tees after coming our of the furnace)		
Diameter of tubes (inside)	dcr	mm		Given	54.0
Total number of tubes				Given	10.0
Average length of tubes		m		Given	8.0
Number of bends		90		Given	2.0
Total flow cross section area	Aur	m²	$10*3.14*$ $(54/1000)^2*/4$		0.0229
Average inclination angle		degree		Given	60
Height of unheated riser	Hur			Given	3.4

Parameter	Symbol	Units	Equation	Calculation	Value
Downcomer					
Diameter of tubes (inside)		mm		Given	95.0
Number of tubes				Given	2.0
Total height of downcomers		m		Given	19.6
Number of elbows		90		Given	2.0
		60		Given	3.0
Flow cross section area		m^2		$2*3.14*(95/1000)^2/4$	0.0142
Data for Thermal Calculations					
Pressure in the drum	P	atm		Given	36.0
Saturation temperature	Ts	C		From steam table	243.0
Enthalpy of water at saturation	h'	kcal/kg		From steam table	251.3
Specific weight of water	row	kg/m^3		From steam table	810.0
Specific weight of saturated steam	ros	kg/m^3		From steam table	17.7
Latent heat of evaporation	L	kcal/kg		From steam table	418.1
Enthalpy of saturated steam	h''	kcal/kg		From steam table	669.4
Temperature of water entering drum	T_1	C			230.0
Enthalpy of water entering drum	h	kcal/kg		From steam table	236.5
Ratio of increase of enthalpy of water to increase of pressure at 36 atm	dh/dP	kcal/atm		From steam table	1.8
Mean heat flux on riser heating surface	qm	$kcal/m^2h$			85,000.0
Total heat absorption of circuit	Q	kcal/s	$qm*Ar$	$85000*14.8/3600$	348.8
To take into account the nonuniform heat flux in the furnace the furnace is divided into two zones with two average heat fluxes					
Coefficient of heat absorption nonuniformity	y			upper quarter	0.75
Specific heat absorption of upper water wall	q_{up}	kcal/s	$q_m^* y$	$348.8*.75$	261.6
Total heat absorption in upper quarter	Q_{up}		$q_{up}/4$	$261.6/4$	65.4
Heat absorption per 1 m of upper quarter				$65.4/(9.72/4)$	26.9
Heat absorption in bottom part of wall	Q_{bot}	kcal/s	$Q - q_{up}$	$348.8-65.4$	283.4
Absorption per 1 m of bottom part	q_{bot}		Q_{bot}/H_{bot}	$283.4/(9.72*3/4)$	38.9
Circulation Calculations					
Steam generated in the whole water wall	G''	kg/s	$Q/(h''-h)$	$343.6/(669.4-236.5)$	0.806
Steam in upper section of the water wall	G''_{up}	kg/s	$Q_{bot}/(h''-h')$	$65.4/(669.4-251)$	0.156
Steam generated in bottom section	G''_{bot}	kg/s	$G''-G''_{up}$	$0.803-0.2$	0.649

Parameter	Symbol	Units	Equation	Calculation	Value	Trial 1	Trial 2	Trial 3
TRIAL FOR CIRCULATION								
Assume velocity of circulation	W_o	m/s			0.8	1	1.2	0.7
Total volume of water through the circuit	V	m³/s	$W_o^* A_t$	.0458*.8	0.037	0.046	0.055	0.032
Total mass + B88 of water through the circuit	G	kg/s		0.037*810	29.7	37.1	44.5	26.0
Circulation ratio	K		G/G″	29/7/.834	36.8	46.0	55.2	32.2
Heat required to bring water to saturation	h_{ud}	kcal/kg	$(h' - h)/K$	(251.3–236.5)/35.6	0.40	0.32	0.27	0.46
Velocity of water in downcomer	W_{dc}	m/s	V/Ad	.037/.0142	2.58	3.23	3.88	2.26
Hydraulic resistance								
Downcomer								
Entrance to drum	fen				0.5	0.5	0.5	0.5
Exit from drum to header	fex				3.0	3.0	3.0	3.0
Elbows	fel			0.2*2 + 0.1*3	0.7	0.7	0.7	0.7
Friction (from table for fully rough flow)	fr		f·l/d	0.0194*19.6/.095	4.0	4.0	4.0	4.0
Total resistance factors	K_{dc}		$f_r + f_{en} + f_{ex} + f_{el}$	.5 + 3.0 + .7 + 4.0	8.2	8.2	8.2	8.2
Pressure drop in downcomers	DP_{dc}	kg/m²	$K_{dc}^* W_{dc}^* 2.row/(2g)$	8.2*2.58^2*810/19.62	2262.5	3535.2	5090.7	1732.2
Head drop in downcomer		m	DP_{dc}/row	2262.5/810	2.8	4.4	6.3	2.1
Bottom section of riser								
Reduced Saturated enthalpy for downcomer level	D_{hp}	kcal/kg	$dh/dp^* row(H - DP_{dc})^* 9.81/101.320$		1.5	1.3	1.0	1.6
Height of economizer section in riser	H_{ec}	m	$(h_{ud} + D_{hp})^* G/q_{bot}$	(.4 + 1.5)*29.7/38.9	1.46	1.53	1.45	1.38
Height of evaporative section in lower furnace	Hb	m	$3/4^* H_r - H_{ec}$	3/4*9.72 – 1.42	5.83	5.76	5.84	5.91
Reduced velocity of steam at exit of lower section	$W_{o.bot}''$	m/s	$G_{bot}''/(Ar^* ros)$	.649/(17.7*.0458)	0.80	0.80	0.80	0.80
Mean reduced velocity in bottom evaporator	$W_{o.av}''$	m/s	$(W_{o.bot}'' + 0)/2$	(0.8 + 0)/2	0.40	0.40	0.40	0.40
Ratio of reduced steam and circulation velocity	V_{ratio}		$W_{o.av}''/W_o$	.42/.8	0.50	0.40	0.33	0.57
Factor $W_o^* 2/d$	F_r			.8^2/.054	11.9	18.5	26.7	9.1
Value of FI from figure 7-6 against F_r & V_{ratio}	F_i			Figure 12-3a	0.22	0.19	0.18	0.24
Correction factor for pressure	K_p			Figure 12-4	1.0	1	1	0.99
Corrected motive head at bottom section	P_{mb}		$F_i^* K_p^* H_b^* (row - ros)$	.22*1.0*5.83* (810 – 17.7)	1016	867	833	1113
Resistance for riser economizer section								
Entrance	f_{en}			Table 12-3	1.0	1.0	1.0	1.0
Elbow	f_{el}			Table 12-3 for 90°	0.2	0.2	0.2	0.2
Tube (assuming fully rough turbulent flow)	f_r			Table 12-2	0.023	0.023	0.023	0.023
Total resistance factor	K_{bec}			1 + 0.2 + .023* (1.3 + 1.85)/.054	2.37	2.40	2.37	2.34
Resistance in economizer section	DP_{ec}	kg/m²	$K_{bec}^* W_o'^{} 2^* row/2g$	2.36*0.8^2/ (2*9.81)	62.7	99.1	140.8	47.3
Resistance for riser evaporator section								
Friction in evaporator section	K_{bg}		$fr^* H_b/d$	.0231*5.44/.054	2.49	2.49	2.49	2.49
Two phase resistance factor	K_{bm}		$row + W_{o.av}'' (row - ros)/W_o$	810 + 0.4* (810 – 17.7)/0.8	1206.8	1127.5	1074.6	1263.5
Resistance in evaporator section	DP_{in}	kg/m²	$K_{bg}^* K_{bm}^* W_o^{} {}^2/2g$	2.51*1207*.8^2/ (2*9.81)	98.2	143.3	196.7	78.7

Parameter	Symbol	Units	Equation	Calculation	Value	Trial 1	Trial 2	Trial 3
Upper section of riser								
Height of upper riser	H_{up}	m	$H_r/4$	9.72/4	2.4	2.4	2.4	2.4
Reduced steam velocity at entrance of upper riser	$W''_{0.en}$	m/s	$= W''_{0.bot}$		0.8	0.8	0.8	0.8
Reduced steam velocity at exit or upper riser	$W''_{0.ex}$	m/s	$G''/(Ar^* ros)$	.806/(.0458* 17.7)	1.0	1.0	1.0	1.0
Mean reduced velocity	$W''_{0.av}$	m/s	$(W''_{0.en} + W''_{0.ex})/2$	(1 + 0.8)/2	0.9	0.9	0.9	0.9
Velocity ratio	V_{ratio}		$W''_{0.av}/W_0$	.9/0.8	1.1	0.9	0.7	1.3
Factor $W_0^{\wedge}2/d$	F_r		$W_0^{\wedge}2/d$	$.8^{\wedge}2/.054$	11.9	18.5	26.7	9.1
Value of Fl from figure 7-6 against F_f & V_{ratio}	F_i		For F_r & velocity ratio	Figure 12-3a	0.38	0.33	0.31	0.42
Correction factor for pressure	K_p		For F_r. P & velocity ratio	Figure 12-4	1.01	1.00	1.00	1.01
Corrected motive head at bottom section	P_{mu}		$F_i^* K_p^* H_{up}^*$ (row − ros)	0.38* 1.01* (810 − 17.7)	738.9	635.3	596.8	816.7
Friction in tube (for fully rough turbulent flow)	f_r		Rough zone turbulence	Table 12-2	0.023	0.023	0.023	0.023
Friction in upper evaporator section	K_{bg}		$f_r^* H_{up}/d$	.0231*2.4/.054	1.04	1.04	1.04	1.04
Two phase resistance factor	K_{bm}		row + V_{ratio}^* (row − ros)	810 + 1.1* (810 − 17.7)	1699.3	1521.4	1402.8	1826.3
Resistance in evaporator section	DP_u	kg/m²	$K_{bg}^* K_{bm}^* W_0^{\wedge}2/2g$	1.04* 1699* 0.8^2/ (2*9.81)	57.6	80.6	107.0	47.4
Total motive head in riser	P_m	kg/m²	$P_{mu} + P_{mb}$	905 + 777.8	1755.3	1502.7	1429.4	1929.5
Total resistance in riser	P_r	kg/m²	$DP_u + DP_{ec} + DP_{in}$	58 + 62.2 + 98.8	218.5	323.1	444.6	173.4
Available head in riser	$P_{av.r}$	kg/m²	$P_m − P_r$	1755 − 218	1536.8	1179.6	984.8	1756.1
Water wall to drum section								
Reduced steam velocity in unheated riser	$W''_{0.ur}$	m/s	$G''/(Aur^* ros)$	.806/(.0229* 17.7)	1.99	1.99	1.99	1.99
Velocity of circulation	W_0	m/s	$G/Aur^* row$	29.7/(810*0.0229)	1.60	2.00	2.40	1.40
Velocity ratio	V_{ratio}		$W''_{0.ur}/W_0$	1.99/1.6	1.24	0.99	0.83	1.42
Factor $W_0^{\wedge}2/d$	F_r			$1.6^{\wedge}2/.054$	47.4	74.1	106.7	36.3
Value of Fl from figure 7-6 against F_f & V_{ratio}	F_i		Figure 12-3a		0.46	0.44	0.4	0.47
Correction factor for pressure	K_p		Figure 12-4		1.00	1.00	1.00	1.00
Correction factor for 60 inclination of tubes	K_a		Fr = 46%. 36 atm, 60°	Figure 12-5	0.90	0.91	0.92	0.92
Corrected motive head at bottom section	P_{mur}		$F_i^* K_p^* K_a^* H_{ur}^*$ (row − ros)	0.46*0.9* 3.4* (810 − 17.7)	1125	1088	1000	1175
Friction in tube (for fully rough turbulent flow)	fr		Table 12-2		0.023	0.023	0.023	0.023
Friction in upper unheated riser	K_{bg}		$f_r^* lur/d$	.0231*8/.054	3.42	3.42	3.42	3.42
Two phase resistance factor	K_{bm}		row + V_{ratio}^* (row − ros)	810 + 1.24* (810 − 17.7)	1795	1598	1467	1936
Local losses: exit from header + 40elbow + 90elbow + entrance to drum	$f_{ex} + f_{el}$ $f_{elb} + f_{en}$			1.0 + 0.1 0.2 + 1.0	1.1 1.2	1.1 1.2	1.1 1.2	1.1 1.2
Total resistance factors	K_f		$f_{ex} + f_{el} + f_{elb}$ $+ f_{en} + K_{bg}$	1.11 + 1.2 + 3.42	5.7	5.7	5.7	5.7
Resistance in unheated riser	DP_{ur}	kg/m²	$K_f^* W_0^{\wedge}2/2g$	5.7*1.6^2*1795/ (2*9.81)	1340	1864	2464	1106
Available motive head in unheated riser	$P_{av.u}$	kg/m²	$P_{mur} − D_{pur}$	1125 − 1340	−215	−776	−1464	69
Total available head	P_{av}	kg/m²	$P_{av.r} + P_{av.u}$	1536 − 215	1322	404	−479	1825
Available circulation from graphical solution	W_0	m/s			0.7			

	Circulation	Velocity	Available head	Pressure drop in downcomer				
Available head		m/s	kg/m^2	kg/m^2	1740.0			
		0.7	1732.2	1825				
		0.8	2262.5	1322				
		1.0	3535.2	404				
		1.2	5090.7	−479				

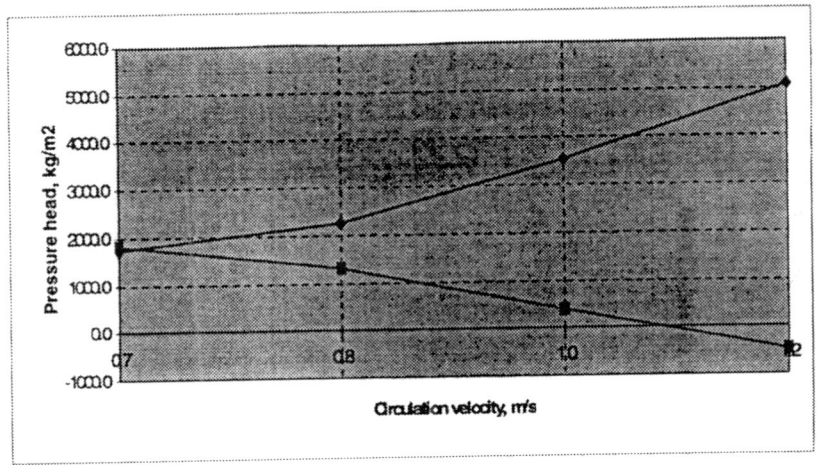

Nomenclature

A	flow cross section of the tube, m^2
d	tube diameter, mm
f	friction coefficient (Eq. 12-16)
G	total flow rate, kg/s
G''	steam leaving the circuit, kg/s
G_1''	flow rate of steam, kg/s
G_2''	flow rate of water, kg/s
H	height of the circuit, m
h'	enthalpy of saturated water kJ/kg
h''	enthalpy of saturated steam, kJ/kg
H_s	vertical height, m
H_{inlet}	enthalpy at the entrance, kJ/kg
H_{sat}	saturation enthalpy, kJ/kg
K	circulation ratio
k	loss coefficient (Eq. 12-16)
L	latent heat of vaporization, kJ/kg
l	tube length, m
m	mass velocity in tube, $kg/m^2 s$
P	pressure, bar
P_{av}	head available to force the mixture through the riser, kg/m^2
P_c	critical pressure, bar
P_m	motive head, kg/m^2
q	heat absorption, kJ/s
q_f	heat flux (Eq. 12-18), MW/m^2
q_{cr}	critical heat flux, MW/m^2
q_m	average heat flux, MW/m^2

R	coefficient (Eq. 12-26)
T_{sat}	saturation temperature, °C
T_w	wall temperature, °C
$W_{m,av}$	reduce average velocity, m/s
W_o	reduced velocity of water in the economizer section, m/s
W_o''	reduced velocity of steam in the evaporator section, m/s
W_o'	reduced velocity of water in the evaporator section, m/s
$W_{o,av}''$	reduced velocity averaged over a length of evaporator section of the tube, m/s
W_s	velocity of steam, m/s
W_w	velocity of water, m/s
x	steam mass fraction in mixture
x_{cr}	critical mass fraction of steam, kg
$x_{cr,low}$	critical vapor mass for the underside of tube, kg
$x_{cr,up}$	critical vapor mass for the crest of tube, kg
x_{in}	steam mass fraction (Eq. 12-22)
Y	coefficient of nonuniformity of heat absorption
Z	total height of the water wall

Greek Symbols

β	volume fraction of steam in the mixture
$\Delta H_{vaporization}$	latent heat of vaporization, kJ/kg
ΔP	pressure drop, Pa
ΔP_{acl}	pressure drop—acceleration, kg/m^2
ΔP_{dc}	flow resistance in downcomer, Pa
ΔP_{en}	pressure drop at the entrance, Pa
ΔP_{ex}	exit pressure loss, Pa
ΔP_{fit}	pressure drop owing to fittings, Pa
ΔP_{fric}	pressure drop—friction, kg/m^2
ΔP_{mix}	pressure drop for steam-water mixture, kg/m^2
ΔP_r	flow resistance, Pa
ΔP_r	lost head because of hydraulic resistance
ΔP_{st}	pressure drop—hydrostatic, kg/m^2
ΔP_w	pressure drop through a section carrying water, kg/m^2
ΔT_{sub}	temperature difference between saturation and water, °C
Δx_{cr}	difference in critical steam fraction between the upper and lower part of a tube
θ	angle of inclination of tube
ϕ	fraction flow area occupied by steam
γ''	specific weight of steam, N/m^3
γ'	specific weight of water, N/m^3
γ_m	specific weight of steam—water mixture ($\rho \cdot g$), N/m^3
ΔP	pressure drop, N/m^2 or Pa
γ_w	specific weight of water in the downcomer, N/m^3
φ	coefficient (Eq. 12-19)

ρ	density, kg/m³
ρ_l	saturated density of water, kg/m³
$\rho_m(z)$	local density of steam–water mixture at a height z, kg/m³
ρ_s	saturated density of steam, kg/m³

Dimensionless Number

Fr'	Froude number, $\dfrac{x_{cr}m}{\sqrt{gd\rho_s(\rho_l - \rho_s)\cos\theta}}$

Subscripts

1	at the entrance of pipe
2	at the exit of the tube
m	properties of steam water mixture

Superscripts

$'$	related to water
$''$	related to steam

References

Alad'yev, I.G., Gorlov, L.D., Dodonov, L.D., and Fedynskiy, O. (1969) Heat transfer to boiling potassium in uniformply heated tubes. Heat Transfer Soviet Research 1(4): 14–26.

Butterworth, D. (1971) "A Model for Prediction Dryout in a Tube with a Circumferential Variation in Heat Flux." AERE-M-2436, Harwell.

CKTI (1965) Standard developed by Boiler Design Institute of former Soviet Union. Doroshuchuk, V.E., Levitan, L.L., and Lantsmann, F.P. (1975) Recommendations for calculating burnout in round tube with uniform heat release. Teploenergetika 2(12): 66–70.

Jens, W.H., and Lottes, P.A. (1951) Analysis of heat transfer, burnout, pressure drop and density data for high pressure water. Argonne National Laboratory Report ANL-4627.

Kefer, V. (1989) Stromungsformen and Warmeubergang in Verdampferrohren Unterschiedlicher Neigung. Dissertation, Technical University of Munich.

Kon'kov, A.S. (1965) Experimental study of the conditions under which heat exchange deteriorates when a steam–water mixture flows in heated tubes. Teploenergetika 13(12): 77.

Perkov, V.I. (1965) "Design of boilers." Course notes on lectures delivered at the Mechanical Engineering Department of Indian Institute of Technology, Kharagpur.

VDI Heat Atlas (1993) Verein Deutscher Ingeieur, ed. J.W. Fullarton, trans. Dusseldorf ISBN 3-18-400915-7, pp. Hbc 1-29.

13
Forced Circulation for Supercritical or Subcritical Boilers

The main objective of a good natural circulation design is to have wet steam at the outlet of all the tubes making up the water walls of the combustion chamber. Provided the dry out and critical heat flux are not reached the boiling inside the tube occurs in a nucleate regime, where the heat transfer coefficient is in the range of 10–40 kW/m^2·C. Owing to such high internal heat transfer coefficients, the temperature of the tube walls remains very close to the saturation temperature of the water. This makes the task of material selection and circulation design easy. Another advantage of this natural circulation is that circulation rate increases either when pressure decreases or when heat flux increases. The former enables inherent flow stabilization during load rejection as long as the level of water is correctly controlled in the drum. The latter allows natural uniformity of flows in adjacent tubes receiving different heat fluxes.

A disadvantage of this natural circulation boiler is that it needs a heavy, thick-walled drum and downcomer as well as large-diameter water wall tubes to allow high water flow rates through them. So the cost of pressure parts is very high. Also, a rapid change in temperature of these pressure parts is difficult to handle and hence restricted. A change any faster than about 50°C/h may make thermal fatigue and cracks appear. So this type of boiler is mainly operated at constant temperature. Since boiling takes place inside water wall tubes, its walls are at saturation temperature. To maintain the walls at constant temperature the boiler must be operated at constant pressure. As a result, to vary the load a natural circulation would use a stop valve at the outlet of the boiler to adjust the flow rate needed by the turbine. Such a throttling action by the steam stop valve can cause vibrations and temperature changes at the inlet of the high pressure (HP) turbine. Also the choking in the stop valve wastes a large amount of feed pump power. Another point is that a gradual heating is required for all these heavy pressure parts. So to initiate the natural circulation during startup the boiler takes a long time and hence wastes fuels.

Finally, for modern high-efficiency plants where design pressure is chosen above 190 bar, natural or even assisted circulation is no longer workable because the circulation rate is insufficient. For this reason, forced circulation has been developed. In the next section, we will present the main characteristics of forced circulation boilers that have to be taken into account while designing or operating this type of boiler at either sub- or supercritical steam conditions. As far as possible, these specific characteristics will be given in comparison with those of a natural circulation boiler.

13-1 General Description

In a forced circulation boiler, unlike natural circulation ones, water in the water walls is fed with water pumps. These pumps can either only be the main water feeding pumps or the main feeding pumps assisted by booster or so-called circulation pumps. This chapter would not discuss assisted circulation boilers, as they are more similar to natural circulation.

Forced circulation boilers also called once-through or monotube boilers, were first developed in the late 1920s by Benson. Here water was fed into the water wall tubes at supercritical pressure irrespective of the load of the plant. In the original Benson-type forced circulation boiler, the heated supercritical fluid after the water walls was expanded through expansion nozzles to the conditions desired for use of steam. So the main purpose of this forced circulation boiler was to avoid the use of heavy pressure parts, i.e., drum and downcomer, which are essential in a natural or assisted circulation one. To allow faster temperature variations was another purpose for this type of boiler. The first design had thick water wall tubes to withstand supercritical pressure and wasted much power in feed pumps at partial loads. Benson and Sulzer improved this design by operating the boiler with sliding pressure (Fig. 13-1). Here the boiler operating pressure varies with the load. This pressure variation following load means that only a slight choking of turbine inlet-valves is needed to allow load increase. The main load variations are carried out by adjusting the water feed and firing rates while keeping the inlet turbine valves

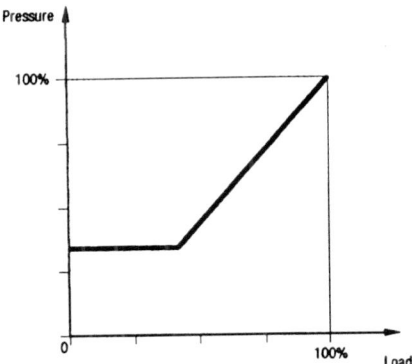

FIGURE 13-1. Sliding pressure variation as a function of load in a typical forced-through boiler (© EDF/SEPTEN/EC)

fully opened. This specific feature protects the HP turbine against mechanical and thermal fatigue and avoids wastage of power in feeding pumps at part load. This feature of the forced circulation boiler greatly helped its development. Today, good availability and efficiency are reported worldwide for this type of boiler.

In the next two sections, we will look at the main criteria to be taken into account for general and circulation design of forced circulation boilers. Then we will discuss in the last section, the main improvements needed to reach advanced steam parameters for both pressure and temperature.

13-2 Design Principles of Forced Circulation Boiler

As seen in the combustion chapter of this book, design of a pulverized coal (PC) fired furnace is mainly dictated by fuel characteristics. The design requires that (i) fuel has to burn completely before leaving the combustion chamber. This depends on volatile matter content, inflammability criteria, and particle diameter especially. (ii) Ash particles must not be molten when reaching the first transverse heat exchangers at the top of the combustion chamber or those in the back pass. This restricts the maximum flue gas temperature at this point. Compared to pulverized coal (PC) boilers, circulating fluidized bed boilers have less constraints.

As can be seen in Figure 13-2, forced circulation boilers can either be erected in single or double pass. The choice for this mainly depends on (i) fuel erosion and

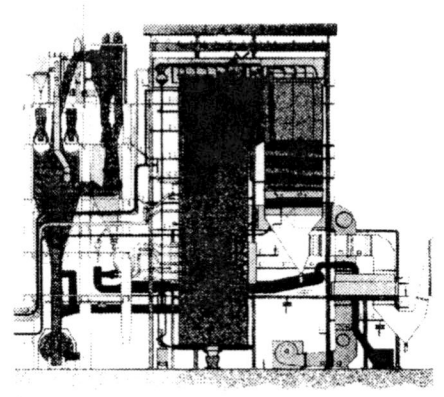

a   b

FIGURE 13-2. Arrangement of two types of boiler: (a) single-pass or open-type furnace or semi-open-type furnace; (b) double-pass or semi-open-type furnace) (© EDF/ SEPTEN/EC)

slagging or fouling characteristics, (ii) operating flexibility, (iii) local conditions for space, wind, and seismic load.

The single-pass boiler does not have a back-pass. All heat exchanger surfaces (except the economizer, whether partially or fully) are stacked inside the furnace enclosed by evaporator water walls. It has several special advantages. The flue gas flows only vertically upward and gas velocity is relatively low in front of the first heat exchangers, where heat transfer is achieved mainly by radiation. As a result of this large spacing and low velocity are allowed between the tubes. Owing to this erosion is low and deposits of ash on tubes are minimal. As all heat exchangers are horizontal, they are easy to drain and they allow larger flexibility for startup procedures. Another advantage is the quasisymmetric general arrangement, which provides more clearly defined flue gas flow, and heat transfer, which enables easier tangential firing and hence (i) higher burn-up with longer coal particle residence time, and (ii) low NO_x emission using both staged combustion and over-fire air. This, however, suffers from the need of a very tall furnace height, resulting in higher costs and less capacity to withstand strong winds or seismic conditions.

Double-pass boilers are mainly used when climatic or seismic conditions do not allow erection of a single pass one. Also these are most commonly used coal fired boilers around the world. Here the suspended heat exchangers (platen wall) cannot be drained, and special care has to be taken to avoid plugged flows and induced thermal stresses in headers during startup procedures. Lack of symmetry in the combustion chamber, makes it difficult to use advanced steam conditions where steam temperature may be close to the maximum allowable material temperature.

Forced circulation boilers are used specially to attain high pressure up to supercritical conditions. Also, for a given power input the enthalpy difference from inlet to outlet of evaporator-tubes decreases with rising pressure (Fig. 13-3). One major goal for advanced steam boiler design is to keep the water wall within the maximum allowable temperature range of the materials.

13-3 Features of Forced Circulation Boilers

A natural or assisted circulation boiler has to be designed to avoid completely vaporized flow inside the tubes enclosing the combustion chamber, whereas in a once-through boiler, a total vaporization of water in the water walls is needed. This is the only way to attain high pressures (which can be supercritical) where saturation enthalpy (or heat of vaporization) can be very low (zero for supercritical pressures); and yet a given amount of heat must be absorbed by the walls to cool the flue gas temperature to a certain level at the outlet of the furnace. So special care must be taken first not to exceed the allowable temperature of the tube walls at any load and second not to allow large temperature differences between parallel membrane tubes which are welded together. As a result, the main design considerations for circulation inside the water walls for the entire range of operation of the plant are—

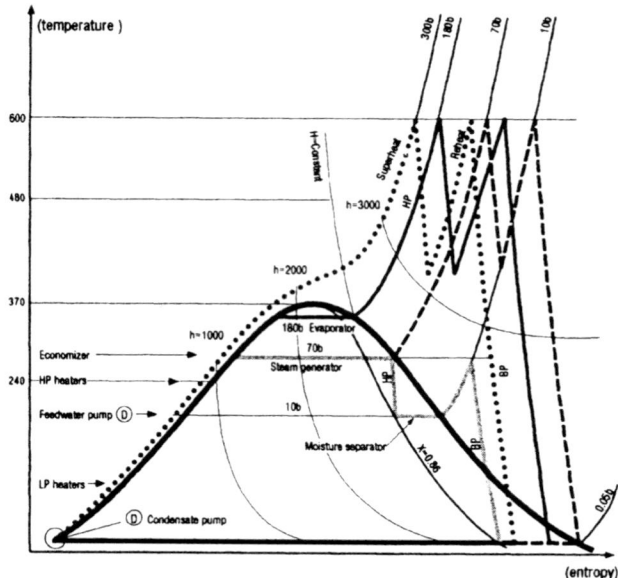

FIGURE 13-3. Temperature/entropy diagram showing three different types of water/steam circuits (© EDF/SEPTEN/EC)—Gray solid lines: Typical french PWR plant; Black dot lines: one reheat supercritical boiler (full load); Black broken lines: one reheat supercritical boiler (low load); Black solid lines: one reheat subcritical boiler (full load)

1. To ensure adequate flow within individual tubes regardless of (i) the load (i.e., heat flux and pressure inside the tube) or (ii) the fouling of outside the tube and oxidation inside the tubes.
2. To maintain uniform temperatures between the different tubes welded together. To attain this the designer must either have uniform heat flux on each tube or compensate for the differences in these heat fluxes by different flow rates.

The plant at times requires fast response to load variations or rejection or trip of the turbine. This generally requires the installation of by-passes on the turbine to prevent sudden disturbance in the operation of the boiler, and also a quick and accurate instrumentation and control (I & C) system to ensure security of the plant. Both circulation features and I & C of forced circulation boilers are discussed below.

13-3-1 Forced Circulation and Water Wall Arrangement

At a given flow rate of water, the first criteria to be fulfilled is to cool the tubes sufficiently during a dry out so that the wall temperature does not exceed the allowable metal temperature. This means that very careful attention must be paid to the design of the circulation system and water wall arrangement. There are many parameters affecting the so-called dry out or departure from nucleate boiling (DNB) or critical heat flux (CHF) (see Chapter 12 on Circulation). The key parameters that influence the increase in temperature during dry out are steam pressure, mass flux, and quality of steam (i.e., wetness); heat flux along the combustion chamber and circumferential to the tube; tube geometry inlet profile, internal and external diameters, bends, inclination, and surface (smooth or riffled); and inside fouling of the tubes. So depending on all these parameters, CHF for the given circulation flow rate in individual tubes (in $kg/m^2 \cdot s$) will be calculated. For large-capacity boilers where the volume (and thus the perimeter) of the furnace has to be increased, the cross section area of the tubes may become too large to get sufficient circulation flow rate. Then tubes have to be inclined to decrease flow area and to increase the circulation flow rate. Such inclined tubes are also called spiral tubing. These also largely improves heat flux homogeneity.

As the tubes making up the furnace walls are joined together with intermediate welded fins, there is a second criterion to be met. It is to avoid temperature differences between these tubes to keep stresses within admissible limits. For this either the heat flux received by each tube has to be identical or if some nonuniformity in heat flux is present between different tubes, then this heat flux has to be compensated by different flow rates to bring a temperature uniformity. Except in the case of low (lignite f.ring in PC) or relatively uniform (CFB for any fuel) heat flux, this criterion is very difficult to fulfill. So, efforts must be made to achieve some uniformity in the heat absorption by different tubes.

For a corner fired boiler using international steam coal, a three dimensional combustion computation has been carried out for one of the 600 MWe units erected in France in the 1980s. Figure 13-4 shows the distribution of heat flux on different

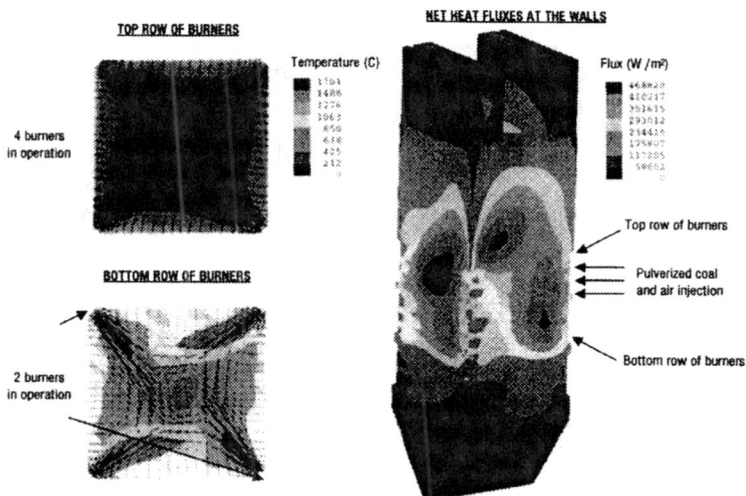

FIGURE 13-4. 3D view of combustion of a 600 MWe corner fired boiler (© EDF/DER/LNH)

walls. As can be seen in this figure large variations of the heat flux occur along both height and width of the furnace. To achieve uniformity in the heat absorbed by the water passing through different tubes, spiral tubing going up around the furnace greatly helps. A major disadvantage of spiral tube walls is that it cannot carry the weight of the total wall together with the ash hopper. Thus, a more complex mechanical design is needed, which will be discussed in the next section.

The other technique to attain this homogeneity is to use orifices adding a pressure drop at the inlet of the tubes overriding any difference in pressure drop in subsequent section of the tube. This method suffers from some shortcomings: e.g., (i) a given orifice cannot levelize the system whatever load or firing rates are (i.e., different burners in service, fouling); and (ii) pumping power is wasted.

Another alternative in a new Benson design, is to use vertical tubing with riffled tubes (Fig. 13-10). Depending on the design of these tubes (diameter, length) natural circulation characteristics of the flow inside the tubes can be attained. This means that when the heat flux increases, the flow rate increases too and then tends to level temperature in the different tubes. Another advantage of riffled tubes is to increase the allowable heat flux without reaching dry out, allowing minimum load to be minimized where recirculation is needed.

As will be seen in Chapter 17, mechanical criteria govern material selection, thickness, and general tube arrangement. Heat flux also imposes some limit on the fin length. So, the diameter and internal surface of the tubes are allowed to some extend to change to fulfill the circulation needs. Other criteria that have to be taken into account are, of course, the economical and mechanical ones that have to make the furnace competitive and withstand expansion problems. Once tube characteristics have been chosen, designers have to check that the maximum temperature allowable by the tubes is not exceeded under any operating conditions (i.e., load variations and rejection, internal tube fouling, different fuels).

With the heat fluxes shown in Figure 13-4, a one-dimensional circulation calculation has been carried out and is given in Figure 13-5 for three different water flow rates inside the tubes.

Figure 13-5 shows that when water flow rate decreases, the dryout level comes down into the tube, and tube temperature increases. This shows that in a forced circulation boiler when load decreases the flow rate of water decreases almost proportionally to this load, but the heat flux in the furnace does not decrease at that rate. So at some load, circulation rate in the tubes would not be large enough to avoid dry out and critical heat flux would be reached. So at some load more water than that needed for steam production has to be fed into the water walls in order to lower the tube temperature. This can be done either by the main feed pump (Fig. 13-6a) or by a booster pump (Fig. 13-6b). Subsequently, for low loads water flowing out of the evaporator has to be separated from steam in order not to feed the first superheater with wet steam. This is achieved in a water–steam separator or cyclone. Thus, for low loads, the forced circulation operation becomes very similar to that of a natural circulation boiler, but with a very low circulation ratio. Each of the arrangements given in Figure 13-6 have their own advantages and disadvantages.

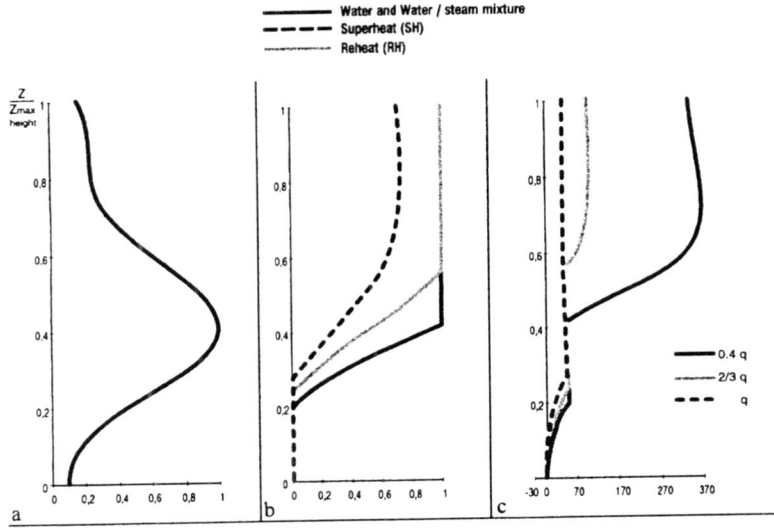

FIGURE 13-5. Results of thermohydraulic calculation for one membrane-tube along the height of the boiler with (© EDF/SEPTEN/EC) (a) heat flux integrated from Figure 13-4 for one spiral-tube; (b) wetness of the steam in the spiral-tube for three different flow rates of steam; (c) steam temperature minus inlet temperature along the height

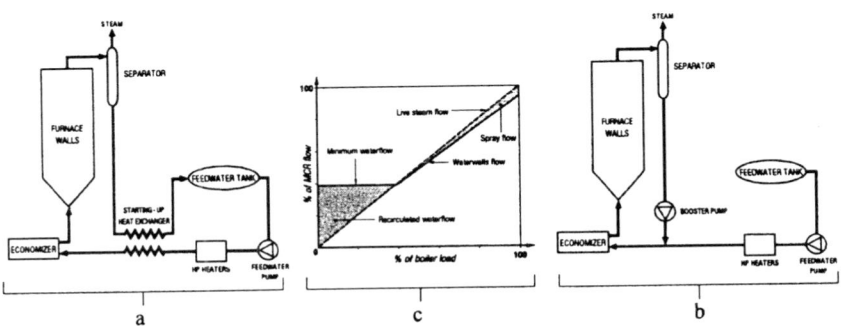

FIGURE 13-6. (a) Recirculation system for low load using a starting-up heat exchange system; (b) recirculation system for low load operation using a booster pump; (c) typical water/steam flow versus load (© EDF/SEPTEN/EC)

13-3-2 Control of Forced Circulation Boilers

Figure 13-7 shows the general arrangement of a natural circulation boiler. As previously discussed, this type of boiler is especially operated at constant pressure. Load following is carried out through variations in inlet valves of the turbine. As

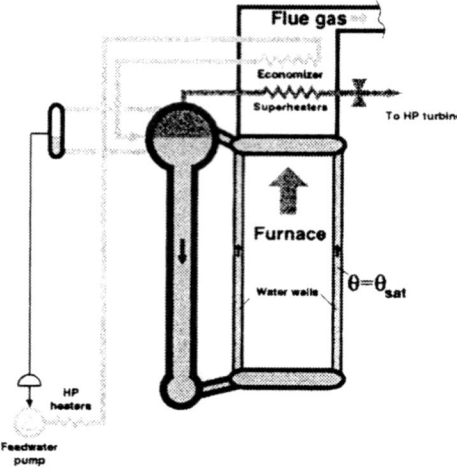

FIGURE 13-7. Drum-type boiler control system (© EDF/SEPTEN/EC)

there is no problem with the water wall temperature as long as water level is controlled within the drum, when load increases, inlet valves open, pressure tends to drop and the firing rate can be raised to adjust the pressure. This increased firing rate tends to lower the level of water in the drum which in turn can be maintained by controlling the feed water pump flow rate. So natural circulation boilers have two main independent variables (pressure and level of water), which require two different control parameters (firing rate and feed pumps).

Figure 13-8 gives a simple description of a forced circulation boiler, which does not use an inlet valve to adjust steam flow rate. In a forced circulation boiler operating at subcritical pressure, the level of water is not maintained at a predefined level. Yet, the temperature of the water walls has to be maintained at a level below the allowable metal temperature. To do this, temperature is measured at some point in or after the first superheater, to find what is equivalent to the level of water in a natural circulation boiler. Later on, this level will be called the pseudo-level of water.

To avoid any fluctuation in this pseudo-level owing to some variation in load or firing rate, one has to adjust simultaneously firing and feeding water rate. These features of a forced circulation boiler make it very different from a natural circulation one as far as control system is concerned. In this case, both firing and feeding rates have to be adjusted simultaneously while taking into account load demand and pseudo-level of water. As all these measurements and controlling loops have different delay times, all of them have to be adjusted very carefully. Finally, these control loops must be closely connected with those that adjust (1) the sliding pressure with the load and (2) the water injection for controlling temperature of superheated steam. The water that is directly injected in between superheaters is drawn from the evaporator before the water wall tubes. This withdrawal of water

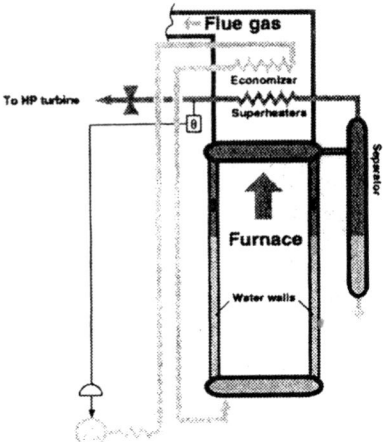

FIGURE 13-8. Once-through boiler type control system (© EDF/SEPTEN/EC)

from the tube again influences the pseudo-level of water in the tube. Finally, to be able to have some load jump, inlet turbine valves are always installed to use the thermal stored energy of the boiler. This again interacts with the previous loops during very quick load changes. These valves operate in a small range of load (about 10%) for quick load adjustments.

Control of superheat steam temperature often uses water injection in between superheaters. Control of reheat steam temperature also uses water injection for fast response to avoid any problem in reheaters. To avoid any loss in efficiency, other systems are proposed for reheat steam temperature control that use (i) burner tilting, (ii) flue gas recycling or excess air while burning different fuels, (iii) triflux system that heats the reheat steam simultaneously with superheated steam and flue gas, and (iv) two passes of flue gas while burning oil or gas. In a CFB, different systems are also used either with by-passing some part of reheater or installing external heat exchangers EHE (Fig. 13-9).

13-4 Supercritical Boilers

Supercritical boilers operate above the critical pressure of 22.089 MPa, and critical temperature of 647.29 K. The higher the steam temperature and pressure, the higher the thermodynamic efficiency of the steam power plant. The efficiency of a steam plant increases with pressure and temperature. Such an increase in the efficiency or drop in heat rate is very significant in reducing the greenhouse gas CO_2 from fossil fuel fired plants. It also helps reduce the cost of power generation.

Above the critical temperature and pressure no latent heat of vaporization is required to evaporate water into steam. There is continuous transition from water

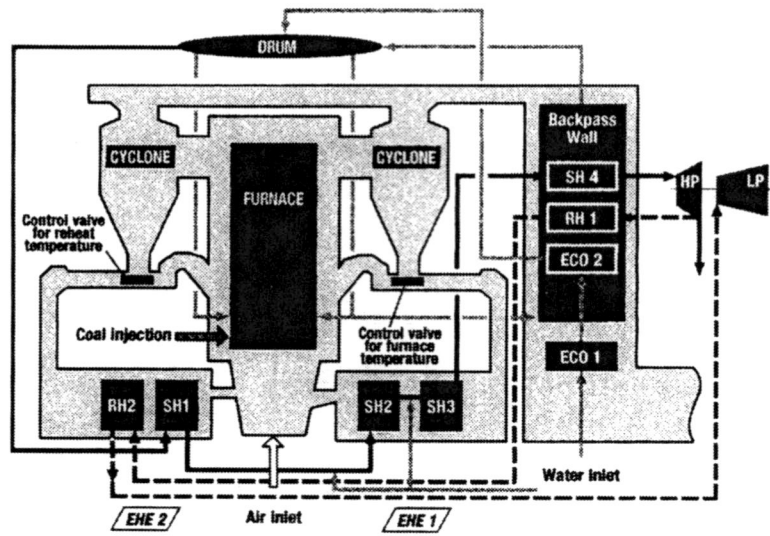

FIGURE 13-9. Control valve is used in external heat exchanger (*EHE* 2) to control reheat temperature in reheater 2 (*RH 2*)—Black solid lines: water and water/steam mixture; Black broken lines: superheat (SH); Gray solid lines: reheat (RH)

to steam. This requires a very high purity of the feed water and very uniform heat flux. These stipulations impose some stringent design conditions for conventional PC, oil, or gas fired supercritical boilers. Circulating fluidized bed boilers owing to their intrinsic furnace temperature uniformity and relatively low peak heat flux better meet the above stipulations of supercritical operation.

13-4-1 Supercritical Circulating Fluidized Bed Boiler

The heat flux and furnace temperature vary widely in a pulverized coal (PC) fired boiler, so a spiral-wound horizontal evaporative tubes have to be used. Both heat flux and temperature in a circulating fluidized bed boiler are relatively uniform. So it does not need the spiral-wound evaporator tubes. Relatively low and uniform heat flux around the furnace perimeter obviates the need for horizontal tubes needed in PC boilers.

A circulating fluidized bed (CFB) boilers has its highest heat flux in the lowest section of the furnace, where the water is coolest. So there is no need for riffled tubes (Fig. 13-10) to avoid DNB in this region. This reduces unnecessary flow resistance. In the upper section where evaporation occurs one may use riffling. The vertical evaporator tubes of supercritical CFB boilers should have orifices for flow equalization, because lengths of some tubes are different owing to the roof and exit section of the furnace wall.

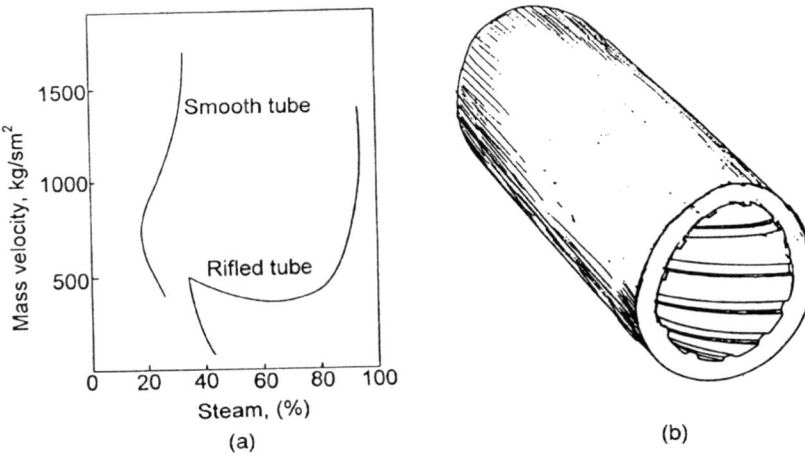

FIGURE 13-10. Riffled tube

13-4-2 Startup and Low-Load Operation of Supercritical Boilers (Fig. 13-11)

1. During startup the steam–water separator (4) collects the excess water, which is drained to the low-pressure condenser (12) or the deaerator (11) after passing through feed water heater (9). During low load, the feed water heater is by passed. The steam, separated in the separator (4), passes through the superheater (5) on its way to the high pressure turbine (HP). However, the heat absorption may not be adequate to have the right steam quality. So the steam may be used either for heating of the turbine component or by-passed to the reheater by the high pressure by pass valve (8). This steam then goes to cool the reheater (6). The reheated steam may go to the low-pressure turbine (LP). If the steam quality is still below the required temperature it exits to the lower pressure condenser (12) through the operation of the low-pressure by-pass valve.

13-4-3 Operating Mode

A supercritical boiler can operate in one of the following two ways:

- Constant operating pressure over the entire range of load
- Sliding pressure operation at which the steam generator operates at constant supercritical pressure from 100% to 80% load and then the pressure will decrease to subcritical as the load reduces further. Eventually the pressure stays constant below about 30% load (Fig. 13-1).

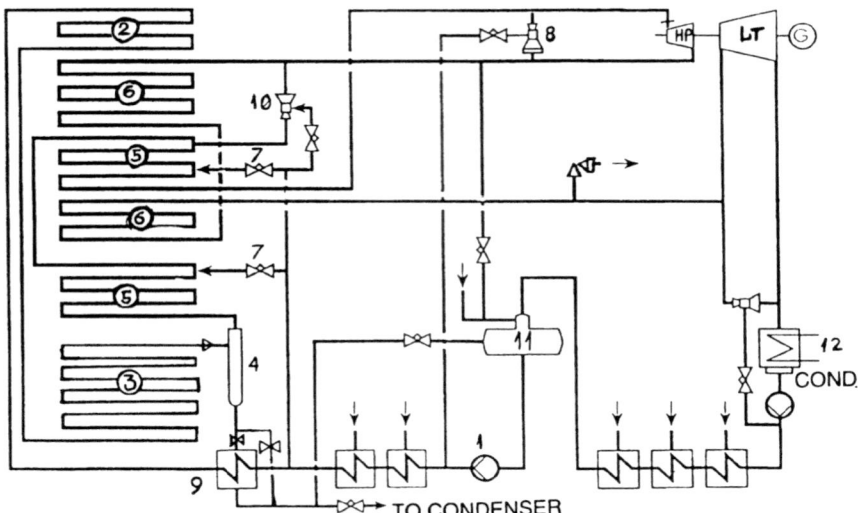

FIGURE 13-11. Schematic of a supercritical steam plant. (1) Feed pump; (2) Economizer; (3) Evaporator; (4) Steam-water separator; (5) superheater; (6) Reheater; (7) attemperator; (8) Turbine by-pass valve; (9) Feed water heater; (10) Reheater by-pass valve; (11) Deaerator; (12) Condenser

The sliding pressure operation enjoys certain advantages as follows:

1. Since the steam temperature remains unchanged during the load change, there is less thermal stress on the turbine components.
2. At low load, the pressure is low, and the specific volume of steam is high. So, it is possible to maintain a good distribution of steam flow in the superheater and reheater.
3. Reduced pressure operations may extend the life of boiler components.
4. It provides an extended range of reheat temperature control.

In addition, there is less external piping and valves. Also, it has a simpler startup.

14
Corrosion and Fouling of Heat Transfer Surfaces

One of the major causes of outage of a modern utility boiler is failure of its boiler tubes. Over the period from 1959 to 1991, boiler tube failures have been ranked as the number one equipment problems in fossil fuel fired power generating plants in the USA. Equivalent unavailability factor owing to tube failure for plants larger than 200 MW capacities was 2.7%. (DOE. 1998). Leading causes of boiler tube failures in the USA arranged in order of availability loss (MWth) are shown below (DOE, 1998):

1. Corrosion fatigue
2. Fly ash erosion
3. Under deposit mechanism (hydrogen damage and acid phosphate corrosion)
4. Long-term overheating/creep
5. Short-term overheating
6. Soot blower erosion
7. Fireside corrosion of water wall, superheater, and reheater

The above ranking may differ from one country to the other depending on the types of fuel fired and operating procedures of the boilers. However, it is apparent that corrosion is one of the most common reasons for tube failure.

A tube can corrode either from inside or from outside. The internal corrosion is largely driven by water chemistry, while the external corrosion is driven by combustion conditions. The former corrosion is known as *internal corrosion*, while the latter is called *fireside corrosion*. The fireside corrosion may take place in either the high- or low-temperature zone, each having its distinct mechanism.

In addition to corrosion, two other undesirable things, viz., fouling and slagging, occur on the outside of the tubes. In fouling and slagging, inorganic materials from the fuel deposit on the surface of the tube. Such deposits reduce the heat absorption capacity of tubes, which increases the downstream flue gas temperature and results in drop in steam output. An increased flue gas temperature may accelerate the corrosion process in the downstream section of the boiler.

This chapter gives an overview of the mechanisms of the corrosion and fouling process and discusses some preventive measures. Erosion is discussed in detail in Chapter 15.

14-1 High-Temperature Corrosion of External Surfaces

High-temperature corrosion takes place on the tube surfaces in the furnace. Boiler furnaces in general operate with excess oxygen in the furnace. At typical oxidizing conditions an oxide layer of Fe_3O_4 and or Fe_2O_3 is formed on the tube surface (Fig. 14-1). This hard oxide layer protects the tube from metal wastage. However, inefficient combustion at times creates localized reducing conditions or allows direct contact of chlorine and carbon with the oxide layer. At high temperature and high heat flux, complex chemical reactions take place, leading to the formation of iron sulfide, which owing to its higher molar volume may crack the protective layer. Also, low melting point compounds may be formed, wasting the metal. If chlorine is present in the flue gas, it diffuses through the oxide layer, making the protective layer porous and therefore less protective.

The high-temperature corrosion of boiler tubes may be caused by one the following five main mechanisms.

1. Formation and melting of sulfate
2. Formation of SO_3 and destruction of oxide layer

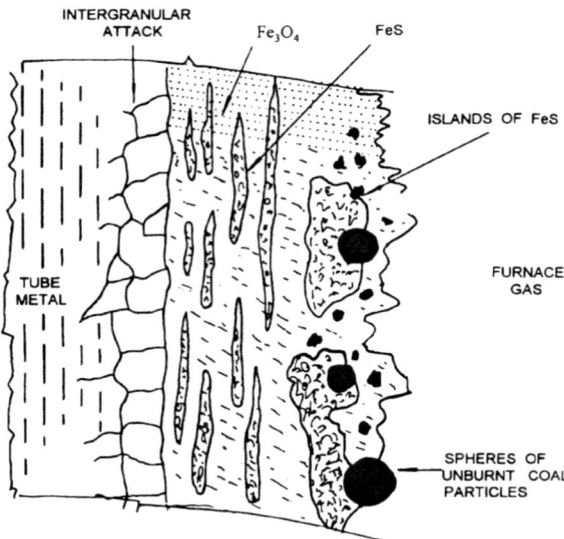

FIGURE 14-1. Nonprotective fast-growing scales are caused by reducing conditions

3. Formation of H_2S and generation of FeS
4. Formation of HCl and its reaction with iron and iron oxides
5. Corrosion due to formation of sodium-vanadium mixture

Details of the above mechanisms are given below.

14-1-1 Formation of Sulfate Deposit and Mechanism of Sulfate Type Corrosion

In pulverized coal (PC) flames, sodium and potassium of fuel volatilize and react with SO_3 in the flue gas to produce Na_2SO_4 and K_2SO_4. The dewpoint of this product is around $1150°K$. Thus, when the gaseous Na_2SO_4 and K_2SO_4 compounds hit relatively cold tube surfaces, they condense on the oxide film of tube walls. Both the chlorine content of coal and high furnace temperature favor the deposition rate of sulfate.

The process of corrosion by sulfate proceeds mainly in the following two ways: (a) sulfate melting and (b) pyrosulfate melting.

a) Sulfate Melting

The first process involves melting of sulfate on heating surfaces which absorbs Fe_2O_3 and forms compound sulfate $(Na,K)_3Fe(SO_4)_3$ (Lai, 1990):

$$Na_2S_2O_4 + Fe_2O_3 \rightarrow Na_3Fe(SO_4)_3$$

$$K_2S_2O_4 + Fe_2O_3 \rightarrow K_3Fe(SO_4)_3 \tag{14-1}$$

The sulfate compound $(K_3Fe(SO_4)_3/Na_3Fe(SO_4)_3)$ cannot form a stable protective layer on tubes like Fe_2O_3 does. When the mole ratio of K and Na is between $1:1$ and $4:1$, the melting point of the above compound falls to $825°K$. Thus, when the deposit thickness of sulfate increases, surface temperature rises to the melting point, the Fe_2O_3 protective membrane is dissolved by compound sulfate, which causes corrosion of tube walls. This is called *sulfate melting corrosion*.

b) Pyrosulfate Melting

Another type is pyrosulfate melting corrosion of alkaline metal: i.e.,

$$FeS_2 \rightarrow FeS + [S] \tag{14-2}$$

When alkaline pyrosulfate is present in an adhesive layer, it will melt at normal wall temperatures because of its relatively low melting point.

Tests show that the corrosion by way of melting of sulfate deposits on metal surface is much faster than that in the gas phase corrosion. One can see from Figure 14-2 that, when the temperature is at about $680°C$, the corrosion rate of melting sulfate is four times faster than that in the gas phase.

When the temperature is about $720°C$, the corrosion rate of sulfate melting is same as that of corrosion in gas phase. When the temperature is higher than $720°C$, the gas phase corrosion will be faster than sulfate melting process.

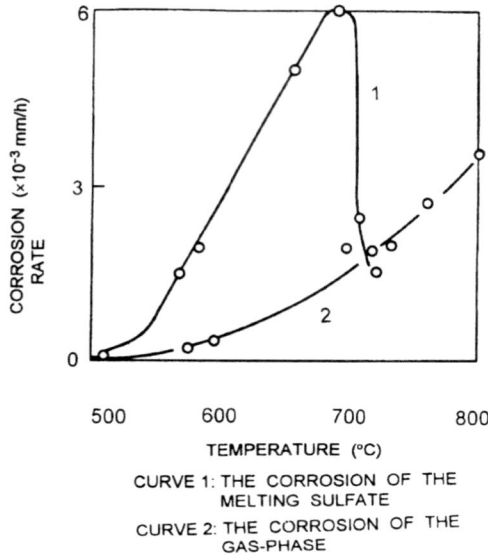

CURVE 1: THE CORROSION OF THE
MELTING SULFATE
CURVE 2: THE CORROSION OF THE
GAS-PHASE

FIGURE 14-2. Corrosion increases with gas temperature, but the melting sulfate corrosion drops sharply after about 680°C

14-1-2 Formation of SO_2/SO_3 and Their Influence on Metal Corrosion

The sulfur in fossil fuels produces sulfur dioxide in course of combustion; but only a small part of the sulfur dioxide may be converted into sulfur trioxide (SO_3). This conversion may take place in one of the following three ways: (i) combustion reactions, (ii) catalytic action of vanadium pentoxide, and (iii) sulfate dissociation.

a) Combustion Reaction

In combustion reactions sulfur dioxide reacts with the atomic oxygen dissociated at high temperature forming sulfur trioxide.

$$SO_2 + [O] = SO_3 \qquad (14\text{-}3)$$

The rate of reaction (Eq. 14-3) depends on the concentration of oxygen atoms. This reaction is first order in atomic oxygen. It is also affected by

• Flame temperature
• Excess air
• Operating conditions

b) Catalysis

When the high-temperature flue gas flows over the convective surfaces, SO_2 is converted into sulfur trioxide by the catalytic action of V_2O_5 and Fe_2O_3 of the ash deposit. The catalytic effect of V_2O_5 on the deposit of dust on the surface of superheater is very marked. The catalytic effect is observed between 425°C to 625°C, but the effect is greatest at 550°C. This temperature is within the range of the superheater surface temperature. The catalysis reaction follows the process:

$$V_2O_5 + SO_2 = V_2O_4 + SO_3$$
$$2SO_2 + O_2 + V_2O_4 = 2VOSO_4$$
$$2VOSO_4 = V_2O_5 + SO_3 + SO_2 \qquad (14\text{-}4)$$

The catalytic formation of SO_3 can be reduced by the following means.

a) Frequent soot blowing
b) Limiting the temperature of tube surfaces below 550°C
c) Reduction in the leakage of air in the furnace and in the area of the superheater

c) Sulfate Dissociation

The quantity of sulfur trioxide released directly from sulfate dissociation is not very high. So, dissociation is a minor factor in the formation of sulfur trioxide. In a small boiler, about 3.2%–7.4% of sulfur dioxide in the flue gases is oxidized to sulfur trioxide; the amount is 0.5%–4.0% in a large-scale boiler. The concentration of sulfur trioxide in the flue gases is in the range of 5–50 ppm. The main source of sulfur trioxide is the combustion reaction. Operating experience shows that the combustion intensity in the boiler and a change in the excess air will obviously affect the concentration of sulfur trioxide in flue gases. The contact time between the flue gas and the surface deposit of the superheater is too short for carrying out the catalytic reaction successfully.

The increase in sulfur trioxide near the water cooling wall is one of the main causes of high-temperature corrosion at the wall surface. Figure 14-3 shows that the rate of corrosion owing to sulfur oxide is very small below 500°C, but when the temperature exceeds 550°C, the corrosion increases very rapidly.

When sulfur trioxide is released the sulfate reacts with the metal oxide film (Fe_2O_3) to form atomic Fe and SO_4^{-2}. This reaction is reversible. The sulfate and the oxide layer on the pipe surface mix and dissolve to form a layer of corrosion. This layer is considered as a kind of undergrowth phenomenon from a chemical and mineral point of view. Thus, it is impossible to remove them by means of any mechanical method.

14-1-3 Formation of H_2S and Its Influence on Metal Corrosion

Another major contributor to the corrosion of the water-cooled wall is hydrogen sulfide in the flue gas. The rate of H_2S-driven corrosion steadily increases with

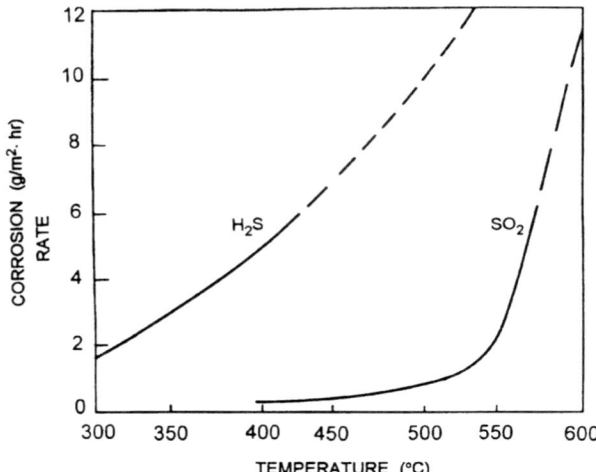

FIGURE 14-3. The rate of corrosion due to SO_3 increases with temperature in the case of carbon steel

temperature (Fig. 14-3). In case of coal combustion, hydrogen sulfide is produced in the combustion zone under reducing conditions. Hydrogen sulfide directly reacts with iron to form iron sulfide, and then the iron sulfide reacts with the pure metal to produce the low melting point censospheres. The hydrogen sulfide penetrates the loose layer of ferric trioxide, and then it reacts with iron oxide in the dense oxide layer of ferric trioxide:

$$H_2S + Fe \rightarrow FeS + H_2 \qquad (14\text{-}5)$$

$$H_2S + FeO \rightarrow FeS + H_2O \qquad (14\text{-}6)$$

The hydrogen sulfide reacts with the metal of the heating surface to produce iron sulfide, and then iron sulfide forms iron oxide. The iron sulfide and iron oxide, being porous, do not protect the tube from corrosion.

14-1-4 Formation of HCl and Its Influence on Metal Corrosion

The combustion of high-chlorine fuels produces hydrogen chloride in boilers, and it greatly enhances the corrosion. The possible corrosion reactions of hydrogen chloride on the tube wall are

$$Fe + 2HCl = FeCl_2 + H_2$$
$$FeO + 2HCl = FeCl_2 + H_2O$$
$$Fe_2O_3 + 2HCl + CO = FeO + FeCl_2 + H_2O + CO_2 \qquad (14\text{-}7)$$

The hydrogen chloride destroys the oxide film on the tube wall, producing iron chloride. The evaporation point of iron chloride being very low, it volatilizes as soon as it is formed. Therefore, the metal of the boiler tube is directly corroded. At the same time, due to the destruction of the oxide film, hydrogen sulfide reaches the surface of the metal further accelerating the corrosion. Tantalum, used in steam lines is the only metal resistant to hydrochloric acid in most concentrations and temperatures (Jones, 1992).

The corrosion potential of hydrogen chloride on metal is appreciable when the partial pressure of hydrogen chloride is above 9.8 Pa. For low-chlorine content fuels, the partial pressure of the hydrogen chloride in flue gas is only 0.0098 Pa. So, the corrosion is not significant. However, for high chlorine content coal it is significant and can be estimated for austenitic tube materials from the following equation (BEI, 1990):

$$r = A \times B\left(\frac{T_g}{E}\right)^m \cdot \left(\frac{T_m - C}{M}\right)^n (Cl - D) \tag{14-8}$$

where r = corrosion rate, nm/h
T_g = gas temperature, °C
T_m = surface metal temperature, °C
Cl = chlorine content of fuel, %
A = tube position factor (dimensionless)
B = material factor (dimensionless)

E, M, D, m, n are constants.

14-1-5 High-Temperature Corrosion Owing to Alkaline Earth Metal and Vanadium

The alkaline metals react with silicate to form low-melting-point eutectic mixtures, which evaporate at high temperature and then condense on low-temperature heating surfaces. These adhesive compounds react with the metal and the flue gas to form coking sulfate and the complex alkaline ferric sulfate. Some types of low-grade fuels contain vanadium pentoxide. So, in the case of oil firing a sodium-vanadium eutectic mixture with a low melting point is produced. This mixture induces melting of the deposit at 600°C. The corroding reaction on the tube wall is:

$$2Na + SO_2 + O_2 \rightarrow Na_2SO_4$$
$$V_2O_5 \rightarrow V_2O_4 + [O]$$
$$V_2O_4 + 1/2O_2 \rightarrow V_2O_5$$
$$V_2O_5 + SO_2 + O_2 = V_2O_5 + SO_3 + [O] \tag{14-9}$$

Both reactions involving V_2O_5 produce highly corrosive oxygen. The ratio of vanadium and sodium in the deposit has an important effect on the corrosion. The corrosion increases with this ratio Fig. 14-4. When the vanadium–sodium ratio

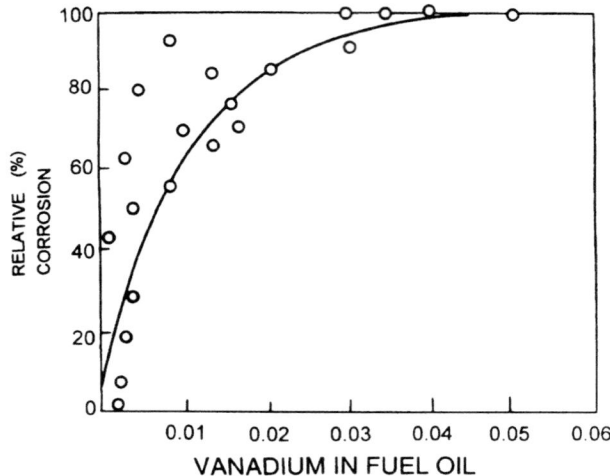

FIGURE 14-4. The corrosion increases with the vanadium content of the fuel oil

exceeds a threshold limit, the corrosion is serious for different metals like low alloy ferritic to austinitic at metal temperatures at or above 660°C (BEI, 1990).

14-2 Prevention of High-Temperature Corrosion

The following measures can be taken to reduce the risk of fire-side corrosion in a boiler.

14-2-1 Low Oxygen Combustion

The use of low excess air can reduce the formation of sulfur trioxide and increase the formation of sulfur dioxide. It also helps oxidization of vanadium to vanadium trioxide instead of more corrosive pentoxide. In case of oil firing, the low oxygen combustion may also reduce the amount of vanadium pentoxide formed. Thus two main contributors to corrosion (reduction in vanadium pentoxide and sulfur trioxide) are reduced by low excess air. However, low excess air may lead to loss in combustion efficiency.

14-2-2 Distribution of Combustion Air

Though the excess air in the burner is greater than 1.0, some parts of the furnace may have excess air less than 1 owing to nonuniform air distribution. Consequently, the generation of sulfur dioxide, the carbon monoxide and the hydrogen sulfide all increase in those parts. Such conditions increase the corrosion rate. Through adjustments of air fuel mixture in each burner, the high-temperature corrosion can be reduced.

14-2-3 Distribution of Coal among Burners

Equal distribution of coal powder amongst coal burner nozzles is important but difficult to achieve at the same time. Each boiler unit has several pulverizers to supply the pulverized coal. The length of the air pipe and the numbers of bend are not necessarily the same. Also, the resistance in every section cannot be kept uniform in all pipes. So an equal distribution of pulverized coal among all burners is difficult to achieve.

The following measures can be taken to achieve uniform distribution of coal powder in all burner nozzles:

- Reduce the length and the number of bends. Try to ensure equal flow resistance in all burners.
- Try to have as long a straight run of the common pipe as possible before it branches out to pipes for individual burners.
- Use flow guide vanes to reduce segregation of coal–air mixture in bends.
- Avoid complete stoppage of a burner nozzle for load reduction. If possible reduce the load on all burners simultaneously down to 50%–60% of their full load capacity.

14-2-4 Control of Particle Size in the Coal Powder

Coarse coal particles may make the flame hit the opposite wall, causing high-temperature corrosion and fouling. Also, some coal particles may not burn completely. Experience (Cen, 1996) shows that when the size of the coal powder is $R_{90} = 8.5\%$–13.5%, the corrosion of the exterior surface of water tubes is much greater than that when $R_{90} = 6\%$–8%.

14-2-5 Avoiding Local Hot Spots

The adverse effect of localized high temperature of the heating surfaces was described earlier. So, efforts should be made to restrict the highest temperature in the flame region of the boiler, and to reduce the heat flux in this region. The high-temperature corrosion of the water wall is greatest near the flame zone, where both temperature and heat flux are greatest.

14-2-6 Protective Air Film near the Water Wall

The corrosion of the fireside of the water wall tubes of a slagging combustor can be prevented by a jet of hot air injected in the corrosion region. An air curtain (Fig. 14-5) washes off the corroding media off the water wall. Also, the rich oxygen atmosphere reduces the sulfur corrosion.

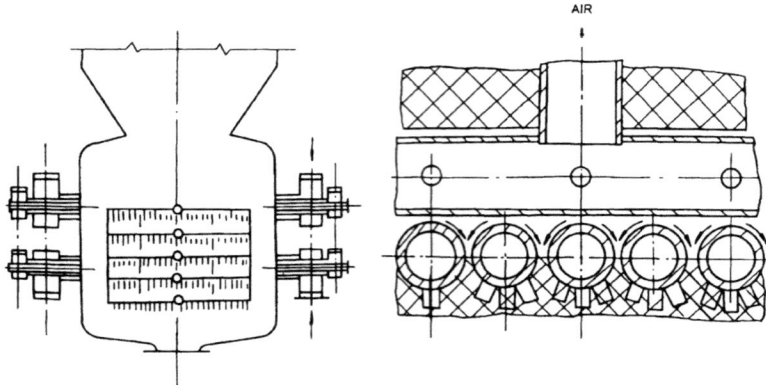

FIGURE 14-5. Air curtain on the side walls of a down jet fired PC boiler; this can reduce corrosion in it

14-2-7 Additives

The addition of metal magnesium or dolomite (Diamant, 1971) can reduce the corrosion in oil fired boilers. The magnesium sulfate or a mixture of magnesium oxide and vanadium oxide form high-melting-point deposits of the tubes. Thus the high-temperature corrosion owing to the melting of the deposit is reduced.

14-2-8 Control of Design Temperature at the Furnace Exit

The furnace exit temperature should also be lower (Table 14-1) for coal with a higher propensity of corrosion. The furnace heat release rate may be restricted to reduce corrosion.

14-2-9 Use of Corrosion Resisting Alloys

High chromium steel, which has superior anticorrosion properties, can be used in furnaces firing corrosive coals. However, high chromium steel is more expensive than carbon steel. So a compromise between the initial cost and the service life expectation should be made when selecting the tube material. The limits of operating temperatures of several types of steel and alloys are given in Tables 14-2 and 14-3.

TABLE 14-1. The corrosion tendency of coal depends on the furnace exit gas temperature.

Corrosion tendency of coal	Furnace exit gas temperature (°C)
Low	1300–1350
Moderate	1250–1300
High	1200–1250

TABLE 14-2. Maximum safe operating temperature for steel alloys.

Type of steel	ASTM Standard	Temperature, °C
Carbon steel	SA178, SA210	450
1/2 Mo	SA209 T1	510
1/2 Cr-1/2 Mo	SA213 T2	550
1.25 Cr-1/2 Mo	SA213 T11	565
2.25 Cr-Mo	SA213 T22	580
12 Cr		750
9 Cr-1 Mo	SA213 T9	650
18 Cr-8 Ni	SA213, TP304H	760
21 Cr		900
29 Cr		1125

TABLE 14-3. Maximum working temperatures of different types of steel for continuous service*.

Metal	Composition (%)						Max. working temperature, °C
	C	Si	Mn	Cr	Ni	Mo	
Carbon steel	0.10	0.15	0.5				450
1% Cr, 1/2% Mo	0.12	0.20	0.5	1.2	—	0.5	510–540
2.25% Cr, 1% Mo	0.10	0.25	0.5	2.3	—	1.0	650
12% Cr, 1% Mo	0.20	0.40	0.55	11.5	1.0	1.0	675
316 Steel	—	0.6	1.7	17.5	13.5	2.7	925
310 Steel		1.2	1.8	21.5	20.5		1150

* Compiled from (1) Modern Power Plant, Central Electricity Generation Board, V-5, 1971, p. 459, (2) ASME Pressure Vessel Code, Part D, Table 1A; and (3) Metal Handbook, V-1, ASM International, 1990, pp. 617, 878.

14-2-10 Flue Gas Recycling

Flue gas recycling may reduce the corrosion of a high-temperature heating surface because

(1) It reduces the maximum flame as well as the furnace exit temperature.
(2) It can reduce the sulfur trioxide content of the gas.

14-2-11 Avoiding High-Temperature Gas near High-Wall-Temperature Zones

For a given wall temperature the corrosion rate is higher for higher gas temperatures (Fig. 14-2). Also, if the wall temperature is lower than 550°C, the corrosion rate will be reduced greatly. So the gas path designs of a heat exchanger should be made such that the metal temperature and gas temperature combination does not give excessive corrosion. Figure (14-6) shows boundaries of desired design limits of these temperatures.

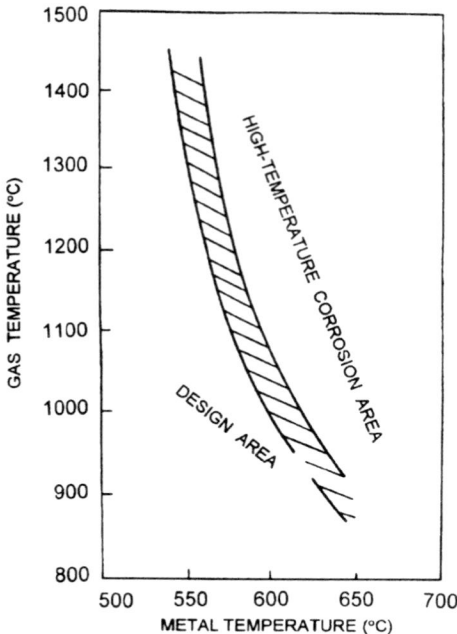

FIGURE 14-6. Design limits of metal temperature for given gas temperatures

14-2-12 Furnace Liner for Protection

The use of a furnace liner (Bradford, 1993) may reduce the corrosion damage to a certain extent, but it would substantially reduce the heat absorption by the wall. This will increase the furnace exit temperature, resulting in other adverse effects.

14-2-13 Protective Shielding on Corrodible Area

Figure 14-7 shows a protective shield on the water wall to reduce high-temperature corrosion (Breuker & Swoboda, 1985). There is a gap between the shield and the tube wall. This shield will decrease radiation from the hot gas. In Figure 14-7 we note that the high-temperature corrosion rate, after peaking at 650°C, drops considerably near 800°C.

14-2-14 High-Temperature Flame Spray on Tubes

The technique of the flame spraying has been studied since 1960s. Aluminum-ferric alloy is most commonly used in the flame spray. Other materials like METCO44, CE2148, and CE2185 are also used to extend the life of the tubes. One could also use 50/50 NiCr, which has good high-temperature resistance and erosion protection properties.

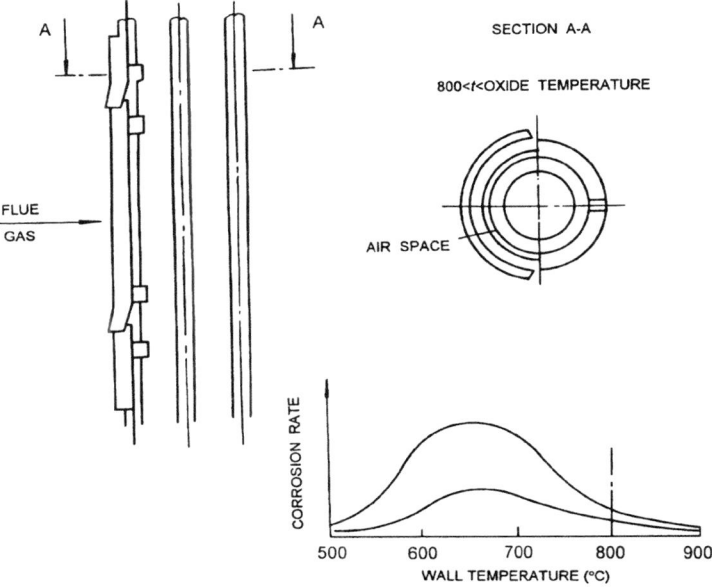

FIGURE 14-7. Protective shields on tubes to reduce the high temperature of corrosion

14-3 Low-Temperature Corrosion on External Surfaces

Heating surfaces of the air preheater and economizer are subjected to relatively cold temperatures. Yet the corrosion of these surfaces can be serious. Sometimes it can be as high as 1 mm per year. This is called *low-temperature corrosion*. Its mechanism is different from that of high-temperature corrosion described so far. The low-temperature corrosion cannot be avoided, but its magnitude can be reduced. To do that one needs to understand the mechanism of corrosion in this part of a boiler.

The main cause of low-temperature corrosion is the condensation of sulfur trioxide below its dew point. The condensation follows this reaction:

$$SO_3 + H_2O \rightarrow H_2SO_4 \qquad (14\text{-}10)$$

A part of the sulfur dioxide, formed from the sulfur in coal, is converted into SO_3. The SO_3 raises the dew point temperature of the acid. When the gas temperature is below the dew point, it forms sulfuric acid solution on the metal surface. The SO_3 can also react with alkaline dust and metal, which leads to fouling and corrosion. The reaction often occurs in the low-temperature segment of the boiler air heater. It is also called low-temperature fouling and corrosion. The SO_3 of the flue gas comes from the sulfur in the fuel, but its formation depends on the combustion condition in the furnace. Thus the low temperature corrosion can be reduced by

- desulfurization of the fuel
- improved combustion condition that reduce the SO_3 generation
- raising the tube wall temperature above the acid dew point
- using corrosion-resistant coating on the tubes

Several empirical relations are available for calculation of the dew point temperature of the flue gas. Here is one such equation:

$$T_{dew} = T_{\text{moist}} + \frac{125\sqrt[3]{S_w}}{1.05\mu A_w} \,^\circ C \qquad (14\text{-}11)$$

where $S_w = 1000 \, (S/LHV) \% \, \text{kg/MJ}$, $A_w = 1000(A/LHV)$.

Additionally, at the flue gas exit section of the air preheater, the deposited H_2SO_4 solution dissolves the oxide film and metal on the tube wall and reacts with fly ash to produce acidulous cohesive dust.

The corrosion reduction measures can be applied in design stage or in the operation stage. The measures are elucidated below.

14-3-1 Design Measures for Control of Corrosion

The following measures can be taken to reduce the potential for low-temperature corrosion:

1. The flue gas temperature in the lowest section of the boiler can be raised.
2. Alternative gas and air arrangement can be made in the air preheater. Figure 14-8 shows several arrangements of the coldest section of the boiler. Here we can see the flue gas and air temperature along with the average tube wall temperature in different sections. It is, thus, possible to control the wall temperature to some extent by changing the design of gas and air path; the corrosion, especially at low load, cannot be avoided entirely.
3. To preheat the air, one could use the bleed steam from the turbine. If the air can be heated to 100°C before it enters the heat exchanger, the low-temperature corrosion can be avoided to some extent.
4. The use of a horizontal tubular air preheater can be helpful. It can attain a higher heat transfer coefficient on the flue gas side than on the air side. As a result the wall temperature will be closer to the hotter flue gas than to the colder air. A horizontal air preheater can have wall temperatures 10–20°C higher than that in vertical types. Furthermore, these are easier to clean as well.
5. In a regenerative rotary air preheater, the surfaces are alternatively exposed to heating and cooling. Acid deposited by the flue gas during the heating process is partially evaporated by the air during the cooling process. So this may also help.
6. Corrosion-resistant material is an effective alternative. Tubular air heaters can be made of glass tubes or glass covered steel tubes. Enamel-coated tubes can also be used. The enamel tubes are, however, difficult to clean. Ceramic tubes can be used in the cold end of the air preheater.

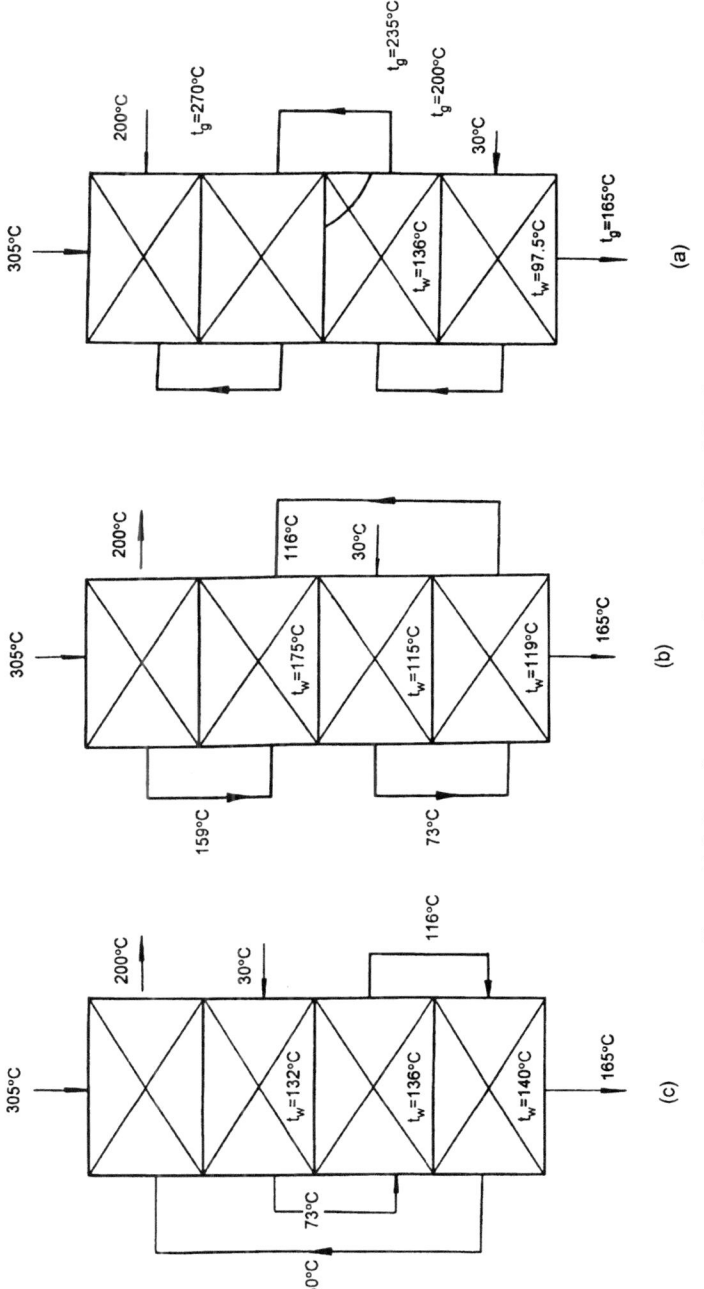

FIGURE 14-8. Several arrangements of gas/air circuit in a PC boiler

399

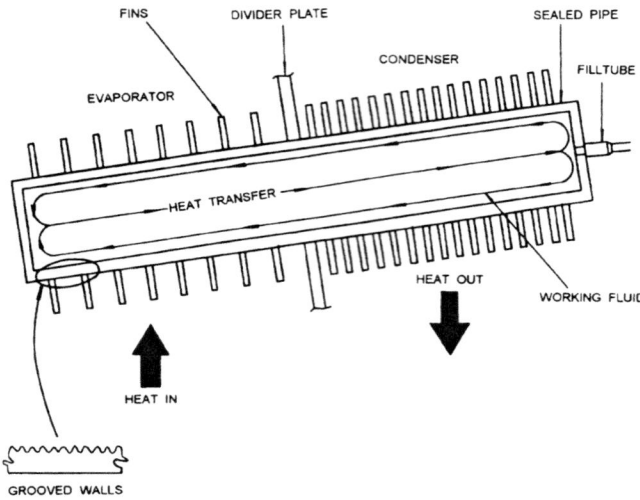

FIGURE 14-9. The working principle of a heat pipe used in an air preheater

7. Indirect heating of the cold air using hot steam or hot water. This type of air heater avoids low-temperature corrosion, as the cold air section does not come in contact with the corrosive flue gas.

8. In a low-temperature and low-H_2SO_4 environment, copper tubes carrying steam show good corrosion resistance. The resistance is, however, no different in low-temperature high acid concentration fields. In the case of the economizer, cast iron tubes show better corrosion resistance than steel tubes. But the strength of cast iron being low, they can be used only in low-pressure non-boiling-type economizers.

9. Figure 14-9 shows a system using heat pipes to heat cold air. If the medium inside the heat pipe is water, the wall temperature of heat pipes of the flue gas side is equal to about the saturation temperature of water. Thus it can alleviate the low temperature problem.

10. It is easier to clean deposits from horizontal tubes. Furthermore since corrosion is likely to occur in one part of the flue gas path, only tubes of that section need to be replaced instead of replacing all the tubes as required for vertical tube air heaters.

14-3-2 Operating Measures

The following operational measures can be taken to reduce corrosion.

1. **Burner capacity:** Experience shows that for a given boiler load, the larger the capacity of a single burner, the higher the dew point. With the increase in the burner capacity the distribution of the flame in the furnace is poorer. Nonuniform

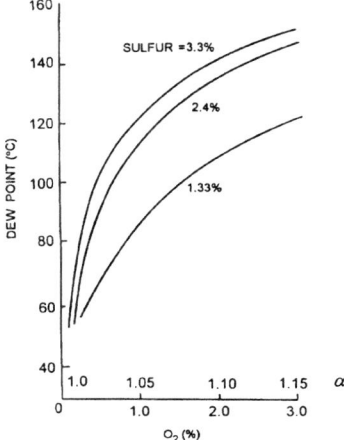

FIGURE 14-10. The dew point drops when the excess air decreases for a fuel with a given amount of sulfur

gas mixture and local high temperature increases the formation of SO_3. Thus high burner throughput is detrimental to corrosion.

2. **Low excess air combustion:** When the CO_2 content of the flue gas from a furnace increases from 10% to 13% the dew point falls from 174°C to 149°C. When the excess air coefficient falls to 1.01–1.02 the dew point falls to 42–52°C (Fig. 14-10). Tests carried out in a 220 t/h boiler (Cen et al., 1994) show that when the excess air coefficient was reduced to 1.03, the loss in combustion efficiency was 0.3%, but the gain in thermal efficiency was 1%, and there was substantial reduction in the corrosion rate.

3. **Control of furnace temperature:** When the temperature, especially at the tip of the flame, increases, the SO_3 increases. A cooler temperature would naturally reduce the SO_3 generation. This can be done by decreasing air preheat or intentionally blowing cold air into the center of the combustion zone.

4. **Avoidance of soot formation in oil firing:** Incomplete combustion forms soot or carbon black. These deposits are captured by the condensed H_2SO_4 on the colder regions. SO_2 and O_2 are absorbed, forming SO_3, which is converted into sulfuric acid. When the particle size of the carbon black increases beyond a size, the deposits lose strength and are dislodged from the tube surface. The flue gas then carry these highly acidic products to the atmospheric as air pollutants.

14-3-3 New Materials

A range of new materials has been developed that allows generation of steam at much higher pressure and temperature without risking corrosion and subsequent failure. Such developments have opened the door for increasing the efficiency of

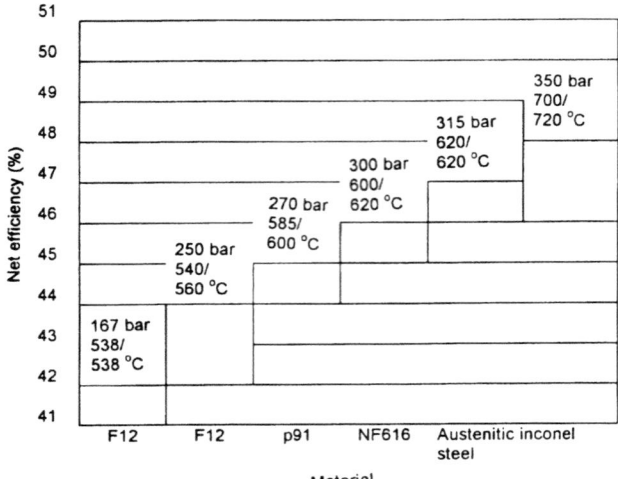

F IGURE 14-11. Higher steam temperature gives higher cycle efficiency on LHV basis, but it requires more expensive alloys

steam plants to higher level. Siemens (Klein et al., 1996) projected that with inconel material it may be possible to increase the cycle efficiency of conventional steam plant to 47% based on lower heating value using steam at 350 bar at 700/720°C (Fig. 14-11).

14-4 Corrosion and Scaling of Internal Surfaces

The interior of tubes carrying steam or water is also subjected to corrosion and or deposits. The corrosion is called water/steam side corrosion while the deposition of solids on the interior is called scaling. Much of the boiler tube failure can be traced to the occurrence of one of these two conditions inside the tubes.

14-4-1 Corrosion

The corrosion, discussed in previous sections, relates to the fire side or the external surfaces of tubes. The mechanism of corrosion inside the tubes is different from above. Water and steam flowing through tubes react with the metal and its alloy constituents, forming ferrous and ferric oxide (Callister, 1991).

$$Fe + 1/2O_2 + H_2O \rightarrow Fe(OH)_2$$
$$2Fe(OH)_2 + 1/2O_2 + H_2O \rightarrow 2Fe(OH)_3 \qquad (14\text{-}12)$$

Thus the metal is consumed at a steady rate if the fresh tube wall is exposed to the water. Fortunately, the oxide layer forms a protective barrier. In the case of chrome alloys the oxide layer is thin but highly protective. For low alloy steels the iron oxide layer thickens, and finally it tends to spall. The breakage of the protective layer occurs often due to differential expansion during cooling. Upsets in water chemistry may also lead to the breakdown of the protective oxide layer.

The CO_2 often remains dissolved in the condensate as carbonic acid (H_2CO_3), which attacks piping and fittings, producing deep channeling, thinning, and penetration of pipe threads. The following reactions may take place:

$$Fe + 2H_2CO_3 = Fe(HCO_3)_2 + H_2$$
$$Fe(HCO_3)_2 = FeO + 2CO_2 + H_2O \qquad (14\text{-}13)$$

The dissolved oxygen of water is also a major contributor to the corrosion. The reaction that follows is

$$2Fe(HCO_3)_2 + \tfrac{1}{2}O_2 = Fe_2O_3 + 2CO_2 + H_2O \qquad (14\text{-}14)$$

Some major types of water or steam side corrosion follow:

1. Corrosion fatigue
2. Pitting
3. Caustic attack
4. Hydrogen embrittlement
5. Stress corrosion
6. Galvanic attack
7. Intergranular corrosion

Corrosion Fatigue

Corrosion fatigue has been the leading single cause of boiler outage in the USA over the last 20 years (DOE, 1998). It is a discontinuous process involving the initial breakdown of the protective layer of Fe_3O_4 scale. When the critical fracture strain for magnetite is exceeded, there is repetitive breakdown of the protective scale. The German boiler standard (TRD301) stipulates that the applied strain levels should be less than 0.1%. The dissolved oxygen has an important effect on the corrosion fatigue. Below 135°C, the dependence is given by (DOE, 1998)

$$\text{Number of cycles to initiate crack} = 2445DO - 0.5256 \qquad (14\text{-}15)$$

where DO is the dissolved oxygen in ppb.

Details of other types of internal corrosion are shown in Table 14-4.

14-4-2 Scaling

Scaling is the deposit of thermally nonconducting solids inside the tube. Scaling is objectionable because it interferes with normal heat flow through the boiler metal and may lead to overheating. This may result in bulging or actual failure

TABLE 14-4. Remedial actions for different types of internal corrosion with brief description of their causes (B&W, 1992).

Types	Reason for occurrence	Location of occurrence	Mechanisms of corrosion	Remedial actions
Corrosion fatigue	Occurs because of fluctuating stress owing to thermal cycling.	Cracking of waterwall tubes near welded attachments		Minimize number of stops and starts, reduce constraints on tubes, and lower dissolved oxygen on starts
Pitting	If the protective oxide layer breaks for some reason, electrochemical reaction initiates between the cracked interior (anode) and the oxide layer (cathode). The interior is corroded forming pits while the cathode remains uncorroded.	Oxygen pitting in economizer Pitting of undrained tubes during shutdowns		Lower O_2 in feed water & auxiliary deaerated during startup Use nitrogen blanket during shutdown
Caustic attack	Here dissolved salts like NaOH concentrate within the oxide deposit, increasing the pH of the solution, and causing accelerated corrosion by continuously dissolving and removing the protective iron oxide layer	Grooving in water wall tubes under deposits		Chemically clean the boiler & control water chemistry; avoid steam blanketing
Hydrogen embrittlement	Hydrogen produced during the corrosion reaction (Eq. 14-13) diffuses into the steel metal, where it meet the carbon in steel. The hydrogen reacts with it forming methane gas forming small bubble. As the methane builds up the pressure on the cavity thus formed increases, eventually forming a fissure	Water wall tubes		Remove deposits, prevent low pH, limit chlorine, etc.

TABLE 14-4. (Continued)

Types	Reason for occurrence	Location of occurrence	Mechanisms of corrosion	Remedial actions
Stress corrosion	Occurs with specific combinations of materials, stress, and environments. These cracks occurs generally in austenitic steels.	Stainless steel superheater & reheater tubes		Flush after cleaning, avoid water carry over
Galvanic attack	Higher-pressure boilers often use more expensive metals in hotter sections of the superheaters. When the junction of such two dissimilar materials is exposed to a corrosive environment an electric potential is developed across two metals. This moves metal from one (anode) to the other (cathode).	Superheater and reheater sections where two tube materials are joined		Proper welding
Acid attack	Here a low pH solution developed under the scale as was the case for acid attack. This allows accelerated corrosion of the metal surface	Water wall tubes		Control water chemistry
Intergranular	Chromium carbide formation in grain boundary may starve the surface of the metal of chromium. Thus the protective layer of chromium oxide is not formed			

of tubes (ASME, 1995). The metal temperature increases when a layer of insulating scale is formed. For example, the temperature of the outer wall of a 22.2 mm thick tube, carrying a heat flux of 300 kW/m^2, is 346°C. If a 0.78 mm thick scale is formed on its inner wall, the temperature of the outer wall will rise to 457°C.

In addition to potential tube failure, scales may create other problems like

- Waste of fuel
- Loss of boiler output
- Maintenance problems for removal of scales

Mechanism of Scale Formation

If the boiler makeup water is not softened, the dissolved bicarbonates break down to carbonate scales

$$Ca(HCO_3)_2 + Heat = CaCO_3 + H_2O + CO_2 \quad \text{(calcium bicarbonate scale)}$$
$$(14\text{-}16)$$
$$Mg(HCO_3)_2 + Heat = MgCO_3 + H_2O + CO_2 \quad \text{(magnesium bicarbonate scale)}$$
$$(14\text{-}17)$$

The formation of scale can be prevented by proper treatment of boiler water. Once formed, it can be removed by chemical cleaning or by mechanical means during maintenance.

14-5 Fouling and Slagging

Fouling and slagging are major problems, especially in coal fired boilers. Fouling is the formation of bonded deposits on the tube. Slagging generally occurs when the deposits are molten. Both slagging and fouling cover heating surfaces, reducing the heat absorption in the region. Such a reduction in heat absorption not only reduces the thermal output of the boiler, it also increases the gas temperature in the downstream region, increasing the corrosion potential there. The deposits on tubes also initiate fireside corrosion. In the extreme case, the gas passage may be restricted, which affects the boiler fluid dynamics.

14-5-1 Effect of Fuel Properties

Fuel is the most important contributor to fouling deposits. When coal burns in air the inorganic and organic mineral matter in the coal are left as solid residues or ash. The ash contains oxides of silicon, aluminum, iron, titanium, calcium, magnesium, manganese, potassium, sodium, sulfur, phosphorous, as SiO_2, Al_2O_3, Fe_2O_3, TiO_2, CaO, MgO, Mn_3O_4, V_2O_5, SO_3, and P_2O_5. In addition to these, some of the elements also reside in complex compounds like silicates, alumino-silicates, sulfates, etc. The composition of ashes from several types of coal are given in Table 14-5. However, as the coal composition varies with its geographical

TABLE 14-5.   Ash composition for some U.S. coals and lignite.

Rank	Low-volatile bituminous			High-volatile bituminous		Sub-bituminous	
Seam	Pocahontas No. 3	No. 9	No. 6	Pittsburgh		Antelope	Lignite
Location	West Virginia	Ohio	Illinois	West Virginia	Utah	Wyoming	Texas
Ash, dry basis, %	12.3	14.1	17.4	10.9	17.1	6.6	12.8
Sulfur, dry basis %	0.7	3.3	4.2	3.5	0.8	0.4	1.1
Ash analysis, by wt %							
SiO_2	60.0	47.3	47.5	37.6	61.1	28.6	41.8
Al_2O_3	30.0	23.0	17.9	20.1	21.6	11.7	13.6
TiO_2	1.6	1.0	0.8	0.8	1.1	0.9	1.5
Fe_2O_3	4.0	22.8	20.1	29.3	4.6	6.9	6.6
CaO	0.6	1.3	5.8	4.3	4.6	27.4	17.6
MgO	0.6	0.9	1.0	1.3	1.0	4.5	2.5
Na_2O	0.5	0.3	0.4	0.8	1.0	2.7	0.6
K_2O	1.5	2.0	1.8	1.6	1.2	0.5	0.1
SO_3	1.1	1.2	4.6	4.0	2.9	14.2	14.6
P_2O_5	0.1	0.2	0.1	0.2	0.4	2.3	0.1

origin, the composition of its ash also varies. The ash composition may even vary from one seam to another even in the same mines. The slagging is a strong function of the composition of the mineral matter in the coal ash.

a) Fouling

Fouling generally occurs in the colder section of the boilers. The volatile inorganic elements like alkali metal are evaporated by the high flame temperature of PC boilers. The flue gas carries these over to the cooler convective section of the boiler. As these come in contact with cooler heat absorbing surfaces they condense as compounds. Ash and these alkali compounds form adhesive solid compounds. Unlike slagging, fouling substances are not molten. Some of them may just be sintered.

b) Slagging

Slags hardly ever form on a clean tube surface. So, a deposition process like fouling needs to precede this. In a PC boiler the temperature in the flame is high enough to melt the ash. When the molten ash hits a relatively cold tube surface it is resolidified on it. Some parts of the solidified ash remains on the surface as deposits (Figure 14-12a). This deposit grows in thickness and join together. The temperature of the external surface of the deposit farthest from the cooling media (steam and water) inside the tube increases slowly, eventually exceeding the melting point of at least some of the constituents of the ash. This process gradually accelerates, and nearly anything that hits the deposit is stuck there. Thus the surface temperature continues to increase, and the molten deposit start to flow (Fig. 14-12b). At times the deposit becomes so heavy that it falls off the tube under its own weight.

(a)   (b)

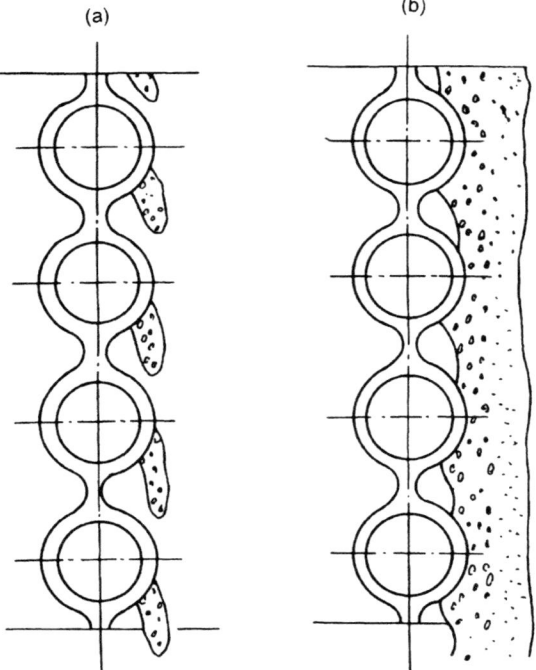

FIGURE 14-12. The process of slag formation on the water wall

c) Ash Fusibililty

The temperature, at which ash melts or starts softening, gives a good indication of the slagging and other behavior of ash at elevated temperatures. In the United States a procedure is outlined in ASTM standard D 1857 (fusibility of coal and coke ash) for measurement of these properties. Other countries have their standards for measurement of these properties.

According to the ASTM method, coal is burnt in an oxidizing condition at 800–900°C to produce the ash. The ash is then pressed into a triangular pyramid [19 mm high (H) × 6.35 mm triangular base (B)]. It is then heated slowly at the rate of 8°C/min. Then one observes the shape of the cone to define the temperature corresponding to the shapes shown in Figure 14-14. They are as follows:

- Initial deformation temperature   Rounding of the pyramid tip
  (IDT)   (Figure 14-13)
- Softening temperature (ST)   Spherical shape of tip, i.e., H = W
- Hemispherical temperature (HT)   Hemispherical lump, i.e., H = W/2
- Fluid temperature (FT)   Molten to nearly flat, i.e., H = 1.59 mm

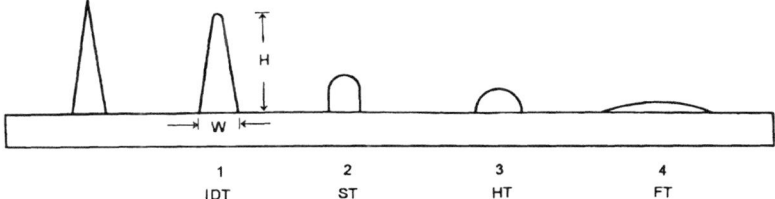

FIGURE 14-13. The shapes of ash cone indicate the initial deformation (IDT), softening (ST), hemispherical (HT), and flowing (FT) temperatures of the ash sample

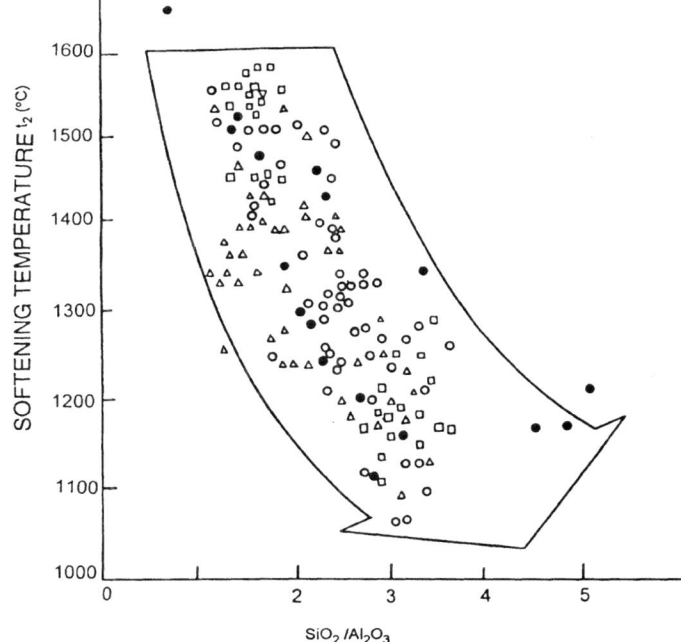

FIGURE 14-14. Melting point of ash drops as the ratio of silica to iron oxides ratio increases

14-5-2 Effect of Mineral Matter of Ash

The higher the alkaline mineral matter content of ash, the greater its slagging tendency and the lower is its softening point. At high flame temperature the sodium compounds are likely to evaporate and react to form vapors containing oxides, chlorides, hydroxides, or sulfates. Then they condense on low-temperature heating surfaces where they react with the metal to form viscous pyrosulfates and complex alkaline ferrous sulfates. This covers the tube surface. Alkali salts like NaCl and KCl contribute to the formation of alkali sintered ashes.

In bituminous coal ash $[Fe_2O_3 > (MgO + CaO)]$, iron has a dominant effect on the slagging characteristics. The iron generally remains as pyrite (FeS_2). In an oxidizing furnace environment the pyrite is converted into hematite.

$$2FeS_2 + 11/2O_2 \rightarrow Fe_2O_3 + 4SO_2 \qquad (14\text{-}18)$$

This form of iron raises each of the four characteristic temperatures of the ash. However, in a reducing region of the furnace it forms FeS, FeO, or Fe. These reduce all characteristic temperatures, and hence increase the slagging propensity.

The slagging propensity may also be described by another index called *the slagging index, Fs* (BEI, 1992, p. 4) derived from characteristic ash temperatures.

$$F_s = \frac{4 \cdot IDT + HT}{5} \; °C \qquad (14\text{-}19)$$

The value of this index is 1230–1340°C for medium slagging, 1050–1230°C for high slagging, and < 1050°C for severe slagging.

14-5-3 Factors Affecting the Melting Point of Ash

Slagging is linked with melting of the ash, which depends on the following factors.

a) Ratio of Silicon to Aluminum Oxides

The ash fusion temperature increases with both silicon and aluminum content of ash, but it decreases with the increasing ratio (SiO_2/Al_2O_3). Figure 14-14 shows a remarkable drop in the softening temperature (ST) with the increase in this ratio (He and Zhang, 1990).

b) Acid-to-Base Ratio

The major components of coal ash can be divided into acidic (SiO_2, Al_2O_3, TiO_2) and basic $(Fe_2O_3, CaO, MgO, K_2O, Na_2O)$. The ratio of basic to acidic components has an important effect on the softening temperatures and, therefore, on the slagging tendencies of ash. The ash fusion temperature decreases with an increase in this ratio of basic to acidic components and reaches a minimum value when the ratio is about 50%. A typical variation of ash fusion temperature with the base-to-acid ratio is shown in Figure 14-15. Here, we notice that an increasing amount of basic component in an acidic ash decreases the fusion temperature in the same way as the increasing acidic components in a basic ash depresses the fusion point (BEI, 1990). The base-to-acid ratio, R, is

$$R = \frac{Fe_2O_3 + CaO + MgO + K_2O + Na_2O}{SiO_2 + Al_2O_3 + TiO_2} \qquad (14\text{-}20)$$

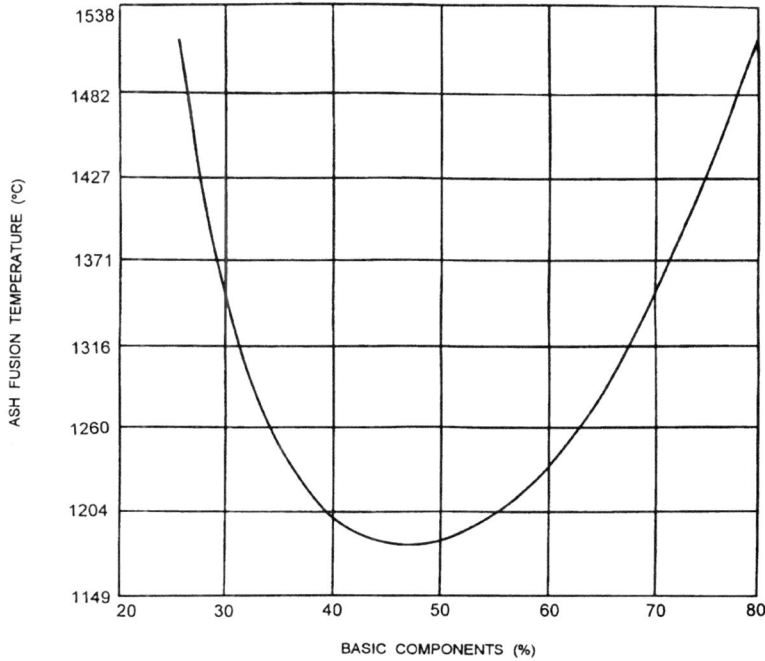

FIGURE 14-15. The variation of ash fusion temperature with the base constituents of the ash

c) Iron Oxide

For a fixed base/acid ratio, when the iron oxide content increases, all fusion temperatures decrease, and therefore, the slagging propensity increases (Fig. 14-16). The slagging propensity is classified into three categories (BEI, 1990, p. 5): high, medium, and negligible when the percentage of iron oxide (Fe_2O_3) is within the range 15%–23% , 8%–115% and 3%–8%, respectively.

Ash content also influences the characteristic temperatures: they drop with ash content down to about 20%. Beyond this value the temperature will increase again. Thus, a minimum value is reached at about 20% ash.

14-5-4 Empirical Relations for Characteristic Temperatures of Ash

The Chinese Coal Research Institute (Chen and Jiang, 1996) developed a number of correlations for prediction of characteristic temperatures of coal ash on the basis of experiments on Chinese coal. These values are yet to be tested against coal from other parts of the world. So their use may be restricted to preliminary assessment

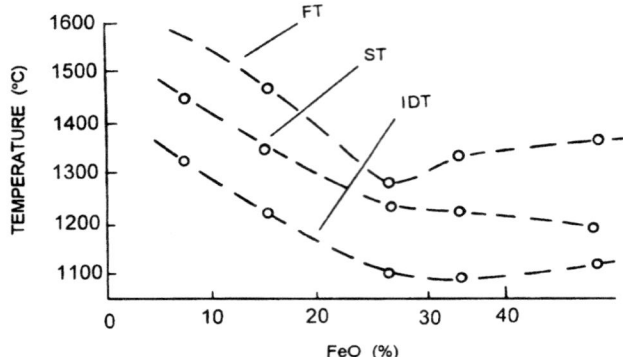

FIGURE 14-16. The variation of ash characteristic temperature (FT = flow temperature; ST = softening temperature; IDT = initial deformation temperature) with the iron oxide in the ash

in cases where experimental values are not available.

For $(CaO + MgO + Fe_2O_3 + (KNa)_2O) < 30\%$

Flow temperature, $FT = 700 + 16Al_2O_3 + 5SiO_2\,°C$ accuracy $\pm 50°C$

$$(14-21)$$

For $(CaO + MgO + Fe_2O_3 + KNaO) > 30\%$

$$FT = 200 + (2.5B + 20Al_2O_3) + (3.3B + 10SiO_2)°C \qquad (14-22)$$

The American Society of Mechanical Engineers (ASME) gave the following empirical relations for softening temperatures (Sondereal et al., 1975)

For bituminous coal:

$$ST = 0.55(2782.67 - 6.9SiO_2 + 0.1Al_2O_3 - 4.2Fe_2O_3$$
$$+ 8.5CaO - 128TiO_2 + 14.9MgO - 8.7NaO_2$$
$$- 8.7NaO_2 + 80K_2O - 5.1SO_2)\,°K \qquad (14-23)$$

For lignite:

$$ST = 0.55(1620.67 + 12.1SiO_2 + 18.8Al_2O_3 + 7.2Fe_2O_3 + 2.0CaO$$
$$+ 128TiO_2 - 11.6MgO - 13.7NaO_2 - 22.3K_2O)\,°K \qquad (14-24)$$

14-6 Calculation of Soot and Ash Deposition

Besides fouling, soot and ash are another form of deposits on tubes that hinder heat absorption in the boiler.

14-6-1 Formation of Fly Ash in the Combustion Process

Fly ash with diameters between 0.1 and 10 μm are too fine to be captured by normal collection devices. As a result they escape through the stack, causing particulate emission. These also contribute to fouling by slag. The fly ash builds up on heating surfaces, causing the surface temperature to increase, which eventually leads to the melting of the ash layer or slagging. The fly ash can come from any or all of the following sources:

- Volatile matter released from coal may condense to microparticles during the cooling period. These particles are typically 0.1 μm in size.
- During the process of crushing and pulverization, coal particles finer than 30 μm are produced. During combustion, some of them may produce spheres with diameters measuring less than 10 μm.
- During the process of combustion the coal particles may undergo fragmentation and attrition, producing fine fly ash particles smaller than 0.1 μm.

14-6-2 Factors Influencing Soot and Ash Deposition on the Heating Surface

Ideally, fouling cannot occur until some component of the ash has a large enough viscosity to adhere to the tube surface. If that occurs, further layers of ash stick to it. Consequently, the temperature of the upper surface of the deposit increases, which eventually leads to the melting of the ash. This process was illustrated in Figure 14-12. Some of the factors that influence the slag fouling are as follows:

1. *Gas velocity*: For a given heat flux, q, the temperature difference, ΔT, between the tube surface and fly ash layer of thickness, x, is

$$\delta T = q(x/\lambda) \tag{14-25}$$

where λ is the thermal conductivity of ash layer. Since the heat flux, q, is fixed at a point, the slagging potential would depend on the ratio x/λ, because a higher value of this ratio will give a higher surface temperature. This ratio, being an important index of slagging and fouling, is termed the *fouling index*. Fouling index decreases with velocity (Fig. 14-17) reducing the slagging potential. Finer ash ($R_{30} = 52.5\%$) shows lower fouling index for both in line and staggered arrangements.

2. *Flow direction*: The flow direction and its magnitude have an important effect on the deposition of ash on tubes. Figure 14-18 shows an example of how the flow direction and gas velocity affect the deposit.

3. *Particle temperature*: The flue gas receives heat from the hotter-burning coal particles. The extent of hotness of burning particles depends on the coal size, burning rate, and the local heat transfer condition. Figure 14-19 shows computed

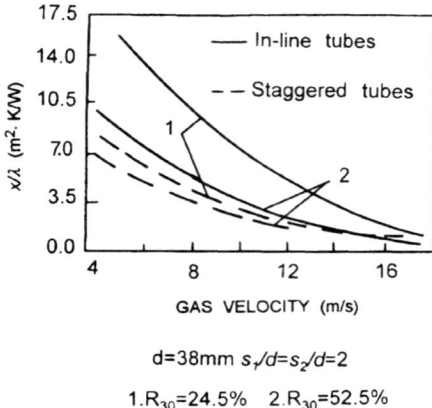

d=38mm $s_1/d=s_2/d=2$

1.R_{30}=24.5%   2.R_{30}=52.5%

FIGURE 14-17. The slagging index (x/λ) decreases with gas velocity, and it is higher for in-line arrangement

results of temperature of burning coal particles. The coarser particles could be as much as 200°C hotter than the gas. Thus, even if the gas is below the softening temperature some burning particles may still soften and stick to the surface.

4. *Excess air coefficient*: In some special situations the local temperature in the flame may exceed the adiabatic flame temperature. This may be due to local preheating of the combustion air. Figure 14-20 shows the measurements in a 300 MW boiler, showing how the ratio of the maximum local temperature increases with the excess air. Higher temperatures increase the fouling potential. The local oxygen concentration is different from its average value, which is responsible for the ratio being greater than one. The gas–solid suspension density does not have any effect on the ash deposition.

5. *Tube geometry*: The fouling potential increases with increasing tube diameter. In a staggered arrangement (Fig. 14-21), when the tube diameter decreases from 76 to 25 mm, the fouling index drops to 25% its original value. In in-line arrangements when the diameter was changed from 38 to 25 mm, the fouling index decreased to 40% of its original value.

6. *Other parameters*: A higher boiler output means higher peak flame temperature. Higher heat release rate also increases the furnace temperature. Both factors positively contribute toward the degree of fouling.

14-6-3 Estimation of Thickness of Ash or Soot Deposits

Based on the hypothesis that the soot deposit on the walls of a furnace is caused by the transverse turbulent fluctuating force in the turbulent boundary layer, Cen (1995) derived the following approximate expression for the amount of deposit

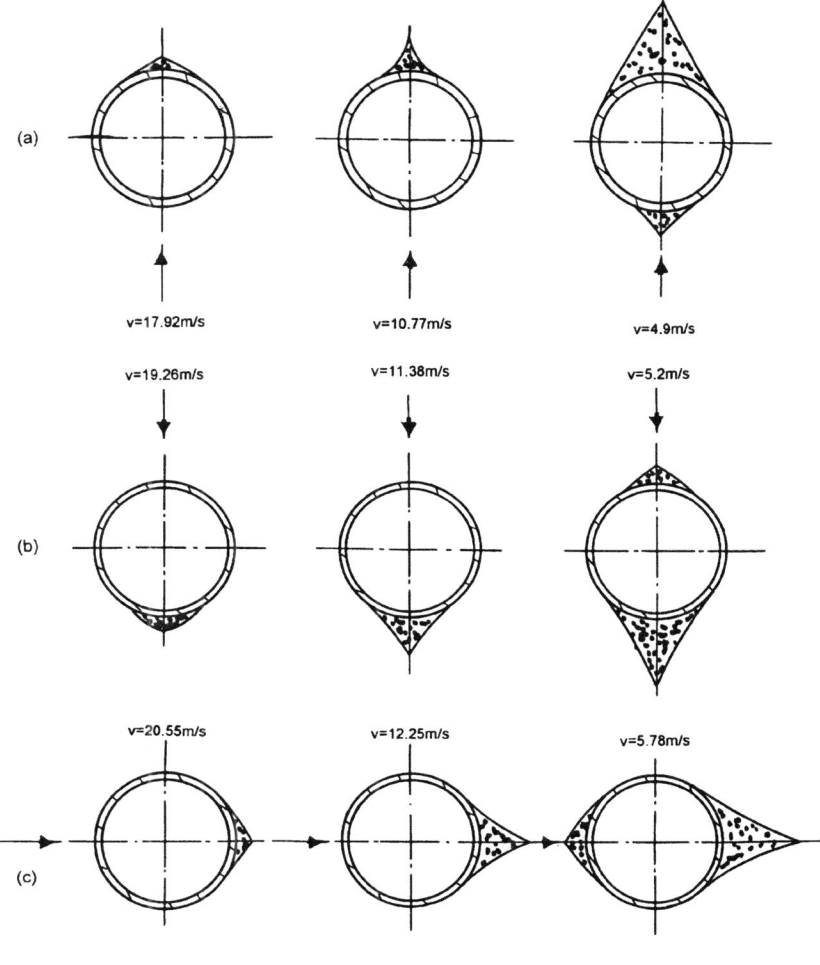

d=38mm,s_1/d=2,s_2/d=2

FIGURE 14-18. Ash and soot deposition on tubes for different direction of flow of gas over tube banks for in-line arrangement

per unit area of the tube surface:

$$G = \frac{K_0 \cdot d_{tube}}{10W^{n-1} Re^{0.333}\nu} \ \text{g/m}^2 \cdot \text{s} \qquad (14\text{-}26)$$

where $K_0 = 90, 415$ and 1500 when $R_{10} = 78, 60$, and 54 μm, respectively; d_{tube}-diameter of tube; ν = kinematic viscosity; Re = Reynolds number; $n = 3.5$; and W = velocity, considering the impact of particles.

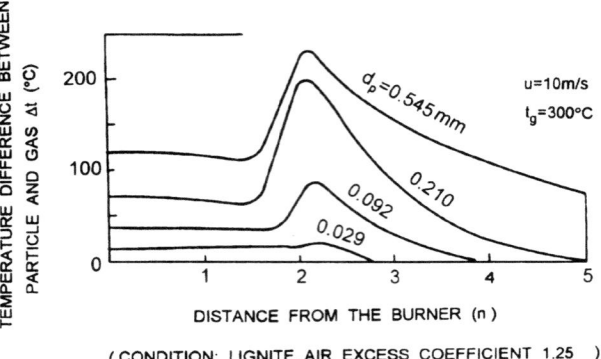

(CONDITION: LIGNITE, AIR EXCESS COEFFICIENT 1.25)

FIGURE 14-19. The variation of computed temperature of the burning coal particles in excess of that of the gas with distance from the burner

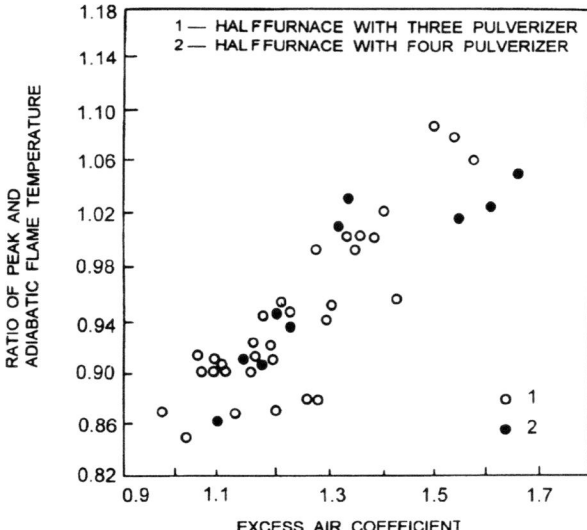

FIGURE 14-20. The effect of excess air on the peak flame temperature reached in the furnace

14-7 Prediction of Slagging Potential

For the purpose of slagging the coal ash can be divided into two categories: bituminous and lignitic (B & W, 1992, p. 20.13).

$$\text{Bituminous} \quad Fe_2O_3 > CaO + MgO$$
$$\text{Lignitic} \quad Fe_2O_3 < CaO + MgO \qquad (14\text{-}27)$$

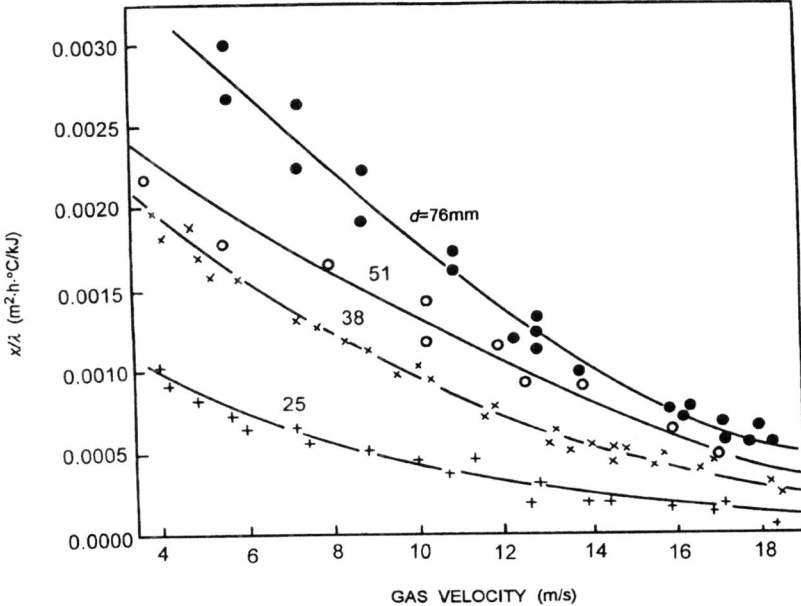

FIGURE 14-21. The effect of tube diameter on the slagging index, x/λ (note the different unit) against gas velocity (staggered arrangement)

14-7-1 Slag Ash Viscosity

Ash viscosity is an important criterion for determining the suitability of a particular ash for slag tap furnace. The viscosity depends on the temperature of the deposit and on the composition of the ash. For slag to flow freely it must have a viscosity lower than 250 poise. The temperature at which ash reaches this viscosity (T_{250}) is an important indicator of slagging potential of the ash. This value depends on the composition of the ash, temperature, and the environment—oxidizing or reducing. This value can be measured in a high-temperature viscometer. A problem with this method is that it is a difficult and expensive process. A crude alternative is to use an empirical correlation developed on the basis of large number of experimental data on different types of bituminous and lignitic coal ashes. The slag viscosity, μ, in poise is obtained by calculating the temperature, T_μ, corresponding to viscosity μ from the equation below:

$$T_\mu = \sqrt{\frac{107m'}{\log^{\mu-c}}} + 150\,°C \qquad (14\text{-}28)$$

where $m' = 0.00835SiO_2 + 0.00601Al_2O_3 - 0.109$ and $c = 0.0145SiO_2 + 0.0142Al_2O_3 + 0.0192MgO + 0.0276Fe_2O_3 + 0.016CaO - 3.92$.

TABLE 14-6. Slagging potential of different types of coal.

Types of ash →		Bituminous/Lignitic	Bituminous	Lignitic
Slagging rate ↓	Index. (Rs)→	$\dfrac{T_{250.oxid} - T_{10.000.red}}{54.1fs}$ °C	R (from Eq. 14-20)S^a %	$\dfrac{(HT_{max} + 4IT_{min})}{5}$ °C
Low		$Rs < 0.277$	$Rs < 0.6$	$Rs < 0.5$
Medium		$0.277 < Rs < 0.55$	$0.6 < Rs < 2.0$	$0.5 < Rs < 1.0$
High		$0.55 < Rs < 1.11$	$2.0 < Rs < 2.6$	$1.0 < Rs < 2.0$
Severe		$1.11 < Rs$	$2.6 < Rs$	$2.0 < Rs$

^{a} S is the sulfur in coal.

The composition of the ash should give $SiO_2 + Al_2O_3 + MgO + Fe_2O_3 + CaO = 100$. For the purpose of slagging the coal ash can be divided into two categories: Bituminous and lignitic (B & W, 1992, p. 20.13)

$$\text{Bituminous} \quad Fe_2O_3 > CaO + MgO$$
$$\text{Lignitic} \quad Fe_2O_3 < CaO + MgO \qquad (14\text{-}29)$$

The Table 14-6 gives the slagging potential of these two types. Here S is the sulfur weight percentage in coal on dry basis, HT_{max} is the higher of reducing or oxidizing hemispherical temperature, and IT_{min} is the lower of reducing or oxidizing initial deformation temperature. B&W (1992) used viscosity as a better means of judging the slagging rate. They based it on two characteristic ash temperatures, $T_{250,oxide}$ and $T_{10,000,red}$. These are temperatures of the ash with a viscosity of 250 poise in oxidizing atmosphere and 10,000 poise in reducing atmosphere, respectively. It is shown in Table 14-6.

A correction factor, fs, is used, which is a function of the average temperatures, corresponding to 2000 poise in oxidizing and reducing environments. Values of this factor are given in Table 14-7.

14-7-2 Prediction of Slagging Using the Ash Deformation Temperature

The deformation of the ash cone is used to measure the ash deformation temperature, or ITD. Different countries have different standards for measurement of this value. While this value may not depend on the country standard for preparation of sample, heating rate, and others, it would definitely depend on the environment. The cone height of the sample changes as the sample is heated slowly in oxidizing

TABLE 14-7. Correction factor, fs, for different average oxidative and reducing temperatures corresponding to 2000 poise viscosity of the ash.

°C	1038	1093	1149	1204	1260	1316	1371	1427	1482	1538
f_s	1.0	1.3	1.6	2.0	2.6	3.3	4.1	5.2	6.6	8.3

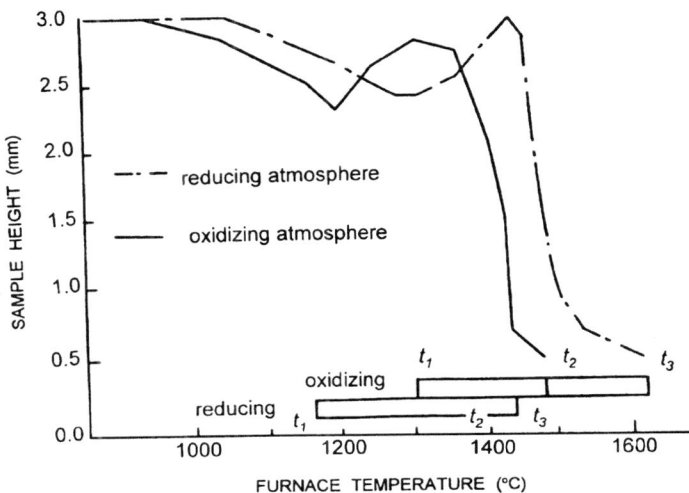

FIGURE 14-22. The change in the height of the test cone of ash depends on the environment in which it is heated

and then in reducing atmosphere (Fig. 14-22). Sometimes the softening temperature (ST) is also used for characterizing the slagging tendency. The Table 14-8 gives values specified by different standards on the slagging potential of ash.

14-7-3 Prediction of Fouling on the Basis of Individual Mineral Constituents

a) Alkali

Among the alkali compounds Na_2O has the greatest influence on the deposition on tubes. Table 14-9 presents a guide to the determination of fouling tendency from a knowledge of the alkali component of the ash. Here the other alkali oxides are

TABLE 14-8. The norm for prediction of slagging tendency as used in different countries on the basis of characteristic temperatures of the ash.

Slagging Type ↓	ITD (°C) Chinese	ITD (°C) CE/USA	ST (°C) Chinese	ST (°C) Japanese
None	>1289	>1371		
Mild			>1390	>1230
Normal	1108–1288	1093–1204	1260–1390	
Severe	<1107		<1260	<1350

TABLE 14-9. Assessment of slagging tendency from the composition of coal-ash.

Slagging tendency	Bituminous ash		Lignitic ash	Others	
	Na_2O (%)	Fe_2O_3/CaO ratio	Na_2O (%)	Fe_2O_3 (%)	SiO_2/Al_2O_3 ratio
None		<0.3 or >3.0		<8	
Weak	<0.5		<2.0		<1.87
Middle	0.5–1.0	0.3–3.0	2–6	8–15	2.65–1.87
Strong	1.0–2.5	0.3–3.0	6–8	>15	
Stronger	>2.5	0.3–3.0	>8		>2.65

to be converted into equivalent sodium oxide using the following relation

$$Na_2O \text{ equivalent} = (Na_2O + 0.659K_2O)ASH/100 \qquad (14\text{-}30)$$

where ASH is the ash percentage and 0.659 is the mole equivalent rate.

b) Iron Oxide

The ferric oxide increases the characteristic temperature because its ITD is highest. The influence of the ferric oxide on the slagging tendency is shown in Table 14-9.

c) SiO_2/Al_2O_3

The ratio of silicon and aluminum oxide also influences the slagging tendency. This is shown in Table 14-9.

d) Fe_2O_3/CaO

This ratio is also an important factor that influences the slagging of bituminous-type coal ash (Table 14-9).

e) Silica Rate

The silica percentage shows some influence on the slagging propensity (Table 14-10). The silica percent in this context is defined as follows:

$$[SiO_2/(SiO_2 + CaO + MgO + Fe_2O_3 + 1.11FeO + 1.43Fe)].$$

TABLE 14-10. Dependence of slagging tendency on the silica content of ash.

Slagging tendency	Chinese	USA/UK	France
	Silica content = $\dfrac{SiO_2}{SiO_2 + CaO + MgO + [Fe]^a}$		
Weak	>78.8	72–80	>72
Normal	66.1–78.8	65–72	65–72
Severe	<66.1	50–65	<65

[a] $[Fe]$ is the equivalent iron = $Fe_2O_3 + 1.11FeO + 1.43Fe$.

A higher percentage of silica results in higher ash viscosity and a higher ash deformation temperature.

14-8 Design Measures for Reduction of Fouling and Slagging

The following design considerations are important for minimization of fouling and slagging in a boiler.

a) **Choice of ash discharge method:** A furnace can be either dry bottom or wet bottom. In a dry bottom the ash does not melt. It is mostly extracted from the fly ash. In a wet bottom ash melts and it is extracted from the furnace as slag. The temperature corresponding to a viscosity of 250 poise must be exceeded for the discharge port for slag to flow freely. Also, the flow temperature (FT) of the ash should be less than 1450°C.

b) **Proper furnace exist gas temperature:** The most economic temperature range for the flue gas exiting the furnace of a pulverized coal fired boiler is 1200–1400°C. To avoid fouling of downstream or back-pass tubes the exit temperature should not be any higher. In boilers where screen tubes are not used at the exit of the furnace, the outlet temperature of the furnace should be a little lower than the initial deformation temperature (ID) of the ash. If the difference between ash temperature and ID is less than 100°C, the exit temperature should be no more than $(ST - 100)$°C subject to $=<1150$°C.

   The temperature behind the screen should be no more than $(ID - 100)$ or $(ST - 150)$°C. The temperature of the screen should be less than 1250°C for weakly slagging coal, less than 1000°C for strongly slagging coal, and less than 1200°C for normal slagging coal.

c) **Choice of heat release rate for different types of coal:** Both volumetric and grate heat release rates influence the temperature of the furnace, which in turn influences the slagging rate of the boiler. A higher temperature can lead to a greater adhesion of slag to the tubes, a higher temperature gradient across the slag layer, and therefore the higher slagging rate. So a small change in the furnace temperature near the critical zone may make a significant change in slagging. Tests show that a 50°C drop in the furnace temperature can decrease the slagging fivefold. A proper choice of the volumetric and grate heat release rate is important to prevent excessive slagging in the boiler.

   Table 14-11 shows average values of volume heat release rate of boilers in China. Modern boilers have a higher heat release rate. For example, a modern cyclone furnace would use a heat release rate of 0.414–0.828 MW/m^3 as opposed to the 0.523–0.698 given in Table 14-11.

   Table 14-12 presents the statistical values of grate heat release rate of dry-bottom boilers in China. The recommended grate heat release depends on the slagging propensity of the coal. The lower limit of the figures are for coal with strong slagging tendencies.

TABLE 14-11. Statistical values of volumetric heat release rate load used in Chinese boilers.

Coal type	Dry bottom (dry ash)	Wet bottom (or molten ash) boiler		
		Open boiler	Half open boiler[a]	Cyclone furnace No.
Anthracite	0.110–0.140	<=0.145	<=0.169	0.523–0.698
Poor coal*	0.116–0.163	0.151–0.186	0.163–0.198	0.523–0.698
Bituminous	0.140–0.198	<=0.186	<=0.198	0.523–0.64
Lignite	0.093–0.151	—	—	—

[a] High ash, low-volatile coal.

TABLE 14-12. Grate or furnace cross section heat release rate of dry-bottom boiler used (MW/m^2).

Evaporation capacity of boiler →		220 t/h	400, 410 t/h	670 t/h
Tangential	Lignite	2.1–2.56	2.9–3.36	3.25–3.71
Tangential	Bituminous	2.32–2.67	2.78–4.06	3.71–4.64
Tangential	Anthracite, Poor coal	2.67–3.48	3.02–4.52	3.71–4.64
Opposed wall & front wall firing		2.2–2.78	3.02–3.71	3.48–4.06

TABLE 14-13. Range of burner zone volumetric (q_{rv}) and projected surface (q_{rf}) heat release rates used in Chinese designs of boilers for coals with different degrees of severity of fouling.

Fouling trend of coal	q_{rv} (kw/m^3)	q_{rf} (kw/m^2)
Low	500–600	410–440
High	500–550	380–410
Severe	450–500	350–380

d) **Heat release rate in the burner zone:** The heat release rate in the burner zone includes the projected surface heat duty, q_{rf}, and the volume heat duty, q_{rv}. These represent the heat flow rate and temperature level in the burner level. The temperature near the burner is higher and therefore it is easiest to slag. Table 14-13 gives values used in Chinese boilers. The projected area heat release rates used in USA and Europe are higher than these. For example, a 500 MW boiler would use q_{rf} 552, 545, and 505 kW/m^2 for opposed wall firing, front firing, and corner firing, respectively (BEI. 1991, p. 18).

e) **Tangential firing:** The bigger the firing circle diameter, the easier it is for the coal air suspension to hit the wall directly, causing slagging. So its diameter should be small.

f) **Distance between the burner and the wall:** In opposed jet firing, the distance between burners in opposite walls must maintain a certain minimum. Otherwise, it will cause fouling.

g) **Proper choice of tube material and gas temperature:** Low alloy carbon steel is often used in economizer tubes. It is generally heat treated to eliminate deposition on the surface and to allow development of the protective iron oxide layer on it. This iron oxide allows the formation of a strong ash layer. On the other hand, it is difficult to form a strong ash layer on austenitic steel, which is often used to prevent corrosion in high-temperature reheaters and super-heaters.

Nomenclature

A	tube position factor (Eq. 14-8)
A_w	$A_w = \frac{1000A}{LHV}$, %kg/MJ
ASH	ash percentage, %
B	material factor (Eq. 14-8)
C	constant °C (Eq. 14-8)
c	parameter in Eq. (14-28)
Cl	chlorine content of fuel, %
D	constant in Eq. (14-8)
d_{tube}	diameter of tube, m
E	constant in Eq. (14-8)
F_s	slagging index, °C
FT	flow temperature, °C
fs	correction factor used in Table 14-6 and Table 14-7
G	amount of soot or dust deposit, g/m^2 s
HT	hemispherical temperature, °C
HT_{max}	higher of reducing or oxidizing hemispherical temperature, °C
IT_{min}	lower of reducing or oxidizing initial deformation temperature, °C
ITD	initial deformation temperature, °C
K_0	coefficient (Eq. 14-26)
M	constant in Eq. (14-8)
m'	parameter in Eq. (14-28)
m	exponent in Eq. (14-8)
n	exponent in Eq. (14-8)
P_{H_2O}	partial pressure, Pa
P_{O_2}	partial pressure, Pa
P_{S_2O}	partial pressure, Pa
q	heat flux, W/m^2
q_{rf}	projected surface heat release rate, kW/m^2
q_{rv}	volumetric heat release rate, kW/m^3
r	corrosion rate, manometer/h
R	base-to-acid ratio
Re	Reynolds number
R_s	slagging index
R_{10}	size of particles less than 10%, μm
S	sulfur content in fuel, %
S_w	$S_w = \frac{1000S}{LHV}$, %kg/MJ
ST	softening temperature, °C

T_{dew}	dew point temperature, °C
T_g	gas temperature, °C
T_m	surface metal temperature, °C
T_{moist}	condensation temperature of water vapor at given partial pressure, °C
T_μ	temperature at which slag will have viscosity μ, °C
$T_{250,oxide}$,	Temperatures at which the ash viscosity is 250 poise in oxidizing
$T_{10000,red}$	atmosphere and 10,000 poise in reducing atmosphere, respectively, °C
ΔT	temperature difference, °C
W	velocity, m/s
x	thickness of ash layer, m

Greek Symbols

α	exponent in Eq. (14-26)
ΔT	temperature difference between tube surface and ash deposit, °C
λ	thermal conductivity, W/m °C
μ	dynamic viscosity, Ns/m^2
ν	kinematics viscosity, m^2/s

Dimensionless Number

Re	Reynolds number, Wd/ν

References

ASME International (1995) "Recommended Rules for Care of Boilers." ASME Boiler and Pressure Vessel Code, Section VII, Subsection C8.410.

Babcock & Wilcox (1992) "Steam." S.C. Stultz and B. Kitto, eds. pp. 42-20,22.

Bradford, S.A. (1993) "Corrosion Control." Van Nostrand Reinhold, New York, p. 224.

Breuker, H., and Swoboda, (1985) "Problem bei der Verfeuerung Salzhaltiger Braunkohlen-Versch lackung, Verschmutzung, Korrosion and Verschleiss," VGB Kraftwerks-technik, Hett 5.

BEI, (1990) British Electricity International, Modern Power Station Practice, Pergamon Press, Oxford.

British Electricity International Ltd. (BEI) (1990) "Modern Power Station Practice," Vol. B, 3rd ed. Pergamon Press, London, p. 16.

Callister, Jr., W.D. (1991) "Materials Science and Engineering," 2nd edition, John Wiley & Sons, Inc., New York, p. 565.

Cen K.F., Fan, J.R., Chi, Z.H. and Shen, L.C. (1994) "The Principles and Calculation of Erosion, Fouling and Corrosion of Boiler and Heat Transfer." Science Press, Beijing.

Cen, K.F. (1996) Personal communication. Institute of Thermal Power Engineering, Zhejiang University, Hangzhou, China.

Diamant, R.M.E. (1971) "The Prevention of Corrosion." Business Books Limited, London, pp. 191-192.

Department of Energy (1998) Materials & Components News Letter. 134:1.

He, P.A. and Zhang, Z.X., (1990), "Experiment Study on Slag Fouling Characteristics of the Chinese Power Coal," Report of Harbin Power Appliance Institute.

Jones, D.A. (1992) "Principles and Prevention of Corrosion." p. 390.

Lai, G.Y. (1990) High-temperature corrosion of engineering alloys. ASME International, p. 152.

15
Erosion Prevention in Boilers

Erosion is a process of wear in which materials are removed from a solid surface by the action of solid particles impinging on it. This type of wear is common in many industrial devices, including boilers. The combustion products of coal contain fly ash particles, which impinge on boiler tubes or fan blades and erode them. Fly ash erosion is the second most important cause for boiler tube failure (DOE, 1998). Tube failures by erosion in some cases account for about one-third of all tube failures in a boiler. Studies on erosion show that the average rate of reduction in metal thickness of eroding tubes in pulverized coal fired (PC) boilers varies from 2.0×10^{-5} μm/s to 15×10^{-5} μm/s. At the highest rate of erosion, tube failures may occur after 16,000 h in service. This chapter explains the mechanism of erosion in fossil fuel fired boilers, presents models for estimation of the extent of erosion, and discusses methods for prevention of erosion.

15-1 Theory of Erosion of Heating Surfaces

Finnie (1958) explained many aspects of the erosion of ductile materials impacted by streams of angular-shaped rough particles. A hard angular particle impinging upon a smooth surface at an angle of attack α will cut into the surface much like a sharp tool. The ductility of the attacked material determines its ability to undergo plastic deformation during the cutting process. By finding the volume removed from the target by the mechanical action of a single particle of mass, m, inferences can be made about the wear arising owing to a larger quantity of simple cutting bodies of total mass, M.

Suppose a gas carrying ash particles of mean diameter d_p is scouring the tubes with a velocity v_p. The impulsive force, F, on a point (Fig. 15-1) of the tube surface would have two components: normal force, F_N, and radial force, F_T. So we can write

$$F_N = F \sin \alpha; \quad F_T = F \cos \alpha \qquad (15\text{-}1)$$

The normal force, F_N, creates pits on the surface and produces shavings. The tangential force, F_T, then removes the shavings. In this way, scouring or erosion

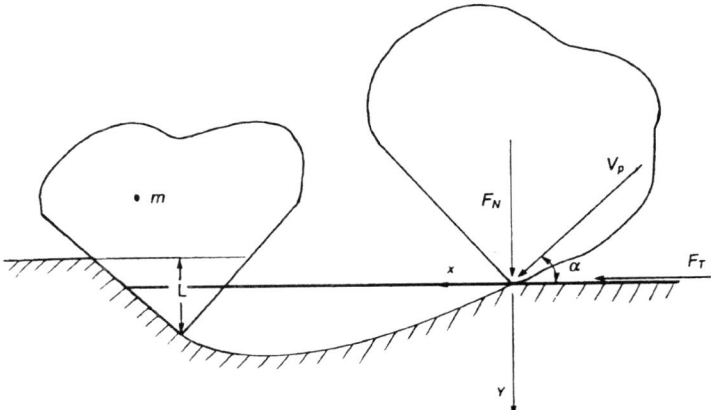

FIGURE 15-1. Mechanism of erosive cutting of a ductile material by the impact of an abrasive particle at an angle α

occurs. The larger the α, the deeper the pit, but the force F_T is reduced. When a tube is scoured uniformly by ashes, the wear differs with the position on the circumference. The metal removal is greatest when α is around 45°. It is reduced as α moves to toward 90° or 0°. An impingement at a very small angle or normal to the surface does not cause much erosion. For this reason in a circulating fluidized bed where the ash particles travel parallel to the wall, the erosion is lower than that in a PC boiler.

According to the theory of metal cutting, the force needed to cut away the metal shavings is

$$F = kbz^\varepsilon \text{ N} \tag{15-2}$$

where ε and k are constants based on metal properties and b and z are width and depth of the shaving chip in mm. The above equation is dimensional, so note the units.

It is assumed that ash particles do not deform and shatter on impact on the tube surface. We also assume that the shape of the pits in the metal surface is square. If N is the number of times of the ash particles hit the tube surface per unit time, the volume of metal removed per second, V is then

$$V = bzlN \tag{15-3}$$

where l is length of the chip and z is the depth of the pit or the thickness of the chip.

The depth of chip, z, depends on the normal component of the particle velocity, $v_p \sin \alpha$ in m/s, impact hardness of metal, H_d, in kg/mm² and diameter, d_p, of the particle in mm. It is given as

$$z = k_1 \frac{d_p v_p \sin \alpha}{\sqrt{H_d}} \text{ mm (Note: the equation is dimensional)} \tag{15-4}$$

where k_1 is a proportionality constant, and v_p is the particle impingement velocity.

Some typical values of impact parameters, H_d, k, and ε are given below:

TABLE 15-1. Some erosion parameters of typical metals.

	Carbon steel	Heat-resisting chrome nickel steel	Aluminum	Lead
H_d	240 kg/mm^2	313 kg/mm^2	26 kg/mm^2	80 kg/mm^2
k	117.55	142.5	79.5	32.9
ε	0.785	0.666	0.8	0.66

Using Eqs. (15-2) and (15-4), we get

$$F = kb\left[\frac{k_1 d_p v_p \sin\alpha}{\sqrt{H_d}}\right]^\varepsilon \tag{15-5}$$

The kinetic energy of the particle in the tangential direction is expended in removing the shaving from the surface. So this kinetic energy, KE, is calculated on the basis of the velocity component ($v_p \sin\alpha$) of the ash particles in the tangential direction:

$$KE = \frac{m v_p^2 \cos^2\alpha}{2} \tag{15-6}$$

This energy is expended in scouring over a length, l, which is the length of the chip. So

$$KE = Fl \tag{15-7}$$

Solving for l,

$$l = \frac{m v_p^2 \cos^2\alpha}{2kb(k_1 d_p v_p \sin\alpha/\sqrt{H_d})^\varepsilon} \tag{15-8}$$

The number of ash particles, N, hitting a tiny area, f, on the surface of a tube per unit time is

$$N = \frac{\mu}{mg} v_p f \sin\alpha \tag{15-9}$$

where m is the mass of one particle, μ is the weight (not mass) concentration (kg/m^3) of fly ash in the gas, and v_p is the velocity of particle impingement on the surface.

In the case of a tube, if the angle of impact to the plane normal to the gas flow is α, the elemental area, f, covered by an elemental angle $d\alpha$ at the center of the tube is

$$f = Lr\,d\alpha \tag{15-10}$$

where L is the length and r is the radius of the tube. By substituting Eqs. (15-4), (15-5), (15-9), and (15-10) into Eq. (15-3), we could get the elemental volume,

dV, of the metal remove d per unit time as follows:

$$dV = \frac{k_1 v_p^4 d_p L r \mu \sin^2 2\alpha}{8g\sqrt{H_d}\left(\dfrac{k_1 v_p d_p}{\sqrt{H_d}}\right)^\varepsilon k \sin^\varepsilon \alpha} \, d\alpha \text{ mm}^3/\text{s} \qquad (15\text{-}11)$$

For a given parallel stream of particles the impact angle α on the tube surface would change around it. The erosion of the surface facing the gas is found by integrating Eq. (15-11) between $-90°$ and $+90°$. Assuming symmetric erosion, we get.

$$V = 2\int_0^{\pi/2} dV = \frac{L \cdot r (7 - \varepsilon) k_1^{1-\varepsilon}}{4g\left(H_d^{1-\varepsilon}(3-\varepsilon)(5-\varepsilon)\right)^{0.5}} \frac{\mu d_p^{1-\varepsilon} v_p^{4-\varepsilon}}{k} \qquad (15\text{-}12)$$

The eroding surface area of the tube is $2\pi r L$. So the weight of the metal removed per unit surface area per second is

$$G = \frac{V\rho}{2\pi r l} = \frac{(7-\varepsilon)k_1^{1-\varepsilon}\rho}{8\pi g\left(H_d^{1-\varepsilon}[3-\varepsilon][5-\varepsilon]\right)^{0.5}} \frac{\mu d_p^{1-\varepsilon} v_p^{4-\varepsilon}}{k} \qquad (15\text{-}13)$$

where ρ is density of the metal. All ash particles do not impinge on the tube equally. If we introduce a frequency factor η to take the randomness of the impact of ash particles into account, the metal removal rate is

$$G = 0.5\eta c M' \mu d_p^{1-\varepsilon} \gamma_p^{4-\varepsilon} \qquad (15\text{-}14)$$

where

$$c = \frac{k_1^{1-\varepsilon}}{8\pi g},$$

and

$$M' = \frac{(7-\varepsilon)\rho}{\left[H_d^{1-\varepsilon}(3-\varepsilon)(5-\varepsilon)\right]^{0.5} k}$$

The depth of erosion, E, of the tube is found as

$$E = \frac{G}{\rho} = 0.5\eta c \frac{M'}{\rho} \mu d_p^{1-\varepsilon} v_p^{4-\varepsilon} \qquad (15\text{-}15)$$

Some average values of the frequency factor, η, are given in Table 15-2. Generally, the exponent ε is chosen as 0.7–0.9. So the dependence of the depth of erosion on the particle velocity is given as

$$E \propto v_p^{4-\varepsilon} \approx v_p^{3.1-3.3} \qquad (15\text{-}16)$$

TABLE 15-2. Average values of frequency factor of randomness of impact, η.

Gas temperature	300°C	600°C
Tube diameter = 32 mm	0.25–0.35	0.2–0.3
Tube diameter = 42 mm	0.2–0.3	0.15–0.2

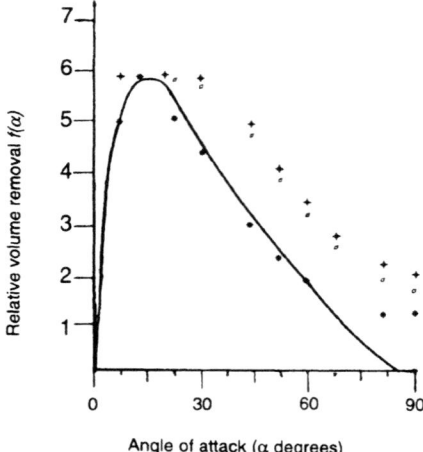

FIGURE 15-2. Predicted variation of volume removed by the impact of an abrasive particle at angle α. Experimental points are for erosion of particles copper (Δ), SA 1020 steel (+), aluminum (o)

Equation (15-11) gives the erosion rate of an elemental area for an impact angle α. To find the maximum erosion rate we use

$$\frac{d(dV)}{d\alpha} = 0$$

Taking the differential of Eq. (15-11), we get

$$\tan^2 \alpha = \frac{2 - \varepsilon}{2} \qquad (15\text{-}17)$$

Substituting $\varepsilon = 0.78$ for carbon steel in Eq. (15-17) we find that the position of maximum erosion is at the $\alpha = 38°$. Figure 15-2 shows how the predicted as well as the experimental wear volume changes when α changes.

Following Eq. (15-15), the Chinese Boiler Thermal Standard (1973) gives the following empirical equation for estimation of the maximum erosion rate.

$$E_{max} = aM\mu k_\mu \tau (k_v V_g)^{3.3} R_{90}^{2/3} \left(\frac{1}{2.85k_p}\right)^{3.3} \left(\frac{S_1 - d}{S_1}\right)^2 C_t \eta \qquad (15\text{-}18)$$

where C_t is a factor taking account of strike efficiency, τ is the time of erosion of the tubes, μ is the density of the fly ash in the gas in the calculation section of the tube banks (g/m^3), k_p is the ratio of gas velocities at designed and average running load. Here, k_μ and k_v are coefficients accounting for the nonuniformity of the fly ash density and gas velocity field, respectively. The gas velocity in the narrowest section of the tube is V_g, and a is the erosion coefficient of fly ash of gas [mm $\cdot$ s^3/(g $\cdot$ h)]. R_{90} is the percentage of the fly ash smaller than 90 μm; M is the abrasion-resistance coefficient of the tube material.

For a double-pass arrangement, $k_v = 1.25$ and $k_\mu = 1.2$. When the gas turns 180° in front of the tube banks, $k_v = k_\mu = 1.6$. For a boiler with a steaming capacity of 120 t/h, $k_p = 1.0$ at designed load, but when the average load is 50–75 t/h, $k_p = 1.4$–1.3.

$M = 1$ for carbon steel tube, $M = 0.7$ for steel with alloy. If v_g is the average velocity of gas before it enters tube banks, then the gas velocity between tubes, $V_g = S_1 v_g/(S_1 - d)$. For the tubes in the first line, we could take $V_g = v_g$. The erosion coefficient a is related to the type of coal. If appropriate test data are not available, it is advisable to choose $a = 14 \times 10^{-9}$ mm $\cdot$ s^3/(g $\cdot$ h).

The design equation recommended in the Chinese Boiler Standard (1973) assumes that ash particles travel approximately at the same velocity as the gas. Therefore, V_g is substituted for V_p in Eq. (15-18). The effect of change in fly ash particle size is taken into account by R_{90}. The effect of change of pitch of the tube banks is taken into account by the factor for the tube bank:

$$\frac{S_1 - d}{S_1}$$

Equation (15-18) gives the highest erosion volume, with the frequency factor of ash particles $\eta = 1.0$. This means that all the ash particles impinge on the tube at the greatest force. But, in fact, some particles travel past the tube without hitting it. So the above equation uses a factor, strike efficiency η_s for a precision calculation. Section 15-4 describes it later. The value of η is a function of the *Stokes' number*, St, which is defined as

$$St = \frac{\rho_p d_p^2 v_p}{\rho v D} \tag{15-19}$$

where ρ_p and ρ are densities of ash particles and gas, respectively, and D is the diameter of the tube. The value of η can be determined according to the value of St. Ash particles have a wide size distribution, each size with its distinct frequency factor. So we divide the size into intervals, and solve them to find the average frequency factor of ash particle's η:

$$\eta_{av} = \frac{\eta_1 \Delta R_1 + \eta_2 \Delta R_2 + \cdots + \eta_n \Delta R_n}{100} \tag{15-20}$$

where $\Delta R_i = R_i - R_{i+1}$ is the volume between two successive sieves whose aperture sizes are i and $(i + 1)$.

For a typical fly ash gas flow, the average value of η is given in Figure 15-3. The higher values of η in the table correspond to higher gas velocities. In addition, changes in the surface temperature of the convection tube bundle will also affect the wear. This effect can be accounted for in the coefficient C_t in Eq. (15-18).

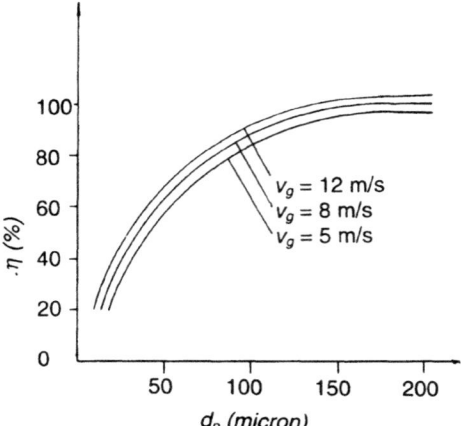

FIGURE 15-3. The frequency of ash particle impact on the boiler tube increases with velocity and particle size

15-2 Example

The economizer tube banks of a boiler (double-pass π construction) are eroded by flue gas entering the economizer, $T_g = 300°C$. The tube surface temperature is $T_w = 50°C$. The excess air coefficient is 1.3. The fly ash parameters are $R_{90} = 13.5\%$; $R_{200} = 1\%$; fly ash concentration μ is 25.6 g/m^3. The gas velocity is 7 m/s. The ash scours the tube banks in cross flow. The tube diameter, $d = 32$ mm; $S_1/d = 2$. $C_t = 0.51$.

a) Determine the depth of erosion into the wall of the tube in the second (row) after one year if the boiler operates 7000 h per year,
b) Find the expected life of the tube if the maximum loss in the tube wall thickness allowed is 2 mm.

SOLUTION

(a) From Eq. (15-18) we get

$$E_{max} = aM\mu k_\mu \tau (k_\upsilon v_g)^{3.3} R_{90}^{2/3} \left(\frac{1}{2.85kp}\right)^{3.3} \left(\frac{S_1 - d}{S_1}\right)^{3.3} \eta C_t$$

Take $a = 14 \times 10^{-9}$ mm·s^3/(g·h) from above, $M = 1$ for carbon steel tube, $k_\mu = 1.2$, $k_\upsilon = 1.25$ for π shape. Since the boiler runs at the design load, $k_p = 1.0$.

As the flow is through tube banks we use the gas velocity through the narrowest section.

$$V_g = \frac{v_g S_1}{S_1 - d} = \frac{7 \times 64}{64 - 32} = 14 \text{ m/s}$$

In addition, we can get $\eta = 0.3$, $C_t = 0.51$ (given), when $T_g = 300°C$. Substituting all the values in the above equation, we get

$$E = 14 \times 10^{-9} \times 25.6 \times 1.2 \times 7000 \times (1.25 \times 14)^{3.3} \left(\frac{1}{2.85}\right)^{3.3}$$

$$\times 13.5^{2/3} \left(\frac{64 - 32}{64}\right)^2 \times 0.3 \times 0.51 = 0.261 \text{ (mm/year)}$$

(b) If the wear depth of the surface is allowed to be 2 mm, the time the tube will last is

$$\frac{2}{0.261} \times 7000 = 53,640 \text{ (h)}$$

which is nearly 8 years in case of 7000 running hours per year. The tubes can fulfill the design requirements.

15-3 Factors Influencing Tube Erosion

The extent of surface erosion by impingement of abrasive particles depends on the following factors:

- Operating conditions (such as particle impinging velocity, impact angle, particle concentration, properties of the carrier gas)
- Nature of target tube material (such as material properties, tube orientation and curvature, and surface condition)
- Properties of impacting particles (such as particle type and grade, mechanical properties, size and sphericity)

Not all of these factors are controllable or even easily measurable during a test but one can attempt to estimate their relative importance. The following sections discuss the effects of these parameters.

15-3-1 Velocity

Particle velocity is by far the most important parameter affecting the erosion process. The erosion rate is proportional to the particle impingement velocity, v_p. It is given as

$$E \propto v_p^n \tag{15-21}$$

The exponent $n = 2.3$–2.7 for ductile materials and 2–4 for brittle materials.

TABLE 15-3. Average values of exponent, n.

Investigator	Japanese	Mitsui	Laitone (1979)	Cen Kefa
Average	3–3.5 ($v_g = 10$–20 m/s)	3.52	3	3.78 ($d_p = 50 \, \mu$m)
value of n	4.2–4.3 ($v_g = 30$–40 m/s)	($v_g = 8$–30 m/s)		3.3 ($d_p = 100 \, \mu$m)
				3.15 ($d_p = 200 \, \mu$m)

Different investigators gave different values of n (Table 15-3). The particles do not truly follow the flow path of the carrying gas stream. As a result their actual impact velocity and angle are often different from that of the gas. A particle (with inertia number less than 0.5), when carried by a gas stream normal to a flat surface, will hit the surface at an angle less than 90°. Laitone (1979) presented the relation between the incidence speeds of the gas and the particle:

$$v_p \propto V_g^m \qquad (15\text{-}22)$$

where the exponent m is a function of the average fluid velocity, V_g, and varies between 1 and 2.13. At very high fluid velocities, $m = 1$, regardless of the value of the inertia number. However, it decreases with velocity. For each particle size there is a critical gas velocity below which $m = 2.13$. The result is important because it gives a material independent explanation for the gas velocity exponents being greater than 2 observed in many ductile erosion experiments. To obtain this result, we substitute Eq. (15-22) into (15-21) and following Finnie (1958), take $n = 2$:

$$E \propto v_p^n = \left(V_g^{2.13} \right)^2 = (V_g)^{4.26} \qquad (15\text{-}23)$$

An important point to note is that the erosion is driven by the local velocity rather than the average design velocity. Thus the cure for erosion lies in reduction of the peak local velocities, which can be two to three times the average velocity. Thus, the rule of thumb suggestion of maximum bulk design velocity of 15 m/s is based on 30 m/s local velocities (DOE, 1998). Cold air velocity techniques (DOE, 1998) have been used to detect regions of high local velocities. Once these are detected diffusion or screens are installed to reduce the velocity peaks.

Figure 15-6 shows the variation of erosion of the first row of a bank of steel tubes impacted by three different types of particles at three temperatures. It shows that the values of the exponent n in the relation $E \propto V_g$ vary between 3.15 and 4.50.

15-3-2 Turbulence in the Flow

The influence of turbulence on erosion by particles was investigated by Dosanjh and Humphrey (1985). Erosion is found to decrease with increasing turbulence intensity. This finding is partly explained by the fact that both particle impact speed and particle flux to the surface decrease with increasing turbulence.

The position of maximum erosion, which is usually 25° with respect to the surface, is significantly displaced toward the jet symmetry axis with increasing turbulence intensity.

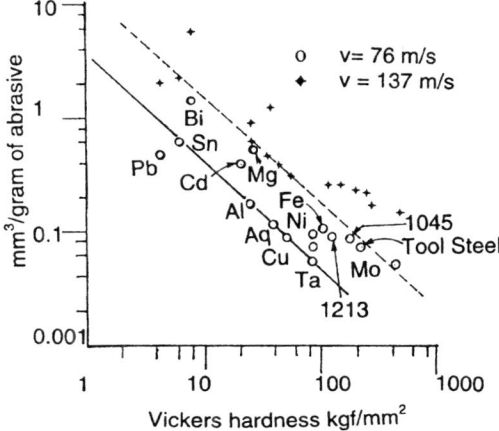

FIGURE 15-4. The volume of metal removed decreases with the Vickers hardness

15-3-3 Size of Ash Particle

Erosion is markedly influenced by the size of impacting particles. The rate of erosion is insignificant when impacting quartz particles are <5 μm. Small particles have insufficient momentum to impact the target. Also, they are carried around the object by the gas flow. The erosion rate increases when the particle size increases from 10 to 100 μm. One interesting aspect of erosion is that for particles larger than about 100 μm the volume removed by a given mass of abrasive grains is independent of particle size. The physical reasons for this size effect are still not clear.

15-3-4 Material of Tube

The effect of surface hardness on erosion is demonstrated in Figure 15-4 for various metals. Here metal surfaces were impacted by silicon carbide particles at an angle closer to the maximum angle of attack. Higher Vickers hardness gives a lower erosion rate. Farthest up from the lower solid line in Figure 15-4 lies data for the body-centered cubic metals (Fe, Mo, and three steels) that eroded much more than expected from their hardness versus erosion rate curve. This can be attributed to the role of small amounts of impurities.

15-3-5 Flue Gas Temperature

Erosion is a mechanical process. Therefore, the enhancement of erosion with temperature may be small, but localized melting owing to high-velocity impact can significantly increase erosion. In the latter case metal softening and localized shear deformation facilitate the formation of deep craters with extended edges from which metal pieces break off easily during subsequent impacts. Melting and re-solidification can also take place at the particle–metal interface. So when the particle rebounds it removes solidified metal material that has adhered to its surface. This mechanism is favored when target surfaces are already at high temperatures.

High temperature also reduces erosion (Yong and Ruff, 1977) owing to the increase in the metal ductility. This facilitates particle embedding, and the production of protective oxide films.

15-3-6 Fly Ash Concentration

Erosion of the tube material increases with increasing ash concentration. In general, the erosion rate is in direct proportion to the fly ash concentration, μ.

15-3-7 Physical Properties of Fly Ash

Fly ash concentration alone does not determine the extent of erosion. The constituent of ash is also important. Quartz ash occurs in coal in the form of discrete large particles, but the pyrite ash remains dispersed in the coal substance. So although the pyrite minerals can be as hard as quartz, the abrasion- and erosion-wear damage caused by pyrite minerals in coal is significantly less than that caused by the same quantity of quartz. A direct relationship between the rate of erosion and the quartz content of the coal was found (DOE, 1992). This is because the abrasiveness of pyrite component is between 20% and 50% of the quartz component of coal. Thus the overall abrasiveness of coal mineral, I_c, is

$$I_c = [C_q + (0.2 \text{ to } 0.5)C_p]I_q \qquad (15\text{-}24)$$

where I_q is the abrasiveness of quartz, and C_q, C_p are concentrations of quartz and pyrite in coal mineral matter, respectively.

The crystallographic structure of the quartz also plays an important role. For example, the α-quartz form is significantly more abrasive than other forms of the quartz in coal. A coal bearing α-quartz could cause significantly more erosion than one with plain quartz.

More angular quartz particles exhibit higher rates of erosion (DOE, 1992). However, sometimes quartz particles are coated over by softer materials that make them more rounded. This has been observed in fluidized bed boilers where alkali-induced low-melting-point eutectics resulted in the coating through complex chemical reactions in limestone beds.

15-4 Analyses of Erosion of Tube Banks in Cross-Flow

To predict the erosion of a tube one may use an empirical correlation, and calculate the mass erosion parameter for a given particle target material combination. Then by using computation results on impact velocities and impingement angles for a large number of particles, one can get the final result. The following section first gives an empirical relation of erosion rate in terms of particle velocity and angle of impact, then discusses results of computation of particle velocities and flow direction from fundamental equations.

15-4-1 Semi-Empirical Equation of Erosion

By impacting coal ash particles on a 304 stainless steel surface at different velocities, Tabakoff et al. (1979) developed the following empirical relation for the erosion mass parameter (E_r), which is the ratio of the eroded mass of the target material (mg) to the mass of impinging particles (g).

$$E_r = K_1 \left[1 + C_k \left(K_2 \sin \left[\frac{90}{\alpha_0} \alpha \right] \right) \right] v_p^2 \cos^2 \alpha \left(1 - R_1^2 \right) + K_3 (v_p \sin \alpha)^4, \text{ mg/g}$$

(15-25)

where v_p and α are the impact velocity and impingement angle, respectively

$$R_1 = 1 - 0.0016 V_p \sin \alpha$$
$$\alpha_0 = 25°, \text{ angle of maximum erosion}$$
$$C_k = 1 \text{ for } \alpha \le 3 \alpha_0$$
$$C_k = 0 \text{ for } \alpha > 3 \alpha_0$$

K_1, K_2, K_3 are constants specific to tube materials. For example, for stainless steel subjected to fly ash $K_1 = 1.5 \times 10^{-6}$, $K_2 = 0.296$, and $K_3 = 5.0 \times 10^{-12}$.

The prediction of flow field around the tube row and the particle trajectories as well as the tube mass erosion is complex, but it must be established before the above equation is used. To compute this, one could either use a commercially available computational fluid dynamics (CFD) software package or numerically solve the basic fluid dynamic equations. An example where the flow pattern was computed from solution of basic equations follows:

EXAMPLE

Air at atmospheric pressure and temperature is carrying coal ash particles. The impaction and erosion damage by coal ash particles are to be obtained numerically for tubes in two types of arrangements: (a) in-line tube banks and (b) staggered tube banks.
The following operating conditions are assumed:

Tube diameter = 30 mm	Tube material = AISI 304 stainless steel
Particle material = coal ash	Particle material density = 2450 kg/m³
Particle size = 20, 30, 40, 50, 100, 200 μm	Gas velocities = 6.10 m/s
Air density = 1.23 kg/m³	Tube arrangements $s_1 = s_2 = 2d$
Kinematic viscosity of air = 1.8 × 10⁻⁵ m²/s	

The equations and solution techniques are not discussed for reasons of brevity. They are available in standard textbooks. The flow field of particles and gas has been numerically computed using fluid flow equations. Results are shown in Figure 15-5, and their implications are discussed below.

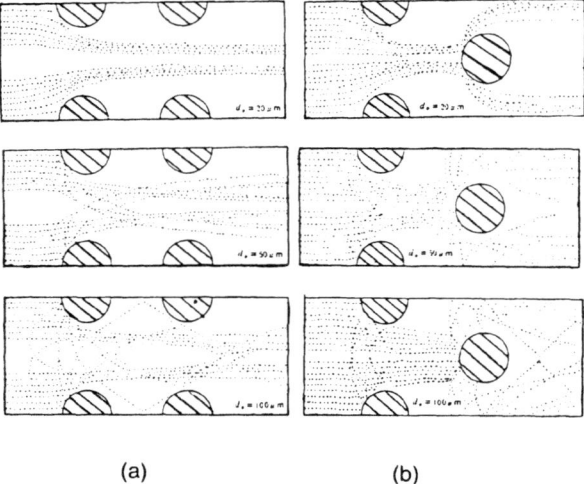

(a)   (b)

FIGURE 15-5. Computed path of solid particle trajectories on tubes of different arrangements

15-4-2 Particle Trajectories

The particle trajectory is affected by the fluid flow path and its properties. Figures 15-5(a) and 15-5(b) show particle trajectories for different particle sizes over tube banks in parallel and in a staggered arrangement, respectively. For the first-row tubes of tube banks, smaller particles deviate away from the tubes. Larger particles, because of their greater inertia, keep to their initial trajectory and impact on the surface of the tubes. They get deflected thereafter. For the second-row tubes of staggered tube banks, almost all particles of various sizes collide with the tube surface. But in the case of parallel tube banks particles rarely strike on the tube surfaces of the second row.

15-4-3 Strike Efficiency

Only a part of the particles initially aimed at a tube impact on it owing to the flow diversion. The strike efficiency, η_s, is defined as the number of particles striking the tube surface as a function of the number initially aimed at it. It depends on the gas velocity and particle sizes. The strike efficiency increases with the increase in particle size or gas velocity. In general, the strike efficiency on the second-row tubes of staggered tube banks is much greater than that on the first-row tubes.

There is a distribution of erosion rate around the tube surface. The erosion distribution function is defined as the erosion of the tube surface per unit center angle per mass of the particle initially aimed at half the tube surface. The strike efficiency distribution function along the tube surface is defined as the number of particles striking the tube surface per unit center angle over the number initially aimed at half the tube surface.

In the first-row tubes of tube banks, higher gas velocities and larger particles cause greater erosion. However, for the second-row tube of staggered tube banks, the smaller particles cause greater erosion than larger particles, because, in this case, the smaller particles have the higher strike efficiency and higher particle impact velocity. On the whole, the erosion damage to the second-row tubes of staggered tube banks is much greater than that to the first-row tubes.

15-5 Permissible Gas Velocity for Safe Operation

The specified service life of convective (back-pass) heating surfaces of a boiler is in the range of $60-100 \times 10^3$ h (the wear less than 0.2 mm/yr). For this service life, the maximum permissible reduction in the tube thickness is 2 mm. For a given coal and operating conditions, the designer could adjust the erosion allowance on the tube thickness to meet a certain service life or vice versa. The key problem here is how to determine the allowable gas velocity at the given condition. The erosion rate of tubes is closely related to the concentration of fly ash, gas velocity, tube geometry, and the type of coal. Figure 15-6 shows how the erosion increases with velocity. The arrangement of heating surfaces, operating conditions, and type of coal also affect the erosion of tube banks. To find the permissible gas velocity and the design parameters for tube banks (S_1/d, S_2/d, allowance on tube thickness), a computer-assisted testing method developed at Zhejiang University, China, may be applied to individual problems. This test technique is presently at an early stage of development. The following is an introduction to this simple semi-empirical method. Before that, we will first sum up the methods used to calculate the principal factors involved in the erosion of tube banks.

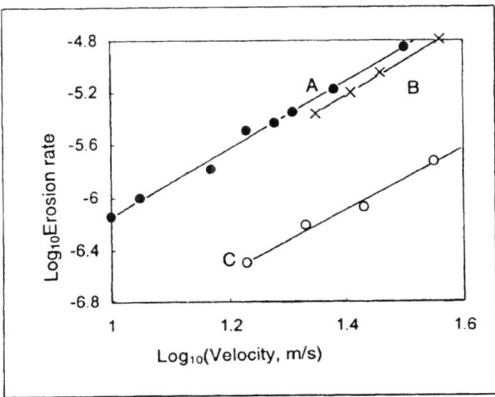

FIGURE 15-6. Effect of particle velocity on erosion of mild steel: the particle is (A) 100 μm quartz at 300°K, (B) quartz of 100 μm at 675°K, and (C) glass sphere at 300°K

Equation (15-18) is rewritten this time with the approximation $v_g = v_p$ to calculate maximum wear of the tube banks as

$$E_{max} = aMk_\mu k_v^n C_d C_\beta C_t v_p^n \mu k_p^{-n} \tau \text{ (mm)} \tag{15-26}$$

where C_d, k_v, k_μ, and n have been discussed earlier. M is the abrasion resistance coefficient ($=1.0$ for carbon steel); and τ is the number of, operating hours. If we know the allowable maximum wear depth of the tubes, E_{max}, and the run time, τ (h), required, we can get the maximum permissible fly ash impact velocity, v_p:

$$v_p \le \frac{k_p}{k_v} \left(\frac{E_{max}}{aMk_\mu C_d C_\beta C_t \mu \tau} \right)^{\frac{1}{n}} \text{ (m/s)} \tag{15-27}$$

Now, we can determine the allowable gas velocity, v_g, in the tube banks or the gas velocity, v_{go}, before it enters into the tube banks.

15-5-1 Velocity Distribution Factor, k_v

As indicated earlier the local velocities are often much higher than the bulk cross section average velocity. Erosion depends on the local velocity. To take account of this increased local velocity the cross section average velocity is multiplied by a velocity distribution factor, k_v. When the gas flow turns by 90°, k_v is chosen as 1.25; when the angle is 180° and the gas scours the tube banks, $k_v = 1.6$. When there is a gap between the tube banks and the enclosing walls, k_v is determined according to the size of the gap, as expressed in Table (15-4).

15-5-2 Concentration Distribution of Fly Ash

The fly ash is also nonuniformly dispersed in the flue gas. As the worst erosion depends on the highest concentration, the average ash concentration, μ, is multiplied by a factor k_μ to account for this variation. A change in direction of gas flow will cause fly ash concentration to increase in some areas. The result is that wear in those area increases by a factor k_μ. When the gas flow direction changes by 90°, $k_\mu = 1.2$, when the angle is 180°, $k_\mu = 1.60$.

15-5-3 Average Size of Fly Ash Particles

The dispersed phase of fly ash contains particles of different sizes, having different kinetic energies. While calculating the erosion volume (Eq. 15-12) we need to use an average diameter to represent the overall kinetic energy of the suspension rather than the mass average diameter of particles. The average diameter (d_{eff})

TABLE 15-4. The relations of gap width to, k_v, in shortcuts.

Gap width (mm)	8	16	24	32	40	48	56	64	72
k_v	1.1	1.4	1.61	1.76	1.88	1.98	2.06	2.12	2.18

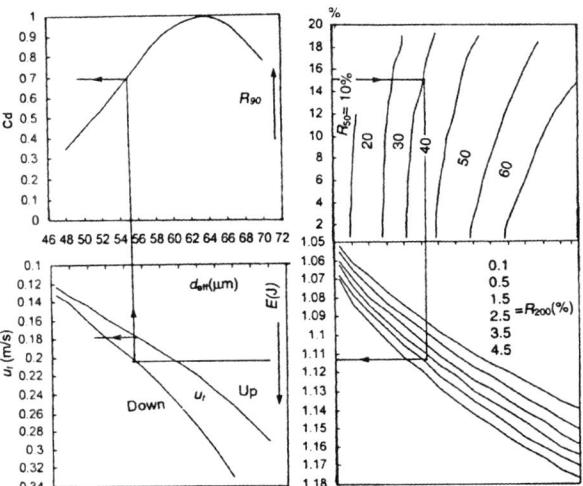

FIGURE 15-7. Nomogram for determination of erosion characteristics of fly ash

would depend on the size distribution or the sieve size analysis of the fly ash. For convenience, we can look up the value of d_{eff} in Figure 15-7 after knowing the size distribution of the fly ash given in terms of residual percent of R_{50}, R_{90}, and R_{200} from the sieve analysis.

15-5-4 Tube Arrangement

We notice from the computed trajectories of particles that in a parallel (in-line) arrangement of tubes, the first row of tubes have the maximum erosion (Fig. 15-5a). For the staggered arrangement of tubes the second-row tubes have the maximum erosion (Fig. 15-5b) provided by the horizontal pitch. If the pitch-to-diameter ratio (S_1/d) is ≥ 2, the peak erosion of tubes will gradually move toward the rear of the bank. The gas–particle flow through tube banks being very complex, we use a simple but safe conservative design approach to the problem.

After the gas enters into the tube banks, its free stream velocity, v_g, increases to V_g owing to the reduced flow area in the tube banks. This is given as

$$V_g = v_g \frac{S_1}{S_1 - d} \qquad (15\text{-}28)$$

where S_1 is the horizontal pitch, and d is the diameter of tube.

The velocity, V_p, of the particles impacting the tube surface is not the same as that (V_g) of the gas carrying it. So we use the following relation of gas velocity and impact velocity of the particles in the tube banks for a first approximation. Suppose the average particle's velocity is V_{po} before it enters into the tube banks:

$$V_{po} = v_g \pm v_t \ (\text{m/s}) \qquad (15\text{-}29)$$

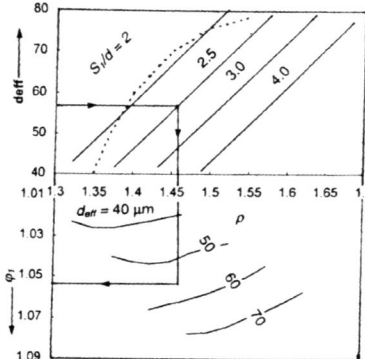

FIGURE 15-8. Nomogram for determination of φ and the coefficient p

where v_t is the terminal settling velocity of the particles whose value could be determined using the equation in Basu and Fraser (1991). In addition, the minus is used in ascending flow, plus in descending flow.

Suppose a particle (Fig. 15-5) misses the first-row tubes and impacts on the second row with a velocity, V_{p1}, and another particle bounces off the first row of tubes and hits the second row tubes with a velocity V_{p2}. Since the impact on the first row of tubes makes the latter particle lose some kinetic energy, $V_{p2} < V_{p1}$. If the average velocity of both types of particles hitting the second row of tubes is V_{pa}, then $V_{p2} < V_{pa} < V_1$, and we may write: $V_{pa} = V_{p1}/\phi_1$, where ϕ_1 is a function of both the diameter of the particles, pitch S_1, S_2, etc. The factor ϕ_1 could be determined from Figure (15.8).

When $S_1/d \leq 2$, the maximum erosion takes place in the second row of staggered tube banks. But when $S_1/d > 2$, some particles may miss both the first and the second row tubes. So they have time to be accelerated which results in $V_p < V_{pa}$. In this way, the maximum erosion moves downstream.

From Eqs. (15-28) and (15-29) we can find the allowable gas velocity over the tubes for a given allowable particle impact velocity, V_p:

$$V_g = \left(\frac{v_p \phi_1 \phi_2}{k'} \pm v_t \right) \frac{S_1}{S_1 - d} \qquad (15\text{-}30)$$

where $k' = V_{p1}/V_{po}$ for $\phi_1 = V_p/V_{pa}$. The relation of ϕ_2 with S_1/d is available from Table 15-5. The function ϕ_1 is found from Table 15-5.

Here the allowable particle impact velocity, V_p, can be obtained from the wear volume calculation (Eq. 15-27); and the coefficient, k', can be determined by the

TABLE 15-5. Values of coefficient ϕ_2.

S_1/d	2.0	2.5	3.0	3.5	4.0	4.5	5.0
ϕ_2	1.0	0.93	0.91	0.91	0.92	0.93	0.94

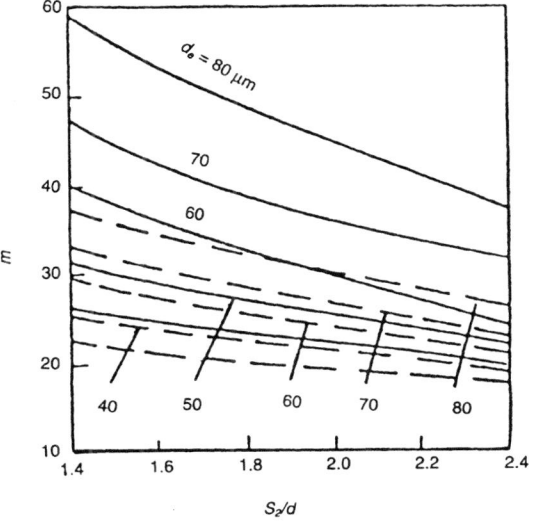

FIGURE 15-9. Graph for determination of m for both downward flow (solid line_____) and for upward flow line (-----x-----)

following formula:

$$k' = f(m, p, S_1/d) \tag{15-31}$$

The value of p can be taken from Figure (15-8), and that of m from Figure 15-9.

15-5-5 Worked-Out Example

Determine the allowable gas velocity in the economizer of a coal fired boiler. The evaporation capacity, $D = 1000$ t/h. The ash percentage in coal is $A_f = 41\%$. The fly ash consists of $R_{50} = 30\%$, $R_{90} = 15\%$, $R_{200} = 3\%$. The gas flows downward. The gas temperature is 700°C when it enters the tube banks. The tube diameter of the economizer is 32 mm, $S_1 = 80$ mm, $S_2 = 60$ mm. The required service life of the economizer is 60,000 h; and the allowable wear is 2 mm.

We get the following values from the above conditions and what we have discussed above; $k_p = 1.15$, $k_v = 1.25$ as the gas turns 90°, $k_\mu = 1.2$ for 90° turns. For steel, $M = 1.0$, and the fly ash density is

$$\rho = \frac{A_f a_y}{w_g} \frac{273}{\tau + 273} = 20 \ (\text{g/m}^3)$$

where a_y is the percentage of fly ash, w_g is the gas volume (nm³/kg), and

$T = 700°C$ is the gas temperature. From Figure (15-8) average diameter $d_{eff} = 58 \ \mu$m, $C_d = 0.05$, wall temperature $T_w = 350°$, and $C_t = 0.55$. The abrasion coefficient, a, is chosen as 6.4×10^{-9}, $C_\beta = 1$, $n = 3.22$.

Taking all the parameters above into Eq. (15-26), we get

$$v_p \leq \frac{1.15}{1.25} \left(\frac{2}{6.4 \times 10^{-9} \times 1 \times 1.2 \times 20 \times 0.05 \times 1 \times 0.55 \times 60000} \right)^{1/3.22}$$

$$= 6.2 \ (\text{m/s})$$

Moreover, from the Figure (15-8) and (15-9) we can look up that $m = 29$, $p = 1.39$, in addition $\phi_1 = 1.055$, $\phi_2 = 0.93$, $v_t = 0.23$ m/s; then we can get the allowable gas velocity in the tube banks on the basis of Eq. (15-28). The value of k is computed as 0.67.

$$v_g = \left\{ \frac{6.2 \times 1.055 \times 0.93}{0.67} + 0.23 \right\} \frac{80}{80 - 32} = 7.2 \ (\text{m/s})$$

15-6 Erosion Protection for the Economizer, Reheater, and Superheater

15-6-1 Reduction in Gas Short Circuiting

The wall region of the convective (back-pass) of a boiler is a common place for erosion. Here there is a gap between the tube banks and the boiler enclosure wall. That gap, which is usually several centimeters in width, allows a gas shortcut (Fig. 15-10). Because the flow resistance in the gap is lower than that in the tube bank, the gas will flow preferentially through it at a higher velocity, and therefore the tube bends exposed to the gap are seriously eroded. To reduce the gas velocity of this gas corridor one can increase the flow resistance of the gas corridor by different means. We summarize below some measures against the erosion:

(1) The velocity in the gas passage or corridor gradually increases along the depth of the bank as more and more gas from the tube bank makes shortcuts through the gas corridor. Furthermore, the greater the number of bends, the higher the difference in flow resistance between the bank and the corridor. So the number of the bends should be kept as low as possible from erosion considerations.

(3) Baffle tubes can be located in front of the economizer (Fig. 15-10). These tubes offer additional resistance to discourage short-circuiting of gas through the gas corridor (space between the tubes and the wall). The diameter, length, and arrangement of the baffle tubes have a power law influence on the velocity dimensionless number of the corridor.

(4) Guard tiles can be arranged (Fig. 15-11). Addition of guard tiles improves the horizontal flow, but it does not change the inlet velocity. The guard tile itself could guard the tube bends against erosion.

(5) A combination of the baffle tubes and guard tiles may improve the velocity distribution of the gas flow in the corridor.

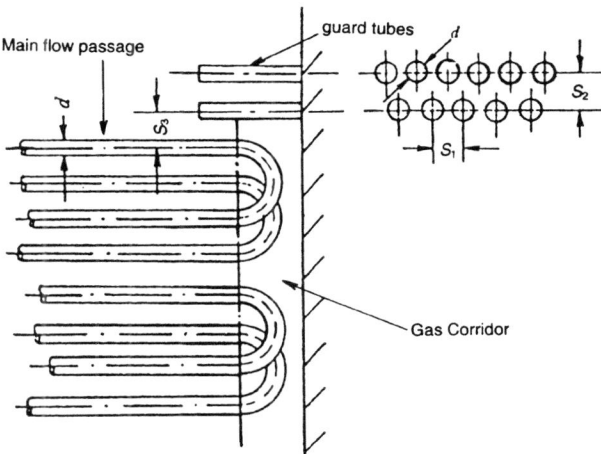

FIGURE 15-10. Higher amounts of gas pass through the gas corridor between the wall and the end of tube bends, this short by flues gases can be avoided by using guard tiles

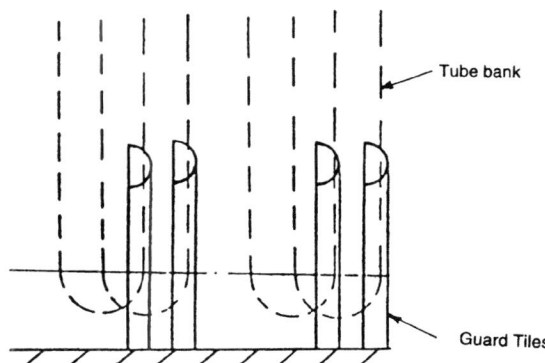

FIGURE 15-11. Guard tiles may protect tube bends by saving them from direct impact as well as by reducing the flow through the passage

(6) A grid resistance between two banks of bends increases the resistance through the gas corridor and thereby reduce the gas short circuiting.

(7) An alternative approach involves reduction in the resistance of the tube banks. This measure will decease the gas velocity in the outlet of the corridor. Following measures can be taken to reduce the resistance:

(a) Use the in-series arrangement of tube banks.
(b) Use the wider horizontal pitch.
(c) Use larger diameter tube for a given S_1/d. So, the $(S_1 - d)$ will increase. Thus the width of the corridor will decrease in comparison to the gap between tubes.

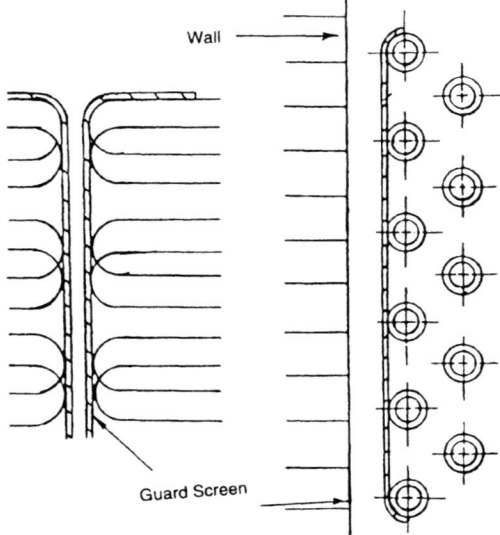

FIGURE 15-12. Guard screens may protect the tubes from direct impact

(9) Fin tubes indirectly reduce the erosion, as these decrease the number of the tube banks. As a result of the smaller number of banks, the resistance differential decreases. Therefore, there is less short-circuiting.

(10) A direct approach is to protect the whole bank of bends in the gas corridor with a guard screen (see Fig. 15-12). This will protect against the erosion of the bends in the corridor, because the guard screen separates the large ash particles that aggregated there because of the centrifugal force when the gas turns direction.

15-6-2 Use of Finned Tube Banks

A new approach to erosion protection, called the finned tube method, was been introduced by Cen et al. (1996). The cross section view of the tube and the flow over finned tubes are shown in Fig. (15-13a,b). Finned tubes reduce erosion because a large proportion of the particles that would have impacted the tube surface impact the fins and rebound from the finned tubes with an altered angle of impact and momentum. Therefore, the wear from impact of the solid particles on the tube wall is reduced. The principle of operation of this technique is explained below.

Figure (15-13) shows typical particle trajectories of particles different sizes. Very small particles ($d_p \leq 20$ μm) with low momentum barely touch the surface of the nonfinned tube. They are completely entrained by the fluid flow. Larger particles ($d_p = >100$ μm) deviate considerably from the gas streamlines because of their high inertia. These particles faithfully follow the original trajectories of the particles until they hit the tube surface or the fins. When fins are added, there are

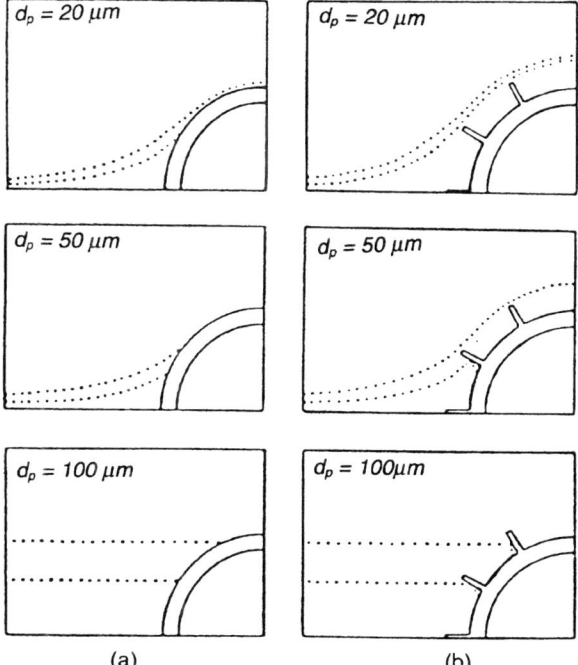

FIGURE 15-13. Fins on tubes help (a) deflect the fine particles away from the tube and (b) reduce the impact velocity of coarser particles

regions of flow recirculation in the vicinity of the fins and gas streamlines which tend to slow down the particles and change their trajectories. A large proportion of the particles, which would originally have hit the tube surface, now impact the fins and rebound from the finned tube surface with reduced momentum. Thus the fins prevent larger particles from impacting the tube surface directly. The addition of fins also changes the local flow patterns. This plays a strong role in reducing the erosion by smaller particles ($d_p \leq 50$ μm), because the flow field's influence is pronounced only for the smaller particles. However, in case of larger particles the erosion is dominated by their impacts.

The freestream velocity affects the erosion of finned tubes. The effect of particle size on the erosion of finned tube is same as that on non-finned tube. Larger particles cause greater erosion damage. However, for $d_p > 100$ μm, erosion damage does not increase with increasing particle size. Greater the fin height, greater is the reduction in erosion damage.

If E_{fin} is the erosion with fin and E_0 is the erosion without it, its ratio (E_{fin}/E_0) will decrease with increasing fin height. There is an optimum fin angle β between $0°$ and $90°$, which minimizes the ratio E_{fin}/E_0. The optimum angle β is around 20–$30°$. The ratio E_{fin}/E_0 decreases with increasing number of fins.

15-7 Erosion in Tubular Air Heaters

Flue gas passes through vertical tubes in an air heater. Thus its erosion mechanism is different from all other sections in the back pass where the flue gas flow is normal to the tubes. The following section discusses the mechanism and prevention of erosion of air heater tubes.

15-7-1 Gas Scouring

The entry section of the tubes is most prone to erosion in a tubular air preheater. The maximum wear occurs between one and three diameters from the tube sheet or the open end. The erosion rate is two to three times greater than its average value for the entire length of the tube.

When the gas enters a tube from an open space, the flow streams contract immediately after the entrance (Fig. 15-14). The minimum diameter of the vena contracta or the contracted flow stream is about $0.8d$, occurring at about $0.4d$ from the entrance independent of the Reynolds number. Owing to the high velocity in the throat or the shrunken area, a negative pressure occurs here. It then hits the wall further down at an angle, as shown in Figure 15-14. Continuous scouring of the surface combined with low pressure inside the tube makes the tube easily collapse in this region.

An interesting phenomenon is that finer particles give greater erosion at the tube inlet. This is because the eddies are formed in the throat or contracted part and carry

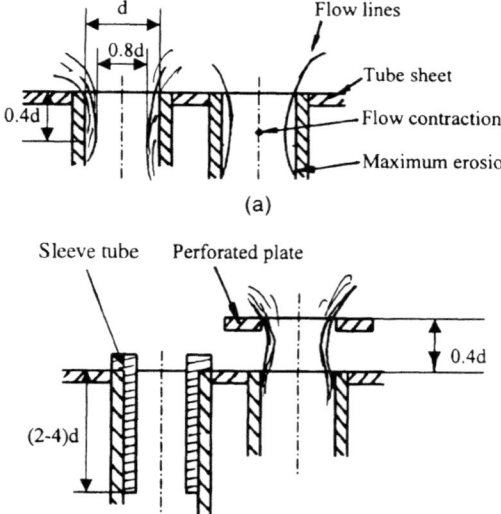

FIGURE 15-14. (a) Flue gas contracts at the entrance to the air heater tube, causing erosion during subsequent expansion, (b) several methods to reduce minimize the erosion owing to flow contraction in tubes

the fine particles more readily than coarser particles. Therefore, the erosion caused by the impact of the accelerated gas beyond the throat is higher for fine particles.

15-7-2 Methods of Calculation of Erosion at the Inlet of Air Preheater

In keeping with Eq. (15-18), the metal removed from a tube wall by gas with ash particles of diameter d is expressed as (CBCS, 1973)

$$E_d = aM\frac{\rho \cdot Y(d)}{100}v_p^{3.22}C_\beta C_d C_t \tau \qquad (15\text{-}32)$$

where ρ is the gas density, $Y(d)$ is the percentage of ash particles of diameter d, C_β is the coefficient taking account of the influence of the incidence angle of particles on erosion, and C_d is a coefficient that accounts for the influence of the diameter of ash particles. The value of C_d is determined by experiments. If the temperature of the tube wall is 150°C, $C_t = 0.73$.

The total erosion is the sum of contribution of each size of particles.

$$\sum_{d=0}^{d=m} E = \sum E_d \qquad (15\text{-}33)$$

The erosion increases along the length of the tube, starting from its inlet. It reaches maximum at $x = 2d$, and then decreases to an asymptotic value further down the tube.

15-7-3 Design for Prevention

Following are some preventive measures for reduction of erosion of the inlet part of air preheater of tubes:

1. A smooth transition section, at the entrance of the air preheater tube, may avoid sudden contraction of flow. This would reduce the formation of eddies which cause fine particles to impact on the walls of the tubes.
2. A sleeve tube, added at the entrance of each air tubes, may avoid erosion inside the tube. The sleeve tubes are usually 2–4d long. They can be changed during major repairs. The main idea of the measure is to let the sleeve tube to bear the erosion.
3. A perforated plate having the same diameter holes as the tube is added at about 0.4d above the tube plate. The gas passing through the hole in the top plate will expand to the size of the tube diameter, and then enter the tube after contraction. Here the flow contraction takes place in the open space outside the tube. As the flow is already contracted there would not be further contraction in the tube. There may still be some high velocity region. So this method is not as effective as adding the sleeve tube.
4. Here, a pore plate, slightly larger in diameter than the tube, is used. Usually it is 0.26d high and 1.2d in diameter, where d is the diameter of the tube. The

intent of the design is to lead gas to contract and decrease swirl in the inlet part of the tube.

15-8 Erosion in Fluidized Bed Boilers

The erosion of boiler tubes in fluidized bed boilers is different from that in pulverized coal (PC) fired boilers. The solid concentration in a fluidized bed boiler is considerably larger than that in a PC boiler. In a bubbling fluidized bed boiler the velocity is low. In circulating fluidized bed (CFB) boilers the velocity is higher, but the solids hit the furnace wall at a very low angle. Owing to the near parallel flow of solids on the wall the erosion is negligible in CFB boilers. In bubbling bed boilers erosion is a major problem. So several design methods have evolved over time. The present section discuss the erosion mechanism in these two types of boilers and then presents methods for erosion protection.

15-8-1 Circulating Fluidized Bed Boilers

The following areas of a CFB boiler are most prone to erosion (Fig. 15-15).

- Water wall to refractory junction on the wall
- Furnace roof
- Target area of cyclone
- Furnace floor where recycled solids enters the furnace

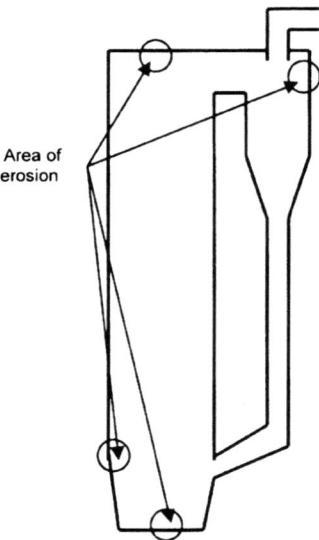

Area of erosion

FIGURE 15-15. Erosion-prone areas of a circulating fluidized bed boiler

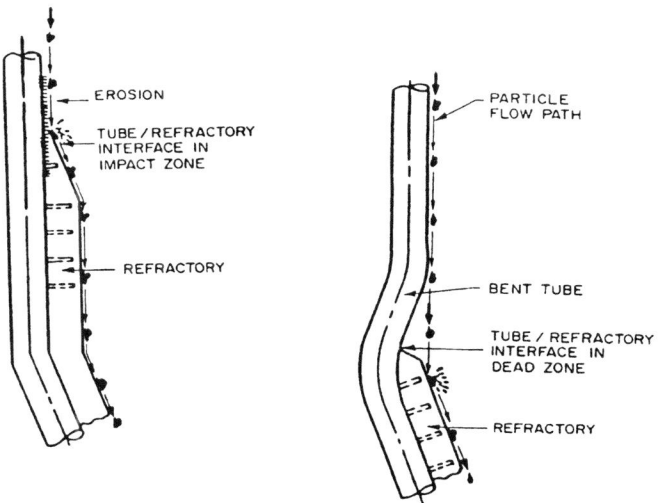

FIGURE 15-16. (a) Mechanism of erosion of boiler tubes in a circulating fluidized bed boilers, (b) The erosion can be reduced by moving the tube away from the point where there is greatest change in momentum of particles due to change in furnace area

In a CFB furnace a layer of slow-moving solids slides down the wall. Except for the regions where they are devoid of this layer, no other regions receive a direct hit of solids from the core of the furnace. When layers of solids flow along the wall, three-body erosion is the dominant mechanism of erosion (Basu & Fraser, 1991). In this system, a single layer of particles flows parallel to the wall. Other particles or agglomerate impact on the wall layer. The wall layer alone cannot cause erosion unless there is a normal force acting on it. In case of a vertical wall of a CFB, this force is absent. However, if there is a sudden change in the cross section of the furnace, the solids are forced to change their flow direction (Fig. 15-16a). This change in direction results in a change of momentum in the lateral direction. The force owing to this change in momentum provides the normal force to erode the wall tubes. For this reason, in most cases, this erosion was found to be maximal just above the change in cross and section.

The areas near the roof corners suffer from both erosion and corrosion owing to the formation of dead-zone reducing conditions. The boundary between high-erosion and high-corrosion areas fluctuates. The metal temperature at the roof is also higher, resulting in a higher rate of erosion and corrosion.

15-8-2 Bubbling Fluidized Bed Boilers

Accelerated wastage is experienced by water-cooled metal surfaces immersed in bubbling fluidized beds, especially when the metal surface temperature is in the

FIGURE 15-17. Erosion protection rings on tubes in bubbling fluidized bed boilers

range of 200–450°C. Motion of bubbles and its frequency on metal surface affect the erosion. Several types of surface treatment and addition of physical shield or armor minimize the problem.

a) Armor

Figure 15-17 shows some of the shields or barrier attached especially to the underside of the tubes to protect them from direct impact of bed solids. The studs or ball are welded directly on the tubes. Sometimes shields in the form of half tube or screen are placed on the edge of the tubes most prone to erosion. These are easy to insert in situ, but may reduce the heat transfer.

b) Coating or Metal Treatment

Nitriding (2.25Cr-1Mo) of bends or parts of the tube most exposed to erosion is an effective means of erosion protection (DOE, 1991). Flame spray with Extendalloy© (tungsten carbide particles in a nickel-base matrix) on the lower side of tubes proved very successful in the 160 MW AFBC boiler in the USA (DOE, 1991). British Coal Corporation found that Cm1 (1Cr-0.5 Mo-balance Fe) improved the wear performance of mild steel tube by 30%; however, weld overlay did not improve the performance much.

15-8-3 Measures Against Erosion in CFB Boilers

In a CFB boilers various protective measures have been attempted. These include

1. welding overlay
2. welding of shelves
3. metal spray
4. kick out

In weld overlay layers of hard metal are welded on the affected or erosion-prone areas of the tubes. Weld overlay may lead to embrittlement of the old overlay and formation of cracks that could propagate into the tube metal underneath (Solomon, 1998). Furthermore, the 4–6 mm thick weld overlay creates ledges that lead to erosion in areas above.

The welding of shelves above the water wall refractory wall junction brings a temporary relief. It directs the problem to other areas, and so other methods are preferred.

TABLE 15-6. Some suggested materials for CFB boilers (compiled from DOE [1991] and Dutheillet et al., [1999]).

Component	Subcritical boiler		Supercritical boiler	
	Refractory on hot end	Tube metals	Refractory on hot end	Tube metals
Furnace	Silica base	Water wall: grade 22	Alumina or SiC	Waterwall: grade 23 with coatings: 502/95XC
Roof	High conductivity SiC based castable	Covered tubes	High conductivity SiC-based castable	Covered tubes
Cyclone uncooled	Dense abrasion resistant super-duty or Mullite brick		Dense abrasion-resistant super-duty or Mulliter brick	
Cyclone water/steam cooled	Thin layer of phosphate bonded plastic refractory	Grade 22	Thin layer of phosphate-bonded plastic refractory	Grade 22 or 23
Cyclone inlet	Dense castable refractory containing stainless steel fibers and organic "burn out" fibers			
External heat exchanger	Silica or alumina base; fused silica for division walls	HTR: Z6CNT 18-12B	none	Casing: grade 91 HTR: super 304H, HR3C
Cyclone dip-leg and loop seal	Vibration cast insulating casable covered by dense shock-resistant fused silica-based castable containing stainless steel fibers and burnout fibers			
Header, pipes		Grade 22– Grade 91		Grade 91 Grade 92/122

Thermal spray involves spraying a thin layer (600 μm) of hard material by arch, flame, or plasma on the tube. Thermal spray does not alter the substrate. A wide range of spray materials are available. A high-velocity oxygen fuel spray of coating materials exhibited the best resistance to erosion–corrosion in CFB boilers (DOE, 1998). For water walls, high-velocity oxygen fuel spray of Cr-Tic/Ni-Cr coatings proved most successful, while Al_2O_3/NiAl coating is recommended for upper furnace sections, roof, and superheater sections (DOE, 1998). Material 502 with composition 20Cr-3 Ti-1.25Si-1Mn and material 95MXC with composition 28Cr-3.7B-1.7Si-Fe (balance).

The kick out proved to be the most effective erosion-preventing measure (Basu & Fraser, 1991). Here, wall tubes are slightly bent outward (Fig. 15-14b), forming a cavity in the region of cross section change. This cavity is filled with hard refractory. Thus, the erosion now occurs on the refractory wall instead of the metal tubes. Refractories are considerably harder than boiler tubes. So they experience minimal erosion.

Refractory failure is a problem in CFB boilers. It may occur owing to thermal differences between the refractories and their anchors. Sometimes the joints between refractory panels or fine cracks in the refractory get filled with bed solids when the furnace is cold. Later, when heated, the panels cannot expand, causing stresses that lead to buckling of large areas and falling chunks of refractory. Table 15-6 presents a suggested list of refractory materials for CFB boilers.

Nomenclature

a	abrasion coefficient, $mm \cdot s^3/g \cdot h$
A_f	ash percentage in coal, %
a_y	percent of fly ash, %
b	width of shaving chips, mm
C_d, C_t, C_β	coefficient (Eq. 15-26)
C_k	coefficient (Eq. 15-25)
C_p	concentration of pyrite in coal ash
C_q	concentration of quartz in coal ash
C_t	factor taking account of strike efficiency in Eq. (15-18)
d	tube diameter, mm
d_{eff}	average diameter, m
d_p	diameter of particle, mm
E_d	erosion of air heater tube, kg/s
E_{fin}	erosion with fin
E_0	erosion without fin
E_{max}	maximum erosion rate, mm/h
E	depth of erosion, mm/s
E_r	ratio of the material removed and mass of impacting solids, mg/g (Eq. 15-25)
E_θ	erosion distribution function
F	impulse force, N
f	elemental area on the surface of the tubes, mm^2
F_N	radial force, N
F_T	tangential force, N
G	solid removed ratio, $g/mm^2 s$
H_d	impact hardness, kg/mm^2
I_c	overall abrasiveness of coal mineral
I_q	abrasiveness of quartz
k	constant (Eq. 15-2)
k'	velocity ratio (Eq. 15-30)
K_0	material constant (Eq. 15-32)
K_1	proportional coefficient of Eq. (15-4)
K_1, K_2, K_3	material constants (Eq. 15-25)
k_μ	coefficient of nonuniformity of the fly ash density
k_p	ratio of gas velocities at designed and average running load, respectively

k_v	coefficient of nonuniformity in gas velocity field
l	length of shaving chip, mm
L	tube length, mm
M	abrasion coefficient
m	exponent in Eq. (15-22)
m	mass of one particles, g
p	exponent (Eq. 15-31)
N	number of times of ash particles impacts on the surface per unit time, s^{-1}
n	exponent (Eq. 15-21)
r	tube radius, mm
R_1	parameter (Eq. 15-25)
R_{50}	percentage of fly ash less than 50 μm
R_{90}	percentage of fly ash less than 90 μm
S_1	horizontal pitch of tubes distance, m
S_2	
T	temperature, °C
T_w	wall temperature, °C
U	stream velocity, m/s
V	volume of metal removed, mm^3/s
V_a	average velocity, m/s
V_g	actual gas velocity between tubes in the tube bank, m/s
V_p	particle velocity impacting the tube surface, m/s
v_{po}	average particles velocity, m/s gas velocity , m/s
v_g	gas velocity in free stream or outside tube bank, m/s
v_p	velocity of particles, m/s
v_{pr}	velocity of ash particles, m/s
v_t	terminal setting velocity of the particles, m/s
W_{pl}	impact velocity, m/s
w_g	volume of gas, nm^3/kg
$Y(d)$	percent of ash particles of diameter d
z	depth of shaving chip, mm

Greek Symbols

α	angle of impact or impingement angle
α_0	angle of maximum erosion
ε	exponent in Eq. (15-2)
β	optimum fin angle, °C
β_0	maximum erosion angle
ϕ_1, ϕ_2	coefficient (Eq. 15-30)
μ	weight concentration of the ash in the gas, g wt/mm^3
ρ	density of gas, kg/m^3
ρ	fly ash density, g/m^3
τ	run time, h
η	frequency factor of ash particles

η_s strike efficiency
ξ coefficient (Eq. 15-14)

Dimensionless Number

St Stokes' number, $\rho_p d_p v_p / \rho v D$

References

Basu, P. and Fraser, S. A. (1991) Circulating fluidized bed boilers—design and operations, Butterworth-Heinemann, Boston.

Bitter, J.G.A. (1963a) A study of erosion phenomena (Part I). Wear 6:5–21.

Bitter, J.G.A. (1963b) A study of erosion phenomena (Part II). Wear 6:169–190.

Cen, K.F., and Fan, J.R. (1988) A numerical model for the turbulent fluctuation and diffusion of gas-particle flows and its application in freeboard of a fluidized bed. Particulate Science and Technology 6(1).

Chinese Boiler Standard Calculation (1973) CBSC.

Dosanjh, S., and Humphrey, J.A.C. (1985) The influence of turbulence on erosion by a particle-laden fluid jet. Wear 102:309–330.

DOE (1991) Materials & components. US Department of Energy Newsletter, 92 (June 1): 5–7.

DOE (1992) Materials & components. US Department of Energy Newsletter 98 (June 1): 1–5.

DOE (1998) Materials & components. US Department of Energy Newsletter 135 (August 1):7–11.

Dutheillet, Y., Dorier, C. and Houzet, S. (1999) Refractory and metallic materials for a 600 Mwe CFB boiler with advanced steam parameters. Proceedings 6th International conference on circulating fluidized beds, Wurzburg, Germany.

Fan, J.R., Zhou, D.D., Jin, J. and Cen, K.F. (1991) Numerical simulation of the tube erosion by particle impaction. Wear 142:171–184.

Fan, J.R., Zhou, D.D., Zeng, K.L., and Cen, K.F. (1992) Numerical and experimental study of finned tube erosion protection methods. Wear 152:1–19.

Finnie, I. (1958) The Mechanism of Erosion of Ductile Metals: Proceedings 3rd U.S. National Congress of Applied Mechanics. Pergamon Press, London.

Finnie, I. (1960) Erosion of surfaces by solid particles. Wear 3:00–00.

Hussian, M.F., and Tabakoff, W. (1974) Computation and plotting of solid particle flows in rotating cascades. Computers and Fluids 2:00–00.

Laitone, J.A. (1979) Erosion prediction near a stagnation point resulting from aerodynamically entrained solid particles. Journal of Aircraft 16:809–814.

Patankar, S.V. (1980) Numerical Heat Transfer and Fluid Flow. McGraw-Hill, New York.

Raask, E. (1969) Erosion by ash impaction. Wear 13:301–315.

Solomon, N.G. (1998) Erosion resistant coating for fluidized bed boiler. Materials Performance, February, pp. 38–43.

Spalding, D.B. (1977). "GENMIX: A General Computer Program for Two-Dimensional Parabolic Phenomena." Pergamon Press, London.

Tabakoff, W., Kotwal, R., and Hamed, A. (1979) Erosion study of different materials affected by coal ash particles. Wear 52:161–173.

Yong, J.P., and Ruff, A.W. (1977) Particle erosion measurements on metals. ASME, Journal of Engineering Materials and Technology 99:121–125.

16
Pressure Drop in Gas and Air Ducts

The purpose of a boiler draft system is to provide the air needed for the combustion of fuel and to discharge the flue gas produced in the furnace. In early days, gas passages in boilers were large. As a result the pressure drop owing to the flow of fluids through these passages was small. So most boilers did not need any external means for moving the fluid through the system. As with an open fire, air was driven through the fuel grate and the boiler furnace simply by the buoyancy force of the hot flue gas. This is called *natural draft*. Modern boilers are much more compact. So the resistance through the system is too large for the natural draft to overcome. A fan is required to force the gas through the system. This type of system is, therefore, called *forced* or *mechanical* draft.

Figure 16-1 shows the schematic of a typical draft system of a modern power boiler. Air from the atmosphere is drawn by forced draft fans and delivered to the air heater. This air passes through the air heater and coal mill. It then enters the furnace, where the fuel burns to produce the flue gas. The flue gas passes through back-pass heat transferring sections, air heater, and air pollution equipment. Finally, an induced draft fan draws the gas and releases it to the atmosphere through a tall chimney. A reasonable assessment of the magnitude of the pressure drop through the system is essential to specify the fans. Furthermore, in most fans the discharge volume varies with the imposed resistance. Thus a knowledge of the system resistance is also important for the assessment of the flow through the entire boiler draft system. The present chapter describes different draft systems and presents methods for calculation of the system resistance.

16-1 Draft Systems

In a modern boiler a large number of heating surfaces are located in a very compact space. As a result a mechanical means, i.e., fan, is required to overcome the flow resistance in its compact flow passages.

In natural draft systems the buoyancy force carries the gas through the system. The mechanical draft uses a fan to draw the air and gas through the system. Mechanical draft can be one of three types: negative, positive, and balanced draft. A forced

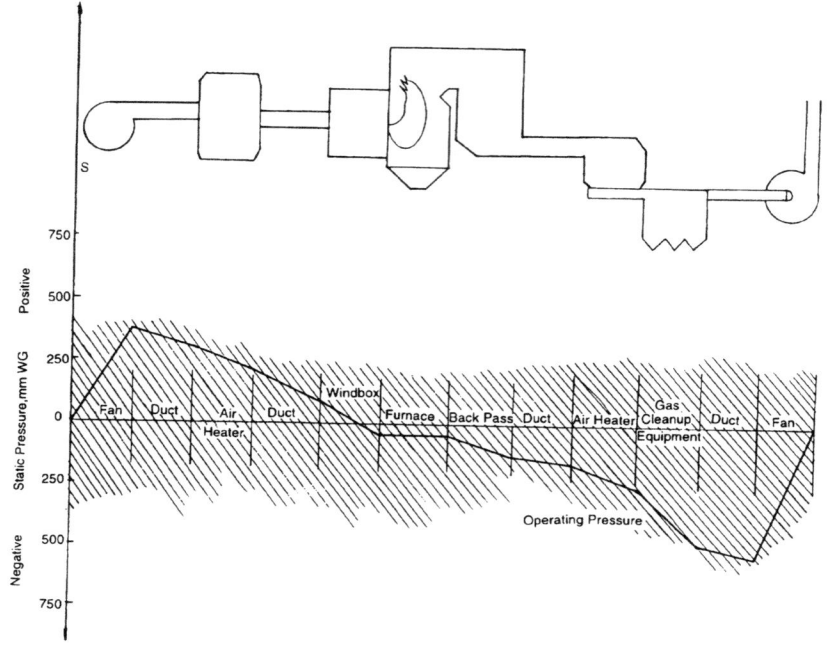

FIGURE 16-1. A typical draft plan of a boiler plant

draft fan draws cold air from the atmosphere and forces it through the boiler draft system. An induced draft fan draws air out of the boiler draft system and forces it back into the atmosphere. A combination of these two gives the balanced draft.

16-1-1 Suction or Negative Draft

In this system, most of the boiler gas and air passage remains at slightly below the atmospheric pressure. An induced draft fan located downstream of the system draws gas and air from the system by overcoming the total resistance of the furnace, air, and gas ducts. Its disadvantage is that a large amount of cold air is drawn into the furnace, increasing stack losses and unburned losses. This reduces the boiler efficiency. Furthermore, the fan power consumption is higher for the negative draft system. Owing to the higher flue gas temperature an induced draft will handle a larger volume of gas than a forced draft fan would pump from the colder end of the draft system.

16-1-2 Pressurized or Positive Draft

Forced draft fans are located upstream of the boiler. These fans pump air to a sufficiently high pressure to overcome the resistance of the air and gas ducts system. The pressure in the furnace is above the atmospheric pressure, and therefore it is called pressurized draft. This type of draft requires that the gas ducts and the furnace, in particular, are gas tight. Otherwise the hot gas, flame, and ash could dangerously leak out of the furnace polluting the environment and endangering the safety of operators. In case of a coal fired furnace, a positive furnace pressure makes it very difficult to keep the boiler house clean. An advantage of the system is that air leakage in the furnace is eliminated. This improves the intensity and efficiency of combustion. In addition, the erosion by fly ash is avoided since there is no induced draft fan to contact the fly ash. Pressurized combustion is used generally in oil and gas fired boilers.

16-1-3 Balanced Draft

Balanced draft uses both forced and induced draft fans. The forced draft fan pumps air into the furnace, and the induced draft fan draws the flue gas out of the furnace. So it is called *balanced draft*. The fans are chosen in such a way that the pressure in the furnace is slightly below the atmospheric pressure. In general, the pressure at the furnace exit is kept 20–30 Pa below the atmospheric pressure. This allows safe operation, keeps the boiler house clean and at the same time reduces excessive leakage of air into the furnace. Furthermore, since the induced draft fan needs to overcome only a small fraction of the system resistance, it does not consume excessive power like the negative draft system. Balanced draft method is widely used in power plant boilers.

16-2 Pressure Drop in Air and Gas Duct Systems

The well-known Bernoulli equation can give the pressure drop across a flow passage. With reference to Figure 16-2, it may be written in a simplified form as

$$\frac{P_1}{\rho_1} + \frac{u_1^2}{2} + gz_1 + \frac{\Delta W}{\Delta m} = \frac{P_2}{\rho_2} + \frac{u_2^2}{2} + gz_2 + \frac{\Delta P}{\rho_2} \qquad (16\text{-}1)$$

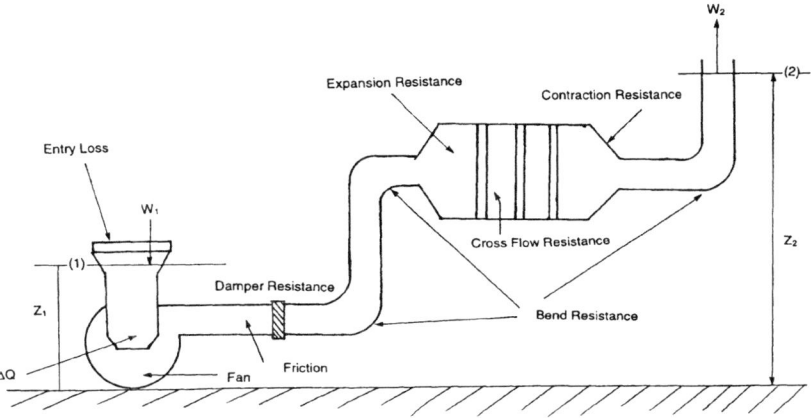

FIGURE 16-2. A schematic of flow through a duct

where P_1, ρ_1, u_1, and z_1, are the pressure, density, velocity, and height, respectively, at section 1. Here, Δp is the pressure drop between sections 1 and 2. Also, $\Delta W/\Delta m$ is the work done by the fan per unit mass of the fluid flowing between sections 1 and 2 to increase its pressure. The pressure drop, Δp, is the sum of the frictional resistance on the walls of the duct and local eddy losses. The latter (local loss) occurs owing to bends, expansion or contraction of the passage, or combination of both. For each of these situations the pressure drop contribution is found separately. In most cases of gas or air flow the term gz is negligibly small compared to other terms, so it is neglected. However, in case of water flow, this term is significant and hence is taken into account. The frictional pressure drop is usually expressed as a factor of the velocity head $(u^2/2)$ as shown below.

16-2-1 Calculations for the Frictional Resistance along the Flow Path

The pressure drop owing to frictional resistance along the flow path, Δp_{mc}, depends on the viscosity and velocity of the fluid (gas, air, steam or water) and the length and roughness of ducts. Neglecting the effect of heat transfer, the frictional resistance is

$$\Delta p_{mc} = f \frac{L}{d_{dl}} \frac{\rho u^2}{2} \text{ Pa} \qquad (16\text{-}2)$$

where
f = coefficient of friction
L = length of the duct, m
d_{dl} = equivalent diameter of the duct, m
ρ = density of air or gas calculated at the mean gas temperature, kg/m^3
u = cross section average velocity of air or gas in the duct, m/s
A = cross section area of duct, m^2

The equivalent diameter for a noncircular duct is given by (4 × flow area/flow perimeter). For a rectangle cross section duct filled with tubes parallel to its axis, the value of d_d is calculated as

$$d_{dl} = \frac{4\left(ab - z\dfrac{\pi d^2}{4}\right)}{2(a + b) + z\pi d} \qquad (16\text{-}3a)$$

where a and b = the length and width of the duct respectively, m
z = tube number in the duct
d = diameter of tubes, m

A more accurate expression of equivalent diameter of rectangular duct is recommended by American Society of Heating, Refrigerating and Air-Conditioning

TABLE 16-1. Roughness (e) of some tube and duct surfaces.

Type of surface	e (mm)
Tubular air heater, platen air heater and seamless steel tube	0.00015
Gas duct welded with steel platen	0.40
Galvanized steel sheet	0.15
Steel tube	0.12
Wrought iron tube	0.046
Pig iron tube and platen	0.80
Steel tube highly eroded	0.70
Duct built by bricks with concrete	0.8–6.0 (mean value 2.5)
Concrete duct	0.3–3.0 (mean value 2.5)
Fiberglass duct	0.0015–0.01 (mean 0.005)

Engineers (ASHARE, 1985),

$$d_{dl} = 1.3 \frac{(ab^{0.625})}{(a+b)^{0.25}} \tag{16-3b}$$

where a and b are sides of the duct, mm.

The coefficients of friction for flow through tubes can be approximated as shown below (Swamee & Jain, 1976):

$$f = \frac{0.25}{\left[\log\left\{\left(\frac{e}{3.7d_{dl}}\right) + \left(\frac{5.74}{Re^{0.9}}\right)\right\}\right]^2}$$

$$\text{for } 5000 < Re < 10^8, 10^{-6} < (e/d_{dl}) < 0.01 \tag{16-4}$$

where e is the roughness of the tube surface (Table 16-1), and $Re = ud_{dl}/\gamma$ the flow Reynolds number. For more reliable values of the tube surface one could use Moody's diagram (Shames, 1992).

16-2-2 Calculations for Local Pressure Drops

In addition to the frictional resistance in the straight passage there is an additional resistance called local or fitting losses. These losses occur whenever there is a sudden change in flow cross section causing eddy dissipation. This loss is written as:

$$\Delta p_l = K \frac{\rho u^2}{2} \tag{16-5}$$

where Δp_l = local pressure drop, Pa
   K = local resistance factor; values for different fittings are given in Table 16-2
   ρ = density of air or gas at the position of the pressure drop calculated, kg/m^3
   u = velocity of air/gas through the fittings (m/s). It is calculated at the position as marked by arrow in the Table 16-2.

A comprehensive list of local resistance factor is given in Idelick (1989).

TABLE 16-2. Local resistance factor, K, where $\Delta P = K u^2/2$.

No.	Name	Sketch	Local resistance factor
1	Inlet (edge fitting with wall)		$K = 0.5$

No. 2 — Inlet (edge out of wall):

For $\delta/d \leq 0.004$ (δ-wall thickness)
  $a/d \geq 0.2$, $K \cong 1.0$
  $0.05 < a/d < 0.2$, $K \cong 0.7$
  $0.02 < a/d \leq 0.05$, $K \cong 0.6$

For $0.004 < \delta/d < 0.04$
  $a/d \geq 0.2$, $K \cong 0.7$
  $0.05 < a/d < 0.2$, $K \cong 0.6$
  $0.02 < a/d \leq 0.05$, $K \cong 0.5$
For $\delta/d \geq 0.04$, $K \cong 0.5$

No. 3 — Inlet with circular edge:

Freestanding	Wall-mounted
For $r/d = 0.05$, $K = 0.4$	For $r/d = 0.05$, $K = 0.22$
For $r/d = 0.1$, $K = 0.15$	For $r/d = 0.1$, $K = 0.12$
For $r/d = 0.2$, $K = 0.03$	For $r/d = 0.2$, $K = 0.03$

No. 4 — Projecting inlet:

K	$1/d = 0.1$–0.2	0.3
$\alpha = 30°$	0.25	0.2
$50°$	0.20	0.15
$90°$	0.25	0.2

K	$1/d = 0.1$	0.2	0.3
$\alpha = 30°$	0.55	0.35	0.2
$50°$	0.45	0.22	0.15
$90°$	0.41	0.22	0.18

No. 5 — Draft duct:

With shield: $K = 0.3$
Without shield: $K = 0.2$

With shield: $\zeta = 0.1$
Without shield: $\zeta = 0.2$

TABLE 16-2. (*Continued*).

No.	Name	Sketch	Local resistance factor
6	Outlet (except chimney)		$K = 1.1$ for contracting outlet $(1 > 20d_{dl})$ $K = 1.0$
7	Inlet under lid		$K = 0.65$
8	Outlet under lid		$K = 0.48$
9	Inlet through orifice or grid		$K = \left(1.707\dfrac{F}{F_1} - 1\right)^2$
10	Inlet through a hole		$F_1/F \leq 0.4$: $K = 2.5\left(\dfrac{F}{F_1}\right)^2$ $F_1/F > 0.4$: $K \approx 2.25\left(\dfrac{F}{F_1}\right)^2$
11	Inlet through two opposite holes		
12	Outlet through orifice or grid		$F_1/F \leq 0.7$: $K \approx 3.0\left(\dfrac{F}{F_1}\right)^2$ $F_1 = $ *area of the hole*
13	Outlet through a hole		$F_1/F \leq 0.7$ $K = 2.6\left(\dfrac{F}{F_1}\right)^2$ $0.7 < F_1/F \leq 1.0$ $K = 3.0\left(\dfrac{F}{F_1}\right)^2$
14	Outlet through two opposite holes		$F_1/F \leq 0.6$: $K \approx 2.9\left(\dfrac{F}{F_1}\right)^2$ $F_1 = $ *area of the hole*

(*Contd.*)

TABLE 16-2. (*Continued*).

No.	Name	Sketch	Local resistance factor
15	Orifice or grid in duct		$K = \left(\dfrac{F}{F_1} - 1 + 0.707 \dfrac{F}{F_1} \sqrt{1 - \dfrac{F_1}{F}} \right)^2$
16	Shield or valve opened		$K = 0.6$ (K relates valve opening and duct shape))
17	Contracting duct		$\alpha = 10\text{–}40°,\quad K = 0.05$ $\alpha = 50\text{–}60°,\quad K = 0.07$ $\alpha = 90°,\qquad K = 0.2$ $\alpha = 120°,\qquad K = 0.25$

18 Throttle valve

Open degree (%)	5	10	30	50	70	90
K	1000	200	18	4	1	0.1

19 Expanding suddenly

F_x/F_d	0.1	0.2	0.3	0.4	0.5	0.6
K	0.81	0.64	0.50	0.36	0.25	0.16

20 Contracting suddenly

F_x/F_d	0.1	0.2	0.3	0.4	0.5	0.6
K	0.45	0.40	0.35	0.30	0.25	0.20

21 Bend

$K = A_1 B_1 C_1$

R/b	0.5	0.7	0.9	1.25	1.5		
A_1	1.18	0.51	0.28	0.19	0.17		

a(°)	0	30	60	90	120	150	180
B_1	0	0.45	0.78	1.00	1.17	1.28	1.40

a/b	0.25	0.5	1.0	1.5	2.0	3.0	6.0	8.0
C_1	1.3	1.17	1.0	0.9	0.85	0.85	0.98	1.0

TABLE 16-2. (*Continued*).

No.	Name	Sketch	Local resistance factor

22 — Welded bend

5	—	0.46	0.33	0.24	0.19	0.09
4	—	0.50	0.37	0.27	0.24	0.18
3	0.9	0.54	0.42	0.34	0.33	0.31

23 — Bend

$K = C_1 K'$, C_1 as (No. 21)

r/b	0.1	0.15	0.13	1.0
K'	0.69	0.57	0.30	0.21

24 — Bend and inside arc

$K = C_1 K'$, C_1 as (No. 21)

r/b	0.1	0.2	0.3	0.5	0.7
K'	0.88	0.70	0.56	0.48	0.43

25 — Three-way tube

F = duct area
V_2, V_1 = air mass flow
$V_2 + V_1 = 1$

Branch resistance factor, K

$K \cdot V_2(F/F_2)$	0.4	0.6	0.8	1.0	1.5	2.0
$\alpha(°)$						
45	0.5 4	0.38	0.29	0.24	0.27	1.0
90	1.0	1.0	1.0	1.0	1.0	1.0

Main duct resistance factor, K, $\alpha = 15–90°$

V_1	0.2	0.4	0.6	0.8	1.0
K	0.26	0.14	0.06	0.02	0

26 — Divergent Outlet of the fan

$K \cdot F_2/F_1$	1.5	2.0	2.5	3.0	3.5
α					
10	0.10	0.18	0.21	0.23	0.24
20	0.31	0.43	0.48	0.53	0.56
30	0.42	0.53	0.59	0.64	0.67

27 — In tube bank

$\alpha = 45°$ $K = 0.5$
$\alpha = 90°$ $K = 1.0$
$\alpha = 180°$ $K = 2.0$

(*Contd.*)

TABLE 16-2. (*Continued*).

No.	Name	Sketch	Local resistance factor
28	Chimney inlet	a) b)	Sharp turn a) $K = 1.4$ Smooth turn b) $K = 0.9$

16-3 Pressure Drop Across Heating Surfaces

Gas and air often flow over tube bundles instead of flowing through them. This is called *cross flow*. The flow resistance in such flows is discussed below.

16-3-1 Pressure Drop Across Tube Bundles

The pressure drop in a gas flow over tube bundles (ΔP_{tb}) is given by the following equation:

$$\Delta p_{tb} = K \frac{\rho u^2}{2} \text{ Pa} \qquad (16\text{-}6)$$

where u is the gas velocity for the minimum cross section of the gas duct along the plane in which the tubes are located, and K is the loss coefficient.

The loss coefficient for cross flow, which depends on the configuration of the tube bundle and the number of tube rows and Reynolds number, can be taken from graphs given by Grimson (1937). Alternately, approximate values may be calculated from the following empirical relations.

a) In-Line Arrangement

The loss coefficient is calculated as

$$K = nK_0 \qquad (16\text{-}7)$$

where n is the number of tube rows along the flow direction; $K_0 =$ loss coefficient for one row of tubes. Its value depends on the $\sigma_1 = s_1/d$, $\sigma_2 = s_2/d$, $\phi = (s_1 - d)/(s_2 - d)$, and the Reynolds number. Here, s_1 is the lateral pitch, s_2 is the longitudinal pitch, and d is the diameter of tubes. The values of K_0 are found as follows

(Perkov, 1965):

$$\text{If } \sigma_1 \leq \sigma_2 \quad K_0 = 1.52\,(\sigma_1 - 1)^{-0.5}\phi^{-0.2}\,Re^{-0.2} \qquad (16\text{-}8)$$

$$\text{If } \sigma_1 > \sigma_2 \quad K_0 = 0.32\,(\sigma_1 - 1)^{-0.5}(\phi - 0.9)^{-0.68}\,Re^{-0.2/\phi} \qquad (16\text{-}9)$$

b) Staggered Arrangement

The loss coefficient is obtained as

$$K = K_0(n + 1) \qquad (16\text{-}10)$$

where K_0 is the coefficient of frictional resistance of one row of tubes. Its value depends on $\sigma_1 = s_1/d$ and $\phi = (s_1 - d)/(s_2' - d)$ and the Re. s_2' is the diagonal tube pitch, which is given by

$$s_2' = \sqrt{0.25 s_1^2 + s_2^2} \qquad (16\text{-}11)$$

and K_0 can be rewritten as the following equation:

$$K_0 = C_s\,Re^{-0.27} \qquad (16\text{-}12)$$

with C_s is a design parameter of the staggered banks.
For $0.17 \leq \phi \leq 1.7$ and $\sigma_1 \geq 2.0$, $C_s = 3.2$, but if $\sigma_1 < 2.0$, then C_s is given by

$$C_s = 3.2 + (4.6 - 2.7\phi)(2 - \sigma_1) \qquad (16\text{-}13)$$

$$\text{For } \phi = 1.7 - 5.2, \quad C_s = 0.44(\phi + 1)^2. \qquad (16\text{-}14)$$

c) Cross-Flow over Finned Tubes

Resistance coefficients for banks of finned or grilled tubes depends on shape and dimension of fins. They are given by the following equations (Perkov, 1965).

(1) In-Line Arrangement

$$K = \left[2.87 + 0.464\sigma_f'^{1.24}(\sigma_f'' - 0.606)\right]\frac{n - 1}{\sigma_1}\,Re^{-0.12} \quad \text{(For round fins)} \qquad (16\text{-}15)$$

where $\sigma_f' = $ (pitch of fin, p_f/diameter of tube, d) and $\sigma_2'' = $ (height of fin, h_f/diameter of tube d), $Re = (u p_f/\nu)$.

$$K = [1.80 + 2.75\sigma_f'']\frac{n - 1}{\sigma_1}\,Re^{-0.12} \quad \text{(for square fins with } \sigma_f'' = 0.33) \qquad (16\text{-}16)$$

(2) Staggered Arrangement

$$K = \left[2.0\,n\sigma_f'^{-0.72}\right]Re^{-0.24} \quad \text{(for round fins } s_1 = s_2 = d + 2h_f)$$

$$K = \left[2.7\,n\sigma_f'^{-0.72}\right]Re^{-0.24}\sigma_f''^{0.45} \quad \text{(for round fins } s_1 = s_2 = 2d) \qquad (16\text{-}17)$$

For square fins substitute $(1.13h_f + 0.065d)$ for h_f in the above equation.

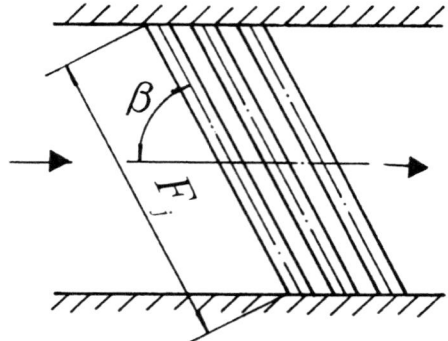

FIGURE 16-3. Cross-flow over inclined tube bundles

For gas flow over inclined tube bundles, the pressure drop can be obtained by the above equation, but the mean velocity, u, should be calculated with respect to the inclined flow cross section area F_j, as shown in Figure 16-3. If the incident angle, β, is less than 75°, the calculated pressure drop is increased by 10% on the basis of the results of cross flow through the tube bundles. Klier (1964) found that the pressure drop in a bundle of cross tubes is roughly the same as that in parallel staggered banks with the identical spacing ratio. Hammeke et al. (1967), however, found that the pressure drop in cross in-line tube bundles could be as much as 50% higher than that in parallel staggered or in-line banks with the same values of lateral and longitudinal pitches.

16-3-2 Pressure Drop in Tubular Air Heaters (for Fluid Inside Tubes)

In this section we try to calculate the gas side pressure drop in an air heater where the flue gas passes through the tube, but the air flows over the tubes. The air side pressure drop can be calculated using the method shown in section 16-3-1.

The pressure drop on the gas side of tubular air heaters is made up of three components: pressure drop at the inlet of tubes owing to flow contraction, wall friction in tubes, and the pressure drop at the outlet of the tubes owing to flow expansion. The combined drop is

$$\Delta p = \Delta p_{mc} + (K_{in} + K_{out})\frac{\rho u^2}{2} \text{ Pa} \qquad (16\text{-}18)$$

where Δp_{mc} is the pressure drop in the tube and K_{in} and K_{out} represent the local resistance factors at the inlet and the outlet, respectively. Their values are taken from Table 16-2 (items 9 and 12) where F_{tube}/F_{sheet} is the ratio of tube cross section and the tube sheet areas. It is calculated approximately by the following

$$\frac{F_{tube}}{F_{sheet}} = \frac{\pi d_{dl}^2}{4s_1 s_2} \qquad (16\text{-}19)$$

where s_1 = lateral pitch, m

$\qquad\qquad s_2$ = longitudinal pitch, m

$\qquad\qquad d_{dl}$ = inside diameter of tubes, m

The pressure drop in the tube, Δp_{mc}, is calculated by Eq. 16-2, where the friction factor, f, is calculated either using the Moody diagram or Eq. (16-4).

16-3-3 Pressure Drop Through Rotary Air Heaters

In a rotary air heater the gas and air pass through a rotating body of corrugated plates. The pressure drop through it is typically within 3–7 in. of water gauge. This is calculated by the equation below

$$\Delta p = f \frac{l}{d_{dl}} \frac{\rho u^2}{2} \qquad (16\text{-}20)$$

where l is the depth of the plate in the direction of the flow and d_{dl} is the equivalent diameter of the flow passage.

The rotary air heater generally uses three types of corrugated heat storage plates (Fig. 16-4). So, the friction factor is calculated using equations appropriate for the basis of type of plate used:

a) Corrugated plate-corrugated setting plate (Fig. 16-4a)

$$Re \geq 2.8 \times 10^3 \quad f = 0.78 Re^{-0.25}$$
$$Re < 2.8 \times 10^3 \quad f = 5.7 Re^{-0.5} \qquad (16\text{-}21)$$

b) Corrugated plate-plane setting plate: (Fig. 16-4b)

$$Re \geq 1.4 \times 10^3 \quad f = 0.6 Re^{-0.25}$$
$$Re < 1.4 \times 10^3 \quad f = 33 Re^{-0.8} \qquad (16\text{-}22)$$

c) Plane plate-plane setting plate: (Fig. 16-4c)

$$Re \geq 1.4 \times 10^3 \quad f = 0.33 Re^{-0.25}$$
$$Re < 1.4 \times 10^3 \quad f = 90/Re \qquad (16\text{-}23)$$

16-3-4 Gas Side Pressure Drop in Finned-Tube Economizers

The flue gas passes over horizontal finned economizers at 90° or in cross flow mode. The pressure drop, Δp, is found in the same way as for tube bundles:

$$\Delta p = K \frac{\rho u^2}{2} \text{ Pa} \qquad (16\text{-}24)$$

where K is the resistance factor or loss coefficient of finned tube bundles.

The resistance factor of the tube bundle of cast iron finned-tube economizers (Fig. 16-5) depends on the ratios (h_f/d), (S_{qp}/d), s_1, and z_1, where h_f = height of fins; d = outside diameter of tubes; s_{qp} = pitch of fins; s_1 = lateral pitch;

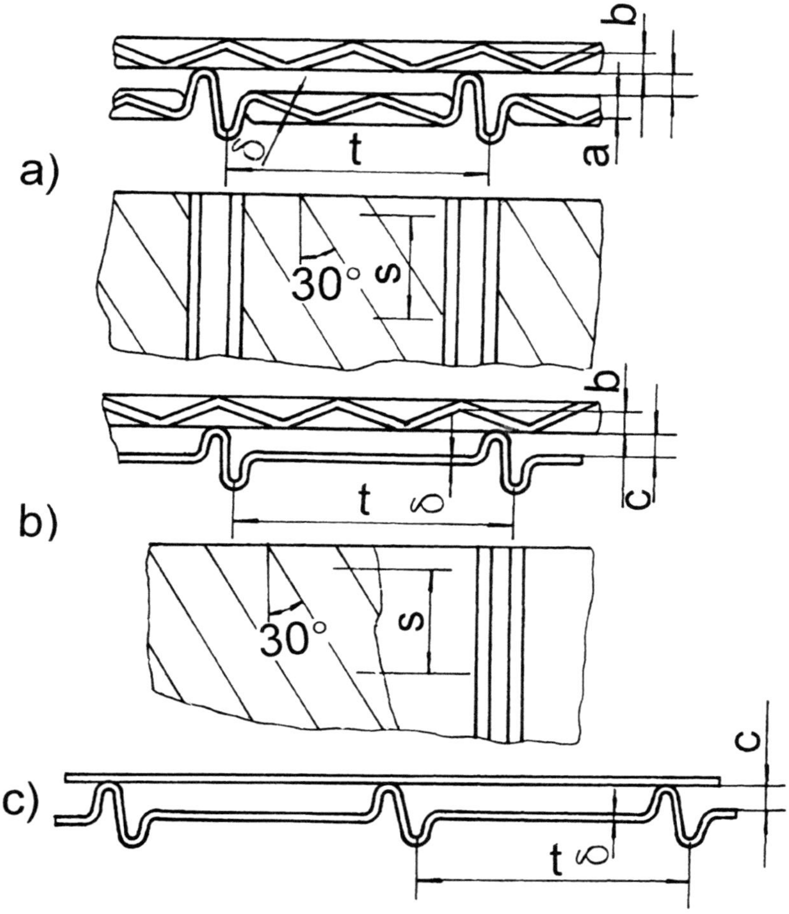

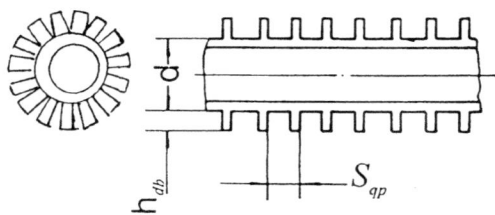

FIGURE 16-4. Rotary air heater and plate types

FIGURE 16-5. A finned tube for an economizer

z_1 = number of tube rows. The expressions for pressure drop across finned tubes were given earlier in Eqs. (16-15)–(16-17). In these equations the Reynolds number is defined as

$$Re = \frac{u h_f}{\gamma} \tag{16-25}$$

where u is the gas velocity on the minimum area of the tube bank, m/s
          v is the kinematic viscosity, m^2/s
          h_f is the height of fins, m

16-4 Pressure Drop in Natural Draft Gas Path

The boiler stack is an open system. The gas temperature in the vertical flue passage is higher than that of the ambient air. The gas density is lower at a higher temperature. Therefore, there is a density difference between the gas in the flue and the air outside the stack. This causes a draft called *natural draft* in the flue.

In case of upward flow the head of natural draft helps the gas flow, but in case of downward flow it hinders the gas flow. The head of natural draft is calculated by the formula

$$h_{nd} = \pm(\rho_a - \rho_g)gH \tag{16-26}$$

("+" for upward flow, and "−" for downward flow)

where h_{nd} = head of natural draft, Pa
          ρ_a = ambient air density, kg/m^3
          ρ_g = gas density in the flue, kg/m^3
          H = height difference between the beginning and the end of the section, m

The flue gas density, ρ_g, is calculated by the following:

$$\rho_g = \rho_g^0 \frac{273}{273 + T_g} \tag{16-27}$$

where T_g = gas temperature in the flue °C
          ρ_g^0 = gas density in the flue under the standard atmospheric condition (0°C, 1 atm), Nm^3/kg

It is determined by

$$\rho_g^0 = \frac{1 - 0.01A + 1.302\,\alpha V^0}{V_g} \tag{16-28}$$

where α = excess air ratio in the flue
          V^0 = theoretical air requirement for unit weight of fuel, Nm^3/kgf
          V_g = flue gas produced per unit weight of fuel, Nm^3/kgf
          A = ash content of the fuel, %

For approximation

$$V_g \approx \alpha V^0 \qquad (16\text{-}29)$$

The total natural draft head, H_{nd}, is found by adding up natural draft head in all parts of the flue gas passage

$$H_{nd} = \Sigma h_{nd} \qquad (16\text{-}30)$$

16-4-1 Pressure Drop in Ducts Joining Air Heater and Dust Collector

The pressure drop in the duct between the air heater and the dust collector is calculated from Eqs. (16-2) to (16-4) using appropriate values of velocity for the mass flow rate of flue gas and the temperature of gases. The pressure drop beyond the dust collector is calculated using the mass flow rate and the temperature at the induced draft fan. If there is no dust collector the pressure drop between the air heater and the induced draft fan is still calculated using the values at the induced draft fan.

The volume flow rate of gases at the induced draft fan is determined by the following:

$$V_{gf} = B(V_g + \Delta\alpha \, V^0)\left(\frac{T_g + 273}{273}\right) \qquad (16\text{-}31)$$

where V_{gf} = volume flow rate of gases through the induced draft fan, m^3/s
T_g = temperature of flue gas entering the induced draft fan, °C
$\Delta\alpha$ = leakage air ratio behind the air heater
B = fuel firing rate, kg/s

In the case of $\Delta\alpha \leq 0.1$ the gas temperature at the induced draft fan may be taken as equal to that behind the air heater T_{gf}.

In the case of $\Delta\alpha > 0.1$, T_{gf} is given by

$$T_{gf} = \frac{\alpha_e T_g + \Delta\alpha T_0}{\alpha_e + \Delta\alpha} \qquad (16\text{-}32)$$

where α_e is the excess air ratio in the flue gas at the induced fan
T_0 is the cold air temperature, °C

16-4-2 Pressure Drop Through Convective Section

Flue gas is usually in cross flow over heating surfaces such as superheaters, economizers, and reheaters. The effective gas passage is often reduced owing to deposition of ash over it. So the calculated pressure drop should be multiplied by a correction factor, Ψ, taking from Table 16-3.

Although the mass of gas flow through the gas duct remains unchanged, the velocity may change owing to the change in duct cross section areas. Instead of calculating the velocities in individual sections and finding the friction factors in them, one could use the following simplified approach. Local resistance

TABLE 16-3. Values of correction factor, ψ, for ash deposition in flow passages.

Heating surface	ψ	Heating surface	ψ
Platen heating surface	1.2	*Boiler tube bank*	
Superheater in horizontal flue		a) Multidrum & mixed swept boiler	0.9
(with soot blower):	1.2	b) Small boiler with horizontal flue	1.0
a) Loose ash deposit	1.8	c) Small boiler with combustion	1.15
b) Adhesive ash deposit		chamber	
Superheater at the gas exit:		*Nonstandard ribbed economizer:*	
(*with soot blower*)		a) With soot blower	1.4
a) Fuels except oil & gas	1.2	b) Without blower	1.8
b) Heavy oil and gas fuel	1.0	*Tubular air heater:*	
First-stage, second-stage &		a) Air path shift number >2	1.1
single-stage economizer		b) Air path shift number <2	1.05
a) Solid fuel with loose ash	1.1	c) Gas side	1.1
b) Solid fuel with adhesive ash	1.2	*Rotary air heater:*	
c) Gaseous fuel	1.0	a) Heavy oil	1.5
First-stage and single-stage	1.2	*Rotary air heater:*	
economizer (heavy oil firing)		a) Gas side	1.5
Second-stage economizer (heavy	1.0	b) Air side	1.2
oil firing, with soot blower)			
Finned tube economizer	1.2		
Platen air heater:			
a) Heavy oil	1.5		
b) Other fuels	1.2		

factors for different sections of the gas passage may be converted into a common cross section area, F, and velocity, u, as shown below. For example, two sections with areas F_1, F_2, velocities u_1, u_2, mass conservation for unchanged density gives

$$uF = u_1 F_1 = u_2 F_2 \qquad (16\text{-}33)$$

From above and Eq. (16-5) the local loss, Δp_1, can be written as

$$\Delta p_1 = K_1 \frac{u_1^2}{2} = K_1 \frac{F^2}{F_1^2}\frac{u^2}{2} = K_1' \frac{u^2}{2}$$

Now the total loss is found by adding up the individual losses expressed in terms of common velocity and the modified local loss coefficients

$$\Delta p = \Delta p_1 + \Delta p_2 + \Delta p_3 + \cdots + \Delta p_n = (K_1' + K_2' + K_3' + \cdots + K_n')\frac{u^2}{2} \quad (16\text{-}34)$$

where $K_1' = K_1(F/F_1)^2$, $K_2' = K_2(F/F_2)^2$.

The local resistance owing to expansion and contraction can be neglected when the variation of the cross section area is small. Specifically, for the divergent tube the local resistance can be neglected if the area expansion ratio exceeds 1.3. For the convergent tube and contraction angle less than 45° the resistance can also be neglected.

TABLE 16-4. Typical characteristics of dust collectors.

Type of dust collector	Gas velocity (m/s)	Collection efficiency (%)	Pressure drop (Pa)
1. Settling tank	0.5–1.5	40–60	50–100
2. Inertial separator	10–15	60–65	450–500
3. Cyclone	15–20	70–90	500–1000
4. Wet scrubber	13–18	80–90	300–700
5. Bag house	0.5–1.0	80–99	1000–2000
6. Electrostatic precipitator	1–2	99	100–200

16-4-3 Dust Collector

The pressure drop of dust collector for the gas velocity and the dust collection efficiency are usually given by the equipment supplier in their product manual. However, some typical values are given in Table 16-4.

16-4-4 Stack

The pressure drop through the stack is made up of two components: the pressure drop caused by friction and the exit pressure loss. The stack pressure drop may be estimated by the following

$$\Delta p_{st} = \left(\frac{f + L_{st}}{D} + K_c \right) \frac{\rho u_c^2}{2} \text{ Pa} \qquad (16\text{-}35a)$$

where Δp_{st} = stack pressure drop, Pa
f = friction factor (≈ 0.03)
L_{st} = height of the chimney, m
D = diameter of the chimney, m
K_c = resistance factor at the stack outlet (≈ 1.0–1.1)
ρ = gas density in the stack, kg/m^3 (It is approximately equal to the density at the induced draft fan.)
u_c = gas velocity at the chimney outlet, m/s

For tapered chimney L_{st}/D may be substituted by $(D_1 - D_2)/[(D_1 + D_2)\alpha_{st}]$, where D_1, D_2 are lower and upper diameter of the chimney and α_{st} is the taper of the stack.

While working out the pressure drop through the chimney the natural draft caused by the gas density difference (Eq. 16-27) in chimney must be considered.

16-4-5 Total Gas Side Pressure Drop

The total gas side pressure drop is calculated summing up pressure drops across all individual sections and then compensating for the pressure of the gas and presence of ash in the flue gas (Lin, 1987):

$$\Delta p_{gas} = [\Sigma\Delta p_1(1 + \mu) + \Sigma\Delta p_2]\left[\frac{\rho_{go}}{1.293}\right]\left[\frac{101325}{P_{av}}\right], \text{Pa} \qquad (16\text{-}35b)$$

where $\sum\Delta p_1$ = total pressure from the furnace outlet to the dust collector, Pa
$\Sigma\Delta p_2$ = pressure drop after the dust collector, Pa
μ = ash content in the flue gas, kg/kg
P_{av} = average pressure of the gas, Pa
ρ_{go} = flue gas density at standard condition (Eq. 16-30), kg/Nm^3

The ash fraction in the flue gas is calculated by the following

$$\mu = \frac{A\,\alpha_{fh}}{100\rho_{go}V_g} \qquad (16\text{-}36)$$

where α_{fh} = ratio of fly ash in flue gas to total ash in the fuel
A = ash content of working mass, %
V_g = average volume of gas from furnace to dust collector
calculated from the average excess air ratio, Nm^3/kg of fuel.

The pressure drop from the balance point (where $P = 0$) of the furnace to the chimney base is

$$\Delta H = \Delta P_{exit} + \Delta H_{gas} - H_{nd} \qquad (16\text{-}37)$$

where ΔP_{exit} = pressure drop up to the boiler outlet
H_{nd} = head owing to natural draft (Eq. 16-30)

At a low gas velocity ($u < 10$ m/s) the flow resistance of the platen heating surface arranged at the furnace exit is close to the head gain by natural draft. Therefore, its pressure drop need not be taken into account. However, when gas velocity, u, is greater than 10 m/s, the resistance of platen heating surface, which is arranged at the horizontal flue (cross-over duct connecting the furnace and the back-pass), should be considered, and it can be estimated on the basis of Eq. (16-2). The length L is taken to be equal to that of gas passage in the platen region, and the equivalent diameter d_{eq} of the gas flue is taken as twice the pitch of the platen.

The pressure drop of the slag screen can be neglected if its rows of tubes (n_2) < 5, the gas velocity $u > 10$ m/s or $n_2 < 2$, and $u > 15$ m/s. In other cases the pressure drop is calculated considering cross flow through it.

16-5   Pressure Drop Through Air Ducts

The pressure drop calculations for air ducts is similar to that for gas ducts. It should be done at the designed load of the boiler, but after the thermal calculation and at the designed load of the boiler. The resistance of the air duct can be divided into four parts, the cold air duct, air heater, hot air duct, and combustion equipment.

i) When the resistance of the cold air duct is calculated, the flow of the cold air in the duct can be worked out with the following formula:

$$V_{ca} = BV^0(\alpha_{fu} - \Delta\alpha_{lf} - \Delta\alpha_{lp} + \Delta\alpha_{lah})\frac{273 + T_a}{273} \qquad (16\text{-}38)$$

where V_{ca} is flow of cold air, m³/s; B is fuel feed rate, kg/s; V^0 is theoretical air flow, Nm³/kg; and α_{fu} is excess air coefficient at furnace outlet; $\Delta\alpha_{lf}$, $\Delta\alpha_{lp}$, and $\Delta\alpha_{lah}$ are leakage air coefficients of the furnace, pulverization system, and air heater, respectively. T_a is the temperature of air in °C.

The velocity of air in ducts generally follows the norm given in Table 16-6. If it is less than 10 m/s, the frictional resistance of the cold air duct can be neglected. But when it is between 10 and 20 m/s, the pressure drop should be calculated from Eq. (16-2). The local pressure drop across bends, etc., can be calculated from Eq. (16-5) taking values of resistance factors from the Table 16-2.

ii) The air side resistance and its correction factor for the rotary air heater is calculated in the same way as done for the gas side (Section 16-3-3).

In a tubular air heater, when the air flows across staggered banks, the pressure drop can be calculated using equations for cross flow. In the flow reversal chamber of the air heater air comes out of one section, turns 180°, and then enters another section. So to calculate the pressure drop across flow reversal sections where $a/h < 0.5$ (Fig. 16-6) one can take the resistance factor for 180° bend. The flow velocity changes as air turns through the chamber (Fig. 16-6). So, it should be calculated on the basis of the mean flow area, F_{av}, which is calculated from individual sectional areas F_1 and F_2 and F_3.

$$F_{av} = \frac{1}{\dfrac{1}{F_1} + \dfrac{1}{F_2} + \dfrac{1}{F_3}} \qquad (16\text{-}39a)$$

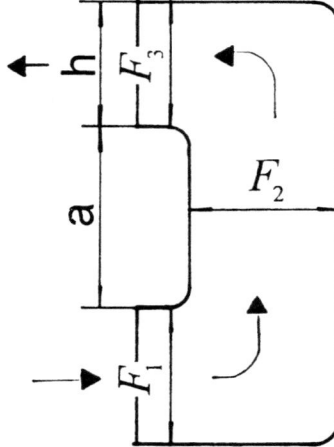

FIGURE 16-6. A 180° bend in the reversal chamber of an air preheater

When $a/h \geq 0.5$, the resistance through the reversal chamber is equal to that for two 90° bends, $K = 2 \times 0.9$. The average flow area for air velocity should be worked out as below:

$$F_{av} = \cfrac{1}{\cfrac{1}{F_1} + \cfrac{1}{F_3}} \qquad (16\text{-}39b)$$

The correction factor for ash deposits, Ψ, for the air side of the tubular air heater can be found in Table 16-3.

iii) The computation of the resistance of hot air ducts is the similar to that for cold air ducts. The volume flow rate of the hot air after the air heater can be calculated from below:

$$V_{ha} = B V^0 (\alpha_{fu} - \Delta\alpha_{lf} - \Delta\alpha_{lp}) \frac{273 + T_{ha}}{273} \qquad (16\text{-}40)$$

where V_{ha} is flow of cold air, m³/s; B is fuel feed rate, kg/s; V^0 is theoretical air flow, Nm³/kg; and α_{fu} is excess air coefficient at furnace outlet; $\Delta\alpha_{lf}$ and $\Delta\alpha_{lp}$, are leakage air coefficients of the furnace and pulverization system, respectively. T_{ha} is the temperature hot air in °C.

The pressure drop across each branch of the air duct can be computed using the flow rate in each of them.

iv) The pressure drop across combustion equipment like a burner, etc., can be calculated from the following equation:

$$\Delta P = K \frac{\rho u^2}{2} \text{ Pa} \qquad (16\text{-}41)$$

where u = velocity of the air at burner exit.
            K = 1.5 for tangential burner
               = 3.0 for swirl burner of gas and heavy oil

For a swirl burner of pulverized coal the K changes with the design of the burner.

For grate fired boilers, the resistance of air flowing through the grate and the coal layer can taken from Table 16-5 (Lin, 1987).

The resistance across the combustor of a bubbling fluidized bed boiler comprises the pressure drop across the distributor plate and that across the fluidized bed. The

TABLE 16-5. Pressure drop through grate and fuel layer in grate fired furnace.

Type of boiler	Type of fuel	Pressure drop (ΔP) in Pa
Grate fired boiler	Bituminous coal, lignite coal	800
	Anthracite	1000
Spreader-stoker fired boiler	Bituminous coal, lignite coal	500
	Oil shale	600
Hand fired boiler	Bituminous coal	800
	Anthracite, lignite coal	1000

grate pressure drop (Δp_g) can be calculated from the Eq. (16-41), where the velocity, u, is taken as the velocity of air through the orifices of the air nozzles. Orifice velocity (u) is generally in the range of 20–45 m/s and the coefficient $K = 1.6$ (Zenz, 1981). The resistance of the fluidized bed of solids, ΔP_b is equal to the weight of the bed solids per unit grate area.

$$\Delta p_b = H_0 \rho_b g \text{ Pa} \tag{16-42}$$

where H_0 is the height of the bed when fluidized, m
         ρ_b is the bed density of the fluidized bed, which is in the range
         1000–1200 kg/m^3

v) The sum of resistance of all parts of the air duct gives the total resistance of the air duct, $\Sigma \Delta p$. If the local elevation above the sea level is higher than 200 m, the pressure drop should be corrected for the local atmospheric pressure. The total pressure drop after this correction ΔP_r is

$$\Delta P_r = \Sigma \Delta p \frac{101,325}{P_{av}} \tag{16-43}$$

where P_{av} is the mean pressure of the air in air duct, Pa
         $\Sigma \Delta p$ is the total resistance of the air ducts system, Pa

When $\Sigma \Delta p$ is less than 3000 Pa, P_{av} is taken to be equal to the mean pressure of the local atmospheric pressure, P; otherwise, P_{av} should be calculated as

$$P_{av} = P + \Sigma \Delta p / 2 \tag{16-44}$$

where P is the average local atmospheric pressure.

Natural draft in air ducts in a boiler can be calculated from

$$P_{nd} = H \cdot g \cdot (\rho_a'' - \rho_a') \text{ Pa} \tag{16-45}$$

where P_{nd} is the natural draft in air ducts, ρ_a'' and ρ_a' are the density of air at the lowest and highest point of the duct, respectively H is the height difference between the beginning and the final sections of the duct. Adding the natural draft component of all parts of the air ducts the total natural draft, ΣP_{nd} is obtained.

The total pressure drop, ΔP_{air}, in the air ducts of the boiler can be obtained by adding

$$\Delta P_{air} = \Delta P_r - \Sigma P_{nd} \text{ Pa} \tag{16-46}$$

If the temperature of the air entering the fan is above that of the ambient temperature, an air inlet higher above the forced draft fan it will the resist while it will aid the flow if the inlet is below the fan level. However, this would be taken care of while calculating the natural draft of individual sections of the air duct using Eq. (16-45).

16-6 Selection of Fans

After the flow and resistance of the gas and air ducts at rated load of the boiler are determined, we can choose the fans. Power stations generally use following two types of fans:

- centrifugal pump
- axial flow pump.

Presently centrifugal pumps are more favored, especially for high head. The fan selection may use the following procedure.

16-6-1 Choice of the Flow and Pressure of the Fan

Since there is always the possibility of a small variation between the actual operation conditions and the theoretical calculation, the fan specifications for both flow and pressure should have safety margins.

a) Choice of Discharge of the Fan

The flow rate, Q_x, of the fan can be calculated using the following equation

$$Q_x = \beta_v V \frac{101,325}{P_j} \text{ m}^3/\text{s} \qquad (16\text{-}47)$$

where V is the flow rate of the gas or air at the rated load, m³/s. It can be calculated using either Eq. (16-31) or (16-40). However, it is customary to leave a 10% design margin for safety. β_v is the safety margin ($\beta \approx 1.1$). Here, P_j is the pressure of the working medium at the entrance of fan, Pa.

b) Choice of the Pressure of the Fan

The pressure head, H_x, of a fan can be calculated from the following equation:

$$H_x = \beta_h \Delta P \text{ Pa} \qquad (16\text{-}48)$$

where β_h is a margin on the pressure ($\beta_h \approx 1.2$). ΔP is the total pressure drop of the gas or air duct system, which can be obtained by Eq. (16-35) or (16-49).

While designing forced fans manufacturers often consider air at standard atmospheric pressure and at 20°C as the working medium. But when they design induced draft fans, they take gas at 200°C and standard atmospheric pressure as the working condition. Therefore, before the fan is chosen, the pressure head of the fan, H_x, should be corrected for the design temperature and pressure.

$$H_o = K_p H_x \text{ Pa} \qquad (16\text{-}49)$$

H_o is corrected pressure head.

Here K_p is the coefficient to correct for the density of the working medium, which can be calculated as below:

$$K_p = \left[\frac{T + 273}{T_i + 273}\right]\left(\frac{101,325}{P}\right) \quad \text{for forced draft fan} \quad (16\text{-}50)$$

$$K_P = \frac{1.293}{\rho_{gn}}\left[\frac{T + 273}{T_i + 273}\right]\left(\frac{101,325}{P}\right) \quad \text{for induced draft fan} \quad (16\text{-}51)$$

where T is the temperature of the working fluid of the fan under rated load, °C
T_i is the design temperature used by the fan manufacturer, °C
ρ_{gn} is the density of the gas under standard conditions, kg/Nm^3
P is the inlet pressure, Pa

16-6-2 Calculations for the Fan Power

The shaft power needed by the fan, W, can be obtained as

$$W = \beta_w \frac{Q_x H_o}{1000\eta_f \eta_c} \text{ kW} \quad (16\text{-}52)$$

where β_w is a safety factor ($\beta_w \approx 1.15$ for forced draft fan; $\beta_w \approx 1.3$ for induced draft fan)
η_f is the fan efficiency (0.9 for efficient fans and 0.6–0.7 for other fans)
η_c is transmission efficiency

Electrical power required for the motor, W_d, can be found as follows:

$$W_d = \frac{W}{\eta_d} \text{ kW} \quad (16\text{-}53)$$

where η_d is the efficiency of the electric motor, usually $\eta_d = 0.9$.

Since fluid handled by a forced draft fan is clean and low-temperature air is used, there is no special requirement for the enclosure of this fan. But the induced draft fan usually transports high-temperature (150–200°C) gases containing ash, which erodes and damages the blades of the fan, leading to fan vibration. Therefore, the rotational speed of the fan should be limited to 960 rpm, and the number of the blades should not be too large. All these measures prolong the life of the fan. In induced fans equipped with water spray separators, we should take measures to avoid acid corrosion.

16-7 Pressure Drop Through Water or Steam Tubes

With the exception of air heaters, the tubes in a boiler generally carry water, steam, or steam–water mixture. The pressure drop through the tubes and intermediate manifolds (or headers) is an important design parameters especially for the natural

TABLE 16-6. Some typical fluid velocities used in boilers.

Fluid pipes	Range of velocities (m/s)
Forced air duct	7.5–18
Induced draft flue gas duct	10–18
Water in economizer	0.75–1.5
Water in natural circulation circuit	0.35–3.5
General service water pipe	2.5–4.0
Steam in high pressure line	40–60
Steam in low pressure line	60–75
Steam in vacuum line	100–200
Steam in superheater/reheater tubes	10–25

circulation calculation. This part has been discussed in details in Chapter 12, Section 12-1-3 and Appendix 12-1. Table 16-6 gives some typical values of fluid velocities used in steam water lines.

Nomenclature

A	ash content of working mass
A_F	cross section area of duct, m^2
a	length, m
b	width, m
B	fuel feed rate, kg/s
D	diameter of the chimney stack, m
d	outer diameter of tube, m
d_{dl}	equivalent diameter of duct, m
e	surface roughness, m
f	coefficient of frictional resistance
F_{tube}	tube cross section area, m^2
F	flow area, m^2
F_{sheet}	tube sheet area, m^2
F_{av}	mean cross section area, m^2
H	height difference, m
H_o	expanded depth of the fluidized bed, m
H_{nd}	sum of head owing to natural draft, Pa
H_x	pressure head of fan, Pa
h_f	height of fin, m
h_{nd}	head of natural draft, Pa
K	loss coefficient of fittings
K_1'	loss coefficient of section 1 based on a common section
K_p	correction coefficient for head (Eq. 16-49)
K	local resistance factor
K_c	resistance factor at the stack outlet
K_{in}	local resistance factor at the inlet
K_o	frictional coefficient of one row of tubes

K_{out}	local resistance factor at the outlet
T_g	flue gas temperature, kg/m^3
T_{gf}	gas temperature at the induced draft fan, °C
l	depth, m
L	length of the duct, m
L_{st}	height of the stack, m
n	number of tube rows
P	static pressure at a section, Pa
Per	perimeter of duct, m
P_{av}	gas average pressure, Pa
P_{nd}	natural draft pressure head, Pa
Q_x	fan flow, m^3/h
s_1	lateral pitch, m
s_2	longitudinal pitch, m
s_2'	diagonal pitch of tubes, m
s_{qp}	pitch of fins, m
T_0	cold air temperature, °C
T_g	flue gas temperature, °C
T_{ha}	temperature of hot air, °C
u	mean velocity of air or gas, m/s
u_c	gas velocity at the stack outlet, m/s
V_{ca}	flow of cold air, m^3/s
V_g	gas volume in the flue, Nm^3/kg
V_{gf}	mass flow rate gases at the induced draft fan, m^3/s
V_{gpj}	average volume of gas, m^3/kg
V^0	theoretically required air volume, Nm^3/kg
W	fan power, kW
W_d	electrical motor power, kW
z	number of tubes
$z_{1(2)}$	height at section (1), m

Greek Symbols

$\Delta\alpha$	leakage air ratio
ΔH	net total pressure drop, Pa
Δh_f	height of fins, m
$\Delta\alpha_{lf}, \Delta\alpha_{lp},$	leakage air coefficients of furnace, pulverization system and air
$\Delta\alpha_{lah}$	heater, respectively
Δh_{nd}	head of natural draft, Pa
Δp	pressure drop, Pa
Δp_{st}	pressure drop in stack, Pa
Δp_{tb}	cross tube bundles pressure drop
Δp_l	local pressure drop, Pa
Δp_f	frictional pressure drop, Pa
Δp_{mc}	pressure drop in the tube, Pa

ΔP_{exit}	pressure drop up to the boiler outlet, Pa
ΔQ	fan work, W
Φ	coefficient (Eq. 16-7)
$\Sigma \Delta h_1$	total pressure from the furnace outlet, Pa
$\Sigma \Delta h_2$	pressure drop after the dust collector, Pa
α	excess air
α_e	excess air ratio in the flue gas at the induced fan
α_{fh}	ratio of fly ash in flue gas to total ash in the flue
α_{pg}	excess air ratio
β_v	safety factor on volume
β_h	safety factor on head
β_h	safety factor on power requirement
η_c	efficiency of the transmission
η_d	efficiency of the electrical motor
η_f	efficiency of the fan
μ	ash content in the flue gas, kg/kg
ρ	density, kg/m^3
$\rho_{1(2)}$	air density at section (1), kg/m^3
ρ_g	flue gas density, kg/m^3
ρ_{gn}	flue gas density under the standard atmospheric, kg/m^3
ρ_k	ambient air density, kg/m^3
$\sigma_{1(2)}$	ratio $s_{1(2)}/d$

Dimensionless Number

Re	Reynolds number $\rho u d / \mu$

Subscripts

1	parameters of section 1
2	parameters of section 2

References

ASHARE Handbook (1985) Fundamentals, Chapter 33. American Society of Heating Refrigerating and Air Conditioning Engineers, Inc., Atlanta, Georgia.

Grimson, E.D. (1937) Correlation and utilization of new data on flow resistance and heat transfer for cross flow of gases over tube banks. Transactions of American Society of Mechanical Engineers 59:583–594.

Hammeke, K.E., Heinecke, and Scholz, F. (1967) Warmeubergangs und Druckverlustmessungen an querangestromten Glattrohrbundeln, insbesondere bei hohen Reynolzahlen. International Heat Mass Transfer 10:427–446.

Idelick, I.E., Fried (1989) "Flow Resistance—A Design Guide for Engineers." Hemisphere Publishing Corp. New York.

Klier, R. (1964) Warmeubergang und Druckverlust bei quer angestromten, gekreuzten Rohrgittern. International Heat Mass Transfer 7:783–799.

Lin, Z., and Zhang, Y. (1987) "Boiler Handbook," (in Chinese). Mechanical Engineering Press, Beijing.

Perkov, V.G. (1966) "Lecture Notes on Boiler Technology." Indian Institute of Technology, Kharagpur, India.

Shames, E. (1992) "Mechanics of Fluids," 3rd edition, McGraw Hill, New York, p. 365.

Swamee, P.K., and Jain, A.K. (1976) Explicit equation for pipe flow problems. Journal of Hydraulic Division, ASCE May:657–664.

VDI Heat Atlas (1993) Association of German Engineers, VDI Society for Chemical Process Engineering. J.W. Fullarton, trans. Dusseldorf, Germany, p. Lb.2.

Zenz, F. (1981) Elements of Grid Design. Proceedings of Tripartite Chemical Engineering Conference. Symposium on Fluidization II, Montreal, pp. 36–39.

17
Mechanical Design
of Pressure Parts

A boiler is classified as a fired pressure vessel. Parts of the boiler that are subjected to high internal pressure of steam or water are referred to here as *pressure parts*. Tubes, drum, and headers are examples of pressure parts. Other components likes burners, etc., are not subjected to such internal pressure. As such they are classified as nonpressure parts. Selection of materials for pressure parts of a boiler and their mechanical design and construction are important aspects of a boiler design. In the early days of the industrial revolution many accidents occurred where boiler vessels burst under pressure, killing many operating personnel. More recently, a failed boiler flange in a ship of the United States Navy killed ten sailors (Peterson, 1997). Such incidents warrant utmost care in the design as well as the operation of pressure parts. The boiler pressure parts are inspected periodically by statutory authorities for safe operation. These authorities set their own guidelines and stipulations that govern the design and operation of boilers in their specific country. The present chapter discusses the principles involved in the choice of materials and in the mechanical design of pressure parts. The following section shows (i) factors affecting the selection of materials, (ii) important mechanical properties, and (iii) how the properties of steel can be improved by altering its composition and by suitable mechanical processing.

17-1 Selection of Materials

The principal pressure parts of boilers, which constitute about 20% of the total boiler weight, include drums, headers, and tubes for superheaters, reheaters, and economizers. These are made from steel. Since 1920 a great variety of steels have been developed. All these steels are essentially iron–carbon alloys; but they differ appreciably in carbon content and in many instances in the amount of other alloying elements present. The properties of steel have been improved considerably by the application of science of metallurgy.

For construction of boiler parts subjected to pressure and their integral attachments, the designer takes into account several factors, including relative cost,

TABLE 17-1. Relative cost of tubes (RC) and their maximum skin temperature.

Material	SA 192	SA210 Gr.Al	SA210 Gr.C	SA209 TI	SA209 TIA	SA213 T11	SA213 T12	SA213 T22	SA213 T91
RC	1.00	1.04	1.08	1.35	1.35	1.50	1.50	1.80	5.00
MST°C	454	454	454	482	482	552	552	579	649
MST°F	850	850	850	900	900	1025	1025	1075	1200

mechanical properties, manufacturing methods, scaling resistance, and maintenance. These factors are described in details below.

17-1-1 Relative Cost

The relative cost and adequate resistance to oxidation indicated by maximum permissible operating skin temperatures (MST) in °C primarily dictate the choice of materials. The ratios of permissible stresses at mean wall temperatures can change the relative cost and must be considered to achieve an economical choice. Further, a rapid rate of load change may dictate the choice of high-strength materials to achieve reduction in wall thickness of components such as drums.

Table 17-1 gives the approximate relative cost of the tubes and conventionally used limitation on the maximum skin temperature for heated tubes to various ASME/ASTM specifications.

17-1-2 Mechanical Properties

Mechanical properties such as ultimate tensile strength, yield strength, creep strength, creep-rupture strength, fatigue strength, ductility, and toughness are required to be considered for boiler construction.

A column supporting a building, being under compression, requires compressive strength and rigidity. It does not require much ductility or toughness. On the other hand, the pressure part of a high-pressure boiler must possess the adequate mechanical properties not only at ambient temperature but also at operating temperature. It should also possess sufficient ductility and toughness to enable it to withstand unavoidable deformations and shocks during fabrication, welding, and in operation, without cracking or fracturing at nominal stresses.

17-1-3 Manufacturing Methods

Welding, cold forming, hot forming, and expanding are some of the modern manufacturing methods used to fabricate a particular material into required shapes. A consideration of these methods is important, as the least expensive fabricated product with adequate properties rather than the cheapest steel should dictate the choice.

17-1-4 Scaling Resistance

Need for scaling resistance at maximum surface temperature, creep properties at mean wall temperature, and availability of welding material with adequate properties must be considered.

17-1-5 Maintenance

Ease of maintenance and compatibility with expected resistance to creep at elevated operating temperatures, as well as corrosive and erosive factors in the service environment also need to be considered.

17-2 Important Mechanical Properties of Various Materials

17-2-1 Common Properties

The following mechanical properties are important at all temperatures.

a) Tensile Strength

The standard tension test provides data on tensile strength, yield strength, and ductility. When metals are pulled with an uniaxially increasing load, the material stretches. Load per unit area is called *stress*, and stretch per unit original length is called *strain*. Until it reaches yield point, the strain is directly proportional to stress, and stress equals strain multiplied by Young's modulus of elasticity. There is no permanent deformation. So the strain vanishes when the load applied is removed.

Continuing to increase the loading beyond the yield point results in plastic strains and eventual breakage of test specimen. The breaking load divided by the original cross sectional area give the *ultimate tensile strength*. Yield point for harder materials such as steel is not easily determined, but the yield strength can be easily determined and is defined as the stress necessary to produce a specific value (0.1% or 0.2%) of plastic strain deformation. It is also called *proof stress*.

Most of the specifications require room temperature tensile testing. The design determines the minimum values of ultimate tensile strength and yield strength, and confirms that the ductility specified is met.

b) Ductility

Ductility is a measure of the amount of plastic deformation the steel will sustain before breaking. It is usually expressed as percentage elongation or reduction in cross section area of test specimen plastic strain exhibited by steel. It is inversely related to the tensile strength. High-strength steels will typically exhibit less ductility than softer, low-strength steels. This property is important with regard to manufacturing and operation. Ductility allows steel to withstand fabrication operations such as bending, swagging, or installation operations like as expanding or welding. Therefore, many *product form specifications*—most noticeably those for pipes and tubes, require various deformation tests such as 1) flattening, 2) flaring, or 3) bending.

c) Malleability

Malleability is the capacity of metals to be permanently extended in all directions when subjected to compressive force.

d) Hardness

Hardness is a measure of resistance to plastic deformation and is related to tensile strength. It is used also as an indicator of machinability and abrasion resistance of steel. In materials where ductility is important for its workability, its maximum hardness is specified.

e) Toughness

Toughness is the ability of a material to resist a brittle mode of failure when subjected to concentrated loads at low temperature. Various impact tests evaluate toughness. The mode of failure in impact tests changes from ductile (shear) to brittle (cleavage) as temperature is lowered. The temperature range at which it occurs is called transition range. Material within or below the transition temperature range may crack if subjected to an impact load or if construction details are such that localized yielding is prevented.

f) Fatigue

Fatigue strength is the magnitude of cyclic stress a material can resist for a specified number of cycles before failure. The fatigue stress a material can withstand under repeated application and removal of load is less than what it can withstand under static conditions. Yield strength can only be used as a guide in designs for materials subjected to static loading. Dynamic and cyclic loadings cause slip and cold working in minute areas localized at grain boundaries and at stress concentrating notches of various types. As sufficient work-hardening develops, microscopic cracks develop and grow until complete fracture results. At sufficiently low stress levels, many materials can tolerate an almost infinite number of cycles.

Areas of peak stress, such as change in section or stress-concentrating notches promote localized concentration of cyclic strains. Tubes can crack at high stress concentration points (such as the end of a flat stud welded to a tube) under flow-induced vibration. Thus they need antivibration restraints to correct the design.

17-2-2 High-Temperature Properties

The following properties are important for high-temperature operations.

a) Creep

Creep is the slow plastic deformation of metals under constant stress. It becomes important in boilers at operating temperatures of 450–650°C. Creep can take place and lead to fracture at an extended static load much lower than that which will cause yielding or breakage when the load is of short duration. The rate of extension varies with the time period at the operating temperature. In ASME boiler code, a creep rate of 0.01% in 1000 h and creep stress to rupture in 100,000 h are considered. No means have been found to predict this behavior quantitatively from short duration tests. It, therefore, becomes necessary to perform tests of creep and stress rupture at several stress levels, various temperatures, and over time periods as long as possible. For design purposes, the ultimate tensile strength, proof stress, and creep rupture strength at operating temperature are, of course desirable.

b) Creep and Fatigue Interaction

When materials are exposed to cyclic loadings while operating at temperatures within their creep range, the creep effect can reduce their fatigue life.

Differences in thermal expansion of superheater terminal tubes, which are welded to the outlet header, and seals and at the furnace enclosure, have often caused circumferential cracks in tubes at the top of the tube-to-header weld. Strain cycling causes these cracks to initiate and propagate in a fatigue-induced manner. Vibration induced cracks are generally transgranular (across grain), which is typical of fatigue. The cracking in the tube at the superheater outlet header is often intergranular (i.e., along grain boundaries). In a longer time creep or stress rupture an intergranular crack is typical. Low-cycle high-strain loading in the creep range is likely to produce intergranular rather than transgranular cracking.

17-2-3 Weldability

Weldability of material is an important manufacturing consideration. Besides parent material, welding electrodes, weld details, and welding techniques also play a part. Welding problems that have to be overcome include

a) solidification cracking
b) heat affected zone liquefaction cracking
c) hydrogen induced cracking
d) lamellar cracking
e) reheat cracking

The nature of the problem and methods to avoid them are discussed below:

(a) Solidification cracking takes place if the ductility of the weld material is lowered by the presence of residual liquid films to a level where it cannot accommodate the strain of contracting of the solidifying crystals. The remedy often involves lowering the strain during welding by improving the fittings, modification of tacking and jigging procedure, and avoidance of preheating the other welds in the vicinity. When welding around a patch or a nozzle, welding in steps instead of welding continuously helps. Basic coated electrodes are more tolerant to sulfur and reduces this cracking incidence.

(b) High peak temperatures induce local melting at grain boundaries in ferritic steels owing to formation of sulfide eutectic. This is the cause of heat affected zone cracking. The accompanying plastic strain may open up microcracks at liqueated boundaries. When the weld cools, these cracks do not heal, and the presence of brittle solidified films reduces the fracture toughness of the heat affected zones to an unacceptable level. The presence of phosphorous and sulfur has an important effect on susceptibility to this problem. Control of the heat input as well as selection and testing of material for welding is the primary remedy.

(c) Hydrogen-induced cracking can occur weeks after welding if no heat treatment is carried out after the welding. Choosing material that does not harden in the heat affected zone is critical. To avoid this type of cracking, one may (a) use dry and controlled hydrogen electrodes, (b) use MIG and a submerged arc process, or (c) select appropriate heat input while welding based on the carbon equivalent concept [carbon equivalent $= C + Mn/6 + (Cr + Mo + V)/5 + (Ni + Cr)/15$].

(d) Lamellar cracking occurs if the through thickness ductility is very low while welding tee joints. Improved weld design, use of forged products, nondestructive testing of plates, use of vacuum degassed plates, and use of low-strength weld metal are some of the remedies.

(e) Reheat cracking occurs when the geometry of the joint and relative properties of the weld material and the heat affected zone are unsatisfactory. Profiling of the toes of the fillet welds by grinding, control of sulfur, use of low-strength weld material, and buttering the weld surface are beneficial.

17-3 Fundamental Metallurgical Operations to Improve Steel Properties

The application of process and physical metallurgy has greatly enhanced the strength and usefulness of steel in the boiler industry. Process metallurgy is concerned with the extraction of metals from ores, while physical metallurgy is concerned with the combination of alloying and heat treatment and processing the metals into useful engineering materials. The following sections introduce some basic concepts of metallurgy that are frequently used in the boiler industry.

17-3-1 Process Metallurgy

The first step in the production of iron and steel is the removal of oxygen from the iron ore with coke and limestone in a furnace. The result is a mixture of molten iron, carbon, and small amounts of other elements as impurities. It is called *pig iron*. It contains about 2–4% carbon and varying amounts of silicon, manganese, phosphorous, and sulfur as impurities.

The pig iron is converted into steel by reduction of excess carbon through oxidation by various processes, such as the open hearth (basic/acid) process, Bessemer basic/acid process, electric process, and LD process (basic oxygen process). Troublesome impurities of phosphorous and sulfur are captured by basic fluxes such as magnesium and calcium. As such the basic processes are important for boiler quality steels, which contain up to about 0.35% carbon, but contain low percentages of phosphorous and sulfur.

17-3-2 Physical Metallurgy

Pure iron is a relatively soft material. It exists in different crystallographic forms at different temperatures. Molten iron solidifies as a body-centered cubic structure (magnetic delta iron) that transforms at about 1400°C to a face-centered cubic structure (austenite—nonmagnetic gamma iron), while below 910°C it reverts to body-centered cubic form as nonmagnetic alpha iron. Below 768°C it converts to body-centered cubic form as magnetic alpha iron. Temperatures at which these changes take place are known as *critical temperatures or critical points*. The solubility of carbon in various forms of iron differs. It is least in alpha iron. The addition of carbon increases the strength of iron in a number of ways.

Carbon has limited solid solubility (0.025%) in alpha iron to produce solid-solution ferrite, which is stronger than pure iron. It can combine with iron to form a compound Fe_3C with 6.67% carbon called *cementite*, which itself is very hard. Carbon at higher temperatures has solubility in austenite up to 2.0%. This difference in solubility is the crux of all heat treatment processes of steel. If a steel containing less than 0.8% carbon is cooled very slowly it transforms to a mixture of ferrite and pearlite, which is a 0.8% carbon containing eutectoid mixture of ferrite and cementite. Pearlite consists of lamellar ferrite and cementite. By changing the rate of cooling, it is possible to change the transformations that take place and produce structures that have a wide range of properties.

As we increase the rate of cooling, the ferrite and pearlite grains become progressively smaller. Smaller grain structure improves both the strength and toughness (grain number is given by the formula $N = 2^{(n-1)}$, where n is the number of grains observed by microscope in square inches at 100 times linear magnification conditions). The resultant ferrite grain size also depends on the initial grain size.

If the cooling rate is further increased, structures called *upper bainite* and *lower bainite* are formed. The upper bainite has more strength than the ferrite/pearlite structure, but it has a relatively poor toughness. Lower bainite has a higher strength and improved toughness.

With further increases in the cooling rate the transformation product becomes very hard and acicular and is called *martensite*. The ductility and toughness of untempered martensite is generally poor, but if reheated and tempered, the steel's strength falls progressively with the increase in tempering temperature, while the ductility and toughness improve. Steels used in pressure vessel and boiler construction include ferrite-pearlite type, bainitic type, and quenched and tempered type. The above general types of transformation also occur in alloy steels.

17-3-3 Heat Treatment

Several physical processes, like heat treatment, rolling, etc., can substantially improve the properties of steel. The following section explains the effect of heat treatment on steel.

a) Annealing (Achieves Softness and Ductility)

Annealing is a process of heating the metal 55°C (100°F) above the upper limit of critical range (760–910°C depending on carbon %) and then allowing it to cool slowly at controlled rate in the furnace. Fully annealed steel will contain ferrite and pearlite, and it is softer.

b) Stress Relieving (Relieves Stress Set Up During Fabrication)

This can be achieved by heating the steel below lower critical range (usually between 540°C (1000°F) and 705°C (1300°F) and holding the components for 1 h per inch thickness. It is then allowed to cool in still air to remove the residual stresses.

c) Normalizing (Relieves Internal Stresses)

Normalizing is a special annealing process wherein steel cools in still air from the upper critical temperature. It relieves internal stresses caused by previous working and produces less softness and ductility. Normalizing is often followed by tempering.

d) Tempering

Tempering removes some of the brittleness induced by the normalizing process. It is done by heating below critical temperature at 205°C (400°F) to 400°C (750°F). The higher the temperature, the softer and tougher is the steel. Some steel becomes brittle when cooled slowly from certain temperatures (called temper brittleness). To overcome this difficulty such steels are cooled rapidly by quenching.

17-3-4 Mechanical Working

Mechanical working can be either cold or hot working. It is usually performed above the transformation temperature. Plates are produced by hot rolling from ingots or slabs. Tubes are manufactured by electrical resistance welding (ERW). Seamless tubes are hot rolled or cold rolled/finished. Pipes are produced by hot rolling or cold rolling or by the extrusion process. Drums are fabricated by cold rolling or hot rolling of plates and welding by the submerged arc or electro-slag welding process. Dished ends are generally pressed from plates and welded to the other cylindrical sections. The quality of steel must be able to meet these mechanical forming requirements. Fully killed and semi-killed steels are often used for these components to ensure their soundness.

17-3-5 Effect of Addition of Different Alloying Elements

Adding small amounts of different elements considerably improves the quality of the steel. Some of those materials are discussed below.

a) Effect of Sulfur Impurity in Steel

Sulfur is usually present in steel as iron-rich sulfides. The sulfides form with iron eutectoids, giving a low melting point. While still liquid, they spread around the boundaries of the metal grain formed and solidify around the grains to give brittle films on further cooling at the grain boundaries. Such steels cannot be hot worked and exhibits brittleness owing to reliquefication of iron sulfides.

b) Effect Manganese in C-Mn Steel

Addition of manganese improves properties of steel. Manganese combines preferentially with sulfur to produce manganese-rich sulfides that have higher melting points that are not in the liquid phase at the hot working temperatures of steel. They remain dispersed and have little effect on toughness. Also, manganese increases the strength of steel by forming a solid solution with iron. Additionally, manganese retards the transformation of austenite to ferrite and pearlite. This means that even in thicker plates and sections that cool more rapidly at the surface than the interior, better and more uniform properties are obtained throughout the thickness. Thus for a given rate of cooling a higher effective strength can be obtained.

Properties of carbon-manganese steel can be further improved in the following ways:

Addition of aluminum to kill the steel refines its grain size to produce greater toughness: Normalizing produces finer grain size to improve the strength and toughness. The process involves heating of steel to sufficiently high temperature to form austenite and then allowing it to cool in still air. Addition of small quantities of niobium or vanadium or nitrogen is frequently made to carbon steels intended for normalizing to ensure improved properties.

Controlled rolling of cast ingots: Hot rolling is generally carried out in the austenite range at which the steel is soft and metal grains are continuously recrystallizing and growing. If rolling temperatures are lowered the steel finally transforms into ferrite and pearlite of finer grain size, giving higher strength and toughness. Small quantities of niobium or vanadium are added in such controlled rolling. By normalizing higher strengths are achieved, combined with superior welding properties and toughness.

Quenching and tempering: Increasing the cooling rate from austenite stage increases the strength of transformed structure. Plates are reheated after rolling and then quenched normally by using high-pressure water sprays. If necessary it can be followed by tempering, resulting in somewhat reduced strength, but with improvement in toughness.

c) Effect of Silicon

It is mainly used as a de-oxidiser and de-gasifier to improve the quality of steel. When added up to 2.5%, silicon increases the ultimate tensile strength without decreasing ductility. If the amount added exceeds 5%, the steel become non-malleable. It increases the resistance to oxidation and surface stability but lowers creep properties.

d) Effect of Chromium

Chromium increases the ultimate tensile strength, hardness, and toughness of steel at room temperatures. It also increases the resistance to oxidation at higher temperatures; and therefore, it is used in high temperature service steels.

e) Effect of Nickel

Nickel increases toughness, resistance to corrosion, and hardness. It improves the creep of low alloy steels slightly, but with chromium it enhances the creep strength considerably.

f) Effect of Molybdenum and Wolfram (Tungsten)

Addition of molybdenum increases the strength, elastic limit, resistance to wear, and impact properties. Molybdenum contributes to red hardness of steel (i.e., hardness is not affected up to red hot temperature). It also restrains the grain growth and the softening of steel. Chromium-molybdenum steels are less susceptible to temper embrittlement and have enhanced creep strength. Wolfram (also known as tungsten) has a similar effect but it is costlier; it is used in tungsten carbide cutting tools.

g) Effect of Vanadium

It is an excellent de-gasifying and deoxidizing agent; but it is seldom used for this purpose because of its high cost. Essentially it is a carbide-forming element that stabilizes the structure especially at high temperatures and increases strength, toughness, and hardness of steel. Excellent creep strengths are achieved with 0.5% to 1.0% molybdenum and 0.1% to 0.5% vanadium.

h) Effect of Aluminum, Titanium and Niobium

Aluminum is an important minor constituent and deoxidizer used to produce killed steel. However, an excessive amount of aluminum has a detrimental effect on creep properties. Titanium and niobium (also called columbium) are carbide-forming elements. They react more readily with carbon than chromium, allowing the latter to remain in solid solution and increase corrosion resistance.

i) Effect of Phosphorous

High phosphorous content has an undesirable effect on properties of carbon steel like resistance to shock and ductility when steels are cold worked. This embrittlement effect is called *cold shortness*. It results from the grain growth and segregation and is encouraged by the phosphorus. The phosphorous content is restricted to a maximum of 0.04% or less. Its presence is objectionable for welding. It increases yield strength and resistance to atmospheric corrosion up to 0.15%. It is used in special structural steel applications.

17-4 Design Methods

There are two main approaches to the design of boiler pressure parts.

- Design by rules
- Design by analysis

These methods are described briefly here.

17-4-1 Design by Rule Method

The experience-based method of design is known as *design by rule*. Design by rule involves determination of design loads, choice of a design formula, and selection of appropriate stress for the material to be used. These are used to calculate the required thickness of the pressure part, which is then fabricated following other design rules and using the construction details permitted by the code. It is not based on detailed stress analysis; instead, the rules generally involve the calculation of primary membrane stresses across the thickness of the walls of vessels. In a few cases where bending stresses caused by imposed loads are explicitly recognized, it is only the combination of primary and bending stress that is limited. This is not to say that codes ignore other type of stresses. In practical designs, rules have been written to limit such stresses to a safe level consistent with experience. The code assigns allowable stresses, with sufficient margin of safety factors to cover the localized stress in normal circumstances. In the USA this method uses Section I of ASME code, which provides the rules for construction of power boilers, heating boilers, and pressure vessels to Section VIII division. 1. Similar codes are available in Germany (Code TRD-series 300), the U.K. (Code BS-1113), India (IBR code), etc. The ASME Section I and ASME Section VIII Division I use maximum stress theory to combine three principal stresses. The TRD rules consider maximum shear stress theory to do so. It is one of the main reasons for differences in the formulae for thickness calculations in different codes. A comparison of the different codes are given in Table 17-3a.

17-4-2 Design by Analysis Method

In this method a detail stress analysis is carried out, which permits the use of higher allowable stresses depending on anticipated loading conditions without reducing the safety and requires rigorous analysis of all types of stresses and anticipated loading conditions. It is thus not necessary to use a single allowable stress limit as given in Table 17-2. One can use appropriate limits based on the location and distribution of the stress in the structure and the type of loading that produces it.

TABLE 17-2a. Maximum allowable stress in N/mm^2 of different materials according to their application (adopted from ASME section II, part D).

	Temperature in °C	38	93	149	204	260	316	343	371	399	427	454	482	510	538	566	593	621	649
	Material specs.																		
Pipes	SA 106 GR A	83	83	83	83	83	83	83	81	74	62	49	34	21	10				
	SA 335 P1	95	95	95	95	95	95	95	95	95	93	90	88	57	33				
	SA 106 GR B	103	103	103	103	103	103	103	99	90	75	54	34	21	10				
	SA 335 P11	103	103	103	103	103	103	103	103	102	99	97	94	64	43	29	19	13	8
	SA 335 P12	103	103	103	103	103	103	103	103	103	101	97	78	50	31	19	12	8	
	SA 335 P22	103	103	103	103	103	103	103	103	103	103	99	94	75	55	39	26	17	10
Tubes	SA 209 T1	95	95	95	95	95	95	95	95	95	93	90	88	57	33				
	SA 210 Gr A1	103	103	103	103	103	103	103	99	90	75	54	34	21	10				
	SA 213 T11	103	103	103	103	103	103	103	103	102	99	97	94	64	43	29	19	13	8
	SA 213 T22	103	103	103	103	103	103	103	103	103	103	99	94	75	55	39	26	17	10
	SA 210 Gr C	121	121	121	121	121	121	121	115	102	83	54	34	21	10				
	SA 213 T91	147	147	146	146	146	143	141	138	134	129	123	115	107	99	89	71	48	30
Plates	A 36	100	100	100	100	100	100	100	96										
&	SA 515 GR 70	121	121	121	121	121	121	121	115	102	83	54	34	21	10				
Flats	SA 516 GR 70	121	121	121	121	121	121	121	115	102	83	50							
	SA 299	129	129	129	129	129	129	129	122	108	83	54	34	21	10				
Rods	SA 105	121	121	121	121	121	121	121	115	102	83	54	34	21	10				

TABLE 17-2b. Stress values (N/mm^2) at different temperatures for chosen material from British standard and 100,000 h design lifetime.

Temperature, °C / Material	250	300	350	390	400	410	420	430	440	450	460
BS3059:part2360S2cat2	111	97	87	79.8	78	77.5	77	76	69	60	52
BS3059:part2620S2	129	129	129	121.8	120	119.2	118.4	117.6	116.8	116	115.
BS3059:part2440S2	134	119	106	102	101	101	100	90	77	65	56
BS3604:part1HFS622cat2	157	153	149	145.8	145	143.4	141.8	140.2	138.6	137	135
BS3059:part2622-490S1	157	153	149	145.8	145	143.4	141.8	140.2	138.6	137	135
BS3602:part1HFS500Nbcat2	163	148	135	127	125	121	105	90	77	65	56
BS1501-271	219	207	201	196.2	195	194.6	194.2	193.8	193.4	193	190
BS3602:part1HFS500Nbcat2		148	135	127	125	121	105	90	77	65	56
BS3604:part1HFS622cat2		153	149	145.8	145	143.4	141.8	140.2	138.6	137	135
BS3059:part2360S2cat2		97	87	79.8	78	77.5	77	76	69	60	52
BS3059:part2440S2		119	106	102	101	101	100	90	77	65	56
BS3059:part2620S2		129	129	121.8	120	119.2	118.4	117.6	116.8	116	115.
BS3059:part2622-490S1		153	149	145.8	145	143.4	141.8	140.2	138.6	137	135
BS1501-271		207	201	196.2	195	194.6	194.2	193.8	193.4	193	190

(Contd.)

TABLE 17-2b. (*Continued*)

Temperature, °C Material	470	480	490	500	510	520	530	540	550	560	570	580	590	600
BS3059:part2360S2cat2	44	36												
BS3059:part2620S2	115.	114	113	112	93	76	62	52	42	33	27			
BS3059:part2440S2	48	42												
BS3604:part1HFS622cat2	133	131	118	105	94	82	72	61	53	45	39	34	29	26
BS3059:part2622-490S1	133	131	118	105	94	82	72	61	53	45	39	34	29	26
BS3602:part1HFS500Nbcat2	48	42												
BS1501-271	181	155	129	107	87	69	54	41	30					
BS3602:part1HFS500Nbcat2	48	42												
BS3604:part1HFS622cat2	133	131	118	105	94	82	72	61	53	45	39	34	29	26
BS3059:part2360S2cat2	44	36												
BS3059:part2440S2	48	42												
BS3059:part2620S2	115.	114	113	112	93	76	62	52	42	33	27			
BS3059:part2622-490S1	133	131	118	105	94	82	72	61	53	45	39	34	29	26
BS1501-271	181	155	129	107	87	69	54	41	30					

TABLE 17-3a. Comparison of some national boiler codes.

SCOPE

German Standard TRD (series 300)	American Standard ASME (section 1)	British Standard BS 113, item 1.1	Indian Standard Indian Boiler Regulation (IBR)
Steam generators and associated economizers, shut-off-type superheaters, steam attemperators, steam and hot water lines and the associated valves/fittings Coverage TRD001 General TRD100 Material TRD2001 Manufacture TRD300 Design TRD401 Equipment and erection TRD500 Inspection and testing TRD520 Licensing TRD601 Operation TRD701 Boilers Gr II HWB TRD801 Boilers Gr I TRD802 Boilers Gr II	Preamble, p. 2—for steam generators ($p >$ 15 psig and/or $t >= 250°$F). Preamble, p. 2—for high temperature water boilers ($p >$ 160 psig and/or $t >= 250°$F). Preamble, PEB—for electric boilers. Preamble, PMB2—for miniature boilers (p < 100 psig and/or gross volume <5ft³ and/or heating surface area <20ft³)	Steam generators (water tube boilers) and steam generator components that are not separated by means of a valve. Reheaters (also separately fired). Separately fired super heaters. Superheaters separated from the boiler by means of a valve. Economizers (also separately fired) separated from the boiler by means of a valve. Valves separating steam from water or other systems.	Boilers (natural & forced circulation capacity >22.75 liters IBR:2c) and associated econo- mizers, superheaters, attemperators, steam and hot water lines and associated valves & fittings. Heat exchangers, steam accumulators, separators & evapo- rators (Chapter XV). Shell-type boilers (chapter XII). Miniature boilers for use in small establishments (pressure <2.11 kg/cm²(g) & heating surface area <20ft²) to be as per Chapter XIV
As per TRD 301, item 1 for all cylindrical shells, (tubes, drums, separating vessels, headers, shells, sections and alike) under internal gauge pressure $da/di <= 1.7$. When da/di up to 2.0 is acceptable where the wall thickness Sv is less than or equal to 80 mm. These rules primarily apply to ductile materials delta phi $>= 14\%$ or also applicable to less ductile materials with higher safety factor and wall thickness $<= 50$ mm as per item 1.3.	External piping is not covered by ASME section I but is covered by design regulations for the materials, calculation, manufactured as per ANSI B.31.1.	Connecting lines, as far they are not covered by another BS standard, as, e.g., BS 806. Piping beyond the stop valve on steam side and beyond non return valve on economizer inlet side are covered by BS 806.	Steam pipes, steam passage up to prime mover and other usage (pressure >3.5 kg/cm²[g]) IBR:2K.

(Contd.)

TABLE 17-3a. (*Continued*)

German Standard TRD (series 300)	American Standard ASME (section 1)	British Standard BS 113, item 1.1	Indian Standard Indian Boiler Regulation (IBR)
Primarily static load of the components to TRD300. Pulsating internal pressure or combined variation of pressure and temperature to TRD301 section.	Fired pressure vessels, Superheater, economizer, and other pressurized parts connected to the boiler, without intermediate valves. Reheaters (regarded as a fired pressure vessel).		The scope covers design regulations for materials, calculations, manufacture, installation and testing as per IBR.
Design rules only consider stress caused by internal pressure. Additional forces and moments of significant magnitude shall be separately considered, indicated and taken care of by the manufacturer.	The scope covers the actual boiler and the certificate as per ASME code.		
Remarks 301, item 1.4. In the case of cylindrical shells without cut-outs (designed as per TRD 301) there is no plastic deformation worth mentioning only in the area of time-independent strength characteristics. In the case of cylindrical shells with cutouts there may be plastic deformations within acceptable			

TABLE 17-3a. (*Continued*) Comparison of some national boilor codes. (Design/calculation pressure)

TRD	ASME (section 1)	BS 113,	Regulation (IBR)
TRD 300, item 7.2 For hot water generators p = permissible total gauge pressure. Design pressure for parts beyond SH. Outlet = Set pressure SH safety valves.	For natural and assisted circulation steam generators design pressure p = maximum allowable working pressure (pp. 21 & 21.2). The maximum allowable working pressure is the pressure determined by employing the allowable stress values, design rules and dimensions designated in the section (p. 21).	For natural and assisted circulation steam generators the highest set safety valve pressure on the drum (Cl 1.3.5) + increased where applicable to take account of the pressure drop and hydraulic head to take care of the most severe conditions (for the yield governed components).	IBR does not use the term *calculation pressure* used by other codes and uses the term *maximum permissible working pressure* (WP) or design pressure and it is taken as equal to maximum overpressure divided by 1.05 when all full lift safety valves are lifted or conventionally the highest set safety valve pressure on the drum for all the pressure parts.
Design pressure to TRD300 7.1. For natural circulation steam generators p = Max allowable working pressure p1 + Hydrostatic pressure >0.05 N/mm². Design pressure = p1 = Highest drum SV pressure [for drum to SH outlet] Design pressure = p1 + geodetic head [for Economizer inlet to evaporator outlets.]	Stresses owing to hydrostatic pressures shall be taken in determining minimum thickness. Additional stresses imposed by the effects other than working pressure or static head that increase the average stress by more than 10% of the allowable working stress shall be taken into account including, the self-weight of the component and method of support as per p. 22.	Or those components whose design stresses are time dependent the lowest set pressure of any superheater or reheater safety valve mounted on the steam outlet increased to take account of pressure drop corresponding to the most severe condition of operation.	
For once-through boilers p = maximum allowable working pressure at the component at MCR. i.e., superheater outlet pressure p1 + losses between the stop valve and the component.	For forced circulation steam generators design pressure p = maximum allowable sustained working pressure of any part not less than the maximum over-pressure reached when one or more over-pressure protection devices are in operation.		
For superheated steam lines downstream of steam boiler outlet design pressure p = maximum pressure safe guarded by safety device against excessive pressure.			

TABLE 17-3a. (*Continued*) Comparison of some national boiler codes. (Design/temperature °C)

Item 8 of TRD	ASME (section 1)	BS1113	IBR
Design temperature = Reference temperature (RT) + at least the temperature allowance (AT).	Design temperature should not be less than the maximum expected mean tube wall temperature. (arithmetical mean temperature on the outer and inner surfaces). For heated tubes the temperature shall be taken as not less than 700°F (371°C).	Cl. 2.2.3 and cl. 1.3.7: Design temperature shall not be less than the highest of the a) maximum mean wall temperature calculated by the manufacturer. In the case of pressure parts carrying superheated steam, the factors specified in cl. 2.2.3.8 shall be taken into account and b) the design temperature specified in cl. 2.2.3.2 to cl. 2.2.3.7.	IBR uses the maximum working metal temperature in place of design temperatures used by other codes. It is taken as given in the next row:
Reference temperature: For water and/or water/steam mixture, RT = saturation temperature at p1 or p, viz., Max. allowable pressure or design pressure.	For unheated tubes the wall temperature equal to the temperature of the working fluid can be taken as a basis. However, this must not be less than the saturation temperature.	As per cl. 2.2.3.2 for unheated drums and headers maximum design temperature (DT) = maximum temperature of internal fluid.	For steam, water and mud drums the saturation temperature corresponding to the working pressure (WP) + 28°C (50°F) for heated parts and saturation temperature + 0°C for unheated parts. For superheater steam drums & headers it is equal to the designed maximum seam temperature +28°C heated and 0°C for unheated or the drums in the third pass.
Temperature allowance for superheated steam: For unheated = 0°C, For heated by radiation = 30°C +3 × wall thickness in mm (or) at least 50°C. For heated by convection = 15°C + 2 × wall thickness in mm (or) 50°C maximum. For screened against radiation = 20°C.		As per cl. 2.2.3.2 for heated drums and headers maximum DT is not less than maximum temperature of internal fluid + 25°C	IBR 338 (a) for tubes: Working metal temperature for economizer (WMT) = maximum fluid temperature +11°C (20°F)

TABLE 17-3a. (*Continued*)

Item 8 of TRD300	ASME (section 1)	BS1113	IBR
Reference temperature: For superheated steam RT = superheated steam temperature.		As per cl. 2.2.3.3 for boiler tubes max. DT shall not be less than saturation temperature at calculation pressure (CP)°C + 50°C for boiler tubes receiving radiant heat from combustion chamber or saturation temperature at CP + 25°C for convection heated boiler tubes	For boiler tubes (WMT) = saturation temperature corresponding to WP + 28°C (50°F). For convection SH and RH tubes (WMT) = maximum anticipated steam temperature + 39°C (70°F) for heated tubes.
Temperature allowance for superheated steam; For unheated = 15°C or at least 5°C if part is after attemperator or mixer. For parts heated by radiation = at least 50°C. For parts heated by convection = at least 35°C. For parts screened against radiation = 20°C.		Heated parts: The maximum fluid temperature + at least 25°C (drums and heaters having wall thickness of >30 mm must not come into contact with waste gases at >650°C without a sufficient cooling)	For radiant SH and RH tubes (WMT) = maximum anticipated steam temperature corresponding to WP + 50°C (90°F). For unheated SH and RH pressure parts WMT = maximum anticipated steam temperature + 0°C.
For once through boilers DT = maximum fluid temperature expected during the operation + allowances as for natural circulation above.		For superheaters and reheaters DT = maximum internal fluid temperature which is taken as anticipated temperature +15°C for variation in flow up to anticipated temperature not above 425°C for header and beyond +50°C surfaces receiving radiation from combustion chamber or +35°C for convective superheater/RH surfaces or +0°C for unheated and +0°C for final SH outlet pipe.	

(*Contd.*)

TABLE 17-3a. (*Continued*)

Item 8 of TRD300	ASME (section 1)	BS1113	IBR
For heated drum DT = RT + 20°C for closely spaced tubes and no significant amount of flue gas flow between the screening tube and the drum.		As per cl. 2.2.3.7, which is applicable for parts beyond final superheaters. The design temperature is further increased by excess over the limits when the normal fluctuation exceeds by 8°C in a year from the rated, abnormal fluctuations in a year exceeds 20°C for 400 h in a year, 30°C for a maximum of 100 h in a year or 40°C for a maximum of 60 h in a year for systems having rated temperature more than 380°C. For systems having rated temperature less than 380°C it will be increased by the excess over 10%.	

TABLE 17-3b. Comparison of calculation methods of US, British and Indian Codes.

Standard	ASME Section I	BSI BS1113	IBR
Permissible stress	Tables, pp. 23.1 and 23.2 —To be taken from tables as maximum allowable stress values in Tables 1A and 1B of Section II, Part D, Appendix 7. The values are generally calculated as: min. set value between the stress a) for temperatures below the creep range —1/4 of the min. tensile strength specified at room temperature —1/4 of the tensile strength at the respective temperature —2/3 of the min. yield strength specified at room temperature —2/5 of the yield strength at the respective temperature b) for temperatures within the creep range —the 100% of the stress causing a creep rate of 0.01% per 1000 c) for temperatures within the transient area —the stress restricted to those being calculated on the basis of a continuous curve connecting the high and low temperature ranges, but being below the curve of 2/3 of estimated min. yield strength at the respective temperature. —Furthermore, the stresses in p. 23.1 are limited to 67% of the mean stress, which would cause the rupture after 100,000 h, or to 80% of the min. 100,000 h rupture stress value	(Table 2.1.2.) Design stress principles described in the Appendix A —For carbon, C-Mn, low-alloy steels (1) Up to and including 250°C use values at 250°C (2) Above 250°C $f = \frac{R_{e(t)}}{1.5}$ or $\frac{R_m}{2.7}$ or $\frac{S_{Rt}}{1.3}$ For austenitic steels $f = \frac{R_{e(t)}}{1.35}$ or $\frac{R_m}{2.7}$ or $\frac{S_{Rt}}{1.3}$ where R_m = minimum tensile strength specified for the grade of material at room temperature $R_{e(t)}$ = minimum value of $R_e 1.0$ or $R_p 0.2$ ($R_p 1.0$ for austenitic steels) at temperature T S_{Rt} = mean value of the stress required to produce rupture in time t at temperature T f = nominal design stress taken as minimum of time-independent nominal design stress	REG 271/338/342/350 (i) For temperatures at or below 454°C. The lower of the following two values: $f = E/1.5$ or $R/2.7$ (ii) For temperatures above 454°. The least of the following three values: $f = E/1.5$ or $S_t/2$ or S_c where E_t = yield point (0.2% proof stress at the design temperature T) R = minimum specified tensile strength at room temperature S_r = the average stress to produce rupture in 100,000 hours at the temperature T and in no case 1.33 times the lowest stress to produce rupture at the temperature S_c = the average stress to produce an enlongation of 1% (creep) in 100,000 h IBR 270/27 for drums and headers IBR 338 for tubes IBR 342 for cylindrical headers IBR 350 for steel pipes

(Contd.)

505

TABLE 17-3b. (*Continued*)

Standard	ASME Section I	BSI BS1113	IBR
Ligament coefficient	Page 27.4, Note 1: pp. 52 and 53 E = ligament coefficient of welded longitudinal seams or of webs between fields and tracks of punched holes, depending on which is the smaller value. E = 1.0 seamless or welded cylinders E = 1.0 for welds, if all weld reinforcements on the longitudinal welds were removed and are flush with the surface of the plates E = 0.9 if position of weld reinforcements is left unchanged E = for webs as per p. 52—for groups of openings forming a uniform pattern and as per p. 53—for groups of openings that do not form a uniform pattern —For tracks of punched holes in parallel the tank axis (PG52) $E = (p - d)/p$ (equal pitch) $E = (pl - nd)/pl$ (unequal pitch) where p = longitudinal pitch between adjacent holes pl = pitch between corresponding holes in one row of symmetrical groups of openings d = diameter of the tube hole n = number of openings in the length pl —For tube holes on a diagonal the ligament coefficient is calculated on the basis of figures on pp. 52.1, 52.6	Ligament efficiency: Cl. (3.3.3.2. to 3.3.3.7.) —For tracks of drilled holes in parallel to the axis $\eta = (s - d)/s$ (equal pitch) $\eta = s_1 + s_2 - \frac{2d}{(s_1 + s_2)}$ (unequal pitch) where s = longitudinal pitch s_1 = shortest pitch s_2 = longest pitch d = diameter of the opening —For tracks of drilled holes on a diagonal: ligament coefficient = 3.3–3.5	IBR 215 E = ligament efficiency $E = (p - d)/p$ (equal pitch) $E = (p - nd)/p$ (unequal pitch) Fig. 15 (diagonal) efficiency & Fig. 14

TABLE 17-3b. (Continued)

Standard	ASME Section I	BSI BS1113	IBR
Allowances	PG 27.4 —Acc. To Note 4: for expanded tubes with $da < 127$ mm and expanded tube ends a thickness addition of 0 to 3.81 mm can be considered for thread cutting and min. structure resistance Note 3: c is for threading allowance only —Corrosion/erosion allowance are to be added if necessary —Acc. To "Note 7": the manufacturing tolerance are pitches as listed on p. 9.	Cl. 3.2.1 For cylindrical shells the min. corrosion allowance = 0.00 mm, unless other agreements were made. Furthermore, the minus tolerance and the manufacturing allowance as per specified standards must be considered.	IBR 270, 338, 350 indicate allowances applicable T = minimum thickness of tubes, i.e., nominal thickness minus the permissible negative tolerance.
Formulae	Page 27.2.1 Tubes up to and including 5 in. outside diameter $t = \frac{p \cdot D}{2s + p} + 0.005D + e$ PG27.4 Note 2-4,8,10 apply —Pipings, drums & headers (based on strength of the weakest section) PG 27.2.2 $t = \frac{p \cdot D}{2s \cdot E + 2y \cdot p} + C$ —for the component with anchorage and not meeting the conditions p. 27.4, Notes 9 and 10. $t = \frac{p \cdot R}{0.882s \cdot E - 0.6 \cdot p}$ t = min. wall thickness required p = max. service pressure D = outer diameter of cylinder η = ligament efficiency S = max. permissible stress C = min. allowance (p. 27·4, Note 3: 1.651 mm for 3.5 in. OD)) e = Thickness factor for expanded tube ends y = temperature coefficient (PG27·4, Note 6)	Cl. 3.3.2 —cylindrical shells $t_m = \frac{p \cdot d_0}{2f\eta + p}$ or $\frac{p \cdot d_i}{2f\eta - p}$ where: t = min. wall thickness required p = calculation or design pressure d_i = internal diameter d_0 = outer diameter f = design stress at design temperature η = ligament efficiency	REG 270 (Eq. 72) for drum and headers: $t = \frac{WP \cdot d_i}{2f - WP} + C_1$ REG 338(a) (Eq. 87) tubes: $t = \frac{WP \cdot d_0}{2f + WP} + C$ REG 350 (Eq. 91) for steam pipes: $t = \frac{WP \cdot d_i}{2f \cdot e + WP} + C_1$ where t = min. wall thickness, mm WP = max. working pressure in kg/cm^2 d_i = internal diameter, mm d_0 = outer diameter, mm f = allowable stress in kg/cm^2 η = ligament efficiency e = efficiency factor = 1.0 for seamless & ERW pipes c_1 = 0.75 mm C = 0.75 mm for working pressure up to and including 70 kg/cm^2; $C = 0$ for $WP > 70$ kg/cm^2

507

Secondary stress developed by constraints is generally self-limiting. It is customary to allow it up to twice the yield strength. Tertiary (local and peak) stress does not cause any noticeable deformation. It is objectionable only as a possible source of a fatigue crack or brittle fracture. For example, the discontinuity local stress and strain at a notch can be several times the yield stress and strain. But this condition is recognized as perfectly safe as long as the material is ductile and the load is not cyclic enough to start a fatigue crack. The average membrane stress in the vessel owing to internal pressure is not allowed to exceed the yield strength. Another example of such stress is thermal stress, which is self-equilibrating within the structure.

The yielding helps the material to accommodate the imposed distortion pattern owing to the thermal strain caused by a nonuniform distribution of temperature or by differing coefficients of expansion within the structure. Such stresses can be safely allowed to exceed the yield strength of the ductile material. The third example is discontinuity stress near the junction of a head and a shell that is produced by internal pressure. In this case, the head and shell expand by different amounts and distortion is required to keep them together at the junction. The stresses associated with this distortion have different significance than the primary stresses required to balance the pressure load. There should be different maximum allowable stresses for different types of the stresses—namely, primary stress, secondary stress, and tertiary (local or peak) stresses and limits appropriately considering the usage. This design by analysis method is not yet widely used in the ASME Section I—power boiler design code. The onus is put on the design organization, who is forced to use the engineering concepts to produce safe design. ASME Section VIII Division 2 and Section III covers the design by analysis method. Other codes have also adopted similar proportions to varying degrees.

17-4-3 Factors Affecting Mechanical Properties of Steels

The thickness of a boiler component (tube, pipe, header, or drum), subjected to internal pressure, depends on the maximum value of the primary stress that can be allowed on it. This permissible limit, called *maximum allowable stress*, is critical to the design of pressure parts. In *rule based* design (see Section 17-4) this value of permissible stress takes care of all forms of loading thermophysical conditions so that the thickness determined from it is safe under all conditions. Thus, the maximum allowable stress depends on a large number of criteria. Different countries have their own national codes or standards for taking account of these factors. For example, in the USA the American Society of Mechanical Engineers (ASME) lists materials in Section II, Part D and their values of maximum allowable stress are given in Table 17-2a. These are based on the minimum of following mechanical properties criteria set:

- One-fourth of specified minimum tensile strength at room temperature
- One-fourth of the tensile strength at elevated temperature
- Two-thirds of specified minimum yield strength at room temperature

- Two-thirds of the yield strength at elevated temperature
- One hundred percent of stress to produce creep rate of 0.01% in 1000 h
- Sixty-seven percent of average stress to produce rupture at the end of 100,000 h
- Eighty percent of minimum stress to rupture at the end of 100,000 h

The relative influence of the different criteria varies with materials as well as with temperature. Carbon steel begins to lose strength at about 370°C (700°F). At 454°C (850°F) its strength is down to about one-half of the room temperature values. The low chromium ferritic alloys start to lose strength above 425°C (800°F) and are down to about half the strength at about 535°C (1000°F).

The strength of austenitic stainless steel declines to an extent from room temperature to 535°C (1000°F) because of a reduction in the yield strength. However, as the temperature rises to 650°C (1200°F) creep drops the strength to half (or less) that at the room temperature. High-temperature strength and the ductility of steel may be strongly affected by grain size, cold working, heat treatment, and other variables and they require consideration in manufacturing and fabrication procedures.

The reason for considering the tensile strength and yield strength both at room temperature and elevated temperatures is probably due to the fact that some of the low carbon steels exhibit marginally higher strength at increased temperature up to about 220°C (425°F).

17-5 Thickness (Scantling) Calculations

In most cases the calculation of the minimum required thickness of the tubes, pipes, and header are based on formulae for "thin shell subjected to internal pressure". The required thickness, t, is directly proportional to the design pressure, P, and outside diameter, d, of the tubes and is inversely proportional to the allowable design stress value, σ_{allow}, for a given material.

$$t = \frac{P \cdot d}{2 \cdot \sigma_{allow}} \qquad (17\text{-}1)$$

Using this basic equation, ASME code specifies for tube thickness the following equation:

$$t = \frac{PD}{2S + P} + 0.005\,D + e \qquad (17\text{-}2)$$

ASME Section I formulae for calculating thickness of tubes and pipes are reproduced below:

	Tubes [Outer dia., $D \le 5$ inch]	Piping, headers, drum
Thickness, t, in in.	$\dfrac{PD}{2S + P} + 0.005D + e$	$\dfrac{PD}{2S.E + 2yP} + C$
Maximum allowable working pressure, P in psi	$P = S\left[\dfrac{2t - 0.01D - 2e}{D - (t - 0.005D - e)}\right]$	$P = \left[\dfrac{2SE(t - C)}{D - 2y(t - c)}\right]$

where e = thickness factor expanded tube ends
            = 0.04 to 0.15 depending upon tube size, etc.
            E = efficiency of longitudinal weld or ligament between openings,
            whichever is smaller
            C = allowance for threading
            y = coefficient of temperature
            D = outside tube diameter, in.
            S = allowable stress in psi

The above the calculation procedure also requires estimation of the following:

(a) Calculation (or design) pressure
(b) Calculation (or design) temperature

Further the procedures also considers:

(c) Choice of the material suitable for operation at the calculation temperature and
    outside skin temperature
(d) Minimum thickness allowance for thinning owing to bending of pipes
(e) Allowance owing to negative tolerance to arrive at the minimum ordering
    thickness

The following items are also checked:

The thickness should conform with the minimum ordering thickness from a manufacturing standpoint (like membrane wall construction), minimum code thickness, and standard thickness of manufactured tubes. The ratio of the chosen thickness to the inside diameter should be less than or equal to 0.25 to satisfy the thin
shell formula limitations. Otherwise, the thick shell formulae should be used to
calculate required thickness.

The chosen thickness should be adequate for external loads and bending loads
over and above the longitudinal pressure stress. Since the pressure-induced circumferential stress is twice the longitudinal stress, the thickness, in most cases, is
governed by requirements of that for withstanding the internal pressure alone. The
tubes and the pipes are generally able to carry external loads without needing to
increase the thickness further. For tubular components carrying very heavy loads
(a double-pass utility boiler, for example), the rear wall sling tubes and horizontal
bank supporting tubes need an increase in thickness by 3 to 4 mm over that required
by the pressure alone. Also, a high-strength material is required for construction.
The reaction safety valve loads on the reheater inlet and outlet header require
that the minimum thickness be about 8% of the outside diameter. An increase in
thickness owing to bending loads is necessary for sling headers and steam drum
of top-supported low-pressure boilers carrying loads.

The calculation method and the design consideration are best illustrated by
worked-out examples. So, the procedure for calculation of the thickness of tube or
header as per BS I 1 13 is described in the worked out example below.

EXAMPLE

> Calculate the thickness of tubes and headers of a two pass natural circulation
> 200 MWe boiler as per the British standard, BS 1113.

17-5-1 Estimation of Design Pressure

Step-I

This step calculates the safety valve set pressures for drum, superheater (SH) outlet, reheater inlet, and reheater outlet for design purposes as per BS1113 C1.1.3.6.

The customer's specification provides data on superheater outlet temperature and pressure. To get the superheater outlet pressure, the boiler designer must add an allowance for pressure drop in piping to the turbine inlet pressure. If no details regarding the pipe layout are available, the pressure drop in the pipeline may be taken as approximately 5%–6% of turbine inlet pressure. The pressure drops through different sections of the superheaters are added to the superheater outlet pressure to find the drum pressure. These pressure drops are determined during the thermal design of superheaters. If these are unknown at the initial thermal design stage, sufficient but not excessive allowance is taken based on the past experience. One should keep in mind that the pressure drops through the downstream components increase design pressures of all the upstream components. The adequacy of the ad hoc pressure drop assumption is checked back after the preliminary thickness of the pressure parts is evaluated.

For the purpose of the present example following values of losses, as given against each item, would be used. Values used here are typical in standard boilers.

SH outlet pressure (SOP) losses $= 137.0$ kg/cm^2 $= 1948.0$ psi.
Losses:

		kg/cm^2	psi
1	Loss through drum internals	0.25	3.5
2	Loss through saturated steam tubes	0.35	5.0
3	Loss through roof tubes and SH enclosure tubes	2.5	35.0
4	Loss through primary SH	1.75	25.0
5	Loss through attemperator 1 & inter connecting pipe	1.5	21.0
6	Loss through platen SH	2.5	35.0
7	Loss through attemperator 2 & inter connecting pipe	1.5	21.0
8	Loss through final SH	3.5	50.0
9	Loss through piping and stop valve	1.15	16.5
	Total losses	**15.0**	**212**

Drum pressure = Steam outlet pressure (SOP) + Losses = 152   2160
Stepping = 0.3   5.0
Subtotal = Drum pressure + Stepping = 152.3   2165
5% margin for safety valve sit back $= 0.05 \times$ Subtotal = 7.7   110

10. **Highest set drum safety valve pressure in drum**
= Margin + Subtotal = 152.3 + 7.7 = **160.0**   **2275**

11. **SH safety valve set pressure**
$= 1.05 \times$ SOP $= 1.05 \times 137 =$ **143.9**   **2046**

12[a]	Static head up to the furnace bottom header	3.5	50.0
13.	Calculation pressure for bottom header = [10] + [12] = 160 + 3.5 =	163.5	2325
14.	Pressure drop through economizer including static head		
	& inter-connecting pipes	5.0	71.0
15.	Calculation pressure for economizer inlet headers		
	= [10] + [14] = 160 + 5 =	165.0	2346
16[a]	Calculation pressure for time independent stress governed primary		
	SH portion = [10] − pressure drop at least load, say, "0" = [10]	160.0	2275
17[a]	Calculation pressure for SH tubes (time dependent stress governed)		
	= [11] + pressure drop up to point at highest evaporation load =		
	143.9 + say (1.1 + 3.5) =	148.55	2112
18[a]	Calculation pressure for SH outlet header = ΔP + Lowest set		
	safety valve pressure for time dependent stress = [7] + [11] =	144.4	2053
19[a]	RH inlet pressure normal —	37.6	535.0
20[a]	RH inlet pressure max. —	43.5	619.0
21[a]	**RH inlet safety valve set pressure = 1.05 × [20]**	**46.0**	**655.0**
22[a]	RH outlet pressure normal —	35.6	506.0
23[a]	RH outlet pressure max. —	41.5	590.0
24[a]	**RH outlet safety valve set pressure = 1.05 × [23]**	**43.6**	**620.0**
25[a]	Calculation pressure for time-independent stress governed		
	RH portion = [21] − pressure drop at least load, say =	46.0	655.0
26[a]	Calculation pressure for RH tubes (time dependent stress governed)		
	= [24] + Pressure drop up to the point at highest evaporation load		
	= 43.6 + say 1.5 =	45.1	641.0
27[a]	Calculation pressure for RH outlet header lowest set safety		
	valve pressure for time dependent stress = [24] =	43.6	620.0

[a] Before estimating the calculation or design pressure it is necessary to estimate the design temperature so as to establish whether the design stress for the part will be a time-independent (yield) criterion or time-dependent (creep) criterion.

17-5-2 Calculation of Design Temperature

Step 2: BS 1113 cl. 2.2.3

Design temperature is defined as the temperature at the middle of the wall of the pressure part. Heat is transferred from the flue gas to the fluid (steam/water) inside heated tubes. Therefore, to arrive at the (I) design temperature the temperature drop across the fluid film inside the tube and half the drop across the metal wall is required to be added to the expected maximum fluid temperature. The design temperature is, therefore, dominantly influenced by the fluid temperature and the drop across the fluid film and the metal by the heat flux. Most codes require the boiler manufacturer to assess the effect of nonuniformity in flow and heat flux to arrive at the design temperature, and to stipulate minimum additional allowances to cover the drop through film and metal wall. One should also add an allowance to cover the variation in fluid temperature from the normally estimated temperatures. For unheated parts beyond the final SH header, it is customary to take design temperature as the rated temperature unless the normal fluctuation in the metal temperature is exceeds the rated temperature by 8°C. In case of abnormal fluctuations, one can take the design temperature as rated temperature if the metal temperature does not exceed by 20°C for a maximum of 400 h in a year or 30°C for a maximum of 100 hours or by 40°C for a maximum of 60 h. And if it does the design temperature shall be increased by the amount of excess. Following values are based on those requirements called for in BS 1113, cl. 2.2.3. The other code requirements will be described in comparison of the code.

The following values are based on those requirements of the code BS II 13, cl. 2.2.3. The other code requirements will be described in comparing the code.

GIVEN

- Design lifetime as required by the customer is 100,000 h
- No continued service review at two-thirds of design life

CALCULATION OF DESIGN TEMPERATURE FOR PRESSURE PARTS

1. *Economizer inlet header (heated or unheated)*
   For heated: Max. temperature of water (**MTW**) = Saturation temperature
   corresponding to design pressure (T_{sat}) + 25°C   = 348 + 25 = 373°C
   For unheated = MTW + 0°C

2. *Economizer tubes (heated or unheated)*
   For heated = MTW + 25°C   = 348 + 25 = 373°C
   For unheated max. temperature of water + 0°C

3. *Economizer outlet header (heated or unheated)*
   For unheated = MTW (T_{sat}) + 25°C
   For unheated = MTW (T_{sat}) + 0°C   = 347 + 0 = 347°C

4. *Economizer outlet pipes (heated or unheated)*
   For heated = MTW (T_{sat}) + 25°C   = 347 + 0 = 347°C
   For unheated = MTW + 0°

5. *Drum (heated or unheated)*
   For heated = (T_{sat}) corresponding to calculation pressure (**CP**),
   calculated in step [10] in Section 17-5.1 + 25°C
   For unheated = (T_{sat}) + 0°C   = 347 + 0 = 347°C

6. *Large-bore downcomers (heated or unheated)*
   For heated = (T_{sat}) corresponding to CP, [13] + 25°C
   For unheated = (T_{sat}) + 0°C   = 348 + 0 = 348°C

7. *Bottom supply pipes (heated or unheated)*
   For heated = (T_{sat}) corresponding to CP, [13] + 25°C
   for unheated = (T_{sat}) + 0°C   = 348 + 0 = 348°C

8. *Furnace bottom headers (heated or unheated)*
   For heated = (T_{sat}) corresponding to CP [13] + 25°C   = 348 + 25 = 373°C
   For unheated = (T_{sat}) + 0°C

9. *Furnace tubes (heated or unheated)*
   For radiant heat from combustion chamber
   (T_{sat}) corresponding to CP [10] + 50°C   = 348 + 50 = 398°C
   For unheated (T_{sat}) + 0°C

10. *Furnace top headers (heated or unheated)*
    For heated = (T_{sat}) corresponding to CP [10] + 25°C
    For unheated (T_{sat}) + 0°C   = 347 + 0 = 347°C

11. *Risers (heated or unheated)*
    For unheated = (T_{sat}) corresponding to CP [10] + 25°C
    For unheated = (T_{sat}) + 0°C   = 347 + 0 = 347°C

12. *Saturated steam tubes (heated or unheated)*
    For heated = (T_{sat}) corresponding to CP [10] + 25°C
    For heated = (T_{sat}) + 0°C   = 347 + 0 = 347°C

13. *Steam cooled furnace roof tubes (heated or unheated)*
For unheated: Maximum expected steam temperature from
thermal design (**MST**) + 0°C
For heated = MST + 15°C (to cover the variations
in steam temperature as per CL 2.2.3.8.1 & 2) + 50° for
radiant heat   = 350 + 15 + 50 = 415°C
However, startup practice requires minimum 485°C. So take   485°C
14. *Primary SH inlet headers (heated or unheated)*
For unheated = MST + 0°C
For heated (as per CL 2.2.3.2) MST + 15°C to cover the
variations in steam temperature (as per CL 2.2.3.8.1 & 2)
+ 25°C = 352 + 15 + 25 = 392°C   392°C
15. *Primary SH bottom bank tubes (heated or unheated)*
For heated = MST + 15°C to cover the variations in
steam temperature (as per CL 2.2.3.8.1 & 2) − 15°C as
the expected normal temperature is not more than 425°C
and the tube is not for subjected to radiation (as per
CL 2.2.3.8.3) + 35°C convective heated (as per CL 2.2.3.2)   = 379 + (15 − 15) + 35 = 414°C
16. *Primary SH intermediate bank tubes (heated or unheated)*
For unheated = MST + 0°C
For heated = MST + 15°C to cover the variations in
steam temperature (as per CL 2.2.3.8.1 & 2) − 15°C as
the expected normal temperature is less than 425°C, and the
tube is not ssubjected to radiation (as per CL 2.2.3.8.3)
+ 35°C for convective heated (as per CL 2.2.3.2) and
= 390 + (15 − 15) + 35 = 425°C   425°C
17. *Primary SH top bank tubes (heated or unheated)*
For heated = MST + 15°C to cover variations in steam
temperature (as per CL 2.2.3.8.1 & 2) − 0°C as the expected
normal temperature is not more than 425°C and the tube is
subjected to radiation (as per CL 2.2.3.8.3) + 35°C for
convective heated (as per CL 2.2.3.2) and 0°C for unheated   = 440 + (15 − 0) + 35 = 490°C
18. *Primary SH outlet header (heated or unheated)*
For unheated = MST + 15°C to cover the variations in steam
temperature (as per CL 2.2.3.8.1 & 2) + 25°C for convective
heated (as per CL 2.2.3.2) and 0°C for unheated   = 440 + 15 + 0 = 455°C
19. *Primary SH outlet pipe and attemperator #1*
For unheated = MST + 15°C to cover the variations in steam
temperature (as per CL 2.2.3.8.1 & 2) + 25°C for heated
(as per CL 2.2.3.2) and 0°C for unheated   = 440 + 15 + 0 = 455°C
20. *Attemperator #1 outlet pipe (heated or unheated)*
For unheated = MST + 15°C to cover variations in steam
temperature (as per CL 2.2.3.8.1 & 2) + 25°C for
convective heated (as per CL 2.2.3.2) and 0°C for
unheated   = 400 + 15 + 0 = 415°C
21. *Platen SH inlet header (heated or unheated)*
For unheated = MST + 15°C to cover variations in steam
temperature (as per CL 2.2.3.8.1 & 2) + 25°C for heated (as
per CL 2.2.3.2) and 0°C for unheated   400 + 15 + 0 = 415°C
22. *Platen SH tubes (heated or unheated)*
For heated = MST + 15°C to cover the variations in steam
temperature (as per CL 2.2.3.8.1 & 2) − 0°C as the expected
normal temperature is less than 425°C and the tube is
subjected to radiation (as per CL 2.2.3.8.3) + 50°C for
radiant heated (as per CL 2.2.3.2) and 0°C for unheated   = 480 + (15 − 0) + 50 = 545°C

23. *Platen SH outlet header (heated or unheated)*
For unheated = MST + 15°C to cover the variations in steam
temperature (as per CL 2.2.3.8.1 & 2) + 25°C for
convective heated (as per CL 2.2.3.2) and 0°C for unheated = 480 + 15 + 0 = 495°C

24. *Platen outlet pipe and attemperator (heated or unheated)*
For unheated = MST + 15°C to cover the variations in steam
temperature (as per CL 2.2.3.8.1 & 2) + 25°C for convective
heated (as per CL 2.2.3.2) and 0°C for unheated = 480 + 15 + 0 = 495°C

25. *Pipe from attemperator to final SH (heated or unheated)*
For unheated = MST + 15°C to cover the variations in steam
temperature (as per CL 2.2.3.8.1 & 2) + 25°C for convective
heated (as per CL 2.2.3.2) and 0°C for unheated = 480 + 15 + 0 = 495°C

26. *Final SH inlet header (heated or unheated)*
For unheated = MST + 15°C to cover the variations in steam
temperature (as per CL 2.2.3.8.1 & 2) + 25°C for convective
heated (as per CL 2.2.3.2) and 0°C for unheated = 480 + 15 + 0 = 495°C

27. *Final SH tubes (heated or unheated)*
For heated = MST + 15°C to cover variations in steam
temperature (as per CL 2.2.3.8.1 & 2) − 0°C as the expected
normal temperature is more than 425°C and the tube is
subjected to radiation (as per CL 2.2.3.8.3) + 35°C for
convective heated (as per CL 2.2.3.4) and 0°C for unheated = 540 + (15 − 0) + 35 = 590°C

28. *Final SH outlet header (heated or unheated)*
For unheated = MST + 15°C to cover variations in steam
temperature (as per CL 2.2.3.8.1 & 2) + 25°C for
heated (as per CL 2.2.3.2) and 0°C for unheated = 540 + 15 + 0 = 555°C

29. *Final SH outlet pipe (heated or unheated)*
For unheated = Rated steam temperature from thermal
design + 0°C (as per CL 2.2.3.7) = 540 + 0 = 540°C

30. *Pipe from HP turbine to Reheater (RH) downstream of
isolating device up to RH inlet header (heated or unheated)*
For unheated = MST + 15°C to cover the variations in steam
temperature (as per CL 2.2.3.8.1 & 2) + 25°C for convective
heated (as per CL 2.2.3.2) and 0°C for unheated = 342 + 15 + 0 = 357°C

31. *RH inlet attemperator (heated or unheated)*
For unheated = MST + 15°C to cover the variations in steam
temperature (as per CL 2.2.3.8.1 & 2) + 25°C for convective
heated (as per CL 2.2.3.2) and 0°C for unheated = 342 + 15 + 0 = 357°C

32. *RH inlet header (heated or unheated)*
For heated = MST + 15°C to cover the variations in steam
temperature (as per CL 2.2.3.8.1 & 2) + 25°C for
heated (as per CL 2.2.3.2) and 0°C for unheated = 342 + 15 + 25 = 382°C

33. *RH tubes bottom bank (heated or unheated)*
For heated = MST + 15°C to cover variations in steam
temperature (as per CL 2.2.3.8.1 & 2) − 15°C as the expected
normal temperature is less than 425°C and the tube is not
subjected to radiation (as per CL 2.2.3.8.3) + 35°C
for heated (as per CL 2.2.3.2) and 0°C for unheated = 360 + (15 − 15) + 35 = 395°C

34. *RH tubes intermediate bank #1 (heated or unheated)*
For heated = MST + 15°C to cover variations in steam
temperature (as per CL 2.2.3.8.1 & 2) − 15°C as the
expected normal temperature is less than 425°C and the
tube is not subjected to radiation (as per CL 2.2.3.8.3)
+ 35°C for convective heated (as per CL 2.2.3.2) and
50°C for radiant unheated = 410 + (15 − 15) + 35 = 445°C

35. *RH tubes intermediate bank #2 (heated or unheated)*
    For heated = MST + 15°C to cover the variations in steam
    temperature (as per CL 2.2.3.8.1 & 2) − 0°C as the
    expected normal temperature is more than 425°C and the
    tube is not subjected to radiation (as per CL 2.2.3.8.3)
    +35°C for convective heated (as per CL 2.2.3.2) and
    0°C for radiant heated   = 465 + (15 − 0) + 35 = 515°C
36. *RH tubes final bank (heated or unheated)*
    For heated = MST + 15°C to cover variations in steam
    temperature (as per CL 2.2.3.8.1 & 2) − 0°C as the
    expected normal temperature is more than 425°C and the
    tube is not subjected to radiation (as per CL 2.2.3.8.3)
    +35°C for convective heated (as per CL 2.2.3.2) and
    0°C for unheated   = 540 + (15 − 0) + 35 = 590°C
37. *RH outlet header (heated or unheated)*
    For unheated = MST + 15°C to cover variations in steam
    temperature (as per CL 2.2.3.8.1 & 2) + 25°C for heated
    (as per CL 2.2.3.2) and 0°C for unheated   = 540 + 15 + 0 = 555°C
14. *RH outlet pipe from RH outlet header up to and including
    isolating device towards IP turbine inlet (heated or
    unheated)*
    For unheated = Rated steam temperature from thermal
    design + 0°C (as per CL 2.2.3.7)   = 540 + 0 = 540°C

17-5-3 Strength Calculation

STEP 3: BS1113

The first step in strength calculation is the selection of materials from Table 2.1.2 of BS 51113. This is done based on design temperatures (mean wall temperatures) calculated in step 2 and by using the following considerations:

Tubes to BS 3059: Part 2

Tubes can be of ERW, CEW, or seamless with a tolerance of S_1 or S_2. Tubes to tolerance S_2 are generally cheaper and are usually used. They have a higher negative tolerance on thickness. S_1 may be preferred where pressure drops are required to be contained such as in final SH banks or RH banks tubes and where parts governed by creep criteria require tubes to have closer tolerance. Followings are some tube materials with their operating limits.

a) Ferritic steel tubes

Carbon and carbon manganese steel tubes to BS 3059, parts 2-360 and 440, can be used up to a maximum design temperature of 480°C. Some regulations, like the Indian Boiler Regulation (IBR), restricts their use up to 454°C.

Alloy steel 243 tubes are carbon molybdenum steel tubes with 0.3% Mo: They can be used up to a maximum design temperature of 540°C. Some regulations (IBR) restrict their use up to 510°C.

Alloy steel 620 tubes are 1%Cr, 0.5% Mo steel tubes: They can be used up to a maximum design temperature of 570°C. However, these tubes are generally used up to 535°C.

Alloy steels 622-490 are 2.25% Cr 1% Mo steel tubes: They can be used up to a maximum design temperature of 600°C. However, these tubes are generally used in the range of 535°C to 600°C. The number 490 indicates the ultimate tensile strength in N/mm² at room temperature.

Alloy steels 629-590 are 9% Cr 1% Mo steels and can be used up to a maximum design temperature of 600°C. However, these tubes are generally used in the range of 535°C to 600°C. The number 590 indicates the ultimate tensile strength in N/mm² at room temperature.

Alloy steels 762 are 12% Cr, Mo, V steel tubes. They can be used up to a maximum design temperature of 620°C. However, these tubes are generally used in the range of 580°C to 620°C.

b) Austenitic stainless steel tubes, BS 3059, Part 2—All cold-finished seamless (CFS)

Steels 304S51 can be used up to a maximum design temperature of 700°C. Steels 316S51,S52 can be used up to a maximum design temperature of 700°C. Steels 321 S51 (I 01 0) can be used up to a maximum design temperature of 650°C. Steels 32IS51 (1105) can be used up to a maximum design temperature of 680°C.

ii) Steel Pipes

Generally headers and pipes are made of hot-finished seamless pipes except for smaller sizes, which may be cold finished seamless.

a) Carbon and carbon manganese steel pipes to BS 3601 and BS 3602: Part 1-360, 430, 500-Nb, can be used up to 480°C. BS 3602-490 can be used up to a maximum design temperature of 350°C.

b) Ferritic alloy steel pipes to BS 3604

BS 3604-620-440 can be used up to a maximum design temperature of 570°C. BS3604-622 can be used up to a maximum temperature of 600°C and -762 can be used up to a maximum temperature of 620°C.

c) Seamless austenitic stainless steel pipes to BS 3605

BS 3605-304S51 can be used up to a maximum design temperature of 720°C; -316S51, S52 can be used up to a maximum temperature of 720°C; and -321S51 (1010) can be used up to a maximum temperature of 660°C.

ii) Plates for Drum to BS 150

BS 1501, part 1, 161-430 (carbon steel, semi- or fully killed) can be used up to a maximum design temperature of 480°C, but generally its use is limited to below 454°C.

BS 1501, part 1, 224-490 (carbon manganese steel, fully killed, aluminum treated) can be used up to a maximum design temperature of 480°C, but generally its used is limited to below 454°C.

BS 1501, part 2, 271 (low alloy steel—manganese, chromium, molybdenum, and vanadium steel) can be used up to a maximum design temperature of 550°C.

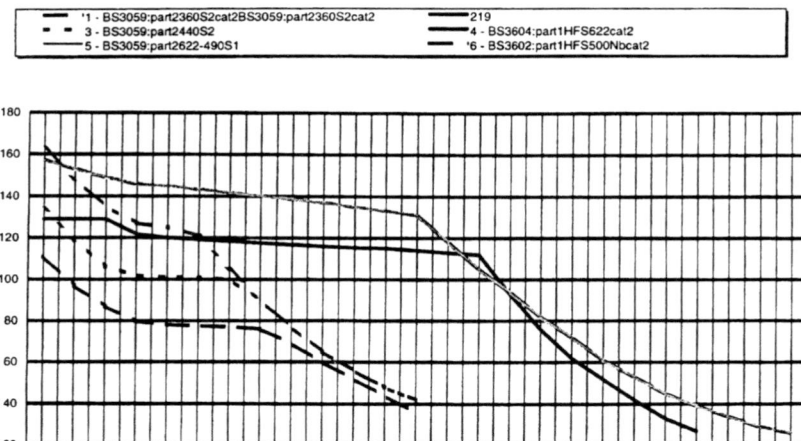

FIGURE 17-1. Plot of allowable stresses for various materials at different design temperatures

It is often used for manufacture of high-pressure drums owing to its high value of allowable stresses for a given temperature. This gives thinner drum construction.

Based on the design temperatures calculated in step 2, the following materials are selected for tubes and headers of different pressure parts of the boiler. The allowable stress for the tube materials at corresponding design temperatures are taken from Table 17-2a as per ASME code, and Table 17-2b as per BS code. This stress is used to calculate the thickness using BS codes (Table 17-3b).

After calculation of tube thickness, appropriate commercially available tube is chosen for the order. For the thickness of ordered tubes the ligament efficiency of various headers are calculated. For the predetermined outside diameter and chosen materials (based on metal temperatures) the thickness is calculated. The suitability of resultant bore is checked from header induced steam unbalance calculations (not a BS1113 requirement) and if necessary inside or outside diameter is increased and the thickness is recalculated. Final results are shown in Table 17-4.

TABLE 17-4. Results of material selection and calculation of thickness of different pressure parts of the 200 Mwe boiler in example.

Sl. No.	Application	Material	Temperature °C	Permissible design stress N/mm²	Outer diameter mm	Ligament efficiency %	Calculation pressure kg/cm²	Calculated thickness mm^a	With bend allowance %^{b}	Selected thickness mm	Reason for increase in thickness
1	Economiser inlet header	BS3602:part1HFS500Nbcat2	373	130	323.9	75.7	165.0	27.3	27.3	36.0	Rationalization
2	Economiser tubes	BS3059:part2440S2	373	103	38	100.	165.0	3.1	3.1	3.6	Close radius bends
3	Economiser outlet header	BS3602:part1 HFS500Nbcat2	347	136	244.5	51.5	165.0	28.2	28.2	36.0	Rationalization
4	Economiser outlet pipes	BS3602:part1 HFS500Nbcat2	347	136	127	100.	165.0	7.9	9.1	12.5	Rationalization
5	Drum	BS1501-271	347	201	1676.4	85.0	160.0	81.7	81.7	90.0	Scaling
6	Large-bore downcomers	BS3602:part1 HFS500Nbcat2	348	135	457	77.3	163.5	36.1	41.3	50.0	Rationalization
7	Bottom supply pipes	BS3602:part1 HFS500Nbcat2	348	135	127	100.	163.5	7.9	9.0	12.5	Rationalization
8	Furnace bottom headers	BS3602:part1 HFS500Nbcat2	373	130	355.6	33.2	163.5	61.8	61.8	75.0	
9	Furnace tubes	BS3059:part2440S2	398	101	63.5	100.	163.5	5.2	5.2	6.3	Load carrying/ bends in opening areas
10	Furnace top headers	BS3602:part1 HFS500Nbcat2	347	13	355.6	33.2	160.0	58.6	58.6	75.0	Rationalization
11	Risers	BS3602:part1 HFS500Nbcat2	347	136	127	100.	163.5	7.9	9.0	12.5	Rationalization
12	Saturated steam tubes	BS3602:part1 HFS500Nbcat2	347	136	127	100.	160.0	7.7	7.7	12.5	Rationalization
13	Steam-cooled furnace roof tubes	BS3059:part262OS2	485	113.5	44.5	100.	160.0	3.2	3.2	6.3	Rationalization

(Contd.)

TABLE 17-4. (Continued).

Sl. No.	Application	Material	Temperature °C	Permissible design stress N/mm²	Outer diameter mm	Ligament efficiency %	Calculation pressure kg/cm²	Calculated thickness mm[a]	With bend allowance %[b]	Selected thickness mm	Reason for increase in thickness
14	Primary SH inlet headers	BS3602:part1 HFS500Nbcat2	392	126	219.1	55.5	160.0	24.5	24.5	32.0	Rationalization
15	Primary SH bottom bank tubes	BS3059:part2440S2	414	100	63.5	100.	160.0	5.1	5.1	6.3	
16	Primary SH intermediate bank tubes	BS3059:part2440S2	425	95	51	100.	160.0	4.2	4.9	5.9	High metal temperature
17	Primary SH top bank tubes	BS3059:part2620S2	490	113	51	100.	160	3.7	4.9	5.4	High metal temperature
18	Primary SH outlet header	BS3604:part1 HFS622cat2	455	136	355.6	82.8	160	25.8	25.8	40.0	High metal temperature
19	Primary SH outlet pipe and attemperator#1	BS3604:part1 HFS622cat2	455	136	355.6	100.	160	21.6	24.3	36.0	To match bore & avoid swage piece
20	Attemperator#1 outlet pipe	BS3602:part1 HFS500Nbcat2	415	113	355.6	100.	160	28.5	32	36.0	To match & bore avoid swage piece
21	Platen SH inlet header	BS3602:part1 HFS500Nbcat2	415	113	355.6	89.3	160	28.5	32	36.0	To match bore & avoid swage piece
22	Platen SH tubes	BS3059:part2622-490S1	545	57	38	100.	151	4.9	4.9	8.0	High metal temperature

TABLE 17-4. (Continued).

Sl. No.	Application	Material	Temperature °C	Permissible design stress N/mm²	Outer diameter mm	Ligament efficiency %	Calculation pressure kg/cm²	Calculated thickness mm[a]	With bend allowance %[b]	Selected thickness mm	Reason for increase in thickness
23	Platen SH outlet header	BS3604;part1HFS622cat2	495	111	355.6	89.3	149.0	27.2	27.2	36.0	High metal temperature
24	Platen outlet pipe and attemperator	BS3604;part1HFS622cat2	495	111	355.6	100.	149.0	24.4	27.5	36.0	High metal temperature
25	Pipe from attemperator to final SH	BS3604;part1 HFS622cat2	495	111	355.6	100.	149.0	28.2	28.2	36.0	To match bore & avoid swage piece
26	Final SH inlet header	BS3604;part1 HFS622cat2	495	111	355.6	85.7	149.0	28.2	28.2	36.0	To match bore & avoid swage piece
27	Final SH tubes	BS3059;part2622-490S1	590	29	44.5	100.	148.55	10.0	10.0	12.7	High metal temperature
28	Final SH outlet header	BS3604;part1 HFS622cat2	555	49.	406.4	88.9	144.4	63.2	63.2	82.0	High metal temperature
29	Final SH outlet pipe	BS3604;part1 HFS622cat2	540	61	406.4	100.	143.9	46.8	52.6	55.0	Rationalization
30	Pipe from HP turbine to reheater (RH)	BS3602;part1 HFS500Nbcat2	357	133	457	100.	46.0	8.4	9.6	11.0	Rationalization
31	RH inlet attemperator	BS3602;part1 HFS500NNbcat2	357	133	457	100.	46.0	8.4	8.4	11.0	Rationalization

(Contd.)

TABLE 17-4. (*Continued*)

Sl. No.	Application	Material	Temperature °C	Permissible design stress N/mm²	Outer diameter mm	Ligament efficiency %	Calculation pressure kg/cm²	Calculated thickness mm[a]	With bend allowance %[b]	Selected thickness mm	Reason for increase in thickness
32	RH inlet header	BS3602:part1 HFS500Nbcat2	382	128	457	88.7	46.0	9.8	9.8	22.2	Safety valve reaction
33	RH tubes bottom bank	BS3059:part2360S2cat2	395	79	63.5	100.	46.0	2.0	2.0	3.6	Bending
34	RH tubes intermediate bank #1	BS3059:part2360S2cat2	445	64	57	100.	45.1	2.1	2.1	3.6	Bending
35	RH tubes intermediate bank #2	BS3059:part2620S2	515	84	51	100.	45.1	1.4	1.4	3.2	Bending
36	RH tubes final bank	BS3059:part2622-490S1	590	29	47.6	100.	45.0	3.8	3.8	5.4	Bending
37	RH outlet header	BS3604:part1 HFS622cat2	555	49	457	91.7	43.6	23.1	23.1	38.0	Safety valve reaction
38	RH outlet pipe from RH outlet header	BS3059:part2620S2	540	52	457	100.	43.6	20.1	22.9	38.0	To match & bore avoid swage piece

[a] Including negative tolerance of 10%. [b] After adding bending allowance of 12.5% for pipes alone.

Nomenclature

C   allowance for threading
d   outside diameter
D   outside diameter, inch
e   coefficient (Eq. 17-2)
E   efficiency
P   pressure (Eq. 17-2), psi
P   pressure, Pa
S   allowable stress, psi
t   thickness (Eq. 17-2), in.
t   thickness, m
y   coefficient of temperature

Greek Symbols

σ allowable stress, Pa

References

ASME Sec. I (1995) American Society of Mechanical Engineers, New York, Part PG, Stress Table, Sub-part 1, Sec. II (1995), Part D, Properties.

Bernstein, M.D. (1988) F.W.E. Corp. Perry Valley, Design criteria for boilers and pressure vessels in the U.S.A. Transaction of the ASME 430 v 110:430–443.

BS 1113 (1992 with updates up to 1994) British Standard Institute, London.

Indian Boiler Regulations (1993) Law Publishers Pvt., Ltd. Calcutta, India.

Peterson, D.G. (1997) Anatomy of a catastrophic boiler accident. National Board Bulletin, Summer 1997:21–26.

Stultz, S.C.D., Kitto, J.B. (1992) "Steam: Its Generation and Use." Babcock & Wilcox Company, Barberton, Ohio pp. 22-1-38.

Technical Rules for Steam Boilers (TRD), English translation by F.D.B.R.E.V. Germany.

18
Tables of Design Data

TABLE 18-1. Specific heat of air, flue gas, and ash at atmospheric pressure.

T (°C)	C_{CO_2} [kJ/(nm³°C)]	C_{N_2} [kJ/(nm³°C)]	C_{O_2} [kJ/(nm³°C)]	C_{H_2O} [kJ/(nm³°C)]	C_k [kJ/(nm³°C)]	C_{CO} [kJ/(nm³°C)]	C_{H_2} [kJ/(nm³°C)]	C_{CH_4} [kJ/(nm³°C)]	$C_{fly\text{-}ash}$ (C_{fh}) [kJ/(nm³°C)]
0	1.5998	1.2946	1.3059	1.4943	1.3188	1.2992	1.2766	1.55	0.7955
100	1.7003	1.2958	1.3176	1.5052	1.3243	1.3017	1.2908	1.6411	0.8374
200	1.7873	1.2996	1.3352	1.5223	1.3318	1.3071	1.2971	1.7589	0.8667
300	1.8627	1.3067	1.3561	1.5424	1.3423	1.3167	1.2992	1.8861	0.8918
400	1.9297	1.3163	1.3775	1.5654	1.3544	1.3289	1.3021	2.0155	0.9211
500	1.9887	1.3276	1.398	1.5897	1.3683	1.3247	1.305	2.1403	0.924
600	2.0411	1.3402	1.4168	1.6148	1.3829	1.3574	1.308	2.2609	0.9504
700	2.0884	1.3536	1.4344	1.6412	1.3976	1.372	1.3121	2.3768	0.963
800	2.1311	1.367	1.4499	1.668	1.4114	1.3862	1.3167	2.4981	0.9797
900	2.1692	1.3795	1.4645	1.6956	1.4248	1.3996	1.3226	2.6025	1.0048
1000	2.2035	1.3917	1.4775	1.7229	1.4373	1.4126	1.3289	2.6992	1.0258
1100	2.2349	1.4034	1.4893	1.7501	1.4499				1.0509
1200	2.2638	1.4143	1.5005	1.7769	1.4612	1.4361	1.3431	2.8629	1.096
1300	2.2898	1.4252	1.5016	1.8028	1.4725				1.1304
1400	2.3136	1.4348	1.5202	1.828	1.483	1.4566	1.359		1.1849
1500	2.3354	1.444	1.5294	1.8527	1.4926				1.2228

TABLE 18-2. Some physical properties of iron, metals, and selected steels.

Material	Identification	C	Si	Mn	Cr	Ni	W	Mo	Heat treatment
				% by Mass					
High-purity iron	A								Annealed
Mild steel	B	0.23	0.11	0.63		0.07			Annealed
Medium carbon steel	C	0.43	0.20	0.69	0.03	0.04		0.01	Annealed
3% nickel steel	D	0.32	0.18	0.55	0.17	3.47		0.04	Annealed
Nickel-chromium steel	E	0.32	0.25	0.55	0.71	3.41		0.06	Annealed, reached to 640°C and furnace-cooled
Nickel-chromium-molybdenum steel	F	0.34	0.27	0.55	0.78	3.53		0.39	Annealed, reached to 640°C and furnace-cooled
13% manganese	G	1.22	0.22	13.00	0.03	0.07			Heated to 1050°C, air-cooled
18/8 chromium-nickel steel	H	0.08	0.68	0.37	19.11	8.14	0.06		Heated to 1100°C, water-cooled
Stainless iron	I	0.13	0.17	0.25	12.95	0.14			Heated to 960°C, tempered 2 hrs at 750°C, air-cooled
Stainless steel	J	0.27	0.18	0.28	13.69	0.20	0.25	0.01	Heated to 960°C, tempered 2 hrs at 750°C, air-cooled
Carbon-chromium steel	K	0.31	0.20	0.69	1.09	0.07		0.01	Annealed

(measurements by the National Physical Laboratory)

TABLE 18-3. Linear thermal expansion of steel: Values of mean coefficient linear expansion $\times 10^6/K^{-1}$.

Type	Mild steel	Medium carbon steel	Nickel-chromium-molybdenum steel	Stainless iron	Stainless steel
Temperature range,°C					
0–100	12.18	11.59	11.63	10.13	9.98
0–200	12.66	12.32	12.12	10.66	10.65
0–300	13.08	13.09	12.61	11.14	11.13
0–400	13.47	13.71	13.12	11.54	11.50
0–500	13.92	14.18	13.50	11.85	11.83
0–600	14.41	14.67	13.79	12.15	12.20
0–700	14.88	15.08	13.45	12.37	12.46
0–800	12.64	12.50	10.67	12.56	12.16
0–900	12.41	13.56	11.99	10.79	10.57
0–1000	13.37	14.45	12.96	11.70	12.18
0–1100			13.62	12.68	13.60

TABLE 18-4. Specific heat capacity of steel: Values of mean[a] specific heat capacity (J/gK).

Type	Mild steel	Medium carbon steel	Nickel-chromium-molybdenum steel	Stainless iron	Stainless steel
Temperature range,°C					
0–100	0.477	0.469	0.477	0.465	0.465
100–200	0.511	0.515	0.515	0.507	0.502
200–300	0.544	0.544	0.548	0.544	0.540
300–400	0.586	0.286	0.595	0.595	0.586
400–500	0.645	0.632	0.653	0.657	0.657
500–600	0.724	0.695	0.745	0.754	0.753
600–700	0.816	0.808	0.938	0.846	0.883
700–800	1.193	1.348	1.151	0.795	0.888
800–900	0.682	0.599	0.632	0.737	0.812
900–1000	0.649	0.628	0.641	0.649	0.649
1000–1100	0.653	0.632	0.636	0.653	0.645
1100–1200	0.662	0.649	0.641	0.649	0.653
1200–1300	0.682	0.647	0.649	0.649	0.670

[a] Specific enthalpy increase divided by temperature increase. The specific enthalpy increase includes that associated with any phase change which may occur in the temperature range.

TABLE 18-5. Electrical resistivity of steel: Values of electrical resistivity at 20°C after water quenching/$\mu\Omega$ cm (for 6.35 mm diameter rods).

Type	Mild steel	Medium carbon steel	Nickel-chromium-molybdenum steel	Stainless iron	Stainless steel
Resistivity	17.9	24.7	37.9	56.2	63.2
% increase	8.5	32.1	30.7	10.6	18.2

TABLE 18-6. Thermal conductivity of steel: Values of thermal conductivity*/(W/m K).

Type	Mild steel	Medium carbon steel	Nickel-chromium-molybdenum steel	Stainless iron	Stainless steel
Temperature, °C					
0	51.9	48.1	33.1	27.2	25.1
100	51.1	48.1	33.9	27.6	26.4
200	49.0	46.5	35.2	27.6	27.2
300	46.1	44.0	35.6	28.1	27.6
400	42.3	41.0	35.6	27.6	27.6
500	39.4	38.5	33.5	27.2	27.2
600	35.6	36.0	30.6	26.4	26.4
700	31.8	31.4	28.1	25.5	25.5
800	26.0	27.2	27.2	25.1	25.1
900	26.4	25.5	27.6	27.2	27.2
1000	27.6	27.2	28.5	27.6	27.6
1100	28.5	28.1	29.7	28.9	28.9
1200	29.7	29.7	30.1	30.6	30.1

*Based on dimension at 20°C.
Note: estimated values are included for steels B to K at above temperatures above 300°C. They are probably correct to within 6%.

TABLE 18-7. Density, heat capacity, and heat conductivity for metals.

Metal	200°C			λ, W/(m°C)									
	ρ kg/m³	c_p J/(Kg°C)	λ W/(m°C)	−100 °C	0 °C	100 °C	200 °C	300 °C	400 °C	600 °C	800 °C	1000 °C	1200 °C
1) Pure Al	2710	902	236	243	236	240	238	234	228	215			
2) Duralium (96Al-4Cu.trace Mg)	2790	881	169	124	160	188	188	193					
3) Aluminium alloy (92Al-8Mg)	2610	904	107	86	102	123	148						
4) Aluminium alloy (8%Al 13Si)	2660	871	162	139	158	173	176	180	118				
5) Beryllium	1850	1758	219	328	218	170	145	129					
6) Pure copper	8930	386	398	421	401	393	389	384	379	366	352		
7) Aluminium bronze (90Cu-10Al)	8360	420	56		49	57	66						
8) Bronze (89Cu-11Sn)	8800	343	24.8		24	28.4	33.2						
9) Brass (70Cu-30Zn)	8440	377	109	90	106	131	143	145	148				
10) Copper-alloy (60Cu-40Ni)	8920	410	22.2	19	22.2	23.4							
11) Gold	19,300	127	315	331	318	313	310	305	300	287			
12) Pure iron	7870	455	81.1	96.7	83.5	72.1	63.5	56.5	50.3	39.4	29.6	29.4	31.6
13) Armco iron	7860	455	73.2	82.9	74.7	67.5	61.0	54.8	49.9	38.6	29.3	29.3	31.1
14) Cast iron (C ∼ 3%)	7570	470	39.2		28.5	32.4	35.8	37.2	36.6	20.8	19.2		
15) Carbon steel (C ∼ 0.5%)	7840	465	49.8		50.5	47.5	44.8	42.0	39.4	34.0	29.0		
16) Carbon steel (C ∼ 1.0%)	7790	470	43.2		43.0	42.8	42.2	41.5	40.6	36.7	32.2		
17) Carbon steel (C ∼ 1.5%)	7750	470	36.7		36.8	36.6	36.2	34.7	34.7	31.7	27.8		
18) Chrome steel (Cr ∼ 5%)	7830	460	36.1		36.3	35.2	34.7	33.5	31.4	28.0	27.2	27.2	27.2
19) Chrome steel (Cr ∼ 13%)	7740	460	26.8		26.5	27.0	27.0	27.0	27.6	23.4	29.0	29.0	
20) Chrome steel (Cr ∼ 17%)	7710	460	22		22	22.2	22.6	22.6	23.3	24.0	24.8	25.5	
21) Chrome steel (Cr ∼ 26%)	7650	460	22.6		22.6	23.8	25.5	27.2	28.5	31.8	35.1	38	
22) Chrome nickel steel (18-20Cr/8-12Ni)	7820	460	15.2	12.2	14.7	16.6	18.0	19.4	20.8	23.5	26.3		

TABLE 18-7. *(Continued)*.

Metal	200°C ρ kg/m³	200°C c_p J/(Kg°C)	200°C λ W/(m°C)	λ, W(m°C) -100 °C	0 °C	100 °C	200 °C	300 °C	400 °C	600 °C	800 °C	1000 °C	1200 °C
23) Chrome nickel steel (17-19Cr/9-13Ni)	7830	460	14.7	11.8	14.3	16.1	17.5	18.8	20.2	22.8	25.5	28.2	30.9
24) Nickel steel (Ni ~ 1%)	7900	460	45.5	40.8	45.2	46.8	46.1	44.1	41.2	35.7			
25) Nickel steel (Ni ~ 3.5%)	7910	460	36.5	30.7	36.0	38.8	39.7	39.2	37.8				
26) Nickel steel (Ni ~ 25%)	8030	460	13.0										
27) Nickel steel (Ni ~ 35%)	8110	460	13.8	10.9	13.4	15.4	17.1	18.6	20.1	23.1			
28) Nickel steel (Ni ~ 44%)	8190	460	15.8		15.7	16.1	16.5	16.9	17.1	17.8	18.4		
29) Nickel steel (Ni ~ 50%)	8260	460	19.6	17.3	19.4	20.5	21.0	21.1	21.3	22.5			
30) Manganese-steel (Mn12–13, Ni ~ 3%)	7800	487	13.6			14.8	16.0	17.1	18.3				
31) Manganese (Mn ~ 0.4%)	7860	440	51.2			51.0	50.0	47.0	43.5	35.5	27		
32) Tungsten (W-5 ~ 6%)	8070	436	18.7		18.4	19.7	21.0	22.3	23.6	24.9	26.3		
33) Lead	11,340	128	35.3	37.2	35.5	34.3	32.8	31.5					
34) Magnesium	1730	1020	156	160	157	154	152	150					
35) Molybdenum	9590	255	138	146	139	135	131	127	123				
36) Nickel	8900	444	91.4	144	94	82.8	74.2	67.3	64.6	69.0	73.3	77.6	81.9
37) Platinum	21,450	133	71.4	73.3	71.5	71.6	72.0	72.8	73.6	76.6	80.0	84.2	88.9
38) Silver	10,500	234	427	431	428	422	415	407	399	384			
39) Tin	7310	228	67	75	68.2	63.2	60.9						
40) Titanium	4500	520	22	23.3	22.4	20.7	19.9	19.5	19.4	19.9			
41) Uranium	19,070	116	27.4	24.3	27	29.1	31.1	33.4	35.7	40.6	45.6		
42) Zinc	7140	388	121	123	122	117	112						
43) Zicronium	6570	276	22.9	26.5	23.2	21.8	21.2	20.9	21.4	22.3	24.5	26.4	28.0
44) Tungsten	19,350	134	179	204	182	166	153	142	134	125	119	114	110

TABLE 18-8. Thermal properties of the saturated water and steam; v_c = sp. vol of water, v_g = sp. vol of steam; h_c = enthalpy of water; h_g = enthalpy of steam (Arranged by temperature)*.

		Specific volume		Enthalpy		Vaporization	Entropy	
		Liquid	Vapor	Liquid	Vapor	heat	Liquid	Vapor
Temperature	Pressure	v_c	v_g	h_c	h_g	h_{cg}	s_c	s_g
T	p							
°C	bar	$\dfrac{m^3}{Kg}$	$\dfrac{m^3}{kg}$	$\dfrac{kJ}{kg}$	$\dfrac{kJ}{kg}$	$\dfrac{kJ}{kg}$	$\dfrac{kJ}{(kg \cdot C)}$	$\dfrac{kJ}{(kg \cdot C)}$
0	0.006108	0.0010002	206.321	−0.04	2501.0	2501.0	−0.0002	9.1565
0.01	0.006112	0.00100022	206.175	0.000614	2501.0	2501.0	0.0000	9.1562
1	0.006566	0.0010001	192.611	4.17	2502.8	2498.6	0.0152	9.1298
2	0.007054	0.0010001	179.935	8.39	2504.7	2496.3	0.0306	9.1035
3	0.007575	0.0010000	168.165	12.60	2506.5	2493.9	0.0459	9.0773
4	0.008129	0.0010000	157.267	16.80	2508.3	2491.5	0.0611	9.0514
5	0.008718	0.0010000	147.167	21.01	2510.2	2489.2	0.0762	9.0258
6	0.009346	0.0010000	137.768	25.21	2512.0	2486.8	0.0913	9.0003
7	0.010012	0.0010001	129.061	29.41	2513.9	2484.5	0.1063	8.9751
8	0.010721	0.0010001	120.952	33.60	2515.7	2482.1	0.1213	8.9501
9	0.011473	0.0010002	113.423	37.80	2517.5	2479.7	0.1362	8.9254
10	0.012271	0.0010003	106.419	41.99	2519.4	2477.4	0.1510	8.9009
11	0.013118	0.0010003	99.896	46.19	2521.2	2475.0	0.1658	8.8766
12	0.014015	0.0010004	93.828	50.38	2523.0	2472.6	0.1805	8.8525
13	0.014967	0.0010006	88.165	54.57	2524.9	2470.2	0.1952	8.8286
14	0.015974	0.0010007	82.893	58.75	2526.7	2467.9	0.2098	8.8050
15	0.017041	0.0010008	77.970	62.94	2528.6	2465.7	0.2243	8.7815
16	0.018170	0.0010010	73.376	67.13	2530.4	2463.3	0.2388	8.7583
17	0.019364	0.0010012	69.087	71.31	2532.2	2460.9	0.2533	8.7353
18	0.020622	0.0010013	65.080	75.50	2534.0	2458.5	0.2677	8.7125
19	0.021960	0.0010015	61.334	79.68	2535.9	2456.2	0.2820	8.6898
20	0.023368	0.0010017	57.833	83.86	2537.7	2453.8	0.2963	8.6674
22	0.026424	0.0010022	51.488	92.22	2541.4	2449.2	0.3247	8.6232
24	0.029824	0.0010026	45.923	100.59	2545.0	2444.4	0.3530	8.5797
26	0.033600	0.0010032	41.031	108.95	2548.6	2439.6	0.3810	8.5370
28	0.037785	0.0010037	36.726	117.31	2552.3	2435.0	0.4088	8.4950
30	0.042417	0.0010043	32.929	125.66	2555.9	2430.2	0.4365	8.4537
35	0.056217	0.0010060	25.246	146.56	2565.0	2418.4	0.5049	8.3536
40	0.073749	0.0010078	19.548	167.45	2574.0	2406.5	0.5721	8.2576
45	0.095817	0.0010099	15.278	188.35	2582.9	2394.5	0.6383	8.1655
50	0.12335	0.0010121	12.048	209.26	2591.8	2382.5	0.7035	8.0771
55	0.15740	0.0010145	9.5812	230.17	2600.7	2370.5	0.7677	7.9922
60	0.19919	0.0010171	7.6807	251.09	2609.5	2358.4	0.8310	7.9106
65	0.25008	0.0010199	6.2042	272.02	2618.2	2346.2	0.8933	7.8320
70	0.31161	0.0010228	5.0479	292.97	2626.8	2333.8	0.9548	7.7565
75	0.38548	0.0010259	4.1356	313.94	2635.3	2321.4	1.0154	7.6837
80	0.47359	0.0010292	3.4104	334.92	2643.8	2308.9	1.0752	7.6135
85	0.57803	0.0010326	2.8300	355.92	2652.1	2296.2	1.1343	7.5459
90	0.70108	0.0010361	2.3624	376.94	2660.3	2283.4	1.1925	7.4805
95	0.84525	0.0010398	1.9832	397.99	2668.4	2270.4	1.2500	7.4174
100	1.01325	0.0010437	1.6738	419.06	2676.3	2257.2	1.3069	7.3564
110	1.4326	0.0010519	1.2106	461.32	2691.8	2230.5	1.4185	7.2402
120	1.9854	0.0010606	0.89202	503.7	2706.6	2202.9	1.5276	7.1310
130	2.7012	0.0010700	0.66851	546.3	2720.7	2174.1	1.6344	7.0281

TABLE 18-8. (*Continued*).

Temperature T °C	Pressure p bar	Specific volume		Enthalpy		Vaporization heat	Entropy	
		Liquid v_c $\frac{m^3}{Kg}$	Vapor v_g $\frac{m^3}{kg}$	Liquid h_c $\frac{kJ}{kg}$	Vapor h_g $\frac{kJ}{kg}$	h_{cg} $\frac{kJ}{kg}$	Liquid s_c $\frac{kJ}{(kg \cdot C)}$	Vapor s_g $\frac{kJ}{(kg \cdot C)}$
140	3.6136	0.0010801	0.50875	589.1	2734.0	2144.9	1.7390	6.9307
150	4.7597	0.0010908	0.39261	632.2	2746.3	2114.1	1.8416	6.8381
160	6.1804	0.0011022	0.30685	675.5	2757.7	2082.2	1.9425	6.7498
170	7.9202	0.0011145	0.24259	719.1	2768.0	2048.9	2.0416	6.6652
180	10.027	0.0011275	0.19381	763.1	2777.1	2014.0	2.1393	6.5838
190	12.552	0.0011415	0.15631	807.5	2784.9	1977.4	2.2356	6.5052
200	15.551	0.0011565	0.12714	852.4	2791.4	1939.0	2.3307	6.4289
210	19.079	0.0011726	0.10422	897.8	2796.4	1898.6	2.4247	6.3546
220	23.201	0.0011900	0.08602	943.7	2799.9	1856.2	2.5178	6.2819
230	27.979	0.0012087	0.07143	990.3	2801.7	1811.4	2.6102	6.2104
240	33.480	0.0012291	0.05964	1037.6	2801.6	1764.0	2.7021	6.1397
250	39.776	0.0012513	0.05002	1085.8	2799.5	1713.7	2.7936	6.0693
260	46.940	0.0012756	0.04212	1135.0	2795.2	1660.2	2.8850	5.9989
270	55.051	0.0013025	0.03557	1185.4	2788.3	1602.9	2.9766	5.9278
280	64.191	0.0013324	0.03010	1237.0	2778.6	1541.6	3.0687	5.8555
290	74.448	0.0013659	0.02551	1290.3	2765.4	1475.1	3.1616	5.7811
300	85.917	0.0014041	0.02162	1345.4	2748.4	1403.0	3.2559	5.7038
310	98.697	0.0014480	0.01829	1402.9	2726.8	1323.9	3.3522	5.6224
320	112.90	0.0014995	0.01544	1463.4	2699.6	1236.2	3.4513	5.5356
330	128.65	0.0015614	0.01296	1527.5	2665.5	1138.0	3.5546	5.4414
340	146.08	0.0016390	0.01078	1596.8	2622.3	1025.5	3.6638	5.3363
350	165.37	0.0017407	0.008822	1672.9	2566.1	893.2	3.7816	5.2149
360	186.74	0.0018930	0.006970	1763.1	2485.7	722.6	3.9189	5.0603
370	210.53	0.002231	0.004958	1896.2	2335.7	439.5	4.1198	4.8031
371	213.06	0.002298	0.004710	1916.5	2310.7	394.2	4.1503	4.7324
372	215.62	0.002392	0.004432	1942.0	2280.1	338.1	4.1891	4.7624
373	218.21	0.002525	0.004090	1974.5	2238.3	263.8	4.2385	4.6467
374	220.84	0.002834	0.003482	2039.2	2150.7	111.5	4.3374	4.5096

Critical values:
* $p_c = 221.15$ bar; $h_c = 2095.2$ kJ/kg; $v_c = 0.003147$ m³/kg; $s_c = 4.4237$ kJ/(kg·C); $t_c = 374.12$°C.

TABLE 18-9. Thermal properties of the saturated water and steam (Arranged by pressure).

Pressure	Temperature	Specific volume		Enthalpy		Vaporization heat	Entropy	
		Liquid	Vapor	Liquid	Vapor		Liquid	Vapor
p	T	v_c	v_g	h_c	h_g	h_{cg}	s_c	s_g
bar	°C	$\dfrac{m^3}{Kg}$	$\dfrac{m^3}{kg}$	$\dfrac{kJ}{kg}$	$\dfrac{kJ}{kg}$	$\dfrac{kJ}{kg}$	$\dfrac{kJ}{(kg \cdot C)}$	$\dfrac{kJ}{(kg \cdot C)}$
0.010	6.982	0.0010001	129.208	29.33	2513.8	2484.5	0.1060	8.9756
0.020	17.511	0.0010012	67.006	73.45	2533.2	2459.8	0.2606	8.7236
0.030	24.098	0.0010027	45.668	101.00	2545.2	2444.2	0.3543	8.5776
0.040	28.981	0.0010040	34.803	121.41	2554.1	2432.7	0.4224	8.4747
0.050	32.90	0.0010052	28.196	137.77	2561.2	2423.4	0.4762	8.3952
0.060	36.18	0.0010064	23.742	151.50	2567.1	2415.6	0.5200	8.3305
0.070	39.02	0.0010074	20.532	163.38	2572.2	2408.8	0.5591	8.2760
0.080	41.53	0.0010084	18.106	173.87	2576.7	2402.8	0.2926	8.2289
0.090	43.79	0.0010094	16.206	183.28	2580.8	2397.5	0.6224	8.1875
0.10	45.83	0.0010102	14.676	191.84	2584.4	2392.6	0.6493	8.1505
0.15	54.00	0.0010140	10.025	225.98	2598.9	2372.9	0.7549	8.0089
0.20	60.09	0.0010172	7.6515	251.46	2609.6	2358.1	0.8321	7.9092
0.25	64.99	0.0010199	6.2060	271.99	2618.1	2346.1	0.8932	7.8321
0.30	69.12	0.0010223	5.2308	289.31	2625.3	2336.0	0.9441	7.7695
0.40	75.89	0.0010265	3.9949	317.65	2636.8	2319.2	1.0261	7.6711
0.50	81.35	0.0010301	3.2415	340.57	2646.0	2305.4	1.0912	7.5951
0.60	85.95	0.0010333	2.7329	359.93	2653.6	2293.7	1.1454	7.5332
0.70	89.96	0.0010361	2.3658	376.77	2660.2	2283.4	1.1921	7.4811
0.80	93.51	0.0010387	2.0879	391.72	2666.0	2274.3	1.2330	7.4360
0.90	96.71	0.0010412	1.8701	405.21	2671.1	2265.9	1.2696	7.3963
1.0	99.63	0.0010434	1.6946	417.51	2675.7	2258.2	1.3027	7.3608
1.2	104.81	0.0010476	1.4289	439.36	2683.8	2244.4	1.3609	7.2996
1.4	109.32	0.0010513	1.2370	458.42	2690.8	2232.4	1.4109	7.2480
1.6	113.32	0.0010547	1.0917	475.38	2696.8	2221.4	1.4550	7.2032
1.8	116.93	0.0010579	0.97775	490.70	2702.1	2211.4	1.4944	7.1638
2.0	120.23	0.0010608	0.88592	504.7	2706.9	2202.2	1.5301	7.1286
2.5	127.43	0.0010675	0.71881	535.4	2717.2	2181.8	1.6072	7.0540
3.0	133.54	0.0010735	0.60586	561.4	2725.5	2164.1	1.6717	6.9930
3.5	138.88	0.0010789	0.52425	584.3	2732.5	2148.2	1.7273	6.9414
4.0	143.62	0.0010839	0.46242	604.7	2738.5	2133.8	1.7764	6.8966
4.5	147.92	0.0010885	0.41392	623.3	2743.8	2120.6	1.8204	6.8570
5.0	151.85	0.0010928	0.37481	640.1	2748.5	2108.4	1.8604	6.8215
6.0	158.84	0.0010009	0.31556	670.4	2756.4	2086.0	1.9308	6.7598
7.0	164.96	0.0011082	0.27274	697.1	2762.9	2065.8	1.9918	6.7074
8.0	170.42	0.0011150	0.24030	720.9	2768.4	2047.5	2.0457	6.6618
9.0	175.36	0.0011213	0.21484	742.6	2773.0	2030.4	2.0941	6.6212
10.0	179.88	0.0011274	0.19430	762.6	2777.0	2014.4	2.1382	6.5847
11.0	184.06	0.0011331	0.17739	781.1	2780.4	1999.3	2.1786	6.5515
12.0	187.96	0.0011386	0.016320	798.4	2783.4	1985.0	2.2160	6.5210
13.0	191.60	0.0011438	0.15112	814.7	2786.0	1971.3	2.2509	6.4927
14.0	195.04	0.0011489	0.14072	830.1	2788.4	1958.3	2.2836	6.4665
15.0	198.28	0.0011538	0.13165	844.7	2790.4	1945.7	2.3144	6.4418
16.0	201.37	0.0011586	0.12368	858.6	2792.2	1933.6	2.3436	6.4187

TABLE 18-9. (*Continued*).

Pressure	Temperature	Specific volume		Enthalpy		Vaporization heat	Entropy	
		Liquid	Vapor	Liquid	Vapor		Liquid	Vapor
p	T	v_c	v_g	h_c	h_g	h_{cg}	s_c	s_g
bar	°C	$\dfrac{m^3}{Kg}$	$\dfrac{m^3}{kg}$	$\dfrac{kJ}{kg}$	$\dfrac{kJ}{kg}$	$\dfrac{kJ}{kg}$	$\dfrac{kJ}{(kg \cdot C)}$	$\dfrac{kJ}{(kg \cdot C)}$
17.0	204.30	0.0011633	0.11661	871.8	2793.8	1922.0	2.3712	6.3967
18.0	207.10	0.0011678	0.11031	884.6	2795.1	1910.5	2.3976	6.3759
19.0	209.79	0.0011722	0.10464	896.8	2796.4	1899.6	2.4227	6.3561
20.0	212.37	0.0011766	0.09953	908.6	2797.4	1888.8	2.4468	6.3373
22.0	217.24	0.0011850	0.09064	930.9	2799.1	1868.2	2.4922	6.3018
24.0	221.78	0.0011932	0.08319	951.9	2800.4	1848.5	2.5353	6.2691
26.0	226.03	0.0012011	0.07685	971.7	2801.2	1829.5	2.5736	6.2386
28.0	230.04	0.0012088	0.07138	990.5	2801.7	1811.2	2.6106	6.2101
30.0	233.84	0.0012163	0.06662	1008.4	2801.9	1793.5	2.6455	6.1832
35.0	242.54	0.0012345	0.05702	1049.8	2801.3	1751.5	2.7253	6.1218
40.0	250.33	0.0012521	0.04974	1087.5	2799.4	1711.9	2.7967	6.0670
50.0	263.92	0.0012858	0.03941	1154.6	2792.8	1638.2	2.9209	5.9712
60.0	275.56	0.0013187	0.03241	1213.9	2783.3	1569.4	3.0277	5.8878
70.0	285.80	0.0013514	0.02734	1267.7	2771.4	1503.7	3.1225	5.8126
80.0	294.98	0.0013843	0.02349	1317.5	2757.5	1440.0	3.2083	5.7430
90.0	303.31	0.0014179	0.02046	1364.2	2741.8	1377.6	3.2875	5.6773
100	310.96	0.0014526	0.01800	1408.6	2724.4	1315.8	3.3616	5.6143
110	318.04	0.0014887	0.01597	1451.2	2705.4	1254.2	3.4316	5.5531
120	324.64	0.0015267	0.01425	1492.6	2684.8	1192.2	3.4986	5.4930
130	330.81	0.0015670	0.01277	1533.0	2662.4	1129.4	3.5633	5.4333
140	336.63	0.0016104	0.01149	1572.8	2638.3	1065.5	3.6262	5.3737
150	342.12	0.0016580	0.01035	1612.2	2611.6	999.4	3.6877	5.3122
160	347.32	0.0017101	0.009330	1651.5	2582.7	931.2	3.7486	5.2496
170	352.26	0.0017690	0.008401	1691.6	2550.8	859.2	3.8103	5.1841
180	356.96	0.0018380	0.007534	1733.4	2514.4	781.0	3.8739	5.1135
190	361.44	0.0019231	0.006700	1778.2	2470.1	691.9	3.9417	5.0321
200	365.71	0.002038	0.005873	1828.8	2413.8	585.0	4.0181	4.9338
210	369.79	0.002218	0.005006	1892.2	2340.2	448.0	4.1137	4.8106
220	373.68	0.002675	0.003757	2007.7	2192.5	184.8	4.2891	4.5748

TABLE 18-10. Thermal properties of unsaturated water and superheated steam at different pressures (Note: The unsaturated water is above the bold line and superheat steam is under the boldline).

T °C	$P = 0.01$ bar $T_s = 6.982$ $v_c = 0.0010001$ $v_g = 129.208$ $h_c = 29.33$ $h_g = 2513.8$ $s_c = 0.1060$ $s_g = 8.9756$			$P = 0.05$ bar $T_s = 32.90$ $v_c = 0.0010052$ $v_g = 28.196$ $h_c = 137.77$ $h_g = 2561.2$ $s_c = 0.4762$ $s_g = 8.3952$		
	v m³/kg	h kJ/kg	s kJ/(kg°K)	v m³/kg	h kJ/kg	s kJ/(kg°K)
0	0.0010002	0.0	−0.0001	0.0010002	0.0	−0.0001
10	130.60	2519.5	8.9956	0.0010002	42.0	0.1510
20	135.23	2538.1	9.0604	0.0010017	83.9	0.2963
40	144.47	2575.5	9.1837	28.86	2574.6	8.4385
60	153.71	2613.0	9.2997	30.71	2612.3	8.5552
80	162.95	2650.6	9.4093	32.57	2650.0	8.6652
100	172.19	2688.3	9.5132	34.42	2687.9	8.7695
120	181.42	2726.2	9.6122	36.27	2725.9	8.8687
140	190.66	2764.3	9.7066	38.12	2764.0	8.9633
160	199.89	2802.6	9.7971	39.97	2802.3	9.0539
180	209.12	2841.0	9.8839	41.81	2840.8	9.1408
200	218.35	2879.7	9.9674	43.66	2879.5	9.2244
220	227.58	2918.6	10.0480	45.51	2918.5	9.3049
240	236.82	2957.7	10.1257	47.36	2957.6	9.3828
260	246.05	2997.1	10.2010	49.20	2997.0	9.4580
280	255.28	3036.7	10.2739	51.05	3036.0	9.5310
300	264.51	3076.5	10.3446	52.90	3076.4	9.6017
350	287.58	3177.2	10.5130	57.51	3177.1	9.7702
400	310.66	3279.5	10.6709	62.13	3279.4	9.9280
450	333.74	3383.4	10.820	66.74	3383.3	10.077
500	356.81	3489.0	10.961	71.36	3489.0	10.218
550	379.89	3596.3	11.095	75.98	3596.2	10.352
600	402.96	3705.3	11.224	80.59	3705.3	10.481

TABLE 18-10. (*Continued*).

	$P = 0.10$ bar			$P = 1.0$ bar		
	$T_s = 45.83$			$T_s = 99.63$		
	$v_c = 0.0010102$	$v_g = 14.676$		$v_c = 0.0010434$	$v_g = 4.6946$	
	$h_c = 191.84$	$h_g = 2584.4$		$h_c = 417.51$	$h_g = 2675.7$	
	$s_c = 0.6493$	$s_g = 8.1505$		$s_c = 1.3027$	$s_g = 7.3608$	
T °C	v m^3/kg	h kJ/kg	s kJ/(kg°K)	v m^3/kg	h kJ/kg	s kJ/(kg°K)
0	0.0010002	0.0	−0.0001	0.0010002	0.1	−0.0001
10	0.0010002	42.0	0.1510	0.0010002	42.1	0.1510
20	0.0010017	83.9	0.2963	0.0010017	84.0	0.2963
40	0.0010078	167.4	0.5721	0.0010078	167.5	0.5721
60	15.34	2611.3	8.2331	0.0010171	251.2	0.8309
80	16.27	2649.3	8.3437	0.0010292	335.0	1.0752
100	17.20	2687.3	8.4484	1.696	2676.5	7.3628
120	18.12	2725.4	8.5479	1.793	2716.8	7.4681
140	19.05	2763.6	8.6427	1.889	2756.6	7.5669
160	19.98	2802.0	8.7334	1.984	2796.2	7.6605
180	20.90	2840.6	8.8204	2.078	2835.7	7.7496
200	21.82	2879.3	8.9041	2.172	2875.2	7.8348
220	22.75	2918.3	8.9848	2.266	2914.7	7.9166
240	23.67	2957.4	9.0626	2.359	2954.3	7.9954
260	24.60	2996.8	9.1379	2.453	2994.1	8.0714
280	25.52	3036.5	9.2109	2.546	3034.0	8.1449
300	26.44	3076.3	9.2817	2.639	3074.1	8.2162
350	28.75	3177.0	9.4502	2.871	3175.3	8.3854
400	31.06	3279.4	9.6081	3.103	3278.0	8.5439
450	33.37	3383.3	9.7570	3.334	3382.2	8.6932
500	35.68	3488.9	9.8982	3.565	3487.9	8.8346
550	37.99	3596.2	10.033	3.797	3595.4	8.9693
600	40.29	3705.2	10.161	4.028	3704.5	9.0979

TABLE 18-10. (*Continued*).

	P = 5.0 bar			P = 10 bar		
	$T_s = 151.85$			$T_s = 179.88$		
	$v_c = 0.0010928$	$v_g = 4.6946$		$v_c = 0.0011274$	$v_g = 0.19430$	
	$h_c = 640.1$	$h_g = 2748.5$		$h_c = 762.6$	$h_g = 2777.0$	
	$s_c = 1.8604$	$s_g = 6.8215$		$s_c = 2.1382$	$s_g = 1.5847$	
T °C	v m³/kg	h kJ/kg	s kJ/(kg°K)	v m³/kg	h kJ/kg	s kJ/(kg°K)
0	0.0010000	0.5	−0.0001	0.0009997	1.0	−0.0001
10	0.0010000	42.5	0.1509	0.0009998	43.0	0.1509
20	0.0010015	84.3	0.2962	0.0010013	84.8	0.2961
40	0.0010076	167.9	0.5719	0.0010074	168.3	0.5717
60	0.0010169	251.5	0.8307	0.0010167	251.9	0.8305
80	0.0010290	335.3	1.0750	0.0010287	335.7	1.0746
100	0.0010435	419.4	1.3066	0.0010432	419.7	1.3062
120	0.0010605	503.9	1.5273	0.0010602	504.3	1.5269
140	0.0010800	589.2	1.7388	0.0010796	589.5	1.7383
160	0.3836	2767.3	6.8654	0.0011019	675.7	1.9420
180	0.4046	2812.1	6.9665	0.1944	2777.3	6.5854
200	0.4250	2855.5	7.0602	0.2059	2827.5	6.6940
220	0.4450	2898.0	7.1481	0.2169	2874.9	6.7921
240	0.4646	2939.9	7.2315	0.2275	2920.5	6.8826
260	0.4841	2981.5	7.3110	0.2378	2964.8	6.9674
280	0.5034	3022.9	7.3872	0.2480	3008.3	7.0475
300	0.5226	3064.2	7.4606	0.2580	3051.3	7.1239
350	0.5701	3167.6	7.6335	0.2825	3157.7	7.3018
400	0.6172	3271.8	7.7944	0.3066	3264.0	7.4606
420	0.6360	3313.8	7.8558	0.3161	3306.6	7.5283
440	0.6548	3355.9	7.9158	0.3256	3349.3	7.5890
450	0.6641	3377.1	7.9452	0.3304	3370.7	7.6188
460	0.6735	3398.3	7.9743	0.3351	3392.1	7.6482
480	0.6922	3440.9	8.0316	0.3446	3435.1	7.7061
500	0.7109	3483.7	8.0877	0.3540	3478.3	7.7627
550	0.7575	3591.7	7.2232	0.3776	3587.2	7.8991
600	0.8040	3701.4	8.3525	0.4010	3697.4	8.0292

TABLE 18-10. (*Continued*).

	$P = 30$ bar			$P = 50$ bar		
	$T_s = 233.84$			$T_s = 263.92$		
	$v_c = 0.0012163$ $v_g = 0.06662$			$v_c = 0.0012858$ $v_g = 0.03941$		
	$h_c = 1008.4$ $h_g = 2801.9$			$h_c = 1154.6$ $h_g = 2792.8$		
	$s_c = 2.6455$ $s_g = 6.1832$			$s_c = 2.9209$ $s_g = 5.9712$		
T $°C$	v m³/kg	h kJ/kg	s kJ/(kg°K)	v m³/kg	h kJ/kg	s kJ/(kg°K)
0	0.0009987	3.0	0.0001	0.0009977	5.1	0.0002
10	0.0009988	44.9	0.1507	0.0009979	46.9	0.1505
20	0.0010004	86.7	0.2957	0.0009995	88.6	0.2952
40	0.0010065	170.1	0.5709	0.0010056	171.9	0.5702
60	0.0010158	253.6	0.8294	0.0010149	255.3	0.8283
80	0.0010278	337.3	1.0733	0.0010268	338.8	1.0720
100	0.0010422	421.2	1.3046	0.0010412	422.7	1.3030
120	0.0010590	505.7	1.5250	0.0010579	507.1	1.5232
140	0.0010783	590.8	1.7362	0.0010771	592.1	1.7342
160	0.0011005	676.9	1.9396	0.0010990	678.0	1.9373
180	0.0011258	764.1	2.1366	0.0011241	765.2	2.1339
200	0.0011550	853.0	2.3284	0.0011530	853.8	2.3253
220	0.0011891	943.9	2.5166	0.0011866	944.4	2.5129
240	0.06818	2823.0	6.2245	0.0012264	1037.8	2.6985
260	0.07286	2885.5	6.3440	0.0012750	1135.0	2.8842
280	0.07714	2941.8	6.4477	0.04224	2857.0	6.0889
300	0.08116	2994.2	6.5408	0.04532	2925.4	6.2104
350	0.09053	3115.7	6.7443	0.05194	3069.2	6.4513
400	0.09933	3231.6	6.9231	0.05780	3196.9	6.6486
420	0.10276	3276.9	6.9894	0.06002	3245.4	6.7196
440	0.1061	3321.9	7.0535	0.06220	3293.2	6.7875
450	0.1078	3344.4	7.0847	0.06327	3316.8	6.8204
460	0.1095	3366.8	7.1155	0.06434	3340.4	6.8528
480	0.1128	3411.6	7.1758	0.06644	3387.2	6.9158
500	0.1161	3456.4	7.2345	0.06853	3433.8	6.9768
550	0.1243	3568.6	7.3752	0.07363	3549.6	7.1221
600	0.1324	3681.5	7.5084	0.07864	3665.4	7.2586

TABLE 18-10. (*Continued*).

	P = 70 bar			P = 100 bar		
	$T_s = 285.80$ $v_c = 0.0013514$ $v_g = 0.02734$ $h_c = 1267.7$ $h_g = 2771.4$ $s_c = 3.1225$ $s_g = 5.8126$			$T_s = 310.96$ $v_c = 0.0014526$ $v_g = 0.01800$ $h_c = 1408.6$ $h_g = 2724.4$ $s_c = 3.3616$ $s_g = 5.6143$		
T °C	v m³/kg	h kJ/kg	s kJ/(kg°K)	v m³/kg	h kJ/kg	s kJ/(kg°K)
0	0.0009967	7.1	0.0004	0.0009953	10.1	0.0005
10	0.0009970	48.8	0.1504	0.0009956	51.7	0.1500
20	0.0009986	90.4	0.2948	0.0009972	93.2	0.2942
40	0.0010047	173.6	0.5694	0.0010034	176.3	0.5682
60	0.0010140	256.9	0.8273	0.0010126	259.4	0.8257
80	0.0010259	340.4	1.0707	0.0010244	342.8	1.0687
100	0.0010401	424.2	1.3015	0.0010386	426.5	1.2992
120	0.0010567	508.5	1.5215	0.0010551	510.6	1.5188
140	0.0010758	593.4	1.7321	0.0010739	595.4	1.7291
160	0.0010976	679.2	1.9350	0.0010954	681.0	1.9315
180	0.0011224	766.2	2.1312	0.0011199	767.8	2.1272
200	0.0011510	854.6	2.3222	0.0011480	855.9	2.3176
220	0.0011841	945.0	2.5093	0.0011805	946.0	2.5040
240	0.0012233	1038.0	2.6941	0.0012188	1038.4	2.6878
260	0.0012708	1134.7	2.8789	0.001648	1134.3	2.8711
280	0.0013307	1236.7	3.0667	0.0013221	1235.2	3.0567
300	0.02946	2839.2	5.9322	0.0013978	1343.7	3.2494
350	0.03524	3017.0	6.2306	0.02242	2924.2	5.9464
400	0.03992	3159.7	6.4511	0.02641	3098.5	6.2158
450	0.04414	3288.0	6.6350	0.02974	3242.2	6.4220
500	0.04810	3410.5	6.7988	0.03277	3374.1	6.5984
520	0.04964	3458.6	6.8602	0.03392	3425.1	6.6635
540	0.05116	3506.4	6.9198	0.03505	3475.4	6.7262
550	0.05191	3530.2	6.9490	0.03561	3500.4	6.7568
560	0.05266	3554.1	9.9778	0.03616	3525.4	6.7869
580	0.05414	3601.6	7.0342	0.03726	3574.9	6.8456
600	0.05561	3649.0	7.0890	0.03833	3624.0	6.9025

TABLE 18-10. (*Continued*).

T °C	P = 140 bar			P = 200 bar		
	$T_s = 336.63$ $v_c = 0.0016104$ $v_g = 0.01149$ $h_c = 1572.8$ $h_g = 2638.3$ $s_c = 3.6262$ $s_g = 5.3737$			$T_s = 365.71$ $v_c = 0.002038$ $v_g = 0.005873$ $h_c = 1828.8$ $h_g = 2413.8$ $s_c = 4.0181$ $s_g = 4.9338$		
	v m³/kg	h kJ/kg	s kJ/(kg°K)	v m³/kg	h kJ/kg	s kJ/(kg°K)
0	0.0009933	14.1	0.0007	0.0009904	20.1	0.0008
10	0.0009938	55.6	0.1496	0.0009910	61.3	0.1489
20	0.0009955	97.0	0.2933	0.0009929	102.5	0.2919
40	0.0010017	179.8	0.5666	0.0009992	185.1	0.5643
60	0.0010109	262.8	0.8236	0.0010083	267.8	0.8204
80	0.0010226	346.0	1.0661	0.0010199	350.8	1.0623
100	0.0010366	429.5	1.2961	0.0010337	434.0	1.2916
120	0.0010529	513.5	1.5153	0.0010496	517.7	1.5101
140	0.0010715	598.0	1.7251	0.0010679	602.0	1.7192
160	0.0010926	683.4	1.9269	0.0010886	687.1	1.9203
180	0.0011167	769.9	2.1220	0.0011120	773.1	2.1145
200	0.0011442	857.7	2.3117	0.0011387	860.4	2.3030
220	0.0011759	0947.2	2.4970	0.0011693	949.3	2.4870
240	0.0012129	1039.1	2.6796	0.0012047	1040.3	2.6678
260	0.0012572	1134.1	2.8612	0.0012466	1134.1	2.8470
280	0.0013115	1233.5	3.0441	0.0012971	1231.6	3.0266
300	0.0013816	1339.5	3.2324	0.0013606	1334.6	3.2095
350	0.01323	2753.5	5.5606	0.001666	1648.4	3.7327
400	0.01722	3004.0	5.9488	0.009952	2820.1	5.5578
450	0.02007	3175.8	6.1953	0.01270	3062.4	5.9061
500	0.02251	3323.0	6.3922	0.01477	3240.2	6.1440
520	0.02342	3378.4	6.4630	0.01551	3303.7	6.2251
540	0.02430	3432.5	6.5304	0.01621	3364.6	6.3009
550	0.02473	3459.2	6.5631	0.01655	3394.3	6.3373
560	0.02599	3485.8	6.5951	0.01688	3423.6	6.3726
580	0.02599	3538.2	6.6573	0.01753	3480.9	6.4406
600	0.2681	3589.8	6.7172	0.01816	3536.9	6.5055

TABLE 18-10. (*Continued*).

T °C	P = 250 bar			P = 300 bar		
	v m³/kg	h kJ/kg	s kJ/(kg °K)	v m³/kg	h kJ/kg	s kJ/(kg°K)
0	0.0009881	25.1	0.0009	0.0009857	30.0	0.0008
10	0.0009888	66.1	0.1482	0.0009866	70.8	0.1475
20	0.0009907	107.1	0.2907	0.0009886	111.7	0.2895
40	0.0009971	189.4	0.5623	0.0009950	193.8	0.5604
60	0.0010062	272.0	0.8178	0.0010041	276.1	0.8153
80	0.0010177	354.8	1.0591	0.0010155	358.7	1.0560
100	0.0010313	437.8	1.2879	0.0010289	441.6	1.2843
120	0.0010470	521.3	1.5059	0.0010445	524.9	1.5017
140	0.0010650	605.4	1.7144	0.0010621	608.1	1.7097
160	0.0010853	690.2	1.9148	0.0010821	693.3	1.9095
180	0.0011082	775.9	2.1083	0.0011046	778.7	2.1022
200	0.0011343	862.8	2.2960	0.0011300	865.2	2.2891
220	0.0011640	951.2	2.4789	0.0011590	953.1	2.4711
240	0.0011983	1041.5	2.6584	0.0011922	1042.8	2.6493
260	0.0012384	1134.3	2.8359	0.0012307	1134.8	2.8252
280	0.0012863	1230.5	3.0130	0.0012762	1229.9	3.0002
300	0.0013453	1331.5	3.1922	0.0013315	1329.0	3.1763
350	0.001600	1626.4	3.6844	0.001554	1611.3	3.6475
400	0.006009	2583.2	5.1472	0.002806	2159.1	4.4854
450	0.009168	2952.1	5.6787	0.006730	2823.1	5.4458
500	0.01113	3165.0	5.9639	0.008679	3083.9	5.7954
520	0.01180	3237.0	6.0558	0.009309	3166.1	5.9004
540	0.01242	3304.7	6.1401	0.009889	3241.7	5.9945
550	0.01272	3337.3	6.1800	0.010165	3277.7	6.0385
560	0.01301	3369.2	6.2185	0.01043	3312.6	6.0806
580	0.01358	3431.2	6.2921	0.01095	3379.8	6.1604
600	0.01413	3491.2	6.3616	0.01144	3444.2	6.2351

TABLE 18-11. Conversion factors.

Multiply	By	To obtain
acre	4046.86	square meter
ampere/centimeter	2.54000	ampere/inch
ampere/inch	39.3701	ampere/meter
ampere/pound (mass)	2.20462	ampere/kilogram
ampere/square foot	10.7639	ampere/square meter
ampere/square inch	1550.00	ampere/square meter
ampere/square meter	0.092903	ampere/square foot
ampere/volt	1.00000	siemens
ampere/volt inch	39.3701	siemens/meter
ampere/weber	1.00000	unit/henry
ampere turn	1.25664	gilbert
ampere turn/inch	39.3701	ampere turn/meter
ampere turn/meter	0.012566	Oersted
atmosphere (kilogram (force)/cm^2)	98.0665	kilopascal
atmosphere (760 torr)	101.325	kilopascal
bar	100.000	kilopascal
barrel (42 U.S. gallons)	0.158987	cubic meter
barrel/ton (U.K.)	0.156476	cubic meter/metric ton
barrel/ton (U.S.)	0.175254	cubic meter/metric ton
barrel/hour	0.044163	cubic decimeter/second
barrel/million std cubic feet	0.133010	cubic decimeters/kilomol
British thermal unit	0.251996	kilogram calorie
Btu (mean)	1.05587	kilojoule
Btu (thermochemical)	1.05435	kilojoule
Btu (39°F)	1.05967	kilojoule
Btu (60°F)	1.05680	kilojoule
btu (I.T.)	1.05506	kilojoule
btu (I.T.)/brake horsepower hour	0.000393	kilowatt/kilowatt
btu (I.T.)/cubic foot	37.2589	kilojoule/cubic meter
btu (I.T.)/hour	0.293017	watt
btu (I.T.)/hour cubic foot	0.010349	kilowatt/cubic meter
btu (I.T.)/hour cubic foot °F	0.018629	kilowatt/cubic meter kelvin
btu (I.T.)/hour square foot	3.15459	watt/square meter
btu (I.T.)/hour square foot °F	5.67826	watt/square meter kelvin
btu (I.T.)/hour square foot °F/foot	1.73074	watt/meter kelvin
btu (I.T.)/minute	0.017581	kilowatt
btu (I.T.)/pound mol	2.32600	joule/mol
btu (I.T.)/pound mol°F	4.18680	kilojoule/kilomol kelvin
btu (I.T.)/pound (mass)	0.555555	kilocalorie/kilogram
btu (I.T.)/pound (mass)	2.32600	kilojoule/kilogram
btu (I.T.)/pound (mass) °F	4.18680	kilojoule/kilogram kelvin
btu (I.T.)/second	1.05487	kilowatt
btu (I.T.)/second cubic foot	37.2590	kilowatt/cubic meter
btu (I.T.)/second cubic foot °F	67.0661	kilowatt/cubic meter kelvin
btu (I.T.)/second square foot	11.3565	kilowatt/square meter
btu (I.T.)/gallon (U.K.)	232.080	kilojoules/cubic meter
btu (I.T.)/gallon (U.S.)	278.716	kilojoules/cubic meter
calorie (I.T.)	4.18680	joule
calorie (mean)	4.19002	joule
calorie (15°C)	4.18580	joule

(Contd.)

TABLE 18-11. (*Continued*).

Multiply	By	To obtain
calorie (TC)	4.18400	joule
calorie (TC)	0.003966	btu (I.T.)
calorie (20°C)	4.18190	joule
Calorie (TC)/gram k	4.18400	kilojoule/kilogram kelvin
calorie (TC)/hour cm^3	1.16222	kilowatt/cubic meter
calorie (TC)/hour cm^2	0.011622	kilowatt/square meter
calorie (TC)/milliliter	4.18400	megajoule/cubic meter
calorie (TC)/pound (mass)	9.22414	joule/kilgram
calorific heat hour	2.64778	megajoule
calorific value	0.735500	kilowatt
calorific heat unit	1.89910	kilojoules
candela/square meter	0.291864	foot lambert
Candela/square meter	0.000314	lambert
centimeter water 4°	0.098064	kilopascals
centipoise	0.001000	pascal second
centistoke	1.00000	square millimeter/second
chain	20.1168	meter
coulomb/cubic foot	35.3146	coulomb/cubic meter
coulomb/foot	3.28084	coulomb/meter
coulomb/inch	39.3701	coulomb/meter
coulomb/meter	0.025400	coulomb/inch
coulomb/square foot	10.7639	coulomb/square meter
coulomb/square meter	0.092930	coulomb/square foot
cubic centimeter	0.035195	ounce fluid (U.K.)
cubic centimeter	0.038140	ounce fluid (U.S.)
cubic centimeter/cubic meter	0.034972	gallon (U.K.)/1000 barrels
cubic centimeter/cubic meter	0.042000	gallon (U.S.) 1000 barrels
cubic centimeter/cubic meter	1.00000	volume parts/million
cubic decimeter/second	2.11888	cubic foot/minute
cubic decimeter/second	0.035315	cubic foot/second
cubic decimeter/metric ton	0.005706	barrel/ton (U.S.)
cubic decimeter/metric ton	0.006391	barrel/ton (U.K.)
cubic decimeter/metric ton	0.268411	gallon (U.S.)/ton (U.K.)
cubic decimeter/metric ton	0.239653	gallon (U.S.)/ton (U.S.)
cubic foot	0.028317	cubic meter
cubic foot	28.3169	cubic decimeter
cubic foot/foot	0.092903	cubic meter/meter
cubic foot/hour	0.007866	cubic decimeter/second
cubic foot/minute	0.471947	cubic decimeter/second
cubic foot/minute square foot	0.005080	cubic meter/second square meter
cubic foot/pound (mass)	62.4280	cubic decimeter/kilogram
cubic foot/pound (mass)	0.062428	cubic meter/kilogram
cubic foot/second	28.3169	cubic decimeter/second
cubic inch	0.016387	cubic decimeter
cubic kilometer	0.239913	cubic mile
cubic meter	6.28976	barrel (42 U.S. gallons)
cubic meter	35.3147	cubic foot
cubic meter	1.30795	cubic yard
cubic meter	219.969	gallon (U.K.)
cubic meter	264.172	gallon (U.S.)

TABLE 18-11. (*Continued*).

Multiply	By	To obtain
cubic meter/kilogram	16.0185	cubic foot/pound (mass)
cubic meter/meter	10.7639	cubic foot/foot
cubic meter/meter	80.5196	gallon (U.S.)/foot
cubic meter/second meter	289870.	gallon (U.S.)/foot
cubic meter/second meter	4022.80	gallon (U.K.)/minute foot
cubic meter/second meter	4831.18	gallon (U.S.)/minute foot
cubic meter/second meter	20114.0	gallon (U.K.)/hour inch
cubic meter/second meter	24155.9	gallon (U.S.)/hour inch
cubic meter/square meter	88352.6	gallon (U.S.)/hour square foot
cubic meter/square meter	3.28084	cubic foot/second square foot
cubic meter/square meter	196.850	cubic foot/minute square foot
cubic meter/square meter	510.895	gallon (U.K.)/hour square inch
cubic meter/square meter	613.560	gallon (U.S.)/hour square inch
cubic meter/metric ton	5.70602	barrel/ton (U.S.)
cubic meter/metric ton	6.39074	barrel/ton (U.K.)
cubic mile	4.16818	cubic kilometer
cubic mile	0.764555	cubic meter
degree Celsius (difference)	(9/5)	degree Fahrenheit (difference)
degree Celsium (traditional)	$(9/5)°C + 32$	degree Fahrenheit (traditional)
degree Fahrenheit/100 feet	0.018227	kelvin/meter
degree Fahrenheit (difference)	(5/9)	degree Celsius (difference)
degree Fahrenheit (traditional)	$(5/9)°F - 32$	degree Celsius (traditional)
degree Rankine	(5/9)	kelvin
degree (angle)	0.017453	radian
dyne	0.000010	newton
dyne/square centimeter	0.100000	pascal
dyne second/square centimeter	0.100000	pascal second
farad/inch	39.3701	farad/meter
farad/meter	0.025400	farad/inch
fathom (U.S.)	1.82880	meter
foot	304.800	millimeter
foot lambert	3.42626	candel/square meter
foot/degree F	0.548640	meter/kelvin
foot/gallon (U.S.)	80.5196	meter/cubic meter
foot/barrel	1.91713	meter/cubic meter
foot/cubic foot	10.7639	meter/cubic meter
foot/day	0.003528	millimeter/second
foot/hour	0.084667	millimeter/second
foot/mile	0.189394	meter/kilometer
foot/minute	0.005080	meter/second
foot/second	0.304800	meter/second
foot poundal	0.042140	joule
foot pound (force)	1.35582	joule
foot pound (force)/gallon (U.S.)	0.358169	kiljoule/cubic meter
foot pound (force)/second	1.35582	watt
foot pound (force)/square inch	0.210152	joule/square centimeter
footcandle	10.7639	lux
gallon (U.K.)	0.004546	cubic meter
gallon (U.K.)/hour foot	4.14306×10^{-6}	cubic meter/second meter
gallon (U.K.)/hour square foot	1.35927×10^{-5}	cubic meter/second . m^2

(*Contd.*)

TABLE 18-11. (*Continued*).

Multiply	By	To obtain
gallon (U.K.)/minute	0.075768	cubic decimeter/second
gallon (U.K.)/minute foot	0.000249	cubic meter/second meter
gallon (U.K.)/minute square foot	0.000816	cubic meter/second square meter
gallon (U.K.)/pound (mass)	10.0224	cubic decimeter/kilogram
gallon (U.K.)/1000 barrels	28.5940	cubic centimeter/cubic meter
gallon (U.S.)	0.003785	cubic meter
gallon (U.S.)/cubic foot	133.681	cubic decimeter/cubic meter
gallon (U.S.)/foot	0.012419	cubic meter/meter
gallon (U.S.)/hour foot	3.44981×10^{-6}	cubic meter/second meter
gallon (U.S.)/hour square foot	1.13183×10^{-5}	cubic meter/second square meter
gallon (U.S.)/minute	0.063090	cubic decimeter/second
gallon (U.S.)/minute foot	0.000207	cubic meter/second meter
gallon (U.S.)/minute square foot	0.000679	cubic meter/second square meter
gallon (U.S.)/pound (mass)	8.34540	cubic decimeter/kilogram
gallon (U.S.)/ton (U.K.)	3.72563	cubic decimeter/metric ton
gallon (U.S.)/1000 barrels	23.8095	cubic centimeter/cubic meter
gauss	0.000100	tesla
gauss/Oersted	1.25664×10^{-6}	henry/meter
gilbert	0.795775	ampere turn
gilbert/maxwell	7.95775×10^{7}	unit/henry
grain	64.7989	milligram
grain/cubic foot	2.28835	milligram/cubic decimeter
grain/100 cubic feet	22.8835	milligram/cubic meter
gram	0.035274	ounce (avoirdupois)
gram	0.032151	ounce (troy)
gram mol	0.001000	kilomol
gram/cubic meter	3.78541	milligram/gallon (U.S.)
gram/cubic meter	0.058418	grains/gallon (U.S.)
gram/cubic meter	0.350507	pound (mass)/1000 barrels
gram/cubic meter	0.008345	pound (mass)/1000 gallons (U.S.)
gram/cubic meter	0.010022	pound (mass)/1000 gallons (U.K.)
grams/gallon (U.K.)	0.219969	kilogram/cubic meter
grams/gallon (U.S.)	0.264172	kilogram/cubic meter
gray	100.000	rad
henry	7.95775×10^{7}	maxwell/gilbert
henry	1.00000	weber/ampere
henry	1.00000×10^{8}	line/ampere
henry/meter	795775	gauss/Oersted
henry/meter	2.54000×10^{6}	lines/ampere inch
horsepower (electric)	0.746000	kilowatt
horsepower (hydraulic)	0.746043	kilowatt
horsepower (U.S.)	0.745702	kilowatt
horsepower (U.S.)	42.4150	Btu/minute
horsepower hour (U.S.)	2.68452	megajoule
horsepower hour (U.S.)	2544.99	Btu (I.T.)
horsepower/cubic foot	26.3341	kilowatt/cubic meter
hundred weight (U.K.)	50.8024	kilogram
hundred weight (U.S.)	45.3592	kilogram
inch	25.4000	millimeter
inch water (39.2°F)	0.249082	kilopascal

TABLE 18-11. (*Continued*).

Multiply	By	To obtain
inch mercury (32°F)	3.38639	kilopascal
inches/minute	0.423333	millimeter/second
inches/second	25.4000	millimeter/second
joule	0.737562	foot pound (force)
joule	23.7304	foot poundal
joule	1.00000	watt second
joule	0.239126	calorie (20°C)
joule	0.238903	calorie (15°C)
joule	0.238662	calorie (mean)
joule	0.238846	calorie (I.T.)
joule	0.239006	calorie (TC)
joule/kilogram	0.108411	calorie (TC)/pound (mass)
joule/mol	0.429923	btu (I.T.)/pound mol
joule/square centimeter	4.75846	foot pound (force)/sq.inch
joule/square centimeter	0.101972	kilogram meter/cm^2
kelvin (degree)	(9/5)	degree Rankine
kelvin (degree)(minus)	−273.16	degree centigrade
kilocalorie (TC)	4.18400	kilojoule
kilcalorie (TC)/hour	1.16222	watt
kilocalorie (TC)/hour m^2	1.16222	watt/square meter°K
kilocalorie (TC)/kilogram°C	4.18400	kilojoule/kilogram°K
kilogram	0.196841	hundred weight (U.K.)
kilogram	0.220462	hundred weight (U.S.)
kilogram	2.20462	pound (avoirdupois)
kilogram meter/second	7.23301	pound (mass) foot/second
kilogram meter/square centimeter	9.80665	joule/square centimeter
kilogram/cubic decimeter	8.34541	pound (mass)/gallon (U.S.)
kilogram/cubic decimeter	10.0224	pound (mass)/gallon (U.K.)
kilogram/cubic meter	0.062428	pound (mass)/cubic foot
kilogram/cubic meter	0.350507	pound (mass)/barrel
kilogram/cubic meter	3.78541	grams/gallon (U.S.)
kilogram/cubic meter	4.54609	grams/gallon (U.K.)
kilogram/meter	0.671969	pound (mass)/foot
kilogram/mol	2.20462	pound (mass)/mol
kilogram/second	7936.64	pound (mass)/hour
kilogram/second	2.20462	pound (mass)/second
kilogram/second	0.059052	ton (mass)(U.K.)/minute
kilogram/second	0.066139	ton (mass)(U.S.)/minute
kilogram/second	3.54314	ton (mass)(U.K.)/hour
kilogram/second	3.96832	ton (mass)(U.S.)/hour
kilogram/second	31037.9	ton (mass)(U.K.)/year
kilogram/second	34762.5	ton (mass)(U.S.)/year
kilogram/second meter	0.671969	pound (mass)/second foot
kilogram/second meter	2419.09	pound (mass)/hour foot
kilogram/second square meter	0.204816	pound (mass)/second square foot
kilogram/second square meter	737.338	pound (mass)/hour square foot
kilogram/square meter	0.204816	pound (mass)/square foot
kilojoule	0.947817	btu (I.T.)
kilojoule	0.943690	btu (39°F)
kilojoule	0.948155	btu (60°F)

(*Contd.*)

TABLE 18-11. (*Continued*).

Multiply	By	To obtain
kilojoule	0.947086	btu (mean)
kilojoule	0.948452	btu (TC)
kilojoule/cubic meter	0.026884	btu (I.T.)/cubic foot
kilojoule/cubic meter	0.004309	btu (I.T.)/gallon (U.K.)
kilojoule/cubic meter	0.003588	btu (I.T.)/gallon (U.S.)
kilojoule/cubic meter	2.79198	footpound (force)/gallon (U.S.)
kilojoule/kilogram	0.429923	btu (I.T.)/pound (mass)
kilojoule/kilogram kelvin	0.238846	btu (I.T.)/pound (mass) °F
kilojoule/kilogram kelvin	0.238846	Btu (I.T.)/pound mol°F
kilojoule/kilogram kelvin	0.239006	calorie (TC)/gram kelvin
Kilojoule/kilogram kelvin	0.239006	Calorie (TC)/gram mol°C
kilojoule/kilogram kelvin	0.239006	kilocalorie (TC)/kilogram °C
kilojoule/kilogram kelvin	0.000278	kilowatt hour/kilogram °C
kilojoule/mol	0.239006	kilocalorie (TC)/gram mol
kilometer	0.621371	mile
kilometer	0.539957	nautical mile
kilometer/cubic decimeter	2.35215	mile/gallon (U.S.)
kilometer/hour	0.539957	knot
kilometer/hour	0.621371	miles/hour
kilomol	1000.00	gram mol
kilomol	2.20462	pound mol
kilomol	836.610	standard cubic foot (60°F, 1 atmosphere)
kilomol	22.4136	standard cubic meter (0°C, 1 atmosphere)
kilomol	23.6445	standard cubic meter (15°C, 1 atmosphere)
kilomol/cubic meter	0.0624280	pound mol/cubic foot
kilomol/cubic meter	0.010022	pound mol/gallon (U.K.)
kilomol/cubic meter	0.008345	pound mol/gallon (U.S.)
kilomol/cubic meter	133.010	standard ft³/barrel (60°F, 1 atmosphere)
kilomol/second	2.20462	pound mol/second
kilomol/second	7936.64	pound mol/hour
kilonewton	0.224809	kip (1000 foot pound)
kilonewton	0.100361	ton (force) (U.K.)
kilonewton	0.112405	ton (force) (U.S.)
kilonewton meter	0.368781	ton force (U.S.) foot
kilopascal	0.010197	atmosphere (kilogram/cm²)
kilopascal	0.009869	atmosphere (760 torr)
kilopascal	0.01000	bar
kilopascal	10.1974	centimeter water (4°C)
kilopascal	4.01474	inch water (39.2°F)
kilopascal	0.295300	inch mercury (32°F)
kilopascal	0.296134	inch mercury (60°F)
kilopascal	7.50062	millimeter mercury (0°C)
kilopascal	20.8854	pound (force)/square foot
kilopascal	0.145038	pound (force)/square inch
kilopascal/meter	0.044208	pound (force)/square inch/foot
kilopascal second	0.145038	pound (force)/square inch
kilowatt	56.8690	Btu (I.T.)/minute
kilowatt	0.947817	Btu (I.T.)/second
kilowatt	1.35962	calorific value
kilowatt	1.34048	horsepower (electric)
kilowatt	1.34102	horsepower (550 foot pound/second)

TABLE 18-11. (*Continued*).

Multiply	By	To obtain
kilowatt	1.34045	horsepower (hydraulic)
kilowatt	0.284345	ton of refrigeration
kilowatt hour	3.60000	megajoule
kilowatt hour/kilogram °C	3600.00	kilojoule/kilogram kelvin
kilowatt/cubic meter	96.6211	btu (I.T.)/hour cubic foot
kilowatt/cubic meter	0.026839	btu (I.T.)/second cubic foot
kilowatt/cubic meter	0.860421	calorie (TC)/hour cm^3
kilowatt/cubic meter	0.037974	horsepower/cubic foot
kilowatt/cubic meter kelvin	53.6784	btu (I.T.)/hour cubic foot°F
kilowatt/cubic meter kelvin	0.014911	btu (I.T.)/second cubic foot°F
kilowatt/kilowatt	2544.43	btu (I.T.)/brake horsepower hour
kilowatt/square meter	0.088055	btu (I.T.)/second square foot
kilowatt/square meter	86.0421	calorie (TC)/hour cm^3
kilowatt/square meter kelvin	0.048919	btu (I.T.)/second square foot °F
kip (1000 foot pounds)	4.44822	kilonewton
kip/square inch	6.89476	megapascal
knot	1.85325	kilometer/hour
lambert	3138.10	candela/square meter
line	1.00000	maxwell
line	1.00000×10^{-8}	weber
lines/ampere	1.00000×10^{-8}	henry
lines/ampere inch	3.93701×10^{-7}	henry/meter
lines/square inch	1.55000	tesla
link	0.201168	meter
lumen/square foot	10.7639	lux
lumen/square inch	1500.00	lux
lux	0.092903	footcandle
lux	0.092903	lumen/square foot
lux	0.000645	lumen/square inch
lux second	0.092903	foot candle second
maxwell	1.00000	line
maxwell	1.00000×10^{-8}	weber
maxwell/gilbert	1.25664×10^{-8}	henry
megagram	1.00000	ton (mass) (metric)
megagram	0.984206	ton (mass) (U.K.)
megagram	1.10231	ton (mass) (U.S.)
megagram/square meter	0.102408	ton (mass) (U.S.)/square foot
megajoule	947.817	btu (I.T.)
megajoule	0.377675	calorific value hour
megajoule	0.372506	horsepower hour
megajoule	0.277778	kilowatt hour
megajoule	0.009478	therm
megajoule	0.102408	ton (mass) (U.S.) mile
megajoule/cubic meter	4.30886	btu (I.T.)/gallon (U.K.)
megajoule/cubic meter	3.58788	btu (I.T.)/gallon (U.K.)
megajoule/cubic meter	0.239006	calorie (TC)/milliliter
megajoule/meter	0.021289	ton (force) (U.S.) mile/foot
megapascal	0.145038	kip/square inch
megapascal	145.038	pound/square inch
megapascal	10.4427	ton (force) (U.S.)/square foot

(*Contd.*)

TABLE 18-11. (*Continued*).

Multiply	By	To obtain
megapascal	0.072519	ton (force) (U.S.)/square inch
megawatt	3.41214	million Btu (I.T.)/hour
meter	0.049710	chain
meter	0.546807	fathom
meter	3.28084	feet
meter	4.97097	link
meter	0.198839	rod
meter	1.09361	yard
meter/cubic meter	0.521612	foot/barrel
meter/cubic meter	0.092903	foot/cubic foot
meter/cubic meter	0.012419	foot/gallon (U.S.)
meter/kelvin	1.82269	foot/°F
meter/kilometer	5.28000	foot/mile
meter/second	3.28084	foot/second
meter/second	196.850	foot/minute
microbar	0.100000	pascal
micrometer	0.039370	mil
micrometer	1.00000	micron
micron	1.00000	micrometer
microsecond/foot	3.28084	microsecond/meter
microsecond/meter	0.304800	microsecond/foot
mil	25.4000	micrometer
mile	5280.00	foot
mile	1.60934	kilometer
mile/gallon (U.S.)	0.425144	kilometer/cubic decimeter
mile/hour	1.60934	kilometer/hour
milligram	0.015432	grain
milligram/cubic decimeter	0.436996	grain/cubic foot
milligram/cubic meter	0.043700	grain/100 cubic foot
milligram/gallon (U.S)	0.264172	gram/cubic meter
millimeter	0.039370	inch
millimeter	0.003281	foot
millimeter mercury (0°C)	133.322	pascal
millimeter mercury (0°C)	0.133322	kilopascal
millimeter/second	283.465	foot/day
millimeter/second	11.8110	foot/hour
millimeter/second	2.36221	inch/minute
millimeter/second	0.039370	inch/second
million Btu (I.T.)/hour	0.293071	megawatt
million electron volt	0.160219	picojoule
million pound (mass)/year	0.014383	kilogram/second
minute (angle)	0.000291	radian
mol/foot	3.28084	mol/meter
mol/kilogram	0.453592	mol/pound (mass)
mol/meter	0.304800	mol/foot
mol/pound (mass)	2.20462	mol/kilogram
mol/square foot	10.7639	mol/square meter
mol/square meter	0.092903	mol/square foot
nautical mile	1.85325	kilometer
newton	1.00000×10^5	dyne

TABLE 18-11. (*Continued*).

Multiply	By	To obtain
newton	0.224809	pound (force)
newton	7.23301	poundal
newton meter	0.737562	pound (force) foot
newton meter	8.85075	pound (force) inch
newton meter	23.7304	poundal foot
newton meter/meter	0.018734	pound (force) foot/inch
newton meter/meter	0.224809	pound (force) inch/inch
newton/meter	0.068522	pound (force)/foot
newton/meter	0.005710	pound (force)/inch
Oersted	79.5775	ampere turn/meter
ohm circular mil/foot	1.66243×10^{-9}	ohm square meter/meter
ohm foot	0.304800	ohm square meter/meter
ohm inch	0.025400	ohm square meter/meter
ohm square meter/meter	6.01531×10^{8}	ohm cirular mil/foot
ohm square meter/meter	3.28084	ohm foot
ohm square meter/meter	39.3701	ohm inch
Ounce (avoidupois)	28.3495	gram
ounce (troy)	31.1035	gram
ounce (fluid) (U.K.)	28.4131	cubic centimeter
ounce (fluid) (U.S.)	29.5735	cubic centimeter
pascal	1	newton/square meter
pascal	10.0000	dyne/square centimeter
pascal	10.0000	microbar
pascal	0.07501	millimeter mercury (0°C)
pascal second	1000.00	centipoise
pascal second	10.0000	dyne second/square centimeter
pascal second	0.020885	pound (force) second/square foot
pascal second	2419.09	pound (mass)/foot hour
pascal second	0.671969	pound (mass)/foot second
picojoule	6.24145	million electron volt
pint (liquid) (U.K.)	0.568262	cubic decimeter
pint (liquid) (U.S.)	0.473167	cubic decimeter
pint (U.K.)/1000 barrels	3.57425	cubic decimeter/cubic meter
pound mol	0.453592	kilomol
pound mol/cubic foot	16.0185	kilomol/cubic meter
pound mol/gallon (U.K.)	99.7763	kilomol/cubic meter
pound mol/gallon (U.S.)	119.826	kilomol/cubic meter
pound mol/hour	0.000126	kilomol/second
pound mol/second	0.453592	kilomol/second
poundal	0.138255	newton
poundal	0.031083	pound (force)
pound foot	0.042140	newton/meter
pound (force)	4.44822	newton
pound (force) foot	1.35582	newton/meter
pound (force) foot/inch	53.3787	newton/meter/meter
pound (force) inch	0.112985	newton/meter
pound (force)/foot	14.5939	newton/meter
pound (force)/inch	175.127	newton/meter
pound (force) square foot	0.047880	kilopascal
pound (force) square inch	6.89476	kilopascal

(*Contd.*)

TABLE 18-11. (Continued).

Multiply	By	To obtain
pound (force)/square inch/foot	22.6206	kilopascal/meter
pound (force)/square foot	47.8803	pascal second
pound (force)/square inch	6.89476	kilopascal second
pound (mass)	32.1740	poundal
pound (mass)	0.453592	kilogram
pound (mass)	1.21528	pound (troy)
pound (mass)/barrel	2.85301	kilogram/cubic meter
pound (mass)/Btu	1.80018	kilogram/kilogram calorie
pound (mass)/cubic foot	16.0185	kilogram/cubic meter
pound (mass)/foot	1.48816	kilogram/meter
pound (mass)/foot hour	0.000413	pascal second
pound (mass)/foot second	1.48816	pascal second
pound (mass)/gallon (U.K.)	0.099776	kilogram/cubic decimeter
pound (mass)/gallon (U.S.)	0.138255	kilogram/cubic decimeter
pound (mass)/1000 gallons (U.K.)	99.7763	gram/cubic meter
pound (mass)/1000 gallons (U.S.)	119.826	gram/cubic meter
pound (mass)/hour	0.000126	kilogram/second
pound (mass)/hour foot	0.000413	kilogram/second meter
pound (mass)/hour square foot	0.001356	kilogram/second m^2
pound (mass)/minute	0.007560	kilogram/second
pound (mass)/mol	0.453592	kilogram/mol
pound (mass)/second	0.453592	kilogram/second
pound (mass)/second foot	1.48816	kilogram/second meter
pound (mass)/second square foot	4.88243	kilogram/second square meter
pound (mass)/square foot	4.88243	kilogram/square meter
pound (mass)/second	0.138255	kilogram meter/second
pound (mass) square foot	0.042140	kilogram square meter
quart (dry) (U.K.)	1.03200	quart (dry) (U.S.)
quart (liquid) (U.K.)	1.136523	cubic decimeter
quart (liquid) (U.K.)	1.136523	liter
quart (liquid) (U.K.)	1.20030	quart (liquid) (U.S.)
quart (liquid) (U.S.)	0.946331	liter
quart (liquid) (U.S.)	0.859370	quart (dry) (U.S.)
rad	0.010000	gray
radian	2.06265	second (angle)
radian	3437.75	minute (angle)
radian	57.2958	degree (angle)
radian/second	0.159155	revolutions/second
radian/second	9.54930	revolutions/minute
radian/second squared	0.159155	revolutions/second squared
radian/second squared	572.958	revolutions/minute squared
revolutions/minute	0.104720	radian/second
revolutions/minute squared	0.001745	radian/second squared
revolutions/second	6.28319	radian/second
revolutions/second squared	6.28319	radian/second squared
rod	5.02920	meter
second (angle)	4.84814×10^{-6}	radian
section	2.58999	square kilometer
siemens	1.00000	ampere/volt
siemens/meter	0.025400	ampere/volt inch

TABLE 18-11. (*Continued*).

Multiply	By	To obtain
square foot	0.092903	square meter
square foot/cubic inch	5669.29	square meter/cubic meter
square foot/hour	25.8064	square millimeter/second
square foot/pound (mass)	0.204816	square millimeter/kilogram
square foot/second	9290.30	square millimeter/second
square foot pound (mass)/second squared	0.042140	joule
square inch	645.160	square millimeter
square kilometer	0.386102	section
square meter	10.7639	square foot
square meter	0.000247	acre
square meter	1.19599	square yard
square meter/cubic meter	0.000176	square foot/cubic inch
square meter/kilogram	4.88243	square foot/pound (mass)
square mile	2.58999	square kilometer
square millimeter	0.001550	square inch
square millimeter/second	1.07639×10^{-5}	square foot/second
square millimeter/second	0.038750	square foot/hour
square millimeter/second	1.00000	centistoke
square yard	0.836137	Square meter
standard cubic foot/barrel (60°F, 1 atmosphere)	0.007518	kilomol/cubic meter
standard cubic foot (60°F, 1 atmosphere)	0.001195	kilomol
standard cubic meter (60°F, 1 atmosphere)	0.044616	kilomol
standard cubic meter (15°F, 1 atmosphere)	0.042293	kilomol
tesla	10000.0	gauss
tesla	64516.0	lines/square inch
therm	105.506	megajoule
ton(force) (U.K.)	9.96402	kilonewton
ton (force) (U.S.)	8.89644	kilonewton
ton (force) (U.S.)	2.71164	kilonewton meter
ton (force) (U.S.) mile	14.3174	megajoule
ton (force) (U.S.) mile/foot	46.9732	megajoule/meter
ton (force) (U.S.)/square foot	0.095761	megapascal
ton (force) (U.S.)/square inch	13.7895	megapascal
ton (mass) (U.K.)	1.01605	megagram
ton (mass) (U.K.)	1.01605	metric ton
ton (mass) (U.K.)	1.12000	ton (mass) (U.S.)
ton (mass) (U.S.)	0.907185	megagram
ton (mass) (U.S.)	0.907185	metric ton
ton (mass) (U.S.)	0.892857	ton (mass) (U.K.)
ton (metric)	1.00000	megagram
ton (mass) (U.K.)/day	0.011760	kilogram/second
ton (mass) (U.S.)/day	0.010500	kilogram/second
ton (mass) (U.K.)/hour	0.282235	kilogram/second
ton (mass) (U.K.)/hour	0.251996	kilogram/second
ton (mass) (U.K.)/minute	16.9341	kilogram/second
ton (mass) (U.S.)/minute	15.1197	kilogram/second

(Contd.)

TABLE 18-11. (*Continued*).

Multiply	By	To obtain
ton (mass) (U.S.)/square foot	9.76486	megagram/square meter
ton refrigeration	3.51685	kilowatt
unit/foot	3.28084	unit/meter
unit/henry	1.00000	ampere/weber
unit/henry	1.25664×10^{-8}	gilbert/maxwell
unit/meter	3.28084	volt/meter
volt/foot	3.28084	volt/meter
volt/inch	39.3701	volt/meter
volume parts per million	1.00000	cubic centimeter/m^3
watt	3.41280	btu (I.T.)/hour
watt	44.2537	foot pound (force)/minute
watt	0.737562	foot pound (force)/second
watt	0.860421	kilocalorie (TC)/hour
watt hour	3.60000	kilojoule
watt/inch	39.3701	watt/meter
watt/meter	0.025400	watt/inch
watt/meter kelvin	0.577789	btu (I.T.) hour square foot °F/foot
watt/meter kelvin	6.93347	btu (I.T.) hour square foot °F/inch
watt/meter kelvin	8.60421	calorie (TC)/hour cm^2 C/cm
watt/meter kelvin	0.002390	calorie (TC)/second cm^2 C/cm
watt/square meter	0.316998	btu (I.T.)/hour square foot
watt/square meter kelvin	0.176110	Btu (I.T.)/hour square foot °F
watt/square meter kelvin	0.860421	kilocalorie (TC)/hour cm^2 C
watt second	1.00000	joule
weber	1.00000×10^8	lines
weber	1.00000×10^8	maxwell
weber/ampere	1.00000	henry
yard	0.914402	meter

Index

556   Index

Mechanical Engineering Series *(continued from page ii)*

Laminar Viscous Flow
V.N. Constantinescu

Thermal Contact Conductance
C.V. Madhusudana

Transport Phenomena with Drops and Bubbles
S.S. Sadhal, P.S. Ayyaswamy, and J.N. Chung

Fundamentals of Robotic Mechanical Systems:
Theory, Methods, and Algorithms
J. Angeles

Electromagnetics and Calculations of Fields
J. Ida and J.P.A. Bastos

Mechanics and Control of Robots
K.C. Gupta

Wave Propagation in Structures:
Spectral Analysis Using Fast Discrete Fourier Transforms, 2nd ed.
J.F. Doyle

Fracture Mechanics
D.P. Miannay

Principles of Analytical System Dynamics
R.A. Layton

Composite Materials:
Mechanical Behavior and Structural Analysis
J.M. Berthelot

Modern Inertial Technology:
Navigation, Guidance, and Control, 2nd ed.
A. Lawrence

Dynamics and Control of Structures:
A Modal Approach
W.K. Gawronski

Electromechanical Sensors and Actuators
I.J. Busch-Vishniac

Nonlinear Computational Structural Mechanics:
New Approaches and Non-Incremental Methods of Calculation
P. Ladevèze

Boilers and Burners: Design and Theory
P. Basu, C. Kefa, and L. Jestin

Printed in the United Kingdom
by Lightning Source UK Ltd.
135307UK00001B/31-45/A